Optimization Methods in Manufacturing Processes

Anand J. Kulkarni
Editor

Optimization Methods in Manufacturing Processes

A Machine-Generated Literature Overview

Springer

Editor
Anand J. Kulkarni
Institute of Artificial Intelligence
Dr Vishwanath Karad MIT World Peace Univ
Pune, Maharashtra, India

ISBN 978-981-96-5256-3 ISBN 978-981-96-5257-0 (eBook)
https://doi.org/10.1007/978-981-96-5257-0

This Springer imprint is published by the registered company Springer Nature Singapore Pte Ltd.
The registered company address is: 152 Beach Road, #21-01/04 Gateway East, Singapore 189721, Singapore

Preface

The manufacturing domain is one of the major drivers of most countries in the world. The manufacturing systems convert raw materials into useful products. The system actively involves several resources, such as human resources, machines, material, transportation, energy, etc. The cost of every resource affects the final price of the product as well as eventually the countries' economy. Moreover, efficient utilization of every resource is very much necessary to sustain in the competitive market. The optimization techniques play a very crucial role in enhancing the efficiency right from the material movement from the mines to the industry and eventually delivery of the final product to customers. In addition, the optimum selection of the machining processes and machining parameter settings decides the quality as well as functionalities of the product. Furthermore, efficient production planning, workforce assignment, scheduling as well as devising systems for reutilization of the waste material is quite possible.

This book provides a machine-generated comprehensive yet classified review of the relevant problems from the manufacturing domain as well as several associated optimization and analytical methods. The methods include deterministic, exact methods as well as metaheuristics. The deterministic methods covered are associated with integer linear programming and branch and bound methods. The metaheuristics covered are genetic algorithms, particle swarm optimization, cohort intelligence algorithm, artificial neural networks, moth search algorithm, Tabu search, simulated annealing teaching–learning-based optimization, Jaya algorithm, grey wolf optimizer, etc. In addition, application of several machine learning methods such as regression analysis, support vector machines, Gaussian regression models, Bayesian classifier, decision tree, random forest, k-nearest neighbour, etc. have also been discussed. The reviews are specifically from the elite Springer publications. The review covers the rationale of the problem being addressed, the associated short literature supporting the contribution, as well as the choice of optimization methods and approaches, the case studies, results, and the significance of the solutions. Such classified review may help the UG and PG students, professors, and researchers from academia as well as industry gain a critical insight at the various systems and aspects of manufacturing systems, and importantly how the optimization methods play a crucial role in utilizing the resources to the maximum. The articles chosen as well as the reviews are necessarily driven by the keywords recommended by the editor of the book as well as generated by the machine based on the suggestions.

Many thanks to the editorial advisors Shrutika Dasari, Chaitra Jawalkoti, Ha-Linh Vu, and Prajana Kate as well as Kamiya Khatter, the Responsible Editor at Springer for their unparallel support in several ways. Their help was very essential in conceptualizing as well as bringing the book to fruition.

Pune, India Anand J. Kulkarni

About the Book

Auto-summaries can be generated by either an abstractive or extractive auto-summarization:

An extraction-based summarizer identifies the most important sentences of a text and uses the original sentences to create the summary. An abstraction-based summarizer creates new text based on deep learning. New phrases are created to summarize the content. The auto-summaries you will find in this book have been generated via an extractive summarization approach.

Each chapter was carefully edited by Dr. Anand J. Kulkarni. The editors selected the papers which were then auto-summarized. The editors have not edited the auto-summaries due to the extraction-based approach and have not changed the original sentences.

You will find the editors' reviews and guidance on the auto-summaries in their chapter introductions. In machine-generated books, editors are defined as those who curate the content for the book by selecting the papers to be auto-summarized and by organizing the output into a meaningful order. Next to the thoughtful curation of the papers, editors should guide the readers through the auto-summaries and make transparent why they selected the papers.

The ultimate goal is to provide a current literature review of Springer Nature publications on a given topic in order to support readers in overcoming information overload and to help them dive into a topic faster; to identify interdisciplinary overlaps; and to present papers which might not have been on the readers' radar.

Please note that the selected papers are not used to train a LLM while the auto-summaries are created.

Contents

Chapter 1
Optimization Methods in Material Handling

Introduction by the Editor

Material handling and distribution is referred to as the movement of material from its very raw form to the final product delivery to the customer or user. The movement of material is expected to be as less as possible as it does not add any value. So, a minimum material handling is very essential to reduce the final cost of the product. The efficient factory layout and route optimization including the scheduling and dispatching of automatic guided vehicles (AGVs), reduced material handling equipment failure rate, etc. play vey essential role in overall minimization of the material handling. This chapter reviews different optimization approaches including Design for Assembly (DFA), goal programming (GP), reinforcement learning (RL), particle swarm optimization (PSO), genetic algorithms (GAs), nonlinear programming (NLP), Artificial Neural Network (ANN), Artificial Immune System, digital twin, multi-objective optimization approaches such as Non-Sorting Genetic Algorithm-II) (NSGA-II), Strength Pareto Evolutionary Algorithm-II (SPEA-II) for material handling in the view point of factory layout, as well as overall supply chain management. In addition, failure mode, effects, and criticality analysis (FMECA), etc. reducing the potential failures of material handling equipment is also discussed along with several cases studies and real-world problems have also been discussed.

The chapter in detail provides the critical review on how reinforcement learning was applied to produce increasingly efficient plans for material handling in representative manufacturing facilities. It necessarily involves defining a formal representation of a realistically complex material handling plan, specifying a set of suitable plans change operators as reinforcement learning actions. As the reinforcement learning algorithms are based on rewards, the algorithm learns the important operators to apply in different situations to improve a material handling plan over the progressive learning process. As there are several material handling problems

A. J. Kulkarni (ed.), *Optimization Methods in Manufacturing Processes*,
https://doi.org/10.1007/978-981-96-5257-0_1

modeled as multi-objective problems, the application of the NSGA-II and SPEA-II have been discussed along with the numerical illustration and sensitivity analysis. The failure mode, effects, and criticality analysis (FMECA), etc. is discussed for showcasing the approach of sufficiency to derive a reliable overhead crane safety failure mode system by reducing the risk of failures. The study concludes that performing FMECA is highly effective in improving efficiency for handling equipment in the industry, and the applicability of its proposed process in this sector was validated.

Furthermore, GA based parameter optimization method for inductive power transfer system based is discussed in details. More specifically, a nonlinear programming method in combination with GA was proposed to obtain the optimal solution of the model. It is used to realize the optimal system parameter design in any given mutual inductance range and reduce the fluctuation of system output during the variance of coupling coefficient. A digital twin approach for the product development covering the product design, validation, and manufacturing is also elaborated. It concludes that the digital validation results can be used to improve the design scheme of the new product and boost the corresponding manufacturing processes. The framework of rapid product development based on digital twin is also discussed. It also highlights the core characteristics and key technologies of digital validation. The review concludes that the ultrahigh-fidelity digital prototype obtained from digital trial manufacturing can not only be used for digital validation and design improvement, but can also accelerate the construction of the digital twin of the new product. A digital twin-based layout optimization method for discrete manufacturing workshop is one of the important contributions. The cases study concludes that according to the sub-framework of digital twin-based workshop partitioning, the workshop partitioning plan of welding production workshop is required to be made based on the layout optimization method.

A review on route optimization for warehouse order picking operations via vehicle routing and simulation is discussed. It necessarily highlights the solution to the routing optimization problem for order picking in a rectangular, multiparallel-aisle warehouse with cross aisles at the middle and end of every aisle. The effect of nature-inspired algorithms and hybrid dispatching rules on the performance of automatic guided vehicles in the flexible manufacturing system is critically reviewed. A simulation methodology is used to compare and analyze the combined effect of four experimental factors, namely four types of priority hybrid dispatching rules, three different nature-inspired algorithms, two levels of loading/unloading times and two levels of machine failures, on the different performance parameters of the FMS.

A unit process life cycle inventory methodology (UPLCI) used to discretize energy consumption and material losses for modeling the L-PBF process is discussed with case study on L-PBF. The objective of the study is to analyze the energy consumption from the production of the part including the support structures. The application of the DFA and goal programming to improve economic efficiency, material handling convenience, and sustainability of a product packaging system is also critically reviewed. The review discusses the development of an effective method to help packaging designers and planners determine the best packaging design solution that balances the multiple objectives of cost, efficiency, and sustainability across an entire supply chain. The simulation-based optimization for material handling systems in manufacturing and distribution industries is reviewed and discussed. The work demonstrates how real-life single and multi-objective optimization problems in manufacturing and distribution industries where material handling system is involved. In addition, it emphasizes on how it can be solved by a simulation-based optimization framework that comprises a simulation tool together with an optimization algorithm based on Artificial Immune System (AIS).

Machine Generated Summaries

Disclaimer: The summaries in this chapter were generated from Springer Nature publications using extractive AI auto-summarization: An extraction-based summarizer aims to identify the most important sentences of a text using an algorithm and uses those original sentences to create the auto-summary (unlike generative AI). As the constituted sentences are machine selected, they may not fully reflect the body of the work, so we strongly advise that the original content is read and cited. The auto generated summaries were curated by the editor to meet Springer Nature publication standards. To cite this content, please refer to the original papers.

Machine generated keywords: handling, manufacturing, material handling, workshop, digital, life cycle, supply chain, manufacturing system, chain, product, cycle, building, layout, development, sustainability.

Applying Reinforcement Learning to Plan Manufacturing Material Handling [1]

This is a machine-generated summary of:

Govindaiah, Swetha; Petty, Mikel D.: Applying reinforcement learning to plan manufacturing material handling [1].

Published in: Discover Artificial Intelligence (2021).

Link to original: https://doi.org/10.1007/s44163-021-00003-3

If you want to cite the papers, please refer to the original.

For technical reasons we could not place the page where the original quote is coming from.

Abstract-Summary "In this work reinforcement learning was applied to produce increasingly efficient plans for material handling in representative manufacturing facilities."

"Doing so included defining a formal representation of a realistically complex material handling plan, specifying a set of suitable plan change operators as reinforcement learning actions, implementing a simulation-based multi-objective reward function that considers multiple components of material handling costs, and abstracting the many possible material handling plans into a state set small enough to enable reinforcement learning."

"This work demonstrates the applicability of reinforcement learning with a multi-objective reward function to realistically complex material handling processes."

Introduction and Motivation

"In order to produce fast and effective changes to MH plans reinforcement learning (RL) was applied instead of mathematical optimization."

"Starting from simple initial MH plans, the RL process produced a sequence of iteratively modified plans."

"The cost of the MH plans calculated by the simulation was used as the reward in the RL algorithm."

"After each plan was generated and tested, the RL algorithm revised it using one of a set of MH-specific plan change operators."

"Based on the rewards, the RL algorithm learned which of the operators to apply in different situations to improve a material handling plan over the iterations of the learning process."

"How can an MH plan be changed in response to a reward to produce a better plan?"

"Can the RL algorithm learn a policy to improve the MH plan?"

Background

"To apply RL to MH, the elements of the factory involved in the MH process must be represented in terms of the elements of an RL algorithm, i.e., environment, states, actions, and rewards."

"The RL algorithm primarily tries to learn a multi-load carrier scheduling policy, rather than learning a policy to improve the MH plan [2]."

"Each of these objectives is associated with an individual reward signal; hence in the RL MH problem the learning agent receives a vector of rewards from the environment after each action."

"Hill climbing is a greedy method, in that the search always moves in the direction that locally improves the objective function value, whereas the standard ε-greedy RL algorithm includes both exploitation (greedy optimization) and exploration (stochastic exploration of non-locally optimal actions)."

Definition of the Problem

"If there is only one delivery in route i or none of the workstations of D_i in route i are running out of parts, then this action tries to remove one of the deliveries from route i and adds it to any other route in plan P which has same type of equipment as route i. This action requires sorting the routes of plan P in descending order of number of different type of parts on the pieces of equipment of each route."

"The reward model maintains the current time and distance lists for each distinct piece of equipment used in the plan P, and it maintains a separate list for the sum of parts stored and sum of non-consumed parts for each workstation and part type in the plan P. These lists record the data at every visit to a particular workstation to drop off a particular type of part."

"The plan is simulated to determine the actual time spent visiting each workstation and to record the sum of stored parts and non-consumed parts, which is used to calculate the reward, i.e., the MH cost."

Learning Process

"To reduce the size of the state space, the RL state was calculated from values of variables in the reward function, rather than using the MH plan itself as the state. (This is quite similar to function approximation in RL using features of the state [3], although in this work the state was calculated from the reward function's output.) Several different ways of doing so, with differing numbers of states, were manually devised based on knowledge of the application and then empirically tested."

"From the No Penalty state, if the agent selects an action that changes the MH plan in a way that the reward model when executed for the changed plan returns a reward with a non-zero Penalty value in the reward vector, then transitions to the Penalty state."

Experiments, Results, and Analysis for Factory Layout 1

"The equipment type combination and equipment limit on number of part types must be checked while moving the delivery from one route to another during the execution of a plan change action to ensure no physical constraints are violated."

"The second column shows how many parts are present in each workstation's temporary parts buffer at the beginning of the shift; the two values in the column, 2.0 and 0.2, indicate 200% and 20% of the workstation's consumption rate of each part type respectively."

"If the value in the column is 1, each route in the plan consists of one piece of the equipment delivering one type of part to one workstation."

"Although the cost stabilized early, and only the first 1000 actions are shown for each initial plan, in total the algorithm was run for 3000 actions so that the learning agent had ample opportunity to explore and improve the plan."

Experiments, Results, and Analysis for Factory Layout 2

"Any details of the factory, the material handling equipment, the initial material handling plans, or of the experimentation not explicitly reported for factory layout 2 should be assumed to be the same as for factory layout 1."

"The majority of the learning occurred within the first 500 actions, after which the cost of the MH plan tended to stabilize."

"Although the cost stabilized early, and only the first 1000 actions are shown for each initial plan, in total the algorithm was run for 3000 actions for each initial plan."

"By comparing initial cost to final cost, the RL process produced a significant reduction in the total MH cost for every initial plan."

Conclusions

"A potentially large set of RL actions was simplified using separate action selection and action parameterization stages; the latter stage exploited the supplemental information produced by the reward calculation."

"One of the contributions of this research is this demonstration that for complex problems such as material handling planning, additional reward calculation information, which we call reward observation, can be useful along with recording the action values of each state for generating and applying an efficient policy."

"Because the cost data and factory layouts used in the experiments were based on a real factory and the discrete event simulation used to evaluate the MH plans is consistent with the discrete event simulations typically used in the manufacturing industry, it is reasonable to expect that an RL algorithm using multiple reward response variables and a two-stage action selection process can be potentially applied to real world MH plan generation in response to changes in demand."

Future Work

"Because MH is a multi-objective optimization problem, action selection methods other than ε-greedy, such as priority-based exploration, may offer better performance."

"Additional response variables, such as the human resources used to perform MH, could be included in the reward function."

"This work considered only centralized and pre-planned MH operations."

"Experiments to determine how to leverage the improved MH plans to benefit other interconnected processes in the manufacturing facility could be valuable."

"After further improving and generalized the RL process for MH planning, processes for applying the process to real-world manufacturing facilities could be developed."

Development of a Supply Chain Configuration Model for New Product Development: A Multi-Objective Solution Approach [4]

This is a machine-generated summary of:

Sinha, Amit Kumar; Anand, Ankush: Development of a supply chain configuration model for new product development: a multi-objective solution approach [4].

Published in: Soft Computing (2021).

Link to original: https://doi.org/10.1007/s00500-021-05761-x

If you want to cite the papers, please refer to the original.

For technical reasons we could not place the page where the original quote is coming from.

Abstract-Summary "We propose a supply chain configuration model under a multi-objective optimization framework."

"We then develop a solution procedure based on a multi-objective evolutionary algorithm, i.e. NSGA-II (Non-Sorting Genetic Algorithm-II), to minimize the total supply chain cost and time-to-market."

"The results obtained from NSGA-II are also compared from SPEA-II (Strength Pareto Evolutionary Algorithm-II)."

Introduction

"In general, some major factors play important role for the success of a new product, especially in a global competitive scenario, for example (1) efficient communication among marketing, design and manufacturing; (2) early involvement of suppliers and customers in new product development; (3) functionalities of product exactly as needed; and (4) timeliness of product development."

"Paper, we propose a multi-objective decision-making model to determine whether consideration of SCM issues in a new product development process is profitable mainly in terms of total supply chain cost and time-to-market."

"To address the above discussed issues of integrated framework of SCN and NPD, this study is carried out with following highlighted objectwise: A systematic integral framework of supply chain management (SCM) and new product development (NPD) has been proposed."

Related Work

"Lamothe et al. [5] identify not only various available solutions for product family and their bill of materials but also a mixed-integer linear programming-based supply chain optimized model while considering supply chain cost has been proposed."

"They have not considered total supply chain cost as well as time-to-market during the modelling which is based on fuzzy stochastic programming of SCCN for NPD."

"Hasan et al. [6] have not developed any methodology for NPD-SCM integration procedure Afrouzy et al. [7] show absence of cost analysis as well as time-to-market or lead time in the proposed modelling Willems [8] has not considered breakage of cost in terms of penalty cost, manufacturing cost, etc. which is not incorporated during supply chain configuration model for NPD."

"Nepal et al. [9] have not considered penalty cost, procurement cost and transportation cost in the modelling Therefore, there is a need for systematic investigation how the modelling could be applied at conceptual design stage in order to design new product with optimal total supply chain costs and time-to-market."

Supply Chain Configuration Model for New Product Development

"The on-hand inventory at stage j is calculated by:Total safety stock cost of the stochastic service model is calculated by:In this supply chain configuration model, design of product has been fixed."

"The two objectives of the supply chain configuration model for new product development have been formulated: one is total supply chain cost whose unit is monetary and the other one is time-to-market whose unit is time."

"This leads us to formulate a multi-objective optimization model where the objectives are incommensurable. (1) Minimize total supply chain cost. (2) Minimize total time-to-market of the supply chain configuration."

"Subject to: Production time constraint Production cost constraintInventory coverage time constraint (since back order is not allowed) Single sourcing constraint (only one supplier is selected at each node) positivity constraint It can be said that the proposed model extends the single-criterion supply chain optimization approach of Graves and Willems [10] in terms of four aspects."

Multi-Objective Optimization for Supply Chain Configuration

"We develop a solution procedure using NSGA-II for supply chain configuration problems because of (1) its computation efficiency, i.e. the computational complexity of NSGA-II is $O(mn^2)$ where m is the number of objectives and n is the number of population sizes [11] and (2) no need of sharing parameters which is generally an important decision factor of evolutionary algorithms."

"The developed procedure produces a random population P(1) of size L. Encoding of each objective functions (total supply chain cost and time-to-market) in terms of each chromosome of P(1) will be the input for the proposed algorithm."

"The main aim of the step is to evaluate the fitness function of total supply chain cost and time-to-market for each possible solution (s) in generation (gen) for all population size because this fitness will be carried out for fast non-dominated sorting."

Numerical Illustration

"Purpose, we used the data in the case study of laptop computer assembly of Graves and Willems [12] which was formulated as a multi-objective optimization problem in the frame of the supply chain configuration model for new product development."

"The original diagnostic model has 34 stages of supply chain with 72 options."

"We have made an attempt to simplify the actual model by appropriately considering only two end items instead of five end items and also by infusing various miscellaneous components in single stage."

Results and Sensitivity Analysis

"In order to efficiently find an optimal set of suppliers/alternatives from a large search space under conflicting criteria, we have proposed a NSGA-II-based solution procedure that reduces the large search space by finding non-dominated solutions over successive generation of NSGA-II."

"The Pareto optimal solutions (i.e. non-dominated solutions) in terms of total supply chain and time-to-market are obtained after 3000 iterations; for this, the population size of 200 was selected."

"To deliver a more clear assessment, we have applied further two well-known heuristics (single-criterion approaches): minimum unit manufacturing cost heuristic (UMC) and minimum time-to-market (MLT) to the Pareto optimal solutions obtained by NSGA-II."

"We can say that robustness of the NSGA-II-based solution procedure is conformed on the basis of SPEA-II and two heuristics UMC and MLT."

"Based on the evaluation of various parameters using NSGA-II algorithm, a trade-off has been obtained such that a bridge between total supply chain cost and time-to-market is established."

Conclusions

"The proposed model has discussed only two objectives, total supply chain cost and time-to-market, while many other important decision factors such as quality, sustainability and flexibility may be further considered."

"The sensitivity analysis of the laptop case study reveals that the total supply chain cost decreases with time-to-market."

"There is a strong correlation exist between total supply chain cost and time-to-market."

"In future further investigation is necessary for establishing correlation between total supply chain cost and time-to-market."

Application of Failure Mode Effect and Criticality Analysis to Industrial Handling Equipment [13]

This is a machine-generated summary of:

Rouabhia-Essalhi, Rachida; Boukrouh, El Hadi; Ghemari, Youcef: Application of failure mode effect and criticality analysis to industrial handling equipment [13].

Published in: The International Journal of Advanced Manufacturing Technology (2022).

Link to original: https://doi.org/10.1007/s00170-022-09099-y

If you want to cite the papers, please refer to the original.

For technical reasons we could not place the page where the original quote is coming from.

Abstract-Summary "It aims to apply one of the most commonly used methods in the industries: failure mode, effects, and criticality analysis (FMECA)."

"FMECA is a technique used to identify critical failure modes and thus improve the reliability of the product or components."

"The study's objective is to reduce potential failures of industrial handling equipment by using FMECA, thereby avoiding potential risks and reducing downtime."

Introduction

"It is a systematic bottom-up failure analysis method at all levels, from component to system or process, to identify failure modes, analyze failure causes, and map failure effects [14]."

"The FMECA is thus an analysis methodology by which all potential failure modes are found, the causes and effects of the failure modes are analyzed, the critical failure modes are selected, and methods to mitigate or remove the effects of the critical failure modes are provided [15–18]."

"Wang et al. [19] proposed an improved FMECA (IFMECA) to overcome the drawbacks of conventional FMECA, which does not consider the opinions of different team members or the relationship between failure modes and causes when assigning criticality."

"In this study, a specialized FMECA procedure for industrial handling equipment was proposed by analyzing several FMECA specifications used in other industries sectors and considering the characteristics and requirements of industrial handling equipment."

Literature Review

"Salvi and Jindal [20] applied the FMEA approach by evaluating the possible causes of failure and their effects on the sub-systems of CNC machines to improve their reliability in the industry and increase the performance of the production rate through the prioritization of the failures occurring in them."

"Since the risk priority number (RPN) calculation method is one of the critical subjects of failure mode and effects analysis (FMEA) research, Wu and Wu [21] recently proposed a fuzzy beta-binomial RPN evaluation method by integrating fuzzy theory, Bayesian statistical inference, and the beta-binomial distribution."

"Other researchers have obtained useful results in improving performance in various industries and avoiding failure by combining FMEA with other improvement methods."

"Peeters et al. [14] proposed a method to perform failure analysis by combining the fault tree analysis (FTA) method and the failure mode and effects analysis (FMEA) method."

Methodology

"It can help managers identify failures and causes/effects and eliminate failures by instituting corrective actions in the risk assessment process [22]."

"There are several types of FMEA/FMECA: Design FMECA (DFMECA): It is used to ensure the reliability of a product by improving its design."

"Process FMECA (PFMECA): It ensures the quality of a product by improving its production operations."

"Concept FMECA (CFMECA): It ensures the availability and safety of production equipment by improving its maintenance."

"The analytical method of FMECA aims to identify potential risks in an area where assessment is performed and score them based on three factors: severity, occurrence, and detection of failures."

Case Study

"We designed a potential FMECA framework for the Pellegrini 10 T as an analytical technique utilized primarily to ensure that the mechanisms of potential failure modes and their associated causes and effects have been considered and addressed."

"We demonstrated that FMECA is sufficient to derive a reliable overhead crane safely failure mode system by reducing the risk of failures."

"For predicting the occurrence of any failure mode on the Pellegrini 10 T overhead crane newly installed, this study is based on the manufacturer's manual of this system and the history of the old overhead crane failures, which have the same characteristics."

"Step 1 Failure data of the Pellegrini 10 T overhead crane were collected."

"The next priority must be given to the gear unit and the electrical cabinet (RPN = 24), mainly due to the degree of severity of the failure, which disrupts the entire movement of the overhead crane."

Conclusions

"Studies have shown the feasibility of implementing a maintenance optimization method as the influence of maintenance activities becomes more and more important in the management of companies."

"This work aims to implement the failure mode effect and criticality analysis (FMECA) on the Pellegrini 10 T overhead crane newly installed in the compactor assembly workshops of an Algerian company to determine the effect of various failures on the system sub-assemblies by ranking and prioritizing each failure according to its risk priority number (RPN)."

"A plan of preventive and corrective actions was established to improve and optimize the machine's performance and ensure its availability."

"The study concludes that performing FMECA is highly effective in improving efficiency for handling equipment in the industry, and the applicability of its proposed process in this sector was validated."

Parameter Optimization Method for Antimisalignment of Inductive Power Transfer System Based on Genetic Algorithm [23]

This is a machine-generated summary of:

Cai, Jin; Wu, Xu-Sheng; Sun, Pan; Sun, Jun; Deng, Qi-jun: Parameter optimization method for antimisalignment of inductive power transfer system based on genetic algorithm [23].

Published in: Journal of Power Electronics (2021).

Link to original: https://doi.org/10.1007/s43236-021-00322-9

If you want to cite the papers, please refer to the original.

For technical reasons we could not place the page where the original quote is coming from.

Abstract-Summary "When conducting dynamic wireless power transfer using an inductive power transfer system, it results in larger fluctuation of output power."

"A parameter optimization method is necessary to improve the stability of inductive power transfer system during dynamic misalignment."

"Genetic algorithm and nonlinear programming were combined to optimize the compensating parameters of the system and to realize minimum fluctuation of output voltage gain of the system within any given range of mutual inductance parameters."

"Test results show that when the mutual induction range is from 29.3 μH to 84.3 μH, the voltage gain of the system varies from 0.67 to 0.77."

"The fluctuation ratio of voltage gain is 6.7%, and the fluctuation ratio of voltage gain under the circumstance of resonance parameters is 40.2%."

Introduction

"In the case of dynamic power transfer, the mutual inductance parameters vary rapidly, where the control is more difficult."

"Literature [24] combined the methods of frequency converting control and phase-shifting control and proposed an optimal variable frequency phase shift control strategy for LCC/S topological system, realizing the zero voltage switching (ZVS) operation of IPT system in wide power range."

"Dynamic regulation has problems, such as communication delay, and is unsuitable to the cases of dynamic power transfer with coupling coefficient changing in real time."

"Literature [25] introduced the method of reducing the sensitivity of system output voltage to mutual inductance by keeping certain detuning rate in the primary and secondary circuits of the system for cascaded compensating topology circuit."

"A parameter optimization method based on genetic algorithm (GA) was proposed in this study to optimize the parameters of the system and to improve the output stability of IPT system in case of the dynamic variation of coupling coefficient."

Model Analysis of S/LCC-IPT System

"The output voltage at equivalent load and its voltage gain relative to output voltage of inverter can be obtained."

"This process is performed to analyze the voltage gain sensitivity to mutual inductance parameter under nonresonance condition."

"The analysis shows that the maximum voltage gain can be obtained at the extreme point and the voltage gain changes gently around the extreme point, suggesting that the extreme point of voltage gain can be set in the given mutual inductance range."

Optimum Design Method of Dynamic Misalignment Resistance

"The parameter optimization solution with extremely stable output can be obtained by solving the above model."

"The voltage gain obtained in the solution set may be extremely small and cannot meet the output requirements of receiving terminal."

"Model [26] is a nonlinear programming model with complicated parameters and multiple constraint conditions, and obtaining the global optima solution of the model by utilizing the traditional equal solution algorithm (such as sequential quadratic programming algorithm and gradient decent algorithm) is difficult."

"Literature [25] introduced the method of reducing the sensitivity of system output voltage to mutual inductance by keeping certain detuning rate in the primary and secondary circuits of the system for cascaded compensating topology circuit Enumeration method is adopted to obtain the optimization parameters of the system."

"This method is extremely cumbersome to implement, requires considerable calculation time, and is difficult to obtain the global optimal solution of the model."

Test Verification

"Under the circumstance of optimization parameters, the theoretical voltage gain range is 0.69–0.89, the measured voltage gain range is 0.67–0.77, and the corresponding fluctuation ratio of voltage gain is 6.9%, which is lower than the theoretical value."

"This condition is because the change incurred by the self-inductance of coil is ignored during the variance of coupling coefficient, and parameter tolerance influences the test results during test."

"The curve trend of measured data is mostly consistent with that of the theoretical curve, suggesting that the proposed parameter design method is effective."

"Under the circumstance of resonance parameters, the range of theoretical voltage gain is 0.73–1.87, the measured voltage gain varies from 0.72 to 1.69, and the corresponding voltage fluctuation ratio is 40.2%, suggesting that the proposed optimization design method can substantially improve the output stability of the system under dynamic misalignment condition."

Conclusion

"A parameter optimization design method based on GA is proposed to improve the dynamic misalignment resistance of IPT system."

"The nonlinear programming method in combination with GA was proposed to obtain the optimal solution of the model."

"This method was used to realize the optimal system parameter design in any given mutual inductance range and reduce the fluctuation of system output during the variance of coupling coefficient."

Toward Digital Validation for Rapid Product Development Based on Digital Twin: A Framework [27]

This is a machine-generated summary of:

Huang, Sihan; Wang, Guoxin; Lei, Dong; Yan, Yan: Toward digital validation for rapid product development based on digital twin: a framework [27].

Published in: The International Journal of Advanced Manufacturing Technology (2022).

Link to original: https://doi.org/10.1007/s00170-021-08475-4

If you want to cite the papers, please refer to the original.

For technical reasons we could not place the page where the original quote is coming from.

Abstract-Summary "Product development should cover product design, validation, and manufacturing."

"In traditional product development, physical validation based on physical trial manufacturing is the key step to confirm the design scheme before physical manufacturing."

"The emergence of digital twin provides an opportunity to accelerate product development by eliminating physical validation toward digital validation in the smart manufacturing era."

"A framework of rapid product development based on digital twin is proposed in this paper."

"During product development, the new product is designed according to the new requirements in the virtual space, in which the existing digital twins of products can be referenced."

"An ultrahigh-fidelity virtual manufacturing system is constructed for digital trial manufacturing based on the digital twin of the manufacturing system and the design scheme of the new product."

"The digital validation results can be used to improve the design scheme of the new product and boost the corresponding manufacturing processes."

Introduction

"Product validation is executed on physical prototypes obtained from physical trial manufacturing to evaluate the feasibility and performance of the design scheme [26], that is, physical validation."

"These technologies can boost the product development process through rapid prototyping and validation in virtual space."

"Physical validation based on physical trial manufacturing is still required to confirm the design scheme before product manufacturing."

"Product validation across two spaces (virtual space and physical space) becomes the major obstacle to rapid product development."

"The digital twin of manufacturing system [28] provides an ultrahigh-fidelity virtual manufacturing system to support the construction of digital trial manufacturing system for the trial manufacturing of new product in the virtual space."

"The digital prototype obtained from digital trial manufacturing is closest to the real product and can achieve complete product validation in virtual space."

Related Works

"Some researchers have attempted to fill the gap between product design and manufacturing based on digital twin [29, 30], extending the investigation and application of digital twin to manufacturing domain [31–35]."

"Digital twin–driven product development synthesizes the study of product design, product validation, and product manufacturing based on digital twin."

"Lo et al. [36] reviewed product design and development based on digital twin."

"Digital twin–driven product validation should also address process planning problem based on product design results, which connects product design and manufacturing system to support product validation through virtual manufacturing."

"Few studies have explored product design, product validation, and product manufacturing based on digital twin together, which is more reasonable and necessary to accelerate the speed of product development."

"To fill this knowledge gap, a framework for rapid product development based on digital twin will be proposed in this paper."

Digital Twin of Product/Manufacturing System

"Although Grieves did not implicitly call it "digital twin," the concept proposed in the presentation has all the elements of digital twin: physical space, virtual space, and the link for data transmission between physical space and virtual space [37, 38]."

"The interaction between physical product and virtual product triggers the vitality of digital twin of product."

"The data flow from physical product to virtual product updates the virtual product to the latest states of the corresponding physical product to maintain its ultrahigh-fidelity characteristic, which can be used to monitor the performance of the corresponding physical product."

"Data feedback from virtual product to physical product can be used to support the performance optimization, maintenance scheduling, and other physical products."

A Framework of Rapid Product Development Based on Digital Twin

"In the virtual space, the new product is designed according to the new requirements, including concept design and detail design, in which the existing digital twins of products and the design suggestions obtained from physical manufacturing processes can be referenced."

"The digital trial manufacturing system is constructed based on the corresponding digital twin of the manufacturing system, which is joint optimization with process planning according to the design scheme of the new product."

"It is a process to map the ultrahigh-fidelity virtual manufacturing system to digital trial manufacturing system according to the design scheme of the new product."

"The digital trial manufacturing system for digital validation can be directly adopted as the initial ultrahigh-fidelity virtual manufacturing system to monitor the manufacturing states of the new product, which increases the construction efficiency of the corresponding digital twin of manufacturing system."

Core Characteristics and Key Technologies of Digital Validation

"The digital trial manufacturing system is adapted from the ultrahigh-fidelity virtual manufacturing system according to the requirements of the new product, which maintains the fidelity of the ultrahigh-fidelity virtual manufacturing system as much as possible."

"The digital trial manufacturing system is an ultrahigh-fidelity virtual manufacturing system for manufacturing the new product as well."

"An ultrahigh-fidelity digital trial manufacturing system is the foundation to eliminate the costly and inefficient physical validation."

"The ultrahigh-fidelity digital prototype makes it possible to skip the product validation on the physical prototype."

"Considering the frequent digital trial manufacturing within the iteration of product design and validation, process planning with high efficiency and validity is crucial for successfully implementing the digital validation based on digital trial manufacturing."

"Intelligent algorithm combining the existing knowledge and the design information of the new product will promote the efficiency and validity of process planning for digital trial manufacturing to a higher level."

Case Study

"The product design, product validation, and product manufacturing of the key part of rotorcraft will be elaborated in this section."

"It is also the most fundamental basis for digital trial manufacturing of the new design of the key part."

"The modified virtual manufacturing system (digital trial manufacturing system) is used for trial manufacturing of new product in the virtual space."

"The product validation of the new product in this case is executed in the virtual space completely, including process planning, trial manufacturing, prototype validation, etc. The goal of the proposed rapid product development framework is achieved through digital validation based on digital twin."

"The digital twin of manufacturing system provides an ultrahigh-fidelity virtual manufacturing system resulting in an ultrahigh-fidelity digital prototype for digital validation, which is really important for eliminating the costly and inefficient physical validation."

Conclusion

"To achieve rapid product development, a framework of product development based on digital twin toward digital validation is proposed in this paper by compensating for the shortcomings of traditional product development."

"The ultrahigh-fidelity digital prototype obtained from digital trial manufacturing can not only be used for digital validation and design improvement, but can also accelerate the construction of the digital twin of the new product."

"The key technologies for boosting the execution of digital validation to support rapid product development are also pointed out."

"It is very promising to eliminate costly and inefficient physical validation based on digital twin, which will significantly promote product development efficiency."

Route Optimization for Warehouse Order Picking Operations via Vehicle Routing and Simulation [39]

This is a machine-generated summary of:

Shetty, Namrata; Sah, Bhawesh; Chung, Sung Hoon: Route optimization for warehouse order picking operations via vehicle routing and simulation [39].

Published in: SN Applied Sciences (2020).

Link to original: https://doi.org/10.1007/s42452-020-2076-x

If you want to cite the papers, please refer to the original.

For technical reasons we could not place the page where the original quote is coming from.

Abstract-Summary "This paper addresses the problem of routing optimization for order picking in a warehouse to minimize the travel time and distance."

"We propose an easy-to-implement vehicle routing based approach in conjunction with the distance matrix for obtaining an optimal route for order picking, which is solved using the off-the-shelf solver Gurobi with the Julia programming language."

"The optimal route is then compared with the routes obtained by other traditional approaches such as S-shape, return, mid-point, and largest gap methods by means of simulation."

"We show that the proposed vehicle routing based approach outperforms the S-shape, return, mid-point, largest gap methods by 36.55%, 44.89%, 46.64%, and 36.9%, respectively, for scenario 1, and by 25.12%, 24.01%, 27.57%, and 26.77%, respectively, for scenario 2."

Introduction

"We aim to contribute to the warehouse operation efficiency by minimizing the order picking travel time and distance by means of the proposed vehicle routing and distance matrix based approach."

"This type of problem, to minimize the order picking travel time, is called the picker routing problem (PRP) and it is usually formulated as a Steiner traveling salesman problem (TSP) [40], a special case of TSP."

"Picker-to-part systems involve the order picker traversing through the warehouse by means of material handling equipment such as forklifts or reach-trucks, physically picking items from their locations to fulfill customer orders."

"While most research in the literature develops a new mathematical model and relevant methodology to tackle the transportation problem in supply chain and logistics, this paper employs both mathematical model and simulation to focus on the order picking routing problem in a warehouse setting with the goal of optimizing the process and improving overall productivity."

Literature Review

"It also explicates various order picking methods and summarizes some of the established routing heuristics such as S-shape, return, mid-point and largest gap policies."

"Dekker et al. [41] present heuristics for the different order picking policies such as S-shape, mid-point, and Combined in conjunction with the storage policy, as the storage policy is closely related to the routing policy."

"For the studies that compare the exact methods and heuristics, Theys et al. [42] present an algorithm based on the NP-hard Steiner TSP to deal with the routing problem for order picking in a conventional multiparallel-aisle warehouse."

"It is worth providing a brief survey on TSP and VRP because these are the basis of our proposed approach to tackle the order picking routing problem."

Model and Methodology

"To examine the impact of different storage policies, we consider the two scenarios: one with uniform pick up locations throughout the warehouse (for random storage policy) and the other with more pick up locations towards the front side (for demand-based storage policy)."

"In order to achieve this, a simulation model is built in which two order demand scenarios (one with uniform distribution and the other with more pick-ups assigned to the areas close to drop-off location) are considered."

"Aisles without picks are not entered and the order picker returns to the drop-off location (depot) from the last visited aisle."

"To evaluate the routing policies by simulation, the inherent random pick locations and random pick amount need to be considered for which two scenarios are generated."

"For the same pick locations and demand quantities, the distances traveled for the alternative policies are calculated."

Results, Analyses, and Discussion

"The VRP results are 4.53 min (1994 ft) and 4.65 min (2044 ft) for scenario 1 and 2, respectively, in terms of travel time (distance), while other routing policy results are more than 6 min for both scenarios."

"The null hypothesis is rejected because the p value is 0.00, which implies that the average travel times are not the same for different routing policies."

"The VRP travel time is significantly smaller than that of the alternative routing policies."

"No significant differences is found between these two routing methods, with the mean time for S-shape being 7.14 min and that for largest gap being 7.18 min in scenario 1."

"The mid-point and return routing policies are significantly more time consuming with the mean times being 8.49 and 8.22 min, respectively, in scenario 1."

Conclusion and Future Work

"This study provides a solution to the routing optimization problem for order picking in a rectangular, multiparallel-aisle warehouse with cross aisles at the middle and end of every aisle."

"A simulation model was generated using VBA based on the actual warehouse layout."

"Two scenarios based on the picking demand distribution within the warehouse were simulated."

"For a set of 15 pick locations, the simulation delivered routes for multiple routing policies: VRP, S-shape, return, mid-point, and largest gap for both scenarios."

"The simulation scenario can be modified for pick locations to accommodate more realistic situation."

"Even though the capacity constraints are not considered in this paper to focus on the main topic of examining the VRP-based method for order picking routing with the aid of simulation, it will be a valuable direction to be sought in the future work."

Effect of Nature-Inspired Algorithms and Hybrid Dispatching Rules on the Performance of Automatic Guided Vehicles in the Flexible Manufacturing System [43]

This is a machine-generated summary of:

Chawla, V. K.; Chanda, A. K.; Angra, Surjit; Rani, Sushma: Effect of nature-inspired algorithms and hybrid dispatching rules on the performance of automatic guided vehicles in the flexible manufacturing system [43].

Published in: Journal of the Brazilian Society of Mechanical Sciences and Engineering (2019).

Link to original: https://doi.org/10.1007/s40430-019-1894-5

If you want to cite the papers, please refer to the original.

For technical reasons we could not place the page where the original quote is coming from.

Abstract-Summary "The application of nature-inspired algorithms and priority hybrid dispatching rules for simultaneous scheduling and dispatching of automatic guided vehicles (AGVs) are observed to be highly significant for the best utilization of the flexible manufacturing system (FMS) facility."

"The aim of this paper is to use the simulation methodology so as to compare and analyze the combined effect of four experimental factors, namely four types of priority hybrid dispatching rules, three different nature-inspired algorithms, two levels of loading/unloading times and two levels of machine failures, on the different performance parameters of the FMS."

"It is observed that the interaction between experimental factors, namely nature-inspired algorithms and loading–unloading time, has a significant effect on the performance measures, namely simultaneous reduction in distance travel and backtracking, mean work center utilization (%) and total production rate of the FMS."

"The interaction between aforesaid experimental factors has no significant effect on mean AGV utilization (%) in the FMS."

"The interaction between priority hybrid dispatching rules and loading/unloading times is found to have significant effect on the mean work center utilization (%) in the FMS facility."

Introduction

"An appropriate combination of heuristic algorithms and PHDRs for the simultaneous scheduling of work centers and AGVs become highly imperative for the sustainable material handling operations in the FMS."

"In order to find mean effect and interaction effect of different experimental factors, namely heuristic algorithms, priority hybrid dispatching rules, loading/unloading times and machine failures on different performance parameters of FMS, in this paper the performance parameters such as simultaneous reduction in distance travel and backtracking of AGVs, total production rate of FMS, mean AGV utilization in FMS and mean work center utilization in FMS are evaluated under four different PHDRs, three different nature-inspired algorithms, two levels of loading/unloading times (10 s and 60 s representing low load/unload and high load/unload, respectively) and two levels of machine failures (0% and 3%) in the ARENA application."

Literature Review

"The scheduling and routing decisions have significant effects on various performance parameters of AGVs and FMS such as simultaneous minimization of distance travel and backtracking of AGVs in the FMS, the total production rate of FMS, mean AGV utilization in FMS and mean work center utilization in FMS."

"In order to yield the best performance from AGVs and FMS, the simultaneous scheduling and routing of FMS and AGVs are generally carried out by the application of priority dispatching rules and heuristic algorithms in combination."

"Review, a research gap to compare and analyze the performance of AGVs and FMS with the application of a combination of nature-inspired algorithms, PHDRs, machine failure rate and loading/unloading time for the simultaneous scheduling of the AGVs and FMS is clearly established."

Problem Definition and Methodology

"For the SPS + SPT (shortest processing sequence + shortest processing time) forming a priority hybrid dispatching rule, the value of "O" can be calculated as Kaban et al. [44]: where k the set of operations; j the set of work centers; PT the time needed for an operation k on work center j; PS the total process sequence for operation k for the completion of the operation."

"To perform a balanced exploration and exploitation of solutions in the solution search space, the modified memetic particle swarm optimization (MMPSO) algorithm is applied so as to harness the characteristic of MA for optimum local search and PSO for the global search of solutions."

"The simulation of the algorithm run with best solution value represents alpha (α) and the second and the third best solutions are considered as beta (β) and delta (Δ), respectively."

Results

"For the performance parameters, namely simultaneous minimization of AGV distance travel and backtracking, the total production rate of the FMS (parts/h), mean work center utilization (%) and mean AGV utilization in the FMS (%), it is clearly evident that the resulting yield of CSA heuristic algorithm outperforms the resulting yield of MMPSO and GWO algorithm."

"The study reflects the effects of appropriate selection of the heuristic algorithm, PHDRs, load/unload time and work center reliability on the total production rate of FMS."

"The two-way interaction between PHDRs and heuristic algorithms; heuristic algorithms and loading/unloading time and work center failures and heuristic algorithms also reflects a significant effect on the mean work center utilization (%) of the FMS."

"The interaction between PHDRs and loading/unloading time, heuristic algorithms and work center failure shows a significant effect on the mean AGV utilization (%) in the FMS."

Conclusions and Future Work

"The effect of four priority hybrid dispatching rules (PHDRs), three nature-inspired heuristic algorithms, two levels of loading and unloading time and two states of work center machining conditions on performance parameters, namely simultaneous minimization of AGV distance travel and backtracking, total production rate of FMS, mean work center utilization (%) in the FMS and mean AGV utilization (%) in the FMS, is obtained."

"The interaction between experimental factors, namely heuristic algorithm and loading/unloading time for FMS, has a significant effect on performance measures such as simultaneous reduction in distance travel and backtracking, mean work center utilization (%) and total production rate of FMS."

"The interaction between aforesaid experimental factors has no significance on mean AGV utilization (%) in FMS, which may be due to fact that FMS loading and unloading time and heuristic algorithms do not have any direct effect on the AGV utilization."

Reusable Unit Process Life Cycle Inventory (UPLCI) for Manufacturing: Laser Powder Bed Fusion (L-PBF) [45]

This is a machine-generated summary of:

Ramirez-Cedillo, Erick; García-López, Erika; Ruiz-Huerta, Leopoldo; Rodriguez, Ciro A.; Siller, Hector R.: Reusable unit process life cycle inventory (UPLCI) for manufacturing: laser powder bed fusion (L-PBF) [45].

Published in: Production Engineering (2021).

Link to original: https://doi.org/10.1007/s11740-021-01050-6

If you want to cite the papers, please refer to the original.

For technical reasons we could not place the page where the original quote is coming from.

Abstract-Summary "The unit process life cycle inventory methodology (UPLCI) was used to discretize energy consumption and material losses for modeling the L-PBF process."

"A reusable approach in terms of materials, parameters, and calculation tools is presented, to estimate the energy consumption and mass loss in practical evaluations of production lines."

"Theoretical equations were also shown to relate the most important parameters of the process with its energy consumption."

"A case study is presented to analyze the UPLCI capacity to improve energy consumption in the manufacturing of medical devices."

Introduction

"Laser powder bed fusion (L-PBF) is a transforming process of mass conservation categorized as a thermal joining process due to the use of laser-sourced energy for the welding of successive layers for building a tridimensional shape."

"Kellens et al. conducted an energy and performance analysis of the L-PBF process by collecting data using an LCI methodology, and by discretizing some of the elements used for machine energy consumption estimates."

"And in order to contribute to a library of UPLCI cases, the elaboration of another UPLCI model is proposed, considering the relevance of additive manufacturing processes nowadays, particularly L-PBF, and to systematically study the electric energy consumption, the material used and recycled, the waste, and the inert gas consumption in a more standardized way."

Methodology for Unit Process Life Cycle Inventory Model

"Inputs such as the material in the form of powder (new or recycled), the inert gas, the energy (electricity), the dry air, the process parameters, and the preparation (set-up) of the machine are introduced into the L-PBF system."

"Input requirements [powder characteristics, inert gas characteristics, energy (electricity), dry air, process parameters, set-up preparation (consumables)]."

"For UPLCI, since it is a methodology dedicated to production mode, the first two steps Generation of laser scanning strategy and set-up by the operator are included to have the overall operation and processing time for this process."

"The L-PBF process produces two typical types of waste streams, gas and solid, and including by-products like the inert gas, the non-used powder and the support material used for the production of parts."

"The inert gas is filtered and reused when the L-PBF machine is processing material, in the meanwhile, the machine uses a constant flow during each fabrication cycle."

Case Study on L-PBF

"We analyze the detailed energy consumption calculations in the production of a tibia implant using an L-PBF machine model Renishaw AM400."

"The objective of the study is to analyze the energy consumption from the production of the part including the support structures."

"Consumed power was quantified through time using an energy logger model Fluke 1732 (Fluke Corporation, Everett, WA)."

"L-PBF process can be divided into three main sections, in agreement with the energy breakdown of the generic model presented above."

"A schematic of the part was added to the graph to detail the energy consumption in three different stages of the manufacturing process where supports and the upper part demanded less energy than the central part's construction."

"Active energy was calculated from the measurements of the power consumed by the laser and the servo motors during the L-PBF process."

Conclusions

"This research paper presented the application of the unit process life cycle methodology (UPLCI) in laser powder bed fusion process, using the calculation of the energy consumption through the process, and the mass and the inert gas losses."

"In the case study presented in this article, a tibia fixation plate was successfully fabricated, and monitored through all the process, from the time in which the operator starts the calibration until the end of the fabrication with a total energy consumption of 62.68 MJ."

"By using UPLCI, users can improve the energy consumption of the process and monitor all the inputs and outputs of the process."

"The analysis of the energy consumed by the laser and active actuators during the process can be helpful to know the right amount of active and idle energy consumed during the process."

Model Construction of Material Distribution System Based on Digital Twin [46]

This is a machine-generated summary of:

Wang, Yunrui; Jiang, Ziqiang; Wu, Yue: Model construction of material distribution system based on digital twin [46].

Published in: The International Journal of Advanced Manufacturing Technology (2022).

Link to original: https://doi.org/10.1007/s00170-022-09636-9

If you want to cite the papers, please refer to the original.

For technical reasons we could not place the page where the original quote is coming from.

Abstract-Summary "Aiming at the problems of poor periodicity of workshop material distribution, difficult prediction of station material demand time node, and redundant distribution route, this paper proposes a model construction method of material distribution system based on digital twin."

"Build the digital twin model of material distribution system, establish the material demand time node dynamic prediction operation mechanism based on LSTM, accurately predict the station material demand time node, establish the material distribution route optimization model with the lowest total cost, and optimize the AGV route."

Introduction

"The introduction of digital twin technology can solve the problems of material demand time node disturbance, material route redundancy, and production data loss caused by the uncertainty of the workshop and comprehensively optimize the material distribution cycle due to the characteristics of virtual and real integration and mutual driving."

"In order to ensure the smooth operation of the control mode of the material distribution system and the orderly connection of each distribution cycle in the replenishment stage, build the digital twin model of the material distribution system, establish the material demand time node prediction model and material distribution route optimization model based on LSTM, and formulate the material distribution scheme with the lowest total distribution cost in line with the actual production of the workshop."

Integration of Digital Twin and Material Distribution System

"Before production, the virtual workshop conducts preliminary simulation based on the highly realistic model with the physical workshop, combined with the material demand time node, road congestion, and AGV usage predicted before production to obtain the preliminary material distribution scheme, and then optimizes the route in combination with the real-time data collected from the physical workshop and simulates and iterates again."

"The core function of MDS is to formulate a real-time material distribution scheme in line with the actual production of the workshop according to the accurate data provided by the twin data platform."

"With the continuous progress of the production process and the continuous interaction between the virtual and real workshops, the data analysis and noise reduction processing are carried out for the prediction results of the material demand time node, and then the material distribution path optimization model is loaded to generate the material distribution scheme and complete the dynamic replenishment stage."

Material Demand Time Node Forecast Based on LSTM

"The dynamic replenishment stage in the full-cycle material distribution is inseparable from the accurate prediction of the material demand time node, and the key factor determining the material demand time node is the material consumption time of the station."

"The accurate prediction of the material consumption time of the station is the basis for the prediction of the material demand time node."

"The historical material consumption time, worker proficiency, fatigue, and other data of the required predicted stations are obtained from the twin data platform."

"If the AGV terminal distribution is inconsistent with the station material demand time window when MDS guides the actual material distribution, feed back the actual material demand time window to the dynamic prediction module, adjust the LSTM prediction model, and update the knowledge base in real time."

"In the established LSTM neural network model, the worker's historical material consumption time, daily working hours, fatigue, and other related parameters can be input to establish the worker's independent prediction model."

Establish the Optimization Model of Material Distribution Route with the Lowest Total Cost

"When the logistics vehicle arrives earlier than the specified time window, the penalty number is T_e *ae i, when the logistics vehicle arrives later than the specified time window, the penalty number is T_l *be i. Set the relevant parameters of the established multi-objective VRPTW model: the production workshop has 1 distribution center and N vehicles for distribution activities."

"The transformed multi-objective function is as follows: In the formula: e—Penalty coefficient per unit time for delivery vehicles arriving earlier than the specified time window."

"In this VRPTW model, the total distribution cost function F can be introduced to convert the multi-objective function with the shortest vehicle travel distance and the smallest time window difference into a single objective function with cost as a single dimension."

Case

"The material BOM of each asynchronous card is not exactly the same, and when the asynchronous card type produced on the production line is switched, the production process is adjusted accordingly, and the material demand quantity, material demand time node, and AGV driving route of relevant stations are also changed."

"The operation mechanism of material distribution in asynchronous line workshop based on digital twin can realize the real-time control of material distribution process."

"If it fails, it will return to the twin data platform to modify the relevant parameters of the virtual workshop according to the real-time data of the asynchronous line workshop for simulation based on actual situation."

"Take the down procedure and test station 1 as an example, retrieve the historical data of station 1, and establish the LSTM material demand time node prediction model for prediction."

"Establish LSTM forecasting models for 9 stations of asynchronous card line to forecast material demand time window."

Conclusion

"This paper studies the application of digital twin technology in the process of material distribution and constructs the control mode of material distribution system based on digital twin and the digital twin model of material distribution system."

"A three-stage full-cycle material distribution mechanism based on material preparation stage, dynamic replenishment stage, and adjacent distribution cycle data transmission is proposed to comprehensively optimize the material distribution cycle."

"Through the example of asynchronous card line workshop, the effectiveness of the proposed method is verified, which provides a practical basis for realizing the material distribution system based on digital twin."

"The full-cycle material distribution mechanism based on digital twin technology needs to be further subdivided, and the high fidelity and synchronization of virtual workshops need to be further studied."

Optimizing Construction Schedules and Material Deliveries in City Logistics: A Case Study from the Building Industry [47]

This is a machine-generated summary of:

Nolz, Pamela C.: Optimizing construction schedules and material deliveries in city logistics: a case study from the building industry [47].

Published in: Flexible Services and Manufacturing Journal (2020).

Link to original: https://doi.org/10.1007/s10696-020-09391-7

If you want to cite the papers, please refer to the original.

For technical reasons we could not place the page where the original quote is coming from.

Abstract-Summary "We consider the problem of designing a logistics system to assure efficient urban construction processes for an innovative urban building area in the city of Vienna."

"We address the challenges of coordinating workers and the timely delivery and storage of material with the objective of optimizing resource-efficiency as well as reducing traffic related to construction tasks."

"The problem is formulated as a hierarchical optimization problem, incorporating tactical construction logistics planning as well as operational construction logistics planning."

"On the tactical level a time schedule of tasks for different construction phases and the corresponding material transports and storage decisions are optimized on a weekly basis, while on the operational level the daily optimization of material transports, modeled as an inventory routing problem, is addressed."

Introduction

"We aim at optimizing construction logistics operations encountered in the development process of newly built areas, comprising multiple construction sites."

"The limited number of transports that are allowed in the construction area, the problem of unproductive time of construction workers due to organizational deficiencies, and the lack of storage facilities on site are main challenges in construction logistics."

"By simultaneously optimizing construction schedules and transports, efficiencies in terms of time and cost, as well as related issues such as quality and environmental considerations are addressed [3]. We present a hierarchical modeling approach, which decomposes the CLOP addressed in the present paper into a tactical and an operational planning level."

"A time schedule of construction tasks for multiple construction sites is determined on the tactical planning level, considering material transports and storage decisions on a weekly basis."

Related Literature

"Artigues et al. [48] present (mixed-)integer linear programming formulations for the resource-constrained project scheduling problem."

"Decision makers regularly face a time-cost problem [49]: carrying out a project on time necessitates crashing certain tasks by exploiting extra resources, which increases the cost of a project."

"Construction scheduling problems are concerned with optimally scheduling tasks over time and allocating resources accordingly for successful completion of construction projects."

"Given precedence and/or resource constraints, the aim of construction scheduling optimization is to find a feasible schedule of tasks (with respect to the constraints) with minimum or maximum objective function value, e.g. shortest project duration or lowest cost [50]."

"In the present paper the hierarchical problem formulation of the tactical and the operational planning problems, where project scheduling and coordinated inventory and routing decisions are made, is addressed."

Problem Description

"The objective of the tactical Construction Logistics Optimization Problem (CLOP) is to determine a schedule for the construction tasks at each construction site within the time horizon and a schedule for transports of construction material to the construction sites, while minimizing total cost."

"Four decisions have to be made: (1) when to start a construction task, (2) how fast to perform a construction task, (3) when to transport material to each construction site, (4) how much material to transport to each construction site, to minimize total cost."

"We therefore face a trade-off between the restriction of transports to be carried out, the amount of material to be stored at the construction sites and the number of workers deployed at the same time to perform a task."

"Material for task p can only be delivered to construction site i, when i is visited at time t, cf."

Case Study

"This is due to the restrictive milestones of this scenario, which force the acceleration of construction tasks in order to avoid penalty cost."

"When acceleration of construction tasks is not possible (scenario NoA), milestones have to be exceeded and high penalty cost incur."

"Transport cost are lowest for scenario MS1, since the restrictive milestones lead to a denser construction schedule, which enables the bundling of material on delivery tours."

"Loose milestones (scenario MS2) lead to an average speed of construction tasks of 0.83, where only one task is accelerated to half of the normal duration."

"The average speed of execution of construction tasks is also comparatively slow, when personnel cost are high (scenario CS2), and obviously when no acceleration is allowed (scenario NoA)."

"It can be observed that the maximum number of transports is very restrictive, especially when milestones limit the construction phases to a short period of time (scenario MS1)."

Conclusion and Managerial Implications

"This work presents a case study from the building industry, focusing on the optimization of construction schedules and material deliveries to reduce total cost and waste of resources."

"We addressed the challenges of coordinating workers and the timely delivery and storage of material with the objective of optimizing resource-efficiency as well as reducing traffic related to construction tasks."

"When storage of material on site is not possible, transports cannot be organized efficiently, i.e. through the bundling of material delivered to several construction sites located close-by."

"The results also show that re-planning during project execution might be necessary due to supply chain uncertainties, i.e. when material is known to be unavailable at the operational planning level."

"Using the proposed hierarchical model, delivery and storage decisions on the operational level can be revised to incorporate dynamic information on material availability, while still following the tactical construction schedule."

A Digital Twin-Based Layout Optimization Method for Discrete Manufacturing Workshop [34]

This is a machine-generated summary of:

Guo, Hongfei; Zhu, Yingxin; Zhang, Yu; Ren, Yaping; Chen, Minshi; Zhang, Rui: A digital twin-based layout optimization method for discrete manufacturing workshop [34].

Published in: The International Journal of Advanced Manufacturing Technology (2021).

Link to original: https://doi.org/10.1007/s00170-020-06568-0

If you want to cite the papers, please refer to the original.

For technical reasons we could not place the page where the original quote is coming from.

Abstract-Summary "This paper focuses on a discrete manufacturing workshop layout optimization based on digital twin, in which the workshop layout problem is solved by twin data fusion, information and physical interaction fusion, and data analysis and optimization."

"A sub-framework of digital twin-based workshop partitioning is established and the workshop partitioning is optimized via simulation analysis."

"A sub-framework of digital twin-based equipment layout optimization is presented, in which equipment layout decisions are made by real-time data collection and value-added processing of twin data."

"A sub-framework of digital twin-based distribution route optimization is developed for the workshop."

Introduction

"The layout optimization of discrete manufacturing workshop can promote the realization of intelligent manufacturing."

"In order to enhance the real-time feedback capability, optimization of the discrete manufacturing workshop layout needs to be combined with the advanced technologies."

"There are a few studies on applying digital twin to the layout optimization of discrete manufacturing workshops."

"This paper puts forward a new workshop layout optimization method based on digital twin to improve the overall quality and efficiency of the discrete manufacturing process."

"In order to solve the above workshop layout problems, a digital twin-based discrete manufacturing workshop layout optimization method is established based on three sub-frameworks of workshop partitioning, equipment layout optimization, and distribution route optimization."

Digital Twin

"Compared with traditional workshop, the physical layer workshop has the ability to perceive and integrate real-time data, as well as the ability to integrate the human-machine-physical-environment elements of the workshop."

"During production, the digital layer continuously accumulates real-time data and knowledge of the physical layer [31]."

"Information fusion layer refers to the optimization through cyber-physical interactive fusion and analysis of twin data."

"Driven by data from the physical and digital layers, the information fusion layer generates a resource allocation plan and initial production plan that meets the task requirements and constraints."

"Before starting production, it revises and optimizes the production plan based on simulation, evaluation, and optimization data of the production plan from the digital layer."

"In the production process, the simulation, verification, and optimization results of the production status in the physical layer and the digital layer are continuously fed back to the fusion layer."

A Digital Twin-Based Layout Optimization Method for Discrete Manufacturing Workshop

"The sub-framework of digital twin-based workshop partitioning operates as follows: physical entities are adjusted according to the layout optimization method, the data collected from the underlying equipment of physical layer is transmitted to the digital layer through the processing of information fusion layer, through the simulation analysis, the digital layer and the simulation results are fed back to the physical layer simultaneously through information fusion layer, and the information fusion layer continues to interact with the physical layer and the digital layer until the problem of unreasonable workshop division is effectively solved [51]."

"The physical workshop and virtual workshop run simultaneously [52]. Real-time data collection, twin data value-added processing, and workshop status monitoring are carried out based on data collection technology. Considering equipment selection, personnel allocation, storage area, etc., the data is continuously and iteratively optimized to obtain optimal layout optimization decisions, so as to further solve the problems of overlong distance between work stations and unreasonable distribution of equipment and facilities."

Case Study

"According to the sub-framework of digital twin-based workshop partitioning, the workshop partitioning plan of welding production workshop is made based on the layout optimization method."

"The twin data from simulation result of virtual workshop is mapped back to the physical workshop to remake the equipment layout optimization plan."

"By applying the digital twin-based discrete manufacturing workshop layout optimization method, we can improve the product quality and production efficiency, improve the level of the on-site management and the operation safety, reduce the costs, and meet the market demands."

"The digital twin-based layout optimization method is applied to optimize the layout of the door line and the front wall line of a welding production workshop."

"The optimization plan implementation effect shows that the digital twin-based discrete manufacturing workshop layout optimization method has a good effect in the welding production workshop."

Conclusions and Future Work

"We focus on a discrete manufacturing workshop layout optimization based on digital twin."

"A sub-framework of digital twin-based equipment layout optimization is presented."

"A digital twin-based discrete manufacturing workshop layout optimization method is established based on the three sub-frameworks."

"There are few researches on the workshop layout optimization using digital twin."

"Future research on this topic needs to be undertaken on better combining digital twin technology with workshop layout optimization and standardizing and personalizing application methods."

Applying DFA and Goal Programming to Improve Economic Efficiency, Material Handling Convenience, and Sustainability of a Product Packaging System [53]

This is a machine-generated summary of:

Chiu, Ming-Chuan; Chung, Wu-Hsun; Lin, Hsin-Han: Applying DFA and goal programming to improve economic efficiency, material handling convenience, and sustainability of a product packaging system [53].

Published in: Research in Engineering Design (2021).

Link to original: https://doi.org/10.1007/s00163-020-00355-4

If you want to cite the papers, please refer to the original.

For technical reasons we could not place the page where the original quote is coming from.

Abstract-Summary "The concept of Design for Assembly (DFA), proved in earlier studies to be effective at improving product manufacturing operations, is applied to the packaging system."

"We formulate the packaging system as a mathematical model for three different objectives."

"To the application of DFA concepts to a packaging system, the contributions of this research lie in the development of a mathematical model (using integer programming and goal programming) for calculating the cost, handling time, and sustainability of the objectives in line with the design needs of the multi-level packaging size from the perspective of a supply chain to provide a range of solutions offering increased options for firms."

Introduction

"Recent studies have suggested that package design should not be focused on the "product's packages" alone but needs to be expanded at the logistics level across the entire supply chain, from manufacturers, to retailers, to customers, or even to the end of product life, including disposal and recycling activities [54–57]."

"Most relevant studies have tended to focus on investigating and showing the influence of packaging on supply chains from the aspects of logistic efficiency or sustainability [54–57]."

"This paper aims to develop an effective method to help packaging designers and planners determine the best packaging design solution that balances the multiple objectives of cost, efficiency, and sustainability across an entire supply chain."

"A goal programming (GP) model based on the IP model was created for use in determining an appropriate packaging design solution at multiple levels in terms of various priorities of the objectives across a supply chain."

Related Literature Review

"Other researchers have used various approaches to investigate the influence of packaging logistics on the environmental performance of supply chains."

"García-Arca et al. [56, 57] conducted a case study involving a retailer in Spain to examine sustainable performance in supply chains when a sustainable packaging logistics plan was implemented."

"According to the reviewed literature, various methods have been developed to evaluate the efficient or sustainable performance of a packaging system in supply chains, but as of yet, no study provides an effective method for determining the design features of a multi-level packaging system to achieve better supply chain performance."

"To address the needs of packaging system design with respect to this matter, the aim of this research is to propose a methodology that meets the requirements of logistic processes to identify a balanced design solution for a multi-level packaging system considering the design requirement of cost, operation efficiency, and sustainability across a supply chain."

Methods

"The performance of the packaging system in a supply chain using this model is threefold: economic performance, efficient performance, and sustainable performance, measured by the cost (C: objective 1), material handling time (D: objective 2), and carbon emission (S: objective 3) accrued in the processes, respectively."

"The objective of the model is to determine the package size and material type in the packaging system that will best minimize overall cost, handling time, and carbon emissions in the germane supply chain activities."

"The constraints associated with DFA in the model are linked to the requirements or feasibility of the logistic processes related to each package level."

"Step 2 Calculate remaining DFA points for assignment to constraints The main criteria related to package design features include weight and length in assembly property (AP) and weight, size, and shape in component property (CP)."

Case Study

"This research demonstrates the proposed model using a box of binder clips as a product example and explores how the characteristics of the packaging system affect the cost, efficiency, and sustainability across the supply chain."

"The box of binder clips (primary package) weighs 0.1 kg (primary package product weight)."

"The DFA index for the secondary and tertiary package constraints are: The DFA points remaining to define the relevant DFA constraints can be calculated as 19.5 (27.5–8 = 19.5)."

"Because the primary package has been clearly defined, we only need to determine the DFA constraints for the secondary and tertiary packaging."

"The DFA constraints relevant to package design features (decision variables) such as the limitation of package weight, size, and shape are listed as follows."

Results and Discussions

"We find the proposed IP model can successfully identify an appropriate package design solution for each individual objective."

"This study's proposed model not only provides a mathematical optimization method to identify a better packaging system design solution from among a large set of package design combinations, but also successfully links package design features (e.g., package sizes at different levels) with the design requirement of a packaging system for reducing handling difficulty during the packaging process via DFA index scoring."

"These goal values derive from the optimal value for each objective of the IP model, which is the minimal value the packaging system can achieve."

"The GP model performs better for identifying a design solution in lower cost, handling time, and carbon emission under the packaging system designers' various priorities of design objectives."

Conclusions

"This study is the first to formulate a multi-level packaging system design using a mathematical IP model."

"Based on the proposed model, a decision maker can easily find an integrated and balanced package design solution at different packaging levels to achieve lower cost, less handling time, and less carbon emission during the packaging process."

"In this work a GP model is developed based on that IP model, which can help package designers determine appropriate packaging design solutions that are dependent on various priorities of the design objectives."

"Packaging designers and planners can adopt appropriate sustainability measures which are able to cover comprehensive environmental impacts and integrate them into the proposed models."

Prediction and Optimization of Abrasive Wear Loss of Ultrahigh Strength Martensitic Steel Using Response Surface Methodology, Harris Hawk and Artificial Neural Network [58]

This is a machine-generated summary of:

Sharma, Varun; Sharma, Sanjay; Verma, Om Prakash; Bhardwaj, Bhuvnesh; Sharma, Tarun Kumar; Pachauri, Nikhil: Prediction and optimization of abrasive wear loss of ultrahigh strength martensitic steel using response surface methodology, Harris Hawk and artificial neural network [58].

Published in: International Journal of System Assurance Engineering and Management (2021).

Link to original: https://doi.org/10.1007/s13198-021-01160-5

If you want to cite the papers, please refer to the original.

For technical reasons we could not place the page where the original quote is coming from.

Abstract-Summary "Three-body abrasion wear problem in real industry application has been significantly reduced by replacing the heat-treatable steels with the newly developed ultrahigh strength martensitic steel."

"The experimental results obtained were used to develop the quadratic model using Response Surface Methodology."

"For the validation of the performance obtained from statistical analysis, experimental data was employed to build the prediction models using Neural Networks (NNs)."

"The proposed improved Generalised Regression-NN (GR-NN) was found to be an efficient and explorative predictive model in comparison to the Levenberg–Marquardt Perceptron-NN and Adaptive Linear-NN."

"The GR-NN was found to be most accurate owing to the minimum error as compared to other prediction models."

"The prediction ability of the GR-NN model with Harris Hawk Optimization was found to be better as compared to quadratic model, which was further validated using Scanning Electron Microscopy (SEM)."

"The proposed model is efficient, accurate and encouraging for the prediction of wear loss in the industrial applications involving abrasion wear conditions."

Introduction

"The wear performance of a newly developed abrasion-resistant steel material, namely Hardox 400, has been compared with the wear plates of chromium cast iron and two cladded plates using statistical approach [59]."

"Different techniques of Design of Experiments (DOE) approach have been also successfully implemented to establish a correlation between the process parameters of three-body abrasion test and wear performance of the materials [60–62]."

"The JFE EH400 steel, which possesses excessively high tensile strength and hardness along with high toughness owing to microalloying additions has neither been extensively investigated nor well documented for its three-body abrasion wear behaviour."

"Pertaining to the above discussion and, attributable to the efficiency of ANNs, the main contributions of this work is to investigate the three-body abrasion wear performance of JFE EH400 steel."

Experimental Procedure

"The metallic wear specimens of length 76.2 mm, width 25.4 mm and thickness 12.7 mm have been extracted for abrasion wear test from 15 mm thick substrate material plate using WEDM (Wire-cut electric discharge machining)."

"Three-body abrasion wear tests have been carried out using silicon carbide (Hardness ≈ 2400 VHN) abrasive particles of size range 150 μm–212 μm evaluated using a sieve analysis method."

"The abrasion wear testing conditions have been developed using design of experiments (DOE) approach, which is defined as a systematic approach for planning of experiments, collection and analysis of data for establishing an empirical relationship between dependent and independent variables of a system by minimizing the number of experiments [63]."

"During the working condition, the abrasive particles have been fed through a chute into the space between the wheel and the test specimen, and abraded for a specific time interval."

Proposed Methodology

"Three ANN based models, Levenberg–Marquardt Perceptron Neural Network (LM-NN), Adaptive Linear Neural Network (ADALINE-NN) and Generalized Regression Neural Network (GR-NN) have been developed to predict abrasion behaviour."

"The computation of number of neurons in the hidden layer has been accomplish by a trial and error approach based on the mean square error, MSE, mean absolute percentage error, MAPE, and root mean square error, RMSE criterion."

"In LM-NN, Levenberg–Marquardt algorithm is used to map the input process parameters including load, revolutions, speed, and flow rate to output as wear loss by minimizing Mean Square Error (MSE) between target and predicted output of network."

"GR-NN is basically a probabilistic neural networks-based function approximation algorithm which predicts the output of given input data."

"GR-NN approach has been employed to estimate the most probable value of output as wear loss which minimizes the MSE."

Results and Discussion

"The most influential factor affecting the abrasion wear loss is the applied load with the percentage contribution of 45.59, followed by the number of revolutions of the rubber wheel (38.72%), a flow rate of abrasive particles (5.18%) and rotational speed of the wheel (4.39%)."

"For load, revolutions and flow rate of abrasives, the curves show a positive slope indicating an increase in wear loss with the rise in values of the variables."

"The demonstrated results reveals that the minimum wear loss of 0.469 has been obtained with maximum desirability of 0.937 at the load of 60 N, 1000 revolutions, speed of 4 m/s and abrasives flow rate of 200 g/min."

"The percentage error between the predicted values and the experimental values of wear loss for GR-NN model optimized through HHO lies within the uncertainty range of 3–4%."

Conclusions

"For effective prediction of wear loss, the HHO has been used for training of GR-NN model by optimizing the weights, biases and number of neurons and attains the minimum value of wear loss (0.4900 g) with the optimum process values within 36 iterations."

"Despite the use of lower number of experimental results, the prediction ability of the GR-NN model with HHO for JFE EH400 steel has been found to be better as compared to the quadratic model obtained through RSM."

"Owing to the uncertainty of only ±6%, both the RSM and GR-NN models can be used effectively to predict the abrasion wear performance of the newly developed ultra-high strength martensitic steels."

"The mathematical models developed in this study helps in predicting the wear performance of JFE EH400 steel only within a specified domain of working environment."

Combining Life Cycle Assessment and Manufacturing System Simulation: Evaluating Dynamic Impacts from Renewable Energy Supply on Product-Specific Environmental Footprints [64]

This is a machine-generated summary of:

Rödger, Jan-Markus; Beier, Jan; Schönemann, Malte; Schulze, Christine; Thiede, Sebastian; Bey, Niki; Herrmann, Christoph; Hauschild, Michael Z.: Combining Life Cycle Assessment and Manufacturing System Simulation: Evaluating Dynamic Impacts from Renewable Energy Supply on Product-Specific Environmental Footprints [64].

Published in: International Journal of Precision Engineering and Manufacturing-Green Technology (2020).

Link to original: https://doi.org/10.1007/s40684-020-00229-z

If you want to cite the papers, please refer to the original.

For technical reasons we could not place the page where the original quote is coming from.

Abstract-Summary "As an established method used in sustainability management, Life Cycle Assessment aims at calculating the environmental impacts from all life cycle stages of a product or system."

"In order to cope with shortcomings of the static character of life cycle models and data gaps this approach combines Life Cycle Assessment with manufacturing system simulation."

"The two life cycles of product and production system are merged to assess environmental sustainability on product level."

"Manufacturing simulation covers the production system and Life Cycle Assessment is needed to relate the results to the final product."

"This combined approach highlights the influences from dynamic effects in manufacturing systems on resulting life cycle impact from both product and production system."

"It can be demonstrate that Life Cycle Assessment can enhance the process simulation and help identify hot-spots along the life cycle."

Introduction and Motivation

"For some types of products (e.g. electronics), the manufacturing stage already dominates the environmental impact from the whole life cycle and combined with increased affluence of the growing population this will lead to higher total impacts of manufacturing [65]."

"These methods and tools have to consider the entire life cycle of a manufacturing system including environmental impacts related to production (e.g. construction of a factory building and production of machines) and end-of-life."

"The environmental impacts related to energy generation are not constant but time-dependent based on the availability of renewable energy sources [66]."

"An approach for combining LCA with simulation is proposed, enabling a holistic evaluation of environmental impacts of a manufacturing system over its life cycle."

"The proposed approach highlights the relevance of the influences from dynamic effects in manufacturing systems on resulting life cycle impacts from both product and production system."

Life Cycle Assessment and Manufacturing System Simulation

"Different methods have approached that problem to quantify the environmental performance of companies (e.g. [67–69]) or manufacturing plants [70] or at more detailed levels in the manufacturing system [71–74], but no model exists which quantifies all relevant environmental impacts holistically, taking a life cycle perspective."

"Some of them use simulation to determine the cumulative amounts of inputs (e.g. energy, materials, supporting media) and outputs (e.g. waste or finished goods) of processes, for example to assess lean and green strategies in manufacturing systems [75, 76]."

"The holistic assessment of interactions between different manufacturing system elements has rarely been done [i.e. process, production equipment and technical building services (TBS)] based on manufacturing simulation."

"This leads most likely to additional environmental impacts of manufacturing due to additional material inputs, and this should be assessed to avoid potential sub-optimization within the manufacturing process (production process vs. equipment vs. TBS)."

"This combined approach highlights the relevance of the influences from dynamic effects in manufacturing systems on resulting life cycle impacts from both product and production system."

Combined Holistic Approach: Life Cycle Manufacturing System Simulation

"Being a combined approach, manufacturing system specific (e.g. state-based behaviour of processes) and LCA specific (e.g. environmental impact of energy supply) parameters are included, as well as combined parameters (e.g. target production levels)."

"The related environmental impact from both energy supply and the production of required equipment (e.g. raw materials to manufacture a machine) is accounted for, as well as a potential positive impact from material recovery."

"Infrastructure: All machinery (e.g. robots, welding guns, installations for compressed air and centralized lubricant system) including their material use and impacts from their manufacturing, maintenance and replacement during the lifetime of the production line, allocated by total estimated production volume of the line over its lifetime HVAC and lighting: All indirect energy demand to keep production running, based on the occupied area of all components including needed space for supply of intermediate products."

Application of New Approach

"For the use phase, upstream processes of energy provision were considered e.g. the production of solar cells or the average environmental impact of the German grid mix."

"Following the process related impacts per product (blue dots), a clear trend of less impact can be identified for the other scenarios due to their increasing share of renewable energy."

"The TBS consumption impacts indicate no change except for the Production Levels Scenarios I–IV where the cycle time per product has been increased."

"The Production Level I Scenario where the throughput has been increased seems to have the least environmental impacts whereas the Battery Storage III Scenario has overall even more impact than the Grid Mix Scenario for all impact categories except GWP."

"Implementing renewable energy can lower the process impact, whereas the Battery Storage Scenarios show an increase in infrastructure impacts and are in total less beneficial."

Conclusion and Outlook

"This leads to the conclusion that an upfront basic test (e.g. if energy data show a high dynamic difference between grid mix and energy mix of on-site VRE supply and demand) might be beneficial to indicate if more detailed modelling is required."

"Applying the integrated LCA/manufacturing system simulation method to a wider range of different product types and production systems (e.g. continuous or batch production processes, dividing and converging material flows) would be beneficial to allow more generalized conclusions on a methodological level."

"From an energy flexibility perspective, including LCA into a structured target search for improving energy flexibility of manufacturing systems (e.g. comparing embodied energy storage, battery storage, wind/solar/CHP supply options) under ecological indicators obtained by LCA is of interest."

"Considering agent-based modelling of manufacturing systems and products, assigning environmental impacts to single product units, depending on system dynamics (i.e. which machine has processed which product utilizing a specific share of renewable energy), would allow a more specific allocation of manufacturing impacts to products."

Production Management for Mass Customization and Smart Cellular Manufacturing System: NSGAII and SMPSO for Factory-Level Planning [77]

This is a machine-generated summary of:
Maalouf, Elie; Daaboul, Joanna; Le Duigou, Julien; Hussein, Bassam: Production management for mass customization and smart cellular manufacturing system: NSGAII and SMPSO for factory-level planning [77].
Published in: The International Journal of Advanced Manufacturing Technology (2022).
Link to original: https://doi.org/10.1007/s00170-022-09188-y

If you want to cite the papers, please refer to the original.
For technical reasons we could not place the page where the original quote is coming from.

Abstract-Summary "This approach is based on three decision levels: factory-level (master planning module), shop floor level (bidding system) dealing with unexpected events, and cell level."

"The approach integrates planning, scheduling, and material handling allocation while considering real-time data from the supply chain."

"A mathematical model for factory-level planning is proposed with two sequence-based resolution approaches implemented on two metaheuristics, NSGAII and SMPSO."

Introduction

"Smart process planning and scheduling are requirements for smart production management which is part of smart factory which in turn is part of smart manufacturing."

"A smart factory has many characteristics among which offering mass customization, having a flexible manufacturing system, being connected to its supply chain, having visibility and optimized decision-making, integrating predictive maintenance, and requiring new planning methods [78]."

"Smart production management is required to highly adapt and change to handle these disturbances and reduce overall costs while satisfying customer needs including order delay."

"Their results show that whereas production scheduling/shop floor control is the activity that most incorporates the new capabilities provided by Industry 4.0, there is still a lack of decision support systems, frameworks, architectures, and models to advance and consolidate smart manufacturing planning and control."

Related Work

"The literature review focused at first on smart and/or production planning and/or scheduling for CMS, then for MC, and finally for dynamic cellular manufacturing systems."

"Half of the works considered integrated process planning and scheduling (4 out of 8), highlighting the importance of this integration for MC."

"Only one work is focusing on production management (and more specifically scheduling) for MC in a CMS."

"This work also contributes to this research field by focusing on production management in CMS for MC and by considering customized customer orders."

"No work included all topics that this research tackles, meaning MC, CMS, integrated planning and scheduling, and considering real-time data from the supply chain."

"This paper contributes to this research field by proposing a method for production management (integrating process planning, scheduling, and resources allocation) for CMS and MC considering real-time data from the factory and the supply chain."

Proposed Framework: Smart Production Management for Mass Customization and Cellular Manufacturing System

"These are (1) the master planning module at the factory-level, (2) the dynamic scheduling module at the shop floor level, and (3) the cell planning module at the cell level."

"The master planning module provides a first schedule for every cell as well as a cell-related job due dates."

"Upon receiving information from the master planning module, the cell planning module disaggregates the information from the master plan and develops a cell-specific schedule respecting the master plan's defined due dates."

"The cell plan will do an integrated process planning and scheduling to define which machine of the cell will do the operation and at which time."

"If a cell is not capable of respecting the master planning module defined due dates, the dynamic scheduling module is triggered."

"An update of the schedule and process plan takes place and is transmitted to the cell planning module."

Master Planning Module

"This problem aims to make the following decisions: Cell level process planning: allocating operations to specific cells Shop floor level scheduling: specifying the process starting time of an operation on its allocated cell Material handling allocation: allocating the material handling equipment for an inter-cell job transfer Notations Input information Demand-related information Production-related information Material handling-related information Supply chain-related information Cost-related information Decision variables Objective functions First objective function: where Second objective function: where f_1 = cost of operation processing: f_2 = cost of setup time: f_3 = cost of transfer within cells: f_4 = cost of using material handling equipment: Subject to: The following equations state the constraints related to an operation starting time."

"In this time plan, the earliest possible time to start operations on the chosen cell is defined, considering the earliest possible slot to move the job using material handling equipment l. These sequences are solutions for metaheuristics algorithms."

Numerical Example

"It is almost equivalent for the first objective function (minimizing tardiness) (32.07 versus 33.18 and 22.85 versus 22.77). (III) Concerning computation time, NSGAII is better performing than SMPSO for both approaches. (IV) Concerning the other performance metrics, NSGAII seems to be better overall performing than SMPSO for the trio-sequence-based approach, and the opposite is true for the duo-sequence approach: (a) For QM, NSGAII is better for the trio-sequence-based approach, whereas SMPSO is better for the duo-sequence approach. (b) For MID, NSGAII is better for the trio-sequence-based approach, whereas SMPSO is better for the duo-sequence approach. (c) For DM, NSGAII is better performing than SMPSO for both approaches. (d) For NPS, NSGAII and SMPSO are almost equivalent for both approaches. (e) Also, it is possible to state that SMPSO is better overall performing for the duo-sequence approach than for the trio-sequence-based approach."

Conclusion and Perspectives

"This paper presented a production management method for MC and CMS based on distributed and smart planning approaches."

"The method is formed of three main modules: the smart master planning module that takes real-time data from the supply chain into consideration for first planning, the cell planning module that focuses on optimizing planning at a cell level, and the dynamic planning module that intervenes if an unexpected event occurs."

"The detailed mathematical multi-objective formulation of the master planning module is presented."

"The results show that the duo-sequence-based approach is better performing."

"Other examples should be run to better analyze the metaheuristics performance."

"These metaheuristics' performance could be improved if combined with other approaches like simulation."

"Future works will focus on further developing the cell planning module and the dynamic scheduling module and testing the full proposition with a numerical example and a case study."

A Heuristic-Based Non-linear Mixed Integer Approach for Optimizing Modularity and Integrability in a Sustainable Reconfigurable Manufacturing Environment [79]

This is a machine-generated summary of:

Massimi, Elisa; Khezri, Amirhossein; Benderbal, Hichem Haddou; Benyoucef, Lyes: A heuristic-based non-linear mixed integer approach for optimizing modularity and integrability in a sustainable reconfigurable manufacturing environment [79].

Published in: The International Journal of Advanced Manufacturing Technology (2020).

Link to original: https://doi.org/10.1007/s00170-020-05366-y

If you want to cite the papers, please refer to the original.

For technical reasons we could not place the page where the original quote is coming from.

Abstract-Summary "To reach these goals, a key paradigm called sustainable manufacturing can be coupled with reconfigurable manufacturing systems (RMSs)."

"The coupling of sustainability concerns with RMS characteristics is a basis to develop a new generation of sustainable production systems."

"This paper outlines sustainability in a reconfigurable environment from an energy consumption point of view."

"A non-linear mathematical model is developed to optimize the energy consumption in an RMS through a redefinition of two of its core characteristics, i.e., modularity and integrability."

"The objective is to minimize the energy consumption of the system by selecting the most suitable modular machines from a set of candidate machines."

Introduction

"To enable this customized flexibility, rapid responsiveness, and reduce the reconfiguration time and recently environmental challenges, a RMS should possess six core characteristics [80, 81]: modularity, the compartmentalization of operational functions into units that can be manipulated among alternate production schemes for optimal arrangements; integrability, the ability to connect modules rapidly and precisely by a set of mechanical, informative, and control interfaces facilitating integration and communication; diagnosibility, the system ability to self-reading its current state to detect and diagnose the root causes of product defects, quickly correcting them; convertibility, the ability to easily transform the functionality of existing systems and machines to suit new production and market requirements; customization, the system and machine flexibility limited to a single product family, thereby obtaining customized flexibility; scalability, the ability to modify easily the production capacity adding or removing resources and changing the system components."

Literature Review

"In this section, the focus will be on RMS characteristics (especially modularity and integrability) and sustainable manufacturing while emphasizing energy consumption and process plan generation."

"Khezri et al. [82] took into consideration the multi-objective single-product process plan generation problem in RMS, where energy consumption is used as one of the criteria to minimize as well as the total completion time and the total production cost."

"The three main challenges facing research on energy efficient RMS are [80]: (1) Define new metrics for energy consumption and greenhouse gas emissions which could be considered during the design of RMS and could thus impact the choice of modules; (2) Propose optimization methods for the reconfiguration of manufacturing systems taking into account energy cost or power peak constraints; (3) Propose reactive methods to dynamically adapt RMS to unexpected changes on power limitations."

Problem Description and Formulation

"The first subsection provides the problem description, where the two characteristics of RMSs (i.e., modularity and integrability) are described considering the core features that enable energy consumption to converge in a reconfigurable manufacturing environment."

"From the energy point of view, modularity is defined as the energy consumption of basic modules, assembling and disassembling auxiliary modules for each configuration as well as the changing machine consumption (Wh)."

"This research work presents a RMS design approach based on a heuristic for selecting the best set of modular machines from a list of candidates as well as the needed set of basic and auxiliary modules that compose these machines."

"The purpose is to design a RMS while minimizing its energy consumption through the redefinition of the two characteristics of integrability and modularity."

Proposed Approach

"All possible combinations of operation sequences (N) that respected the precedence constraints (i.e., feasible solutions) are generated."

"The N sequences of operations are explored one by one, where for each one, all the possible sequences of machines and corresponding configurations are generated."

"The sequence of machines with minimum value of energy consumption is selected and compared with the initial value."

"The heuristic explores and evaluates all the possible solutions (process plans) and rank them based on the energy consumption value."

"The best sequence of operations and the best sequence of machines as well as the needed auxiliary and basic modules are those corresponding to the minimum value of energy consumption stored as E_{min}."

Experimental Results and Analyses

"In each scenario, the number of operations changes as well as the number of machines, auxiliary modules, and basic modules."

"Comparing the two process plans, less energy consumption can be reached by balancing the choice of machine changes with the number of assembling and disassembling auxiliary modules between operations."

"The aim is to identify the choice that leads to a lower energy consumption when an operation cannot be carried out with the same machine configuration as the previous operation; therefore, it is necessary to choose between machine change or auxiliary module change (assembling and/or disassembling)."

"What is noticeable from the results of the third scenario is that even though the total number of interfaces is greater in the best solution, the minimum energy consumption value is obtained by balancing the number of machine changes with the number of modules used and the consequent number of interfaces needed between them."

Conclusions and Future Work Directions

"Work, the problem of sustainable process plan generation to design a RMS is addressed."

"The research work is inspired by the actual industrial scenario while looking towards the challenges that the world will face in the next few years, where two key-concepts have emerged: sustainability and reconfigurability."

"Work, sustainability and reconfigurability of RMSs are evaluated simultaneously by integrating energy consumption while modeling two core characteristics of RMS namely "integrability" and "modularity.""

"Based on the energy minimization model, a heuristic method is developed to solve the optimization problem and generate process plans."

"Through this heuristic, it was possible to generate illustrative examples of low-complexity process plan generation problems to show the applicability of our model."

"Future and challenging research direction on how to include the rest of the other RMS characteristics in the model might be a natural extension of this work."

Exploring the State-of-the-Art Service Composition Approaches in Cloud Manufacturing Systems to Enhance Upcoming Techniques [83]

This is a machine-generated summary of:

Hayyolalam, Vahideh; Pourghebleh, Behrouz; Pourhaji Kazem, Ali Asghar; Ghaffari, Ali: Exploring the state-of-the-art service composition approaches in cloud manufacturing systems to enhance upcoming techniques [83].

Published in: The International Journal of Advanced Manufacturing Technology (2019).

Link to original: https://doi.org/10.1007/s00170-019-04213-z

If you want to cite the papers, please refer to the original.

For technical reasons we could not place the page where the original quote is coming from.

Abstract-Summary "Cloud manufacturing (CMfg) is a new paradigm that has been known as a promising integrated technology in which distributed resources of manufacturing are integrated and transformed into manufacturing services and handled centrally."

"It permits multiple users to request services simultaneously by submitting required tasks to a manufacturing platform."

"The service composition and optimal selection (SCOS) is the fundamental issue of CMfg that aims to select appropriate services from available manufacturing cloud resources to complete the manufacturing task and satisfy the users."

"As far as we know, this paper is the first research that tries to investigate and discuss the CMfg-SCOS methods in a systematic way."

"The open challenging and the lacks within this issue have been discussed which can be applied to upcoming research."

Introduction

"Most of the task execution and composition of hardware services require the users' monitoring and control."

"The experience factors such as the average task execution efficiency, amount of completed tasks, and user rating can be utilized to assess the human resources service [84]."

"The users' requirements in the CMfg are classified into two types including multiservice requirement task (MSRTask) and single-service requirement task (SSRTask) [85]."

"The vital challenge in CMfg is completing a MSRTask optimally through selection and composition of manufacturing services within their accessible time frames by considering restrictions comprising quality of service (QoS) and transportation [86]."

"Finding the optimal manufacturing service composition from available resources to complete the manufacturing task is defined as the service composition and optimal selection (SCOS) issue [87] which is identified as a Np-Hard problem [88–95]."

"The SCOS problem includes combining different services with various capabilities aiming to provide services for highlighting some of the complex manufacturing tasks."

Background

"The CMfg has been developed as a new business paradigm for the manufacturing industries wherein the distributed Mfg resources are transformed into Mfg cloud services and organized and handled in the CMfg platform [85]."

"Various users of various industries could submit their necessities and utilize different sorts of cloud services via the CMfg platform [96]."

"As specified in [97, 98], the resource service composition's life cycle in CMfg can be classified into four stages including design, fabrication, execution, and post-processing which are described in below."

"The users' requested tasks, which are decomposed into abstract composite services and subtasks, are assembled via flow match and function."

"The basic composition structures can be aggregated to obtain the QoS value of any composite CMfg service (CCMfS) [99]."

Related Work

"As another work, authors in [100] have reviewed systematically service composition techniques in the IoT. They have discussed methods in two groups including the methods with focusing on functional feature aspect and the methods with focusing on non-functional feature aspect."

"An evaluation of the service composition methods in cloud computing has been proposed by researchers in [101] in which the selected techniques have been reviewed in regard to the algorithmic taxonomy including agent-based, heuristic-based, and framework-based techniques."

"By analyzing and investigating different methods in each category, this paper has given an overview about the techniques on a bio-inspired algorithm in web service composition and point out future directions."

"A methodical literature analysis of the QoS-aware service composition techniques in regard to the computational intelligence methods in a cloud environment has been proposed by authors in [102]."

"Just four studies have used the systematic literature review (SLR) method for reviewing the service composition techniques."

Research Methodology

"RQ5: Which research, journals, and authors are hot in the field of SCOS in CMfg?"

"The authors have automatically searched keywords and as a result found 1560 articles from the conference, journal papers, and books."

"In order to select more relevant articles, authors have limited the search to the title of papers; consequently, they found 93 papers."

"Authors have conducted a three-phase process to select the papers for a SLR as follows: Phase 1: Automatic search considering the keywords. Phase 2: Collection in regard to the article title. Phase 3: Selection in regard to the databases and type of article In phase 1, in order to find appropriate papers, authors have searched the keywords comprising service selection, service composition, cloud manufacturing, service composition in CMfg systems, and service selection in CMfg systems."

"In this phase, authors have applied the search strings to the article title; subsequently, 93 articles have been recorded."

Service Composition/Selection Mechanisms

"As another research, to resolve the scalable IoT service composition in CMfg, authors in [103] have proposed a method that adopts the genetic algorithm based on empirical knowledge (EK-GA)."

"Although this method could attain a high success rate, good scalability, low cost, increased reliability, and low energy consumption, it suffers from low availability and high response time."

"Although the proposed method improves availability, cost, energy consumption, and execution time, it provides low scalability and low reliability."

"The proposed method achieves high availability, high flexibility, low cost, and low execution time; however, it provides low scalability."

"This method improves cost, execution time, energy consumption, maintenance, and scalability; nevertheless, it provides low reliability."

"The proposed method achieves high reliability, high availability, high reputation, low cost, and low execution time; however, it provides low scalability."

Findings and Discussion

"Fitness function explicitly: This metric is employed to examine the clarification of objective function adopted in the investigated papers."

"It seems that the studies which adopted case study [87, 104–114] or real data [115] are more likely to be applied in real environment such as real manufacturing environments or industries. Penalty function: Some papers adopt penalty function to punish bad solutions."

"Adopted algorithms: This metric only is investigated on Heuristic-based studies."

"Adopted datasets: In order to simulate and investigate the proposed methods, each research has adopted different datasets."

"The investigated studies were examined in case of the adopted dataset."

"Most of the researchers adopted random datasets, while very few researchers used real data such as real manufacturing environment or real manufacturing industry."

"Researchers mostly rely on a study with real data rather than the work with random dataset."

Open Issues and Future Trends

"Task constraints, continuous task arrival, and inventory are some of the parameters that can be taken into account to explore more efficient algorithms and effective techniques, and also establish more accurate models for SCOS problem in CMfg."

"Conducting some proper experiments to collect data and gathering them together as a valid dataset for this issue seems to be interesting and can be useful for upcoming studies in the field of SCOS in CMfg."

"Developing the novel fitness functions design and more efficient resolving algorithm design for the multi-objective service composition and providing a method of automatically QoS-aware web service composition without a definite composition schema are recommended for upcoming studies."

"Current techniques in this field have not been potent enough to handle issues having more than three-objectives; so, developing the multi-objective evolutionary algorithms for coping with three or more objectives SCOS issues can be considered as an interesting open issue for future studies."

Limitations

"The limitation of this research will be mapped out."

"It should be mentioned that this systematic research takes only the articles published in the major international journals into consideration."

"Although it can be asserted that Google scholar is able to offer the mostly related and valid studies based on the considered keywords, selecting all relevant and appropriate studies may not be assured."

"Different causes such as wrong keywords may lead to overlooking some relevant articles."

Conclusion

"Considering the results of SLR, the minimum and a maximum number of published studies belong to the year 2015 (1 paper) and years 2017 (11 papers) and 2018 (11 papers) respectively."

"In case of publishers, Springer has published the maximum number of journal papers among four selected publishers (16 of 31 papers, 52%)."

"Researchers have classified the 31 selected papers into two distinct groups including single-objective and multi-objective."

"It was observed that 61% of selected papers (19 of 31 papers) belong to the single-objective group and remain 39% fit to the multi-objective group."

"Since this research has comprehensively reviewed the state-of-the-art studies and obviously emphasizes their pros, cons, and future trends, the outcomes of this paper can be used as a convenient roadmap for upcoming research in this field."

Simulation-Based Optimization for Material Handling Systems in Manufacturing and Distribution Industries [116]

This is a machine-generated summary of:

Leung, Chris S. K.; Lau, Henry Y. K.: Simulation-based optimization for material handling systems in manufacturing and distribution industries [116].

Published in: Wireless Networks (2018).
Link to original: https://doi.org/10.1007/s11276-018-1894-x

If you want to cite the papers, please refer to the original.
For technical reasons we could not place the page where the original quote is coming from.

Abstract-Summary "Material handling is a vital element of the supply chains, which involves a variety of operations including the movement, storage, protection and control of materials and products throughout the processes of manufacturing and distribution."

"We apply a simulation-based optimization framework for solving several real-life single objective and multi-objective optimization problems."

"The results reveal that the simulation-based optimization framework could become an effective decision making tool for solving both single and multi-objective optimization problems in manufacturing and distribution industries and help management to find near-optimal system parameters to fulfill important objectives."

Introduction

"The ability to analyze and evaluate material handling systems is essential for decision makers to improve material flow management and optimize manufacturing and distribution operations."

"Although it is well-acknowledged that simulation modeling techniques provide an effective means to analyze and visualize the performance of complex systems, the best solution(s) cannot be guaranteed based on the results generated by simulation studies as they often depend on the quality of the simulation model and the experience of the analyst."

"The research reported in this paper is to demonstrate how real-life single and multi-objective optimization problems in manufacturing and distribution industries where material handling system (MHS) is involved can be solved by a simulation-based optimization framework that comprises a simulation tool together with an optimization algorithm based on Artificial Immune System (AIS)."

Simulation-Based Optimization

"The main driving force behind the hybridization is the enhancement of the original algorithm performance, such as enhancement of solution diversity or convergence rate, etc. For example, Leung and Lau [117] proposed a suppression controlled clonal selection algorithm (SCCSA) together with a simulation tool for the optimization of the key parameters in the field of material handling system."

"Another hybrid PSO and GA with population partitioning strategy named Hybrid Particle Swarm Optimization and Genetic Algorithm (HPSOGA) was proposed by Ali and Tawhid [118] for the minimization of the molecular energy function."

"Lau and Leung [119] also developed a hybrid AIS-based algorithm called Suppression-controlled Multi-objective Immune Algorithm (SCMIA) for solving several benchmark numerical optimization problems."

"Liu [120] made use of a newly developed coevolutionary multi-swarm particle swarm optimization algorithm (CMPSODMO) for dealing with dynamic multi-objective optimization problems (DMOPs) in rapidly changing environments."

A Simulation-Based Optimization Framework

"The SCCSA is, in principle, a search algorithm in the solution space initiated by the antigenic affinity, which comprises four main immune operators: cloning operator, hypermutation operator, suppression operator, and selection operator."

"The algorithm makes use of the Pareto dominance for fitness assignment and some common AIS-based algorithm's features for guiding the search process, such as clonal selection and expansion, affinity maturation, antibody concentration, metadynamics and immune memory."

"The suppression operator works on the whole population including the active parent cells and mature clones $A(t) \cup C^c(t)$ to eliminate similar individuals so as to avoid a particular search space being over exploited and acquire the uniformly distributed Pareto front based on the idea of immune network theory [121] such that B-cells are stimulated and suppressed by not only non-self antigens but also the interacted B-cells."

Simulation-Based Optimization Study

"Two sets of experiments are conducted to evaluate the performance of the adopted simulation-based optimization framework incorporating a multi-objective optimizer—SCMIA in the case studies 3 and 4, that is (1) to compare the results of integrating simulation and optimization with the results without using any optimization algorithm for examining the effectiveness of the simulation-based optimization framework, and (2) to benchmark SCMIA against two immune-inspired algorithms—MISA [122] and NNIA [123], and two other evolutionary algorithms—NSGA-II [124] and SPEA2 [125] with respect to the optimality and the diversity under the same framework."

"The results overall demonstrate the capability of the simulation-based optimization framework including the adopted optimizers to act as a decision support tool for helping management for effectively and efficiently finding near-optimal system operating conditions and parameters such as the number of workers, the speed of various kinds of equipment, or other decision variables of interest in order to satisfy multiple objectives and constrains such as cycle time, resource utilization, etc. Consequently, substantial savings in capital, energy, etc., are achieved through the effective and efficient deployment of material handling systems and well-coordinated processing activities based on the optimized results generated from the framework."

Conclusions

"This research applies a simulation-based optimization framework incorporating an AIS-based optimization algorithm (SCCSA for single objective problems and SCMIA for multi-objective problems) for evaluating the optimality of the material handling systems in manufacturing and logistics industries with respect to certain criteria such as system cycle time and resource utilization through simulation modelling."

"Different from other AIS-based multi-objective algorithms, an innovative suppression operator is introduced to help eliminate similar antibodies, hence significantly minimizing the number of unnecessary searches and increasing the population diversity."

"Another feature of SCMIA is that a modified crossover operator originated from the biological evolution is adopted to help further enhance the diversity of the clone population and the convergence of the algorithm because some good genes from the elitist parents can be passed to the offspring for facilitating the search of optimal solutions."

References

1. Govindaiah S, Petty MD (2021) Applying reinforcement learning to plan manufacturing material handling. Discov Artif Intell 1:8. https://doi.org/10.1007/s44163-021-00003-3
2. Chen C, Xia B, Zhou BH, Xi L (2015) A reinforcement learning based approach for a multiple-load carrier scheduling problem. J Intell Manuf 26(6):1233–1245
3. Sutton RS, Barto AG (2018) Reinforcement learning: an introduction, 2nd edn. MIT Press, Cambridge
4. Sinha AK, Anand A (2021) Development of a supply chain configuration model for new product development: a multi-objective solution approach. Soft Comput 25:8371–8389. https://doi.org/10.1007/s00500-021-05761-x
5. Lamothe J, Hadj-Hamou K, Aldanondo M (2006) An optimization model for selecting a product family and designing its supply chain. Eur J Oper Res 169(3):1030–1047

6. Hasan SM, Gao J, Wasif M, Iqbal SA (2014) An integrated decision making framework for automotive product development with the supply chain. Proc CIRP 25:10–18
7. Afrouzy ZA, Nasseri SH, Mahdavi I (2016b) A genetic algorithm for supply chain configuration with new product development. Comput Ind Eng 101:440–454
8. Willems SP (1999) Two papers in supply chain design: supply chain configuration and part selection in multigeneration products. Solan School of Management, MIT, Cambridge
9. Nepal B, Monplaisir L, Famuyiwa O (2012) Matching product architecture with supply chain design. Eur J Oper Res 216:312–325
10. Graves SC, Willems SP (2003) Supply chain design: safety stock placement and supply chain configuration. Handb Oper Res Manag Sci 11:95–132
11. Deb K, Pratap A, Agarwal S, Meyarivan T (2002) A fast and elitist multiobjective genetic algorithm: Nsga-II. IEEE Trans Evol Comput 6(2):182–197
12. Graves SC, Willems SP (2005) Optimizing the supply chain configuration for new products. Manag Sci 51(8):1165–1180
13. Rouabhia-Essalhi R, Boukrouh EH, Ghemari Y (2022) Application of failure mode effect and criticality analysis to industrial handling equipment. Int J Adv Manuf Technol 120:12. https://doi.org/10.1007/s00170-022-09099-y
14. Peeters JFW, Basten RJI, Tinga T (2018) Improving failure analysis efficiency by combining FTA and FMEA in a recursive manner. Reliab Eng Syst Saf 172:36–44. https://doi.org/10.1016/j.ress.2017.11.02
15. IEC (2001) Analysis techniques for system reliability - Procedure for failure mode and effects analysis (FMEA). In: IEC standard. IEC-60812. International Electrotechnical Commission, London
16. MIL (1980) Procedures for performing a failure mode, effects and criticality analysis. In: Military standard. MIL-1629a. US Department of Defense, Washington
17. SAE (2000) Potential failure mode and effects analysis in design (design FMEA), potential failure mode and effects analysis in manufacturing and assembly processes (process FMEA). In: SAE standard. SAE-J1739. SAE, Pittsburgh
18. SAE (2001) Failure modes, efects, and criticality analysisprocedures. In: SAE standard. SAE-ARP5580. SAE, Pittsburgh
19. Wang X, Zhang Y, Shen G (2016) An improved FMECA for feed system of CNC machining center based on ICR and DEMATEL method. Int J Adv Manuf Technol 83:43–54. https://doi.org/10.1007/s00170-015-7551-y
20. Salvi RK, Jindal S (2017) FMEA to enhance quality and efficiency of machines: a case study in valve manufacturing industry. Ind Eng J 10(5):13–19
21. Wu X, Wu J (2021) The risk priority number evaluation of FMEA analysis based on random uncertainty and fuzzy uncertainty. Complexity 2021:8817667. https://doi.org/10.1155/2021/8817667
22. Kim KO, Zuo MJ (2018) General model for the risk priority number in failure mode and effects analysis. Reliab Eng Syst Saf 169:321–329
23. Cai J, Wu X-S, Sun P, Sun J, Deng Q-J (2021) Parameter optimization method for antimisalignment of inductive power transfer system based on genetic algorithm. J Power Electron. https://doi.org/10.1007/s43236-021-00322-9
24. Hu H, Cai T, Duan S et al (2019) An optimal variable frequency phase shift control strategy for ZVS operation within wide power range in IPT systems. IEEE Trans Power Electron 99:1
25. Ren J, Liu YR, Yue PF (2019) Study on anti-misalignment inductive power transfer system based on parameter optimized method. Proc CSEE 39(5):1452–1461
26. Faithfull PT, Ball RJ, Jones RP (2001) An investigation into the use of hardware-in-the-loop simulation with a scaled physical prototype as an aid to design. J Eng Des 12(3):231–243
27. Huang S, Wang G, Lei D, Yan Y (2022) Toward digital validation for rapid product development based on digital twin: a framework. Int J Adv Manuf Technol. https://doi.org/10.1007/s00170-021-08475-4

28. Tao F, Zhang M (2017) Digital twin shop-floor: a new shop-floor paradigm towards smart manufacturing. IEEE Access 5:20418–20427
29. Schleich B, Anwer N, Mathieu L, Wartzack S (2017) Shaping the digital twin for design and production engineering. CIRP Ann 66(1):141–144
30. Negri E, Fumagalli L, Macchi M (2017) A review of the roles of digital twin in cps-based production systems. Proc Manuf 11:939–948
31. Zhuang C, Liu J, Xiong H (2018) Digital twin-based smart production management and control framework for the complex product assembly shop-floor. Int J Adv Manuf Technol 96(1–4):1149–1163
32. Park KT, Lee J, Kim HJ, Do Noh S (2020) Digital twin-based cyber physical production system architectural framework for personalized production. Int J Adv Manuf Technol 106(5):1787–1810
33. Ma J, Chen H, Zhang Y, Guo H, Ren Y, Mo R, Liu L (2020) A digital twin-driven production management system for production workshop. Int J Adv Manuf Technol 110(5):1385–1397
34. Guo H, Zhu Y, Zhang Y, Ren Y, Chen M, Zhang R (2021) A digital twin-based layout optimization method for discrete manufacturing workshop. Int J Adv Manuf Technol 112(5):1307–1318
35. Son YH, Park KT, Lee D, Jeon SW, Do Noh S (2021) Digital twin–based cyber-physical system for automotive body production lines. Int J Adv Manuf Technol 115:291–310
36. Lo CK, Chen CH, Zhong RY (2021) A review of digital twin in product design and development. Adv Eng Inform 48:101297
37. Grieves M, Vickers J (2017) Digital twin: mitigating unpredictable, undesirable emergent behavior in complex systems. In: Transdisciplinary perspectives on complex systems. Springer, Cham, pp 85–113
38. Jiang H, Qin S, Fu J, Zhang J, Ding G (2021) How to model and implement connections between physical and virtual models for digital twin application. J Manuf Syst 58:36–51
39. Shetty N, Sah B, Chung SH (2020) Route optimization for warehouse order picking operations via vehicle routing and simulation. SN Appl Sci. https://doi.org/10.1007/s42452-020-2076-x
40. De Koster R, Le-Duc T, Roodbergen KJ (2007) Design and control of warehouse order picking: a literature review. Eur J Oper Res 182(2):481–501
41. Dekker R, De Koster M, Roodbergen KJ, Van Kalleveen H (2004) Improving order-picking response time at ankor's warehouse. Interfaces 34(4):303–313
42. Theys C, Bräysy O, Dullaert W, Raa B (2010) Using a TSP heuristic for routing order pickers in warehouses. Eur J Oper Res 200(3):755–763
43. Chawla VK, Chanda AK, Angra S, Rani S (2019) Effect of nature-inspired algorithms and hybrid dispatching rules on the performance of automatic guided vehicles in the flexible manufacturing system. J Braz Soc Mech Sci Eng. https://doi.org/10.1007/s40430-019-1894-5
44. Kaban AK, Othman Z, Rohmah DS (2012) Comparison of dispatching rules in job-shop scheduling problem using simulation: a case study. Int J Simul Model 11(3):129–140
45. Ramirez-Cedillo E, García-López E, Ruiz-Huerta L, Rodriguez CA, Siller HR (2021) Reusable unit process life cycle inventory (UPLCI) for manufacturing: laser powder bed fusion (L-PBF). Prod Eng. https://doi.org/10.1007/s11740-021-01050-6
46. Wang Y, Jiang Z, Wu Y (2022) Model construction of material distribution system based on digital twin. Int J Adv Manuf Technol. https://doi.org/10.1007/s00170-022-09636-9
47. Nolz PC (2020) Optimizing construction schedules and material deliveries in city logistics: a case study from the building industry. Flex Serv Manuf J. https://doi.org/10.1007/s10696-020-09391-7
48. Artigues C, Koné O, Lopez P, Mongeau M (2015) Mixed-integer linear programming formulations. In: Schwindt C, Zimmermann J (eds) Handbook on project management and scheduling, vol 1. Springer, Berlin, pp 17–41
49. Doerner KF, Gutjahr WJ, Hartl RF, Strauss C, Stummer C (2008) Nature-inspired metaheuristics for multiobjective activity crashing. Omega 36(6):1019–1037

50. Zhou J, Love PED, Wang X, Teo KL, Irani Z (2013) A review of methods and algorithms for optimizing construction scheduling. J Oper Res Soc 64(8):1091–1105
51. Guo HF, Lian X, Zhang Y, Ren YP, He ZB, Zhang R, Ding N (2020) Analysis of environmental policy's impact on remanufacturing decision under the effect of green network using differential game model. IEEE Access 8:115251–115262
52. Zhao HR, Liu JH, Xiong H, Zhang CB, Miao T, Liu JS, Wang B (2019) 3D visualization real-time monitoring method for digital twin workshop. Comput Integr Manuf Syst 25(6):1432–1443
53. Chiu M-C, Chung W-H, Lin H-H (2021) Applying DFA and goal programming to improve economic efficiency, material handling convenience, and sustainability of a product packaging system. Res Eng Des. https://doi.org/10.1007/s00163-020-00355-4
54. Hellstrom D, Saghir M (2007) Packaging and logistics interactions in retail supply chains. J Pack Technol Res 20(3):197–216
55. García-Arca J, Prado-Prado JC, González-Portela Garrido AT (2012) Strategic packaging logistics: a case study from a supply chain perspective. In: Prado-Prado J, García-Arca J (eds) Annals of industrial engineering. Springer, London, p 374
56. García-Arca J, González-Portela AT, Prado-Prado JC (2014) Packaging as source of efficient and sustainable advantages in supply chain management: an analysis of bricks. Int J Prod Manag Eng 2(1):15–22
57. García-Arca J, Prado-Prado JC, Gonzalez-Portela Garrido AT (2014) "Packaging logistics": promoting sustainable efficiency in supply chains. Int J Phys Distrib Logist Manag 44(4):325–346
58. Sharma V, Sharma S, Verma OP, Bhardwaj B, Sharma TK, Pachauri N (2021) Prediction and optimization of abrasive wear loss of ultrahigh strength martensitic steel using response surface methodology, Harris Hawk and artificial neural network. Int J Syst Assur Eng Manag. https://doi.org/10.1007/s13198-021-01160-5
59. Adamiak M, Górka J, Kik T (2010) Structure analysis of welded joints of wear resistant plate and constructional steel. Arch Mater Sci Eng 46:108–114
60. Basavarajappa S, Joshi AG, Arun KV, Kumar AP, Kumar MP (2010) Three-body abrasive wear behaviour of polymer matrix composites filled with SiC particles. Polym-Plast Technol Eng 49:8–12. https://doi.org/10.1080/03602550903206407
61. Chang BP, Chan WH, Zamri MH, Md Akil H, Chuah HG (2019) Investigating the effects of operational factors on wear properties of heat-treated pultruded Kenaf fiber-reinforced polyester composites using Taguchi method. J Nat Fibers 16:702–717. https://doi.org/10.1080/15440478.2018.1432001
62. Sahin Y (2006) Optimal testing parameters on the wear behaviour of various steels. Mater Des 27:455–460. https://doi.org/10.1016/j.matdes.2004.11.031
63. Selvi S, Rajasekar E (2015) Theoretical and experimental investigation of wear characteristics of aluminum based metal matrix composites using RSM. J Mech Sci Technol 29:785–792. https://doi.org/10.1007/s12206-015-0140-z
64. Rödger J-M, Beier J, Schönemann M, Schulze C, Thiede S, Bey N, Herrmann C, Hauschild MZ (2020) Combining life cycle assessment and manufacturing system simulation: evaluating dynamic impacts from renewable energy supply on product-specific environmental footprints. Int J Precis Eng Manuf-Green Technol. https://doi.org/10.1007/s40684-020-00229-z
65. Kim SJ, Kara S (2014) Predicting the total environmental impact of product technologies. CIRP Ann Manuf Technol 63(1):25–28. https://doi.org/10.1016/j.cirp.2014.03.007
66. Raichur V, Callaway DS, Skerlos SJ (2015) Estimating emissions from electricity generation using electricity dispatch models: the importance of system operating constraints. J Ind Ecol 20(1):42–53. https://doi.org/10.1111/jiec.12276
67. CDP, UN Global Compact, WRI (2015) Science based targets. WWF
68. Martinez-Blanco J, Finkbeiner M, Inaba A (2015) Guideance on organizational life cycle assessment. UNEP/SETAC Life Cycle Initiative

69. Neugebauer S, Martinez-Blanco J, Scheumann R, Finkbeiner M (2015) Enhancing the practical implementation of life cycle sustainability assessment: proposal of a Tiered approach. J Clean Prod 102:165–176. https://doi.org/10.1016/j.jclepro.2015.04.053
70. Favi C, Germani M, Mandolini M, Marconi M (2016) Plantlca: a lifecycle approach to map and characterize resource consumptions and environmental impacts of manufacturing plants. Proc CIRP 48:146–151. https://doi.org/10.1016/j.procir.2016.03.102
71. Duflou JR, Sutherland JW, Dornfeld D, Herrmann C, Jeswiet J, Kara S et al (2012) Towards energy and resource efficient manufacturing: a processes and systems approach. CIRP Ann Manuf Technol 61(2):587–609. https://doi.org/10.1016/j.cirp.2012.05.002
72. Veleva V, Ellenbecker M (2001) Indicators of sustainable production: framework and methodology. J Clean Prod 9:519–549
73. Jayal AD, Badurdeen F, Dillon OW, Jawahir IS (2010) Sustainable manufacturing: modeling and optimization challenges at the product, process and system levels. CIRP J Manuf Sci Technol 2(3):144–152. https://doi.org/10.1016/j.cirpj.2010.03.006
74. Kellens K, Dewulf W, Overcash M, Hauschild MZ, Duflou JR (2012) Methodology for systematic analysis and improvement of manufacturing unit process life-cycle inventory (UPLCI)-CO2PE! initiative (cooperative effort on process emissions in manufacturing). Part 1: methodology description. Int J Life Cycle Assess 17(1):69–78. https://doi.org/10.1007/s11367-011-0340-4
75. Greinacher S, Moser E, Hermann H, Lanza G (2015) Simulation based assessment of lean and green strategies in manufacturing systems. Proc CIRP 29:86–91. https://doi.org/10.1016/j.procir.2015.02.053
76. Diaz-Elsayed N, Jondral A, Greinacher S, Dornfeld D, Lanza G (2013) Assessment of lean and green strategies by simulation of manufacturing systems in discrete production environments. CIRP Ann Manuf Technol 62(1):475–478. https://doi.org/10.1016/j.cirp.2013.03.066
77. Maalouf E, Daaboul J, Duigou L, Julien; Hussein, Bassam. (2022) Production management for mass customization and smart cellular manufacturing system: NSGAII and SMPSO for factory-level planning. Int J Adv Manuf Technol. https://doi.org/10.1007/s00170-022-09188-y
78. Shrouf F, Ordieres J, Miragliotta G (2014) Smart factories in Industry 4.0: a review of the concept and of energy management approached in production based on the Internet of Things paradigm. Int Conf Ind Eng Eng Manag 2014:697–701. https://doi.org/10.1109/IEEM.2014.7058728
79. Massimi E, Khezri A, Benderbal HH, Benyoucef L (2020) A heuristic-based non-linear mixed integer approach for optimizing modularity and integrability in a sustainable reconfigurable manufacturing environment. Int J Adv Manuf Technol. https://doi.org/10.1007/s00170-020-05366-y
80. Benyoucef L (2020) Reconfigurable manufacturing systems: from design to implementation, Springer series in advanced manufacturing. Springer, Cham
81. Bortolini M, Galizia FG, Mora C (2018) Reconfigurable manufacturing systems: literature review and research trend. J Manuf Syst 49:93–106
82. Khezri A, Haddou-Benderbal H, Benyoucef L (2020) Sustainable multi-objective process plan generation in RMS through modelling energy consumption. In: Benyoucef A (ed) Reconfigurable manufacturing systems: from design to implementation, Springer series in advanced manufacturing. Springer, Cham, pp 161–177
83. Hayyolalam V, Pourghebleh B, Kazem P, Asghar A, Ghaffari A (2019) Exploring the state-of-the-art service composition approaches in cloud manufacturing systems to enhance upcoming techniques. Int J Adv Manuf Technol. https://doi.org/10.1007/s00170-019-04213-z
84. Huang B, Li C, Tao F (2014) A chaos control optimal algorithm for QoS-based service composition selection in cloud manufacturing system. Enterp Inf Syst 8(4):445–463
85. Li F, Zhang L, Liu Y, Laili Y (2018) QoS-aware service composition in cloud manufacturing: a gale-shapley algorithm-based approach. IEEE Trans Syst Man Cybern Syst 99:1–12

86. Akbaripour H, Houshmand M, Van Woensel T, Mutlu N (2015) Cloud manufacturing service selection optimization and scheduling with transportation considerations: mixed-integer programming models. Int J Adv Manuf Technol 95(1–4):1–31
87. Zhou J, Yao X (2017) Hybrid teaching–learning-based optimization of correlation-aware service composition in cloud manufacturing. Int J Adv Manuf Technol 91(9–12):3515–3533
88. Zhang M, Liu L, Liu S (2016) Genetic algorithm based QoS-aware service composition in multi-cloud. Int Conf Collab Comput 2015:113–118
89. Chen Y, Huang J, Lin C, Shen X (2016) Multi-objective service composition with QoS dependencies. IEEE Trans Cloud Comput 99:1
90. Zhou J, Yao X (2017) A hybrid approach combining modified artificial bee colony and cuckoo search algorithms for multi-objective cloud manufacturing service composition. Int J Prod Res 55(16):4765–4784
91. Shehu U, Safdar G, Epiphaniou G (2016) Fruit fly optimization algorithm for network-aware web service composition in the cloud. Int J Adv Comput Sci Appl 1(7):1–11
92. Liu Z-Z, Chu D-H, Song C, Xue X, Lu B-Y (2016) Social learning optimization (SLO) algorithm paradigm and its application in QoS-aware cloud service composition. Inf Sci 326:315–333
93. Bharath Bhushan S, Pradeep Reddy CH (2016) A Qos aware cloud service composition algorithm for geo-distributed multi cloud domain. Int J Intell Eng Syst 9(4):147–156
94. Seghir F, Khababa A (2016) A hybrid approach using genetic and fruit fly optimization algorithms for QoS-aware cloud service composition. J Intell Manuf 29:1773–1792
95. Jula A, Othman Z, Sundararajan E (2015) Imperialist competitive algorithm with PROCLUS classifier for service time optimization in cloud computing service composition. Expert Syst Appl 42(1):135–145
96. Xiang F, Xu L, Jiang G (2016) Green manufacturing service composition in cloud manufacturing system: an introduction. In: Industrial electronics and applications 2016 IEEE 11th conference. IEEE, Piscataway, pp 1988–1993
97. Guo H, Zhang L, Tao F, Ren L, Luo YL (2010) Research on the measurement method of flexibility of resource service composition in cloud manufacturing. Adv Mater Res 139:1451–1454
98. Zhang L, Guo H, Tao F, Luo YL, Si N (2010) Flexible management of resource service composition in cloud manufacturing. In: 2010 IEEE international conference on industrial engineering and engineering management. IEEE, Piscataway, pp 2278–2282
99. Bouzary H, Chen FF (2018) Service optimal selection and composition in cloud manufacturing: a comprehensive survey. Int J Adv Manuf Technol 97:795–808
100. Asghari P, Rahmani AM, Javadi HHS (2018) Service composition approaches in IoT: a systematic review. J Netw Comput Appl 120:61–77
101. Vakili A, Navimipour NJ (2017) Comprehensive and systematic review of the service composition mechanisms in the cloud environments. J Netw Comput Appl 81:24–36
102. Jatoth C, Gangadharan GR, Buyya R (2017) Computational intelligence based QoS-aware web service composition: a systematic literature review. IEEE Trans Serv Comput 10(3):475–492
103. Li T, He T, Wang Z, Zhang Y (2018) An approach to Iot service optimal composition for mass customization on cloud manufacturing. IEEE Access 6:50572–50586
104. Zhou J, Yao X, Lin Y, Chan FTS, Li Y (2018) An adaptive multi-population differential artificial bee colony algorithm for many-objective service composition in cloud manufacturing. Inf Sci 456:50–82
105. Zhou J, Yao X (2017) DE-caABC: differential evolution enhanced context-aware artificial bee colony algorithm for service composition and optimal selection in cloud manufacturing. Int J Adv Manuf Technol 90(1–4):1085–1103
106. Lu Y, Xu X (2017) A semantic web-based framework for service composition in a cloud manufacturing environment. J Manuf Syst 42:69–81

107. Xiang F, Jiang GZ, Xu LL, Wang NX (2016) The case-library method for service composition and optimal selection of big manufacturing data in cloud manufacturing system. Int J Adv Manuf Technol 84(1–4):59–70
108. Zheng H, Feng Y, Tan J (2016) A fuzzy QoS-aware resource service selection considering design preference in cloud manufacturing system. Int J Adv Manuf Technol 84(1–4):371–379
109. Que Y, Zhong W, Chen H, Chen X, Ji X (2018) Improved adaptive immune genetic algorithm for optimal QoS-aware service composition selection in cloud manufacturing. Int J Adv Manuf Technol 96(9–12):4455–4465
110. Xiang F, Hu Y, Yu Y, Wu H (2014) QoS and energy consumption aware service composition and optimal-selection based on Pareto group leader algorithm in cloud manufacturing system. CEJOR 22(4):663–685
111. Zhou J, Yao X (2017) Multi-objective hybrid artificial bee colony algorithm enhanced with Lévy flight and self-adaption for cloud manufacturing service composition. Appl Intell 47(3):721–742
112. Akbaripour H, Houshmand M (2018) Service composition and optimal selection in cloud manufacturing: landscape analysis and optimization by a hybrid imperialist competitive and local search algorithm. Neural Comput & Applic 32:10873–10894
113. Li C, Guan J, Liu T, Ma N, Zhang J (2018) An autonomy-oriented method for service composition and optimal selection in cloud manufacturing. Int J Adv Manuf Technol 96(5–8):2583–2604
114. Zhou J, Yao X (2017) Multi-population parallel self-adaptive differential artificial bee colony algorithm with application in large-scale service composition for cloud manufacturing. Appl Soft Comput J 56:379–397
115. Wang Y, Dai Z, Zhang W, Zhang S, Xu Y, Chen Q (2018) Urgent task-aware cloud manufacturing service composition using two-stage biogeography-based optimisation. Int J Comput Integr Manuf 31(10):1034–1047
116. Leung CSK, Lau HYK (2018) Simulation-based optimization for material handling systems in manufacturing and distribution industries. Wirel Netw. https://doi.org/10.1007/s11276-018-1894-x
117. Leung CSK, Lau HYK (2011) An optimization framework for modeling and simulation of dynamic systems based on AIS. In: International federation of automatic control world congress. International Federation of Automatic Control (IFAC), Laxenburg, p 11608
118. Ali AF, Tawhid MA (2017) A hybrid particle swarm optimization and genetic algorithm with population partitioning for large scale optimization problems. Ain Shams Eng J 8(2):191–206. https://doi.org/10.1016/j.asej.2016.07.008
119. Leung CSK, Lau HYK (2016) A hybrid multi-objective immune algorithm for numerical optimization. In: Filipe AJ (ed) The 8th international joint conference on computational intelligence, vol 3. SciTePress, Setúbal, pp 105–114
120. Liu R, Li J, Fan J, Mu C, Jiao L (2017) A coevolutionary technique based on multi-swarm particle swarm optimization for dynamic multi-objective optimization. Eur J Oper Res 261(3):1028–1051. https://doi.org/10.1016/j.ejor.2017.03.048
121. Jerne NK (1974) Towards a network theory of the immune system. Ann Immunol 125(1–2):373–389
122. Coello Coello CA, Cortés NC (2005) Solving multiobjective optimization problems using an artificial immune system. Genet Program Evolvable Mach 6(2):163–190. https://doi.org/10.1007/s10710-005-6164-x
123. Gong M, Jiao L, Du H, Bo L (2008) Multiobjective immune algorithm with nondominated neighbor-based selection. Evol Comput 16(2):225–255. https://doi.org/10.1162/evco.2008.16.2.225

124. Deb K, Agrawal S, Pratap A, Meyarivan T (2000) A fast elitist non-dominated sorting genetic algorithm for multi-objective optimisation: NSGA-II. In: The 6th international conference on parallel problem solving from nature. Springer, Cham, pp 849–858
125. Zitzler E, Laumanns M, Thiele L (2001) SPEA2: improving the strength pareto evolutionary algorithm. In: Technical report 103. Computer Engineering and Communication Networks Lab (TIK), Zurich

Chapter 2
Optimization Methods in Traditional Machining Processes

Introduction by the Editor

Machining processes are at the center of manufacturing processes. They govern the quality and functioning of the final product. The optimization of the parameters associated with the machining process is very essential. The chapter covers machining processes such as, milling, drilling, electrical discharge machining, abrasive waterjet machining, electrochemical machining and associated parameters such as material removal rate and electrode wear rate, multi-step production process, turning force, surface roughness and deviation of the depth of cut, micro-hardness, wear volume, cladding efficiency, input voltage, tool feed rate, several tolerances, etc. with focus on the importance and significance with a view point of optimization. The solution methodologies such as certain variants of cohort intelligence algorithm, ANN, moth search algorithm, particle swarm optimization, conditional design optimization, genetic algorithms, tabu search, simulated annealing, response surface methodology, design of experiment, grey relational analysis, TOPSIS, NSGA, ANOVA, lexicographic method, variations of teaching–learning-based optimization, Jaya algorithm, grey wolf optimizer, principal component analysis have been discussed for the single and multi-objective as well as multi-criteria optimization approaches. The chapter highlights the future directions of the applications of these methods for several traditional as well as non-traditional machining processes.

The review of process modelling and optimization on electrical discharge machining is provided. It reviews the Moth search algorithm (MSA) was used to determine the optimal machining parameters, such that the MRR was maximized and EWR was minimized. It discusses the influence of machining parameters on the electrical discharge machining performances. In addition, optimization of electrical discharge machining process using MSA is also discussed. A systematic approach

A. J. Kulkarni (ed.), *Optimization Methods in Manufacturing Processes*,
https://doi.org/10.1007/978-981-96-5257-0_2

to the optimization of both design and manufacturing variables across a multi-step production process is provided in detail. It surveys process and product design optimization for manufacturing processes. The holistic approach over manufacturing optimization enables the creation of a reliable model for optimizing both economic and technological sides of a NNS manufacturing chain. The work may open doors to the cut surface quality with integrity for utilizing laser microgrooving more effectively and challenging possibilities for exploring effective applications of laser technology. Moreover, it is concluded that the RSM, the second-order quadratic equation is the most common response model, and the approximation of the response function is obtained in the form of predictive variables by establishing the relationship between input parameters and desired responses.

The review discusses various MCDM methods, grey relational analysis (GRA) and TOPSIS have emerged out as the most potent tools for multi-objective optimization of milling processes due to their several added advantages. In addition, it concludes that the applications of MCDM techniques may help in identifying a compromised combination of the milling parameters while simultaneously achieving higher MRR and lower Ra values. The review provides insights that in order to have a high MRR, the voltage and feed rate should be maximum while the ERS should be around the mid-value in the design space such that it improves the machining condition by assisting in the debris removal without causing incomplete sparking. The review effects of selected input factors, i.e., water jet pressure (WJP), stand-off distance (SOD), and traverse speed (TS) on the performance characteristics, namely taper angle (TA), surface roughness (Ra), and the material removal rate (MRR). Moreover, the review underscores that the optimization of dimensional tolerances requires that a cost-tolerance function is evaluated consistently for all the part features involved in a given functional requirement. The properties of the cost-tolerance functions are exploited in a scaling procedure for tolerance allocation, which allows to minimize cost without having to explicitly solve an optimization problem is also highlighted.

The review of the lexicographic method, a multi-objective optimization tool highlights that the determination of the optimal parametric settings for three NTM processes, i.e. EDM, Nd:YAG LBM and USM which have been found to be quite suitable for machining of engineering ceramic materials. It also provides discussion related to the adaptation of the Analysis of variance (ANOVA) to assess the effect of the process parameters on the achieved depth, the material removal rate (MRR) and roughness parameters, while response surface methodology (RSM) and artificial neuronal networks (ANNs) to model the process behaviours. The parametric optimization of μEDM drilling on titanium using principal component analysis is discussed in very details. It concludes that it can be attributed to higher discharge energy of the process resulting from increment in voltage and capacitance, which ultimately removes material at a faster rate due to the higher intensity of discharges occurring between the two electrodes. The selection of parameters in cost-tolerance functions is also reviewed. It reviews general properties, cost-tolerance functions and associated data, methods for parameter selection and concludes that the tolerance allocation problems can be solved with many different cost-tolerance functions.

The algorithm such as multi-cohort intelligence algorithm for solving advanced manufacturing process problems is also critically reviewed highlighting the optimization of the process parameters associated with the abrasive water jet machining (AWJM), EDM and conventional micro-turning, micro-drilling and micro-milling processes in order to achieve the above-discussed responses. The approaches such as follow better, follow best, alienation and roulette wheel and Multi-CI algorithm, have been compared with experimental results from the literature as well as contemporary algorithms such as GA, SA, FA, PSO, BPNN and RSM. The machining characteristics and optimization of process parameters in die-sinking EDM of Inconel 625 have also been reviewed. The important conclusion is ANOVA analysis of the individual parameters reveals that peak current is the most dominating factor for MRR followed by pulse on time while surface roughness is mostly influential by pulse on time followed by pulse off time.

Machine Generated Summaries

Disclaimer: The summaries in this chapter were generated from Springer Nature publications using extractive AI auto-summarization: An extraction-based summarizer aims to identify the most important sentences of a text using an algorithm and uses those original sentences to create the auto-summary (unlike generative AI). As the constituted sentences are machine selected, they may not fully reflect the body of the work, so we strongly advise that the original content is read and cited. The auto generated summaries were curated by the editor to meet Springer Nature publication standards. To cite this content, please refer to the original papers.

Machine generated keywords: machining, edm, composite, mrr, material removal, surface, machining process, electrode, strength, removal, roughness, jet, abrasive, property, laser.

Intelligent Approach for Process Modelling and Optimization on Electrical Discharge Machining of Polycrystalline Diamond [1]

This is a machine-generated summary of:

Ong, Pauline; Chong, Chon Haow; bin Rahim, Mohammad Zulafif; Lee, Woon Kiow; Sia, Chee Kiong; bin Ahmad, Muhammad Ariff Haikal: Intelligent approach for process modelling and optimization on electrical discharge machining of polycrystalline diamond [1].

Published in: Journal of Intelligent Manufacturing (2018).

Link to original: https://doi.org/10.1007/s10845-018-1443-6.

Copyright of the summarized publication:

If you want to cite the papers, please refer to the original.

For technical reasons we could not place the page where the original quote is coming from.

Abstract-Summary "Its applications are restricted because of the PCD material is difficult to machine."

"Electrical discharge machining (EDM) is an ideal method suitable for PCD materials due to its non-contact process nature."

"The performance of EDM, however, is significantly influenced by its process parameters and type of electrode."

"Soft computing technique was utilized to optimize the performance of the EDM in roughing condition for eroding PCD with copper tungsten or copper nickel electrode."

"The EDM experiment was conducted based on the design experimental matrix."

"The effectiveness of EDM on shaping PCD with copper tungsten and copper nickel was evaluated in terms of material removal rate (MRR) and electrode wear rate (EWR)."

"The predictive model of radial basis function neural network (RBFNN) was developed to predict the MRR and EWR of the EDM process."

"Response surface plot was drawn to study the influences of machining parameters of EDM for shaping PCD with copper tungsten and copper nickel."

"Moth search algorithm (MSA) was used to determine the optimal machining parameters, such that the MRR was maximized and EWR was minimized."

Introduction

"Maity and Mishra studied the performance of ANN in predicting the MRR, overcut effect and recast layer thickness of micro EDM during machining of Inconel 718, in which the input parameters of voltage, peak current, pulse on duration and pulse off duration were considered [2]."

"The study of Mohanty et al. [3] on EDM of Inconel 718 super alloy aimed to find the optimal parameter settings of open circuit voltage, discharge current, pulse-on time, duty factor, flushing pressure and tool material, which gave maximum MRR and minimum tool wear rate, SR and radial over cut, using quantum behaved PSO and PSO."

"In terms of simulation modelling, the EDM performance of PCD was predicted using radial basis function neural network (RBFNN); and lastly, the process parameters were optimized using moth search algorithm (MSA) such that the MRR was maximized and EWR was minimized."

Principles of Die-Sinking Electrical Discharge Machining

"EDM is thermal erosion process where controlled electric spark discharge occurring repetitively between the electrode and the workpiece immersed in a dielectric fluid to have the eroding effect on workpiece, in order to produce a replica of tool on workpiece."

"When the electrode reaches the closest point to the workpiece, typically at a very small gap, the plasma channel breakdowns and discharge column is created."

"During the discharging phase, the spark strikes on the workpiece at a temperature in the range of 8000–20,000 °C [4, 5], and causes the melting and evaporation of both electrode and workpiece material."

Experimental Setup and Results

"All experiments on PCD were conducted in die-sinking EDM machine (Sodick AQ55L)."

"The CTB 010 PCD containing 89.7% diamond and 10.3% cobalt, with particle size of 10 μm, was used as workpiece material in this work."

"The copper tungsten (CuW) (75% tungsten, 25% copper) and copper nickel (CuNi) (70% copper, 30% nickel) were used as an electrode tool to erode the PCD workpiece in EDM process."

"The total machine depth or in another word, in-feed of the workpiece was maintained at 0.5 mm."

"The performance of the EDM is influenced by the process parameters."

"MRR and EWR were two important indicators of EDM process, used to monitor the time of finishing the workpiece, and accuracy of machined parts, respectively."

Electrical Discharge Machining Process Modelling Using Radial Basis Function Neural Network

"In the EDM process, any slight variation in the machining parameters may affect the machining performance drastically."

"A variant of ANN model, particularly, the RBFNN model, was used to establish a predictive model that correlates the machining parameters to the machining performance in this study."

"The spread parameter was fixed beforehand and only the synaptic weights were updated iteratively using the pseudoinverse algorithm during the learning process, such that the network model was best fit to the observed data."

"The change of spread parameter influences the generalization capability of RBFNN significantly."

"The value of spread parameter that ensures the best network configuration was chosen."

"From empirical simulation, it was found that the RBFNN was optimized, when the spread parameters were assigned as 5.099, 0.518, 1.318, and 1.79, for the MRR of copper tungsten electrode, EWR of copper tungsten electrode, MRR of copper nickel electrode, and EWR of copper nickel electrode, respectively."

Influence of Machining Parameters on the Electrical Discharge Machining Performances

"During the pulse off time, no discharge energy is supplied to the workpiece material and thus, the higher pulse interval will lead to lower MRR."

"Both surface plots suggest that conducting the EDM process at high pulse interval is less favored, as indicated by low MRR in this region, regardless of type of electrode used."

"Longer pulse duration allows more discharge energy being transferred into the work piece and produces larger craters in the work piece The MRR is dependent on the energy applied to the work material during the pulse duration."

"The excessive pulse duration causes the expansion of discharge column, leading to the reduction of energy density within the discharge spot, which is insufficient to melt and vapor the workpiece [6]."

"With prolonged pulse interval, the energy released from the electrode increases, contributing to lower temperature in the electrode."

Optimization of Electrical Discharge Machining Process Using Moth Search Algorithm

"The moth search algorithm (MSA)—a newly developed bio-inspired optimization algorithm, was used in this regard, such that the optimal machining parameters that maximized MRR and minimized EWR can be identified."

"Such behaviour has been translated into MSA with the assumptions of (1) when approaching the light source, the moths will adjust its flight position, thereby causing the moths to plummet down."

"The moths in subpopulation 1 were considered to be much closer to the light source (higher fitness values) than the moths in subpopulation 2."

"The moths in subpopulation 2, which were considered far from the light source, will fly in a straight line towards the light."

"For each run, the search of the optimized machining parameters was continued until the maximum generation step of 2000 was achieved."

Conclusions

"The findings of this study are as follows: (1) The adequacy of the developed prediction model based on the RBFNN was considered satisfactorily, with the considerable small mean-squared error (MSE) within the range of 6.47×10^{-5} to $7.29 \times$

10^{-6}. (2) In the case of copper tungsten electrode, with the utilization of the MSA as parameter optimizer, the maximum MRR was obtained at the machining parameters of (I, 9.23 A; T_{off}, 13.18 μs; T_{on}, 13.18 μs), while the minimum EWR was obtained at the machining parameters of (I, 8.4 A; T_{off}, 46.82 μs; T_{on}, 32.77 μs)."

"The combination of (I, 8.64A; T_{off}, 13.18 μs; T_{on}, 13.18 μs) was found to achieve maximum MRR and minimum EWR simultaneously. (3) In the case of copper nickel electrode, through the optimization of MSA, the maximum MRR took place at the machining parameters of (I, 15.36 A; T_{off}, 13.18 μs; T_{on}, 39.58 μs), while the minimum EWR was obtained at the machining parameters of (I, 8.63 A; T_{off}, 46.82 μs; T_{on}, 46.82 μs)."

Concurrent Optimization of Process Parameters and Product Design Variables for Near Net Shape Manufacturing Processes [7]

This is a machine-generated summary of:

Marini, Daniele; Corney, Jonathan R.: Concurrent optimization of process parameters and product design variables for near net shape manufacturing processes [7].

Published in: Journal of Intelligent Manufacturing (2020).

Link to original: https://doi.org/10.1007/s10845-020-01593-y.

If you want to cite the papers, please refer to the original.

For technical reasons we could not place the page where the original quote is coming from.

Abstract-Summary "This paper presents a new systematic approach to the optimization of both design and manufacturing variables across a multi-step production process."

"To enable such computation a methodology, named conditional design optimization (CoDeO) is proposed which allows the modelling and simultaneous optimization of process parameters and product design (geometric variables), using single or multi-criteria optimization strategies."

"After investigation of CoDeO's requirements, evolutionary algorithms, in particular Genetic Algorithms, are identified as the most suitable for overall NNS manufacturing chain optimization The CoDeO methodology is tested using an industrial case study that details a process chain composed of casting and machining processes."

Introduction

"In industrial enterprises and academic literature, product design and manufacturing process optimization are linked by predominately qualitative "Design-for-X" methodologies rather than the explicit setting of dimensions and process variables."

"The main aim of the NNS optimization approach proposed here is to minimize cost (or time) and improve technological performances (e.g. minimizing turning force) through systematic variation of product design and manufacturing process parameters."

"Quantify benefits by implement it on a case study The proposed methodology, known as conditional design optimization (CoDeO), is limited to a NNS manufacturing chain: where, a NNS chain is composed of a primary shaping process and a limited number of machining steps."

"In other contexts the idea of concurrent optimization between process parameters, manufacturing systems and design have been explored using different methodologies."

"In the context of NNS manufacturing, the machining steps gives an opportunity to formulate a generic approach to optimize NNS processes in accordance with production requirements and processes dynamics."

Literature Survey: Process and Product Design Optimization for Manufacturing Processes

"Researchers introduce process models (e.g. in turning process: turning power, cutting force, obtainable roughness equations) into optimization algorithm in different ways: using these multi-variable equations (that often take the form of non-linear functions) as an objective function itself (i.e. having a multi-criteria optimization and therefore a Pareto solution), as non-linear constraints or in a combination of the previous."

"Evolutionary algorithms have been found particularity effective in multi-objective optimization [8–11] and, in particular, in manufacturing sector [12, 13]."

"Evolutionary algorithms have been applied to the optimization of several different production systems and manufacturing processes."

"Evolutionary algorithms have been applied to optimize the performances of production (e.g. [14]) and assembly lines (e.g. [15, 16])."

"Sardinas et al. [17] use a micro GA for optimizing a turning process by two objective functions: turning time and tool life."

Concurrent Optimization of Near Net Shape Processes

"For optimizing the product design and process parameters, it is necessary to have a cost model which is a function of both."

"Cost model of the primary shaping process, dependent on process parameters (e.g. forging force or intermediate geometry) and/or product design variables (e.g. diameter) 2."

"Cost model of the machining process dependent on process parameters and/or product design variables 3."

"Support models (i.e. linear or non-linear) for the primary shaping or machining processes (e.g. production rate target or turning force constraint) The presence of support models is dependent on the nature of the processes, production targets and product requirements."

"These models allow the exploration of iterative systematic product and process design variation that can be optimized concurrently by minimizing the process cost (as main target),process technological features and product/production targets."

Case Study

"Literature survey: GA functions and settings for turing process optimization Many different applications for turning process optimization through metaheuristic search based algorithms have been reported in literature, in particular for GA."

"Cus and Balic [18] use a GA search base algorithm for optimizing the turning process of steel."

"Saravanan et al. [19] use a lower number of variables (i.e. feed rate and cutting speed) for testing turning process optimization through GA and SA."

"Sardinas et al. [17] use a micro GA for optimizing a turning process by two objective functions: turning time and tool life."

"D'Addona and Teti [20] use the MATLAB GA optimization toolbox for optimizing the turning process of cast steel blank with an HSS tool, comparing two crossover operators (single point and two-points) and mutation coefficients."

Case Study Results

"To the machining parameters increase, the cutting force increase, remaining into feasible ranges."

"The machining parameters can be set higher than those currently applied ones (except from the feed rate that remaining unchanged)."

"Cutting force and turning power vary congruently with machining parameters 7."

"Surface roughness varies with the machining parameters, satisfying the target requirements."

"Machining time varies with machining parameters, remaining lower than current value (75 min on average)."

"Optimized machining parameters gives up to a 9.5 times higher removal rate."

Discussion

"CoDeO optimize concurrently the manufacturing chain's cost and process feasibility, depending on analysed process characteristics and global cost model."

"With other NNS approaches [21, 22], CoDeO optimization can be adapted to different NNS processes."

"With other concurrent optimization approaches, CoDeO achieves similar optimization targets to the methodologies used for robotic systems [23–26] in terms of process parameters and design (both subject to design constraints)."

"CoDeO is not an "offline" methodology and its scope is limited to the concurrent optimization of the product design and process parameters of NNS manufacturing chains."

"While the Design for Control methodologies are designed for the flexible manufacturing systems, CoDeO application is limited to NNS processes."

"CoDeO can be integrated with this general methodology in order to provide mathematical modelling and concurrent optimization of NNS manufacturing chains."

Conclusions

"The holistic approach over manufacturing optimization enable the creation of a reliable model for optimizing both economic and technological sides of a NNS manufacturing chain."

"CoDeO can be mathematically integrated with other models for generating a holistic production models and concurrent optimization."

"Variable and targets definition are critical for the optimization, however, in the context of a global NNS process chain, the general model developed defines the key outcomes (i.e. objective functions definition) and limit possible search space (i.e. variable selection).Further details of the methodology can be found in Marini [27]."

Parametric Optimization of Nd:YAG Laser Microgrooving on Aluminum Oxide Using Integrated RSM-ANN-GA Approach [28]

This is a machine-generated summary of:

Dixit, Salila Ranjan; Das, Sudhansu Ranjan; Dhupal, Debabrata: Parametric optimization of Nd:YAG laser microgrooving on aluminum oxide using integrated RSM-ANN-GA approach [28].

Published in: Journal of Industrial Engineering International (2018).

Link to original: https://doi.org/10.1007/s40092-018-0295-1.

If you want to cite the papers, please refer to the original.

For technical reasons we could not place the page where the original quote is coming from.

Abstract-Summary "The present work addresses the modeling and optimization study on dimensional deviations of square-shaped microgroove in laser micromachining of aluminum oxide (Al_2O_3) ceramic material with pulsed Nd:YAG laser by considering the air pressure, lamp current, pulse frequency, pulse width and cutting speed as process parameters."

"The performance of the predictive ANN model based on 5.8–8.3 architecture gave the minimum error (MSE = 0.000099) and presented highly promising to confidence with percentage error less than 3% in comparison with experimental result data set."

"The ANN model combined with GA leads to minimum deviation of upper width, lower width and depth value of −0.0278 mm, 0.0102 mm and −0.0308 mm, respectively, corresponding to optimum laser microgrooving process parameters such as 1.2 kgf/cm^2 of air pressure, 19.5 A of lamp current, 4 kHz of pulse frequency, 6% of pulse width and 24 mm/s of cutting speed."

Introduction

"In an another study Kalita et al. [29] performed an investigation with same workpiece material for optimization of process parameters in laser beam micromarking using genetic algorithm (GA) and particle swarm optimization (PSO)."

"Kibria et al. [30] modeled the relationship between the laser microturning process parameters (laser power, pulse frequency, rotational speed, air pressure and feed rate) and quality characteristics using ANN and performed multi-objective optimization using RSM in order to achieve the desired surface roughness and minimum depth deviation of turned part."

"Appropriate combination and utilization, along with proper adjustment of pre-cited machining parameters, are of prime importance for acquiring good grade of microgroove, which is a challenging task as it consumes precious time and effort due to the dynamic behavior of the laser micromachining process."

Experimental Setup and Procedure

"Al_2O_3 flat plate was subjected to microgrooving by multiple laser pulses using Nd:YAG laser treatment with actual peak power moving between 0.7 and 5 kW. The compressed as well as regulated gas (air) is supplied into a fine coaxial nozzle to allow the grooving as per the experimental design."

"Air pressure, lamp current, pulse frequency, pulse width and cutting speed are considered to be the process parameters which affect the response of interest in laser microgrooving operation, namely upper width deviation, lower width deviation and depth deviation."

"The input parameters have predominant impacts on the output criteria of laser microgrooving, as the lamp current combined with the pulse frequency provides the energy to remove the material and achieve the desired depth."

"The range of process parameters setting has been selected after performing some pilot experiments using fixed focal length and also by inspecting the workpiece for a through microgroove of acceptable quality."

Results and Discussion

"In RSM, the second-order quadratic equation is the most common response model, and the approximation of the response function is obtained in the form of predictive variables by establishing the relationship between input parameters and desired responses (output)."

"The variation in ANN and experimental result is under 3%, which avoids the misleading conclusion to consider reliable model, particularly ANN for predicting the responses satisfactorily under pre-cited process parameters in laser microgrooving operations."

"The following points give a generic view of how GAs operate [31]: (1) a population of individuals (solutions) is created that consists of random individuals (initialization); (2) a function or a model (objective function) measures individual performance and determines their ability to survive and reproduce; (3) individuals are ranked (ranking) and the best/fittest individuals (according to a fitness function, which is an objective function transformation) are chosen (selection operator) to mate in pairs (crossover operator) and thus create new individuals and hence a whole new population."

Conclusions

"Design of experiments (DOEs), response surface methodology (RSM), artificial neural network (ANN) and genetic algorithm (GA) have been applied for experimentation, mathematical modeling and multi-response optimization of laser micro-grooving operation on aluminum oxide ceramic workpiece."

"The present research based on GA, ANN and statistically multi-regression analysis (RSM) has demonstrated the ability to optimize and to accurately model the dimensional deviations of microgroove through advances in computer technology."

"The proposed multiple approaches (experimental, evolutional, statistic and stochastic) present reliable methodologies to improve laser microgrooving process, and they can be employed in real-time process monitoring, model predictive control and optimization in several machining processes."

"The present research study of pulsed Nd:YAG laser microgrooving operation on aluminum oxide ceramic will be useful as technological guidelines for further research in the area of laser microgrooving process of structural ceramics."

Multi-Objective Optimization of Coating Properties and Cladding Efficiency in 316 L/WC Composite Laser Cladding Based on Grey Relational Analysis [32]

This is a machine-generated summary of:

Lian, Guofu; Xiao, Shihong; Zhang, Yang; Jiang, Jibin; Zhan, Youji: Multi-objective optimization of coating properties and cladding efficiency in 316 L/WC composite laser cladding based on grey relational analysis [32].

Published in: The International Journal of Advanced Manufacturing Technology (2021).

Link to original: https://doi.org/10.1007/s00170-020-06486-1.

If you want to cite the papers, please refer to the original.

For technical reasons we could not place the page where the original quote is coming from.

Abstract-Summary "This study aims at improving the coating properties and cladding efficiency in composite material laser cladding."

"By altering the input of laser power, scanning speed, gas flow, and tungsten carbide powder ratio in laser cladding setup, their influence on the micro-hardness and wear resistance of the clad as well as the cladding efficiency had been studied."

"This study achieved processing parameter optimization with grey relational analysis by combining multiple objectives."

"To obtain the clad with maximum micro-hardness and minimum wear volume while also attaining the maximum cladding efficiency, grey relational analysis was utilized to combine these multiple objectives in optimization."

"This study verified the applicability and provided the theoretical basis for multiple-objective optimization of the coating properties and cladding efficiency in composite laser cladding."

Introduction

"A single-factor experiment was designed to investigate the variation in the geometrical size, dilution rate, and hardness of cladding layers by the influence of different processing parameters."

"Wu and others investigated the porosity and cracks in Ni60A alloy laser cladding, applied response surface methodology, and obtained the optimized processing parameters to achieve coating with no cracks and low porosity [33]."

"The processing parameter investigation of their influence on multiple objectives simultaneously in laser cladding was rarely discovered."

"This study aims at achieving preferable properties in 316 L/WC composite coating as well as guaranteeing the cladding efficiency by investigating the influence of selected processing parameters with grey relational analysis and orthogonal experimental design."

"This study provides the theoretical basis for the multiple-objective optimization and prediction of the coating properties and cladding efficiency by altering processing parameters in laser cladding."

Experimental Design

"The preparation work before conducting laser cladding included (1) cleaning the substrate surface with acetone and rinsing it with alcohol and drying afterwards and (2) mixing the 316 L and WC powder under a certain weight percentage in a ball mill at 400 rpm speed for 30 min and drying the mixture in a vacuum dryer for another 30 min [34, 35]."

"Laser cladding is a complex process since the process involves the influence of different factors."

"The Taguchi orthogonal design aims at studying the influence of different factors and levels with less experimental runs and obtaining the optimal combination [36–39]."

"After the completion of laser cladding processes, 16 samples were obtained."

"The clad properties selected to be investigated in this study were micro-hardness, wear resistance, and cladding efficiency."

"The wear resistance in this study was represented by the wear volume in the clad."

Results and Analysis

"The processing parameter setup was the 14th run, as a laser power of 1.6 kW, scanning speed of 6 mm/s, gas flow of 800 L/h, and WC powder ratio of 40%."

"It can be seen that the principle order affecting the grey relational grade from the selected processing parameters was WC powder ratio, laser power, scanning speed, gas flow."

"The projected processing parameter setup that could provide the highest GRG was $LP_{level4}SS_{level3}GF_{level4}PR_{level4}$, referring to 1.6 kW laser power, 7 mm/s scanning speed, 1400 L/h gas flow, and 40% WC powder ratio."

"By comparing the setup in the 14th run and the predicted optimal set, it can be seen that the two significant factors (laser power and WC powder ratio) were at the same set point in this two-parameter set, while the difference only existed in the two insignificant factors (scanning speed and gas flow)."

Conclusion

"This study utilized Taguchi orthogonal design and investigated the influence of selected processing parameters (laser power, scanning speed, gas flow, and WC powder ratio) on the clad micro-hardness, wear resistance, and cladding efficiency."

"In the analysis of the micro-hardness and the wear volume, the principle order of processing parameters that affect these two indexes was WC powder ratio, laser power, scanning speed, gas flow."

"In the grey relational analysis, when evaluating the micro-hardness, wear volume, and cladding efficiency evenly, the principle order of the processing parameters was WC powder ratio, laser power, scanning speed, gas flow, with WC powder ratio, and laser power being statistically significant."

"Under the target of achieving maximal micro-hardness, maximal wear resistance, and maximal cladding efficiency, the predicted processing parameter setup was 1.6 kW laser power, 7 mm/s scanning speed, 1400 L/h gas flow, and 40% WC powder ratio."

Plasma Transfer Arc Additive Manufacturing of 17-4 PH: Assessment of Defects [40]

This is a machine-generated summary of:

El Moghazi, Sandy N.; Wolfe, Tonya; Ivey, Douglas G.; Henein, Hani: Plasma transfer arc additive manufacturing of 17-4 PH: assessment of defects [40].

Published in: The International Journal of Advanced Manufacturing Technology (2020).

Link to original: https://doi.org/10.1007/s00170-020-05540-2.

If you want to cite the papers, please refer to the original.

For technical reasons we could not place the page where the original quote is coming from.

Abstract-Summary "The plasma transferred arc additive manufacturing system's ability to produce large samples, compared with other common additive manufacturing techniques, makes it highly desirable in many industrial applications."

"The selected material in this additive process is 17-4 precipitation hardened stainless steel, which is widely used in numerous fields, such as the aerospace, chemical, and mining industries."

"The identified voids were oxide layers and porosity."

"The presence of oxide layers was correlated to the interaction of atmospheric oxygen with the chromium present in the stainless steel."

"The other source of voids was attributed to the porosity in the initial powders and to the choice of welding process parameters."

"Changing the powder supplier and optimizing the process parameters, mainly by increasing the heat input to ensure complete melting of the powders, greatly reduced the amount of porosity in the finished part."

"Hardness was only affected by the aging process, during which copper precipitates formed within the 17-4 precipitation hardened stainless steel matrix."

Introduction

"To increase strength and hardness of the 17-4 PH stainless steel, an aging heat treatment is needed to precipitate the copper."

"Powder bed fusion (PBF) AM processes, including laser and electron beam, have successfully fabricated parts using 17-4 PH stainless steel."

"Porosity is also a common finding when casting 17-4 PH stainless steel and, likewise, can be attributed to the processing parameters during casting."

"It can be concluded that defects found in the AM parts while printing 17-4 PH stainless steel using the PTA-AM system were common findings in other AM processes and not specific to the PTA-AM process."

"The first type of defect observed in fabricated samples of 17-4 PH stainless steel is associated with the formation of oxide layers between the interlayers of AM [41]."

"The formation of oxide layers is not common to all AM processes; it is more specific to direct energy deposition (DED) processes due to the lack of a continuous inert gas environment over the whole AM surface as it is being fabricated."

Experimental Procedures

"The integrated system is able to control most of the parameters in this process, such as movement direction, shield gas, powder gas and center gas flow rates, powder feed rate, current, voltage, and coordinated travel speed."

"The sectioned samples were hot mounted with a thermosetting phenolic powder for hardness testing and optical imaging."

"A 5 × 5 hardness matrix was taken over the cross-section of the samples, with indentations 5 mm apart in width and 2 mm apart in height."

"Obtained from as welded samples, samples with the lowest porosity and amount of oxide layers were heat treated at the optimum heat treatment at a temperature of 480 °C for 1 h. Fiji Image J, Version 1.8.0_172, was used to quantify the porosity in the fabricated parts."

"AM samples were sectioned, mounted, ground, and polished and then imaged in the SEM."

Results and Discussion

"Efforts to print a sample required the raising of the heat input, by increasing current or decreasing travel speed, to ensure complete melting of all the powders and the inhibition of the associated internal porosity and, thus, print a part with minimum porosity [42]."

"With an almost 15% higher heat input, through a combination of increased current and voltage and without the shielding hood, the volume percent of voids dropped from 18 to 8%."

"Using the increased heat input and the shielding hood dropped the porosity from 10 to 2%, due to elimination of the oxide layers and LOF porosity."

"After eliminating the oxide layer and a large number of voids through parameters' optimization and enhanced shielding, the deposits contained approximately 2 vol% porosity."

"There are several advantages to using this gas mix: removing the oxide layer after it has formed and providing higher heat inputs during welding [43–45]."

Conclusions

"Voids were found both at the interface of the interlayers (oxide layers) and within the bulk of a given layer (porosity)."

"The first of these was attributed to the lack of shielding from the atmosphere, resulting in oxygen combining with the chromium present in the 17-4 PH stainless steel to form oxide layers."

"Bulk porosity formation was sensitive to the initial powder quality and process parameters [46]."

"Pore formation in the bulk layers was traced back to the quality of the powders utilized."

"The hardness of the samples was independent of powder quality, process parameters, or layer height."

Experimental Investigation and Parametric Optimization of a Milling Process Using Multi-Criteria Decision Making Methods: A Comparative Analysis [47]

This is a machine-generated summary of:

Kalita, Kanak; Madhu, S.; Ramachandran, M.; Chakraborty, Shankar; Ghadai, Ranjan Kumar: Experimental investigation and parametric optimization of a milling process using multi-criteria decision making methods: a comparative analysis [47].

Published in: International Journal on Interactive Design and Manufacturing (IJIDeM) (2022).

Link to original: https://doi.org/10.1007/s12008-022-00973-3.

If you want to cite the papers, please refer to the original.

For technical reasons we could not place the page where the original quote is coming from.

Abstract-Summary "This paper deals with exploring the influences of cutting speed, feed rate and depth of cut on material removal rate (MRR) and average surface roughness (Ra) during milling operation of aluminum 1100 alloy work material."

"It becomes imperative to deploy multi-criteria decision making (MCDM) tools to identify the most appropriate combination of the considered milling parameters leading to a compromise solution resulting in higher MRR and lower Ra."

"Six popular MCDM techniques in the form of weighted sum model, weighted product model, weighted aggregated sum product assessment, multi-objective optimization on the basis of ratio analysis, evaluation based on distance from average solution and technique for order preference by similarity to the ideal solution are employed and comprehensively assessed here to search out the optimal machining condition for the said process."

"It can be revealed from the detailed analysis that higher MRR can be obtained at an optimal parametric combination of cutting speed = 210 rpm, feed rate = 40 mm/min and depth of cut = 0.4 mm."

"An optimal parametric intermix of cutting speed = 170 rpm, feed rate = 40 mm/min and depth of cut = 0.4 mm would lead to better surface quality of the machined components."

Introduction

"The application potentiality of six popular MCDM techniques, i.e. weighted sum model (WSM) [48], weighted product model (WPM) [49], weighted aggregated sum product assessment (WASPAS) [50], multi-objective optimization on the basis of ratio analysis (MOORA) [51], evaluation based on distance from average solution (EDAS) [52] and technique for order preference by similarity to ideal solution (TOPSIS) [53] in identifying the optimal combinations of three input parameters, i.e. cutting speed, feed rate and depth of cut (DOC), while machining Al 1100 alloy on a vertical milling machine is studied in this paper."

"The contributions of this paper are three-fold, i.e. (a) to perform statistically well-designed milling experiments on Al 1100 alloy with cutting speed, feed rate and DOC [54] as the input parameters, and material removal rate (MRR) and average surface roughness (Ra) as the responses, (b) to apply six well-accepted MCDM tools to derive the optimal parametric combinations for achieving maximum MRR and minimum Ra values, and (c) to perform a comparative analysis of the ranking performance of the considered MCDM tools."

Literature Review

"Among various MCDM methods, grey relational analysis (GRA) [55] and TOPSIS have emerged out as the most potent tools for multi-objective optimization of milling processes due to their several added advantages."

"The TOPSIS and AHP methods were deployed to derive the optimal parametric combination of the said milling process from the Pareto optimal solutions obtained using non-dominated sorting genetic algorithm (NSGA-II)."

"Chakraborty and Chakraborty [56] performed an exhaustive review of the applications of different MCDM techniques for optimizing turning, drilling and milling processes, and concluded that GRA and TOPSIS methods could be adopted for process optimization."

"Considering spindle speed, feed rate and DOC as the input parameters, and MRR and Ra as the responses, Trung [57] presented the simultaneous applications of EDAS, measurement alternatives and ranking according to compromise solution (MARCOS), TOPSIS, proximity indexed value (PIV) and MOORA methods for optimizing a milling process."

Multi-Criteria Decision Making Techniques

"The application of any of the MCDM problems starts with development of a decision matrix D = $[x_{ij}]_{m \times n}$ having m alternatives and n criteria. Where x_{ij} is the performance measure of ith alternative against jth criterion."

"The overall score (S_i^{WSM}) for each of the alternatives is derived after multiplying elements of the normalized decision matrix by the corresponding weight vector."

"All the considered alternatives are arranged in descending order of their AS_i values with the identification of the best alternative having the maximum appraisal score."

"The Euclidean distances of each alternative from the ideal best solution (S_i^+) and ideal worst solution (S_i^-) are computed using the following expressions: Finally, using S_i^+ and S_i^- values, the closeness coefficient (CC_i) is computed."

"In a decision matrix, greater dispersion in criterion values for the alternatives leads to extraction of more information resulting in allotment of higher weight to that criterion [58]."

Materials and Methods

"During Taguchi method-based experiments, different combinations of cutting speed (in rpm), feed rate (in mm/min) and DOC (in mm) are considered, and the corresponding values of MRR (in mm^3/min) and Ra (in μm) are measured as the responses."

"Based on the procedural steps of EDAS method, it identifies experimental trial number 4 as the optimal parametric intermix for the said milling process in simultaneously achieving the targets values of both MRR and Ra."

"Cutting speed is the most influencing milling parameter, that is why trial number 4, having higher cutting speed, and lower feed rate and DOC, is picked up as the best choice by most of the MCDM techniques."

"Since the weight (calculated using entropy method) of Ra is about 4% more than MRR, WSM selects alternative 1 as the ideal solution, given that Ra is the least for alternative trial number 1 among all the experiments."

Discussions

"This is because that in such environment, it is economical to carry out a small batch of pilot experiments and tune the weights assigned to the responses."

"When applications are non-repetitive and/or too complex, the need for pilot experiments to generate a dataset that would help tune the importance to various responses is not economically viable."

"The main challenge arises when there is difference in opinions among the decision makers regarding allocation of response weights in a group decision making environment."

"The relative importance to multiple responses should be decided by a group of experts/decision makers whose opinions may vary depending on their understanding of the needs and objectives of the problem."

"A response may appear to be quite important to a decision maker, but another decision maker may treat it as least significant."

Conclusions

"The paper presents a comprehensive comparison of six different MCDM techniques in ranking of alternative experimental trials and subsequent selection of the optimal milling parameters during machining of Al 1100 alloy."

"Applications of MCDM techniques would help in identifying a compromised combination of the milling parameters while simultaneously achieving higher MRR and lower Ra values."

"These two optimal settings of the milling parameters only differ with respect to the value of cutting speed."

"Although this paper considers six popular MCDM techniques for optimization of the milling process, the application potentialities of other yet to be popular MCDM tools, like multi-attributive real–ideal comparative analysis (MARICA), CoCoSo, simultaneous evaluation of criteria and alternatives (SECA), multi-attributive border approximation area comparison (MABAC) methods etc. may be explored."

Characterization and Parametric Optimization of Micro-Hole Surfaces in Micro-EDM Drilling on Inconel 718 Superalloy Using Genetic Algorithm [59]

This is a machine-generated summary of:

Dilip, Deepak G.; Panda, Satyananda; Mathew, Jose: Characterization and Parametric Optimization of Micro-hole Surfaces in Micro-EDM Drilling on Inconel 718 Superalloy Using Genetic Algorithm [59].

Published in: Arabian Journal for Science and Engineering (2020).

Link to original: https://doi.org/10.1007/s13369-019-04325-4.

If you want to cite the papers, please refer to the original.

For technical reasons we could not place the page where the original quote is coming from.

Abstract-Summary "Drilling of cooling holes in turbine blades on difficult to machine materials like Inconel 718 has been one of the significant applications of micro-EDM."

"Owing to the difficulties in the contact measurement techniques, the actual side wall characteristics of micro-holes were not explored in detail."

"Optical non-contact 3D profilometry is used to characterize the side wall surface."

"The 3D roughness parameter, Sa, has been introduced to measure the side wall surface quality of the holes drilled using micro-EDM."

"Along with side wall roughness, all the significant responses in the micro-hole manufacturing like material removal rate, overcut, and taper angle were also taken as the responses."

"The analysis showed that voltage was the most significant factor influencing all the responses within the design space."

Introduction

"The workpiece and tool electrode selection, electrical and non-electrical process parameters, the process responses like material removal rate (MRR), tool wear rate, dimensional accuracy and surface quality are some of the major areas of concern while hole drilling using micro-EDM."

"Hardly any studies were found for micro-hole drilling using an RC circuit-based micro-EDM with voltage, feed rate and electrode speed as factors."

"Majority of the research in micro-EDM drilling concentrated on the MRR, tool wear rate/electrode wear ratio, overcut, taper angle, and diameter variation between entry and exit (DVEE) whereas very few works have been attempted on studying the inner surface quality of micro-EDM drilled holes."

"A comprehensive study on the inner surface roughness measured by the 3D roughness parameter Sa of the micro-hole created by an RC circuit-based micro-EDM with voltage, feed rate and electrode speed on Inconel 718 is necessitated."

Experimental Methods

"Based on the preliminary experiments conducted and prior knowledge from the literature, gap voltage, tool feed rate and ERS were chosen as the three factors, with side wall roughness (SWR), MRR, overcut and taper angle as the responses."

"The experiments were conducted after randomizing the standard experimental design and the total number of experimental runs obtained as per the design are 17."

"In order to facilitate for the surface roughness measurement, the holes were cut along their axis using Ecocut CNC Wire EDM; Make: Electronica."

"The surfaces obtained were then analyzed using SU6600 Field Emission Scanning Electron Microscope (FE-SEM); Make: Hitachi and the 3D optical profiler."

Results and Discussion

"It can be observed that the overall rate of increase in MRR tends to decrease after 115 V. This can be attributed to the fact that at higher discharge energies, the concentration of debris in the spark zone will be more."

"The increase in feed rate leads to an increase in the sparking rate which in turn results in more material removal and higher MRR."

"It can be observed that in order to have a high MRR, the voltage and feed rate should be maximum while the ERS should be around the mid-value in the design space such that it improves the machining condition by assisting in the debris removal without causing incomplete sparking."

"This increased debris particle concentration results in more side sparking and increased overcut as the feed rate is increased."

"These particles on removal undergo side sparking at the top surface of the hole resulting in increased overcut when the ERS is increased."

Parameter Optimization Using Genetic Algorithm

"This section deals with finding an optimal parameter combination of the variable factors within the design space."

"Find the optimal input parameter combination to, Minimize SWR, overcut and taper angle Maximize MRR Here, the three input parameters, voltage, feed rate and ERS, act as the decision variables."

"In order to optimize the objective functions and find out an ideal input parameter combination, we maximize the sum of weighted objectives given byThe optimization problem is solved using genetic algorithm (GA) in MATLAB."

"The GA has proven to be a suitable optimization technique associated with EDM/micro-EDM [60–64]."

"Due to its advantages, it has been adopted to propose an optimal input parameter combination within the design space."

"The nearness of the optimized result to the lowest surface roughness, overcut and taper angle possible within the design space make this combination suitable for micro-EDM hole drilling on Inconel 718."

Conclusion

"The micro-EDM drilling experiments on Inconel 718 were carried out using the Box–Behnken experimental design for three factors: voltage, ERS and feed rate to study their effect on the surface characteristics of the inner side walls of the hole along with the MRR, diametrical overcut, and taper angle."

"The effect of each factor on all the responses was examined in detail and an optimum combination was proposed for micro-EDM drilling within the design space."

"Higher voltages proved deterrent to the surface characteristics, particularly due to the excessive debris formation, recast layer formation and the increased side sparking."

"Higher feed rates proved advantageous particularly when the material removal is concerned due to the increased sparking rate, but it reduces the surface quality by increasing the recast layer formation and debris deposition on the surface, thereby reducing its effectiveness."

Multi-Objective Optimization of AAJM Process Parameters for Cutting of B_4C/Gr Particles Reinforced Al 7075 Composites Using RSM-TOPSIS Approach [65]

This is a machine-generated summary of:

Kolli, Murahari; Ram Prasad, A. V. S; Naresh, Dasari Sai: Multi-objective optimization of AAJM process parameters for cutting of B_4C/Gr particles reinforced Al 7075 composites using RSM-TOPSIS approach [65].

Published in: SN Applied Sciences (2021).

Link to original: https://doi.org/10.1007/s42452-021-04699-x.

If you want to cite the papers, please refer to the original.

For technical reasons we could not place the page where the original quote is coming from.

Abstract-Summary "The effects of selected input factors, i.e., water jet pressure (WJP), stand-off distance (SOD), and traverse speed (TS) on the performance characteristics, namely taper angle (TA), surface roughness (Ra), and the material removal rate (MRR) are investigated."

"Further, the multi-optimization method "Technique for Order of Preference by Similarity to Ideal Solution" is considered to find out the best combinations of input factors for optimized output factors on the hybrid composite."

"The ANOVA results confirm that among the input factors, WJP and SOD are the most significant factors, and the percentage distribution of input factors are found to be jet pressure (55.21%), stand-off distance (23.36%), and traverse speed (2.56%)."

"The multi-objective optimum conditions of the input factors are WJP (A_1) 210 bar, SOD (B_1), and TS (C_3) 30 mm/min, that produce optimal values of the considered responses, i.e., MRR up to 4.8703 mm^3/min, Ra up to 3.57 μm and TA up to 0.189°."

"Hybrid Al7075/B_4C/Gr composite fabricated through the rotary stir casting technique Experimental planning and designing layouts using Response Surface Methodology scheme and mathematical equations are produced with Design Expert 11.0."

"The best TA was obtained by RSM-TOPSIS approach, found at a lower WJP and SOD and a higher TS."

Introduction

"Thamizhvalavan and others studied the machinability characteristics of Abrasive Aqua Jet (AAJ) on Al 6063/ B_4C/$ZrSiO_4$ composites with 5% of fixed reinforcement particles added to the matrix material."

"WJP, TS and SOD have been selected as machining factors to estimate the response factors like TA, MRR and Ra by adopting the Taguchi-DEAR approach."

"The performance parameters of study considered are Ra, kerf top and kerf bottom against the AAJM input factors, namely SOD, TS and WJP."

"They have also carried out investigations on the abrasive wear behavior of Al/ B_4C composite with a pin-on-disc machine using hybrid particle swarm optimization-artificial neural network (PSO–ANN) technique."

"Shabani and others implemented the ANFIS-PSO algorithm to predict the experimental results and optimize the processing parameters of Al/Al2O3 composites and Al composite."

"The present work investigates the effect of with process parameters of AAJM on hybrid Al 7075/B_4C/Gr composite."

Experimental Procedure

"To calculate the MRR by weight, differences of the before and after machined-sample of the composite are measured by an electronic digital weighing machine (Make: Citizen, India, Model: CY 204) with least count of 0.0001 g. The experiments are designed using RSM-CCD plan to arrive at the best optimal AAJM parameters for Al7075/B_4C/Gr composite."

"The fundamental reason for the adoption of this approach is to investigate the connection between the input parameters and the output measurements on the hybrid Al 7075/B_4C/Gr composite machining."

"An attempt was made to investigate the best multiple degree characteristics of the AAJM process from the RSM-CCD plan through the TOPSIS approach."

"Step 1: Evaluate the RSM-CCD experiments through the performance measurements, namely TA, Ra and MRR."

Results and Discussion

"It testifies that MRR approaches its maxima at the higher levels of SOD, TS and to its minima at WJP."

"Due to proper interactions between WJP and TS, low levels of WJP were sufficient to induce the impact and kinetic energy that is passed on to the target point of the work-piece material that resulted in B_4C and Gr particles being easily pulled out of the composite material."

"It has been confirmed that lower values of WJP (210 bar), higher values of SOD (3 mm), and lower values of TS (10 mm/min) gives the maximum MRR value at 5.3155 mm^3/min."

"Interaction plots revealed that at lower conditions of WJP (210 bar), SOD (2 mm) and TS (10 mm/min), the minimum value of the Ra (2.65 μm) was obtained."

"It was observed that at minimum TA of 0.375 was obtained at the lowest levels of WJP and the medium level and higher levels of SOD and TS."

Conclusion

"The optimum MRR was indicated with ($A_1B_3C_1$) optimum condition concerning lower water jet pressure (210 bar), higher Stand-off distance (3 mm), and lower traversing speed (10 mm/min)."

"The desirable lower Ra was obtained at 2.65 μm and conditions were ($A_1B_1C_1$); 210 bar jet pressure, 2 mm Stand-off distance and 10 mm/min traversing speed and correlating values of R^2 value and Adj."

"The optimum value for taper angle was 0.375°, and dominating input factors are set at ($A_1B_2C_3$): lower jet pressure (210 bar), Stand-off distance (2.5 mm), and higher traverse speed (30 mm/min), corresponding values of R^2 value and Adj."

"The corresponding value recommended through the RSM-TOPSIS method indicated that optimum lower input factors were jet pressure and Stand-off distance and higher factors were traversing speed, it will give the best experimental setting for experimental condition 15 and corresponding values of R^2 value and Adj."

An Extended Form of the Reciprocal-Power Function for Tolerance Allocation [66]

This is a machine-generated summary of:

Armillotta, Antonio: An extended form of the reciprocal-power function for tolerance allocation [66].

Published in: The International Journal of Advanced Manufacturing Technology (2022).

Link to original: https://doi.org/10.1007/s00170-021-08579-x.

If you want to cite the papers, please refer to the original.

For technical reasons we could not place the page where the original quote is coming from.

Abstract-Summary "The optimization of dimensional tolerances requires that a cost-tolerance function is evaluated consistently for all the part features involved in a given functional requirement."

"This is difficult because the parameters of commonly used functions are set using cost data from various sources and on possibly different scales."

"The paper proposes a revised form of one of the available cost-tolerance functions (reciprocal power), which expresses its parameters in empirical relationship with a set of design specifications on the toleranced features."

"Following a previous study based on cost data available in literature, the values and expressions of the parameters are validated and refined using a feature-based method for the estimation of machining cost."

Introduction

"The objective function of the optimization problem is the sum of the costs of the features in the tolerance chain; these are evaluated using a cost-tolerance function for each feature."

"The cost-tolerance functions proposed in literature include parameters whose values are not usually related to the properties of the toleranced features."

"Orientative values for the parameters of the function were calculated using cost-tolerance data collected from various sources and normalized on a common scale."

"The objective of this paper is to validate and refine the above formulation, in order to get a cost-tolerance function that can be used on different types of mechanical assemblies without the need to collect domain-dependent cost data."

"The method is applied to the most common cases of toleranced features to generate consistent cost data, which allow a statistical estimation of the parameters of the extended cost-tolerance function."

Background

"Comparative studies on cost-tolerance functions [67–71] focus on the choice of the correct function as a compromise between accuracy and suitability to specific allocation methods."

"To overcome such limitation, some authors use cost-tolerance curves for combined processes, which were first proposed in [72]; they span a wide range of tolerance values, which correspond to different possible process sequences to machine a feature."

"This seems one of the first attempts to relate cost data with design specifications, thus improving the consistency between the cost estimates of the different features in a tolerance chain."

"Several methods have been used to estimate the parameters of cost-tolerance functions from available cost data."

"The estimation of machining cost [73, 74] is a critical task when developing cost-tolerance functions accounting for the influence of design specifications."

"An existing feature-based method for the estimation of machining time will be used to develop an extended cost-tolerance function with the formulation suggested in [75]."

Methodology

"As the machining cost is roughly proportional to the cutting time, the factors that may have an effect on b_i include the following: the nominal dimension X_i, because a larger dimension will require a longer cutting time; the surface area of the feature, for a similar reason as above; the type of feature (rotational surface, planar surface, hole, etc), because it requires different machining processes that will have different cutting times for the same feature size; the material of the workpiece, because its machinability will require different cutting times for the same feature type and size."

"For all machining operations, the cutting time t [min] for a feature with area A_m [cm^2] is estimated as where Q_A [cm^2/min] is the area removal rate, which depends on the type of operation."

Implementation

"Any type of feature requires a sequence of machining operations depending on the specified tolerance."

"The variable cost of the feature corresponds to the cutting time associated with the tolerance grade of the last operation."

"For an IT7 tolerance the sequence includes only the first two operations, and for an IT10 tolerance it is limited to the first operation."

"This is what actually occurs for a finishing operation, while in a roughing operation only the last pass will be considered (as any previous pass is meant to remove excess material regardless of tolerance); moreover, as discussed above, the multiple passes of grinding can be treated as a single-pass operation with an equivalent removal rate."

"Due to the slight differences in slopes, the intercepts of the regression lines with the line $g = 1$ would not be consistent with the actual proportions of costs in the useful range of tolerance grades."

Results and Discussion

"The proposed form of the reciprocal power cost-tolerance function can be used in any tolerance allocation method that is based on the combined-process assumption, i.e., on the use of a single cost function for the entire allowable tolerance range on a part feature (as opposed to the use of different functions for machining processes with increasing accuracy)."

"The result has the advantage of representing an evaluation of "ease of machining" resulting from a set of factors (material, type of feature, nominal size, influence on the assembly requirement) that had not been considered together in previous approaches to the allocation problem."

"An alternative route for the solution of the allocation problem is the optimal scaling procedure deriving from the properties of the cost-tolerance function."

"The other two dimensions have looser tolerances mainly for reasons related to the material (wheel hub 6) and to the type of feature (width of the supports 3), which tend to increase their machining cost."

Conclusions

"The paper has proposed an extended form of the reciprocal power cost-tolerance function, which evaluates cost parameters according to a set of design specifications (material, type of feature, area, and nominal dimension)."

"The properties of the cost-tolerance functions are exploited in a scaling procedure for tolerance allocation, which allows to minimize cost without having to explicitly solve an optimization problem."

"The results could give the following advantages for the solution of tolerance allocation problems: The extended form of the function allows to evaluate the costs of all dimensions of a tolerance chain in the correct proportions."

"A ready-to-use cost-tolerance function can be of help to future research on tolerance allocation."

"New cost-based optimization methods will be able to benefit from the easy estimation of the parameters of cost-tolerance functions."

"The scope of the cost-tolerance function is still limited to the allocation of dimensional tolerances."

Lexicographic Method-Based Parametric Optimization of Non-traditional Machining Processes for Ceramic Materials [76]

This is a machine-generated summary of:

Das, Partha Protim; Chakraborty, Shankar: Lexicographic method-based parametric optimization of non-traditional machining processes for ceramic materials [76].

Published in: OPSEARCH (2020).

Link to original: https://doi.org/10.1007/s12597-020-00439-8.

If you want to cite the papers, please refer to the original.

For technical reasons we could not place the page where the original quote is coming from.

Abstract-Summary "Various non-traditional machining (NTM) processes have been successfully deployed to generate different complex shape geometries on engineering ceramics with the achievement of the desired dimensional accuracy and surface integrity."

"To achieve the best machining performance of those NTM processes, it has been recommended to operate them at the optimal combinations of their different input parameters."

"Lexicographic method is applied as a multi-objective optimization tool to derive the optimal parametric settings of three NTM processes, i.e. electrical discharge machining, Nd:YAG laser machining and ultrasonic machining which have found wide ranging applications in machining of engineering ceramics."

Introduction

"Due to their excessive hardness, very high strength and low fracture toughness properties which are maintained even at elevated temperatures, machining of engineering ceramics while achieving the desired surface finish and surface integrity has been observed to be a difficult and challenging task."

"These conventional machining processes are being substituted by the non-traditional machining (NTM) processes to effectively machine engineering ceramic materials within the given ranges of dimensional accuracy and surface roughness (SR)."

"Hybrid machining processes, combining the simultaneous advantages of two or more constituent NTM processes, like electrochemical discharge machining (ECDM), abrasive water jet machining (AWJM) etc. have also become popular for efficient machining of engineering ceramic materials [77]."

"A multi-objective optimization tool, in the form of lexicographic method, is applied to achieve the best machining performance of three NTM processes, i.e. EDM, Nd:YAG LBM and USM, while determining a single parametric combination which would conjointly optimize all the process outputs (responses)."

Literature Review

"Bhattacharyya et al. [78] performed experimental studies on the performance of an ECDM process while machining different non-conductive ceramic materials and determined the optimal parametric mix of the considered process parameters."

"Kibria et al. [79] developed RSM-based equations interrelating various input parameters of a pulsed Nd:YAG laser machining process with SR and DD of the machined alumina ceramics."

"Goswami and Chakraborty [80] also employed bird mating optimizer as an optimization tool in deriving the optimal parametric combination of an ECDM process which could be effectively applied for machining of engineering ceramic materials."

"In this paper, lexicographic method, a multi-objective optimization tool, is applied for identifying the optimal parametric mixes for three NTM processes, i.e. EDM, Nd:YAG LBM and USM which have already found to be extremely suitable for machining of engineering ceramic materials."

Lexicographic Method

"It allows the decision maker to provide priority level or target of achievement for each objective and prioritize the order in which the goals are to be achieved, as each of the objectives has its own importance in a particular problem."

"The optimal solution X^* is then derived while minimizing/maximizing the objective functions beginning with the most important one and proceeding according to the order of preference of the objectives."

"The user is able to provide level or target of achievement for each of the objectives and set the priority order for attaining the predefined goals."

"The solution of the first problem is achieved as $f = f_1(X)$ at X. In the same way, the second problem can be stated as follows: The solution of the second problem is attained as $f = f_2(X)$ at X. This optimization procedure is continued until all the k objectives are taken into consideration."

Examples

"The maximum value of MRR that is to be achieved is now included as a new constraint in the second stage optimization problem.subject to the constraints. The solution of the second stage optimization problem provides a minimum value of SR as 12.203 μm at the optimal parametric settings of the four EDM process parameters as discharge current = 4.5 A, pulse-on time = 188.975 μs (~200 μs), duty factor = 0.25 and open discharge voltage = 110 V. Step 3: The last optimization problem deals with minimization of TWR."

"Kibria et al. [79] applied desirability function approach as a multi-objective optimization tool, and determined the optimal settings of the considered Nd:YAG LBM process parameters as average power = 7.81 W, pulse frequency = 5600 Hz, rotational speed = 436 rpm, air pressure = 0.3 kgf/cm^2 and Y feed rate = 0.443 mm/s so as to achieve the minimum values of SR and DD as 5.6277 μm and −0.0002 mm respectively."

Conclusions

"Lexicographic method, a multi-objective optimization tool, is employed for determination of the optimal parametric settings for three NTM processes, i.e. EDM, Nd:YAG LBM and USM which have been found to be quite suitable for machining of engineering ceramic materials."

"The application of this method for parametric optimization of NTM processes while machining ceramic materials sequentially optimizes all the considered responses based on the priority order set by the user."

"It is interestingly observed that for all the three NTM processes, the derived optimal parametric mixes would proffer better response values as compared to those obtained by the earlier researchers while using other optimization approaches."

"It is concluded that lexicographic method can be efficiently adopted for parametric optimization of any of the conventional as well as non-conventional machining processes for achieving better quality characteristics."

An Integrated Approach for the Modelling of Silicon Carbide Components Laser Milling Process [81]

This is a machine-generated summary of:

Leone, Claudio; Genna, Silvio; Tagliaferri, Vincenzo: An integrated approach for the modelling of silicon carbide components laser milling process [81].

Published in: The International Journal of Advanced Manufacturing Technology (2021).

Link to original: https://doi.org/10.1007/s00170-021-07516-2.

If you want to cite the papers, please refer to the original.

For technical reasons we could not place the page where the original quote is coming from.

Abstract-Summary "The paper deals with characterisation and modelling of laser milling process on silicon carbide hard ceramic."

"Square pockets, 5 × 5 mm^2 in plane dimension, were machined at the maximum nominal average power (30 W), under different laser process parameters: pulse frequency, scan speed, hatching distance, repetitions and scanning strategy."

"Analysis of variance (ANOVA) was adopted to assess the effect of the process parameters on the achieved depth, the material removal rate (MRR) and roughness parameters, while response surface methodology (RSM) and artificial neuronal networks (ANNs) were adopted to model the process behaviours."

"Results show that both RSM and ANNs fault in MRR and RSm roughness parameters modelling."

"An integrated approach was developed to overcome the issue; the approach is based on the use of the RSM model to obtain a hybrid Training dataset for the ANNs."

Introduction

"EDM and WEDM are applicable only on electrically conductive materials or require a complex process to machine the insulating ones [82–86]."

"Also laser milling shows some drawbacks, such as low MRR and the presence of thermal damages (recast layer, microfracture, HAZ), and it requires process parameters optimisation."

"Statistical methodologies allow obtaining a robust model that can be also adopted in process optimisation [87–94]."

"There are several studies about laser milling of advanced ceramics [87, 90, 95–103]; however, in the authors' knowledge, there is lack of paper specifically focussed on the laser milling of SiC. The present paper deals with the laser milling of SiC with the aim to detect which and how the process parameters affect the laser beam–material interaction, and to explain the effect of the process parameters on the achieved depth, the MRR and surface quality."

Materials, Equipment and Procedures

"This approach involves different tasks: problem definition; a collection of information about the process by way of bibliographic analysis and relevant background; control factors (i.e. the process parameters) and their level definition, response variables (measured quantities) identification, experimental planning, tests execution and data analysis [104–106]."

"This function automatically develops and trains several network architectures and compute predictions from each by adopting advanced algorithms to determine the selection of inputs (i.e. the control factors to be adopted in the input layer), the number of hidden units, and other key factors involved in the network design."

"It is adopted to stop the training process when the error on the Validation dataset increases because of the over-fitting problems (i.e. the network tends to model the noise present in the Training dataset rather than the overall phenomenon; thus, the Training error is reduced but the Validation and Testing errors increase)."

Results

"More in details, from the figure, a good agreement between the model and the experimental data is observed for the depth, Rz and Rt, since the data fault straddling and close to the 45° line (model response = experimental data), while some scattered points are still visible for MRR, RSm and, in a lesser way, for the Ra parameter."

"A possible solution is to increase the number of points in the training dataset, decreasing the numbers for validation and testing ones; however, this solution would result in a less effective model in performance assessment."

"In order to investigate the possibility to forecast the MRR and RSm parameters by way ANNs, taking also into account the ability of neural networks to work in the presence of noise [107–109], it was decided to develop a new series of networks combining both experimental data and RSM model to obtain a new Training dataset."

Discussion

"About MRR and RSm, the H.ANN shows the best performances with Errors [%] that are 14.9 and 14.3 for the RSM and ANN, respectively."

"A possible reason is that the adoption of the H.ANN is useful mainly when the response of RSM is better than the response of the ANN (i.e. ANN has a lower R[%])."

"When the ANN shows already a response better than the RSM, the adoption of additional data (affected by a certain error) can only get worse the response."

"Either way, results appear encouraging; moreover, the hybrid approach appears as a useful method especially when the regression model (i.e. the RSM) is not fully adequate to predict the response variable (such as in the case of MRR and RSm) or when the number of repetitions is such as not to constitute a sufficiently robust database."

Conclusion

"The roughness parameters Ra, Rz and Rt mainly depend on the f, Ss and R, while the hatching distance and the Strategy show a secondary role."

"Both RSM and ANN can easily model the machine depth and the roughness parameters: Ra, Rz and Rt, allowing a low error in the estimation (<15%)."

"An integrated approach was developed to overcome the latter issue; the approach is based on the RSM model to obtain a hybrid training dataset for the ANN."

"The results show that the ANNs trained with the hybrid dataset show a drastic reduction of the error committed in the evaluation of both the MRR and the RSm."

"Apart from the results obtained for the most critical parameters (MRR and RSm), comparing the H.ANNs and RSM techniques, it must be highlighted that the latter has the advantage of requiring shorter processing times."

Non-dominated Sorting Modified Teaching–Learning-Based Optimization for Multi-Objective Machining of Polytetrafluoroethylene (PTFE) [110]

This is a machine-generated summary of:

Natarajan, Elango; Kaviarasan, Varadaraju; Lim, Wei Hong; Tiang, Sew Sun; Parasuraman, S.; Elango, Sangeetha: Non-dominated sorting modified teaching–learning-based optimization for multi-objective machining of polytetrafluoroethylene (PTFE) [110].

Published in: Journal of Intelligent Manufacturing (2019).

Link to original: https://doi.org/10.1007/s10845-019-01486-9.

If you want to cite the papers, please refer to the original.

For technical reasons we could not place the page where the original quote is coming from.

Abstract-Summary "A non-dominated sorting modified teaching–learning-based optimization (NSMTLBO) is proposed to obtain the optimum solution for a multi-objective problem related to machining Polytetrafluoroethylene."

"A fuzzy decision maker is also integrated with NSMTLBO to determine the preferred optimum machining parameters from Pareto-front based on the relative importance level of each objective function."

"The proposed NSMTLBO is reported to outperform other six peer algorithms due to its excellent capability in generating the Pareto-fronts which are more uniformly distributed and resulted higher percentage of non-dominated solutions."

"The prediction results of NSMTLBO are validated experimentally and it is reported that the performance deviations between the predicted and actual results are lower than 3.7%, implying the applicability of proposed work in real-world machining applications."

Introduction

"Three machining parameters known as discharge current, pulse-on and pulse-off times were optimized by a multi-objective genetic algorithm [111] in order to achieve the maximization of material removal rate and the minimization of tool wear rate simultaneously."

"Three process parameters for the electrochemical machining (ECM) of die tool steel with nanofluid coolant (i.e., tool feed rate, applied voltage and electrolyte discharge rate) were optimized using a multi-objective genetic algorithm developed by Sathiyamoorthy et al. [112]."

"Similar multi-objective Jaya algorithm was further applied by Rao et al. [113] to optimize the process parameters of abrasive waterjet machining process and a multi-objective decision making method was incorporated to select a preferred solution from Pareto front based on the predefined requirements of process planner."

"A multi-objective moth search algorithm was then applied to determine the optimal machining parameters, aiming to maximize the material removal rate and minimize the electrode wear rate simultaneously."

Experimental Design and Modelling of PTFE

"The virgin PTFE rod in 30 mm diameter, prepared by sintering process was used for experimental analysis."

"Three levels of cutting speed (V_c), feed rate (f), depth of cut (ap) and nose radius (N_r) were used in design of experiments (DOE) to design L_{27} orthogonal array (4 × 27)."

"The nose radius of $N_r = 0.4$, 0.8 and 1.2 mm as per the guidance from the cutting tool manufacture was used."

"The experiments were carried out with servo super cut coolant 32 t according to DOE."

"Four trials were done for each experiment and the average of the measurements was recorded."

"Response surface methodology (RSM) is a common modelling technique used to investigate the effects of several input control variables and one or more output response variables."

Proposed Methodology

"The non-dominated sorting modified teaching–learning-based optimization (NSMTLBO) is proposed as an improved posterior approach to solve the multi-objective optimization problems."

"In order to solve the multi-objective optimization problems effectively and efficiently, both of the fast non-dominated sorting approach and the crowding distance mechanism are incorporated into NSMTLBO to determine the ranking of all learners."

"In order to prevent premature convergence, it is crucial for NSMTLBO to maintain the diversity of its Pareto optimal solution sets by choosing the non-dominated solutions from sparse regions as teacher in guiding the search process of remaining learners."

"The existance of mutually contradictory objectives in the multi-objective optimization problems tends to produce multiple non-dominated learners that are able to guide other learners as teachers in searching for the Pareto front."

Experimental Studies

"The proposed NSMTLBO is reported to have the most competitive optimization performance because of its capability to produce the Pareto fronts with larger percentages of non-dominated solutions as compared with the remaining six peer algorithms in all population sizes."

"As compared with MO-ITLBO and NSTLBO, the proposed NSMTLBO has more competitive performance in producing the Pareto fronts with better quality and well distributed solutions."

"The proposed NSMTLBO has better performance than MOPSO and MO-ITLBO for being able to produce the Pareto fronts consisting of higher percentages of non-dominated solutions and more uniformly distributed of solution sets."

"It is observed that the Pareto-fronts generated by MOPSO, MOGWO, MOTLBO, MO-ITLBO and NSTLBO are inferior than that of NSMTLBO because the proposed approach can produce higher percentages of non-dominated solutions, especially those at the boundary regions of objective space."

Conclusions

"The response surface model (RSM) was rendered from the experimental results and the minimization function of surface roughness and maximization function of material removal rate were derived."

"The optimization performance of NSMTLBO was evaluated through extensive simulation studies and it was proven that the proposed approach can outperform the other six peer algorithms due to its excellent capability in generating the Pareto fronts with better uniformity and higher numbers of non-dominated solutions."

"The simulation results obtained by NSMTLBO were validated and it was reported that the performance deviations between the simulated and experimental results of both surface roughness and material removal rates for PTFE are less than 3.70%, implying the applicability of proposed work in machining of PTFE."

Optimization of Machining Parameters for Kerf Angle and Roundness Error in Abrasive Water Jet Drilling of CFRP Composites with Different Fiber Orientation Angles [114]

This is a machine-generated summary of:

Altin Karataş, Meltem; Motorcu, Ali Riza; Gökkaya, Hasan: Optimization of machining parameters for kerf angle and roundness error in abrasive water jet drilling of CFRP composites with different fiber orientation angles [114].

Published in: Journal of the Brazilian Society of Mechanical Sciences and Engineering (2020).

Link to original: https://doi.org/10.1007/s40430-020-2261-2.

If you want to cite the papers, please refer to the original.

For technical reasons we could not place the page where the original quote is coming from.

Abstract-Summary "Carbon fiber-reinforced polymer (CFRP) composites with three different fiber orientation angles (M1: $[0°/90°]_s$, M2: $[+45°/-45°]_s$, and M3:

$[0°/45°/90°/−45°]_s$) were drilled (with and without pilot holes) on an abrasive water jet (AWJ) machine and the effect of the drilling parameters on the kerf angle (K) and roundness error (Re) of selected holes was investigated to determine quality characteristics."

"The first aim of the study was the single-objective optimization of drilling parameters for minimum K and Re individually."

"The second aim was the multi-objective optimization of drilling parameters for the simultaneous minimization of both K and Re."

"Analysis of experimental findings revealed that pilot drilling improved the kerf angle of the hole by 12.4% and the roundness error by 22.87%."

"Minimum kerf angle and roundness error were realized in the AWJ drilling of the M3 CFRP."

"The most effective parameter on kerf angle and roundness error in AWJ drilling of M1, M2, and M3 CFRP materials was water pressure (80.6%, 76.9%, and 73.8%, respectively), followed by stand-off distance (11.7%, 12.0%, and 13.5%), while other drilling parameters remained below 10%."

Introduction

"Doreswamy and others examined the effects of parameters such as jet operating pressure, traverse feed rate, stand-off distance, and abrasive material concentration on kerf width in the AWJ machining of graphite-filled glass fiber-reinforced epoxy composites."

"Of Siddiqui and others, the effect of three main process parameters (water jet pressure, abrasive mass flow rate, and quality level) on kerf quality characteristics (surface roughness and kerf angle) was investigated in the AWJ bilateral cutting of three different types of epoxy composite laminates (aramid, glass, and carbon) manufactured from prepregs."

"In the studies found in the literature, it was observed that drilling parameters such as water pressure, traverse feed rate, stand-off distance, abrasive type and dimension, nozzle diameter, and abrasive mass flow rate were shown to have significant effects on the kerf angle, roundness error, and average surface roughness of the holes in AWJ drilling operations."

Materials and Methods

"To determine optimum drilling parameters and analyze the effects of the determined drilling parameters on kerf angle and roundness error values, the Taguchi method was used, as it decreases experimental study costs and provides a systematic approach."

"Step 3 Variance analysis (ANOVA) was applied to determine the degree of effect of the machining parameters on the quality characteristics following the AWJ drilling operation of the CFRP composites."

"The second objective of this experimental study was to determine the optimum combination of drilling parameters to simultaneously acquire minimum kerf angle and minimum roundness error in the holes in the AWJ drilling of M1, M2, and M3 CFRP composites having different fiber orientation angles."

"The objective of this study was to determine optimum drilling parameters in the pilot hole/no pilot hole AWJ drilling of CFRP composites manufactured with three different fiber orientation angles (M1 CFRP, M2 CFRP, and M3 CFRP) for minimum kerf angle and minimum roundness error."

Experimental Results and Evaluations

"As a general tendency of AWJ machining parameters, the kerf angle increases with an increase in traverse feed rate and stand-off distance and decreases with an increase in water pressure."

"Optimal drilling parameter levels were determined in an effective and systematic way, and minimum K and Re values in AWJ-drilled M1, M2, and M3 CFRP composites with/without pilot holes were obtained by applying all steps of the Taguchi method (single-objective optimization technique), and the quality of the holes was improved."

"According to this, minimum kerf angle and roundness error will be formed in the holes when M1, M2, and M3 CFRP composites are AWJ-drilled with pilot holes at the optimal drilling parameter levels determined as $WP_4F_3L_1D_3$ (WP_4: 5300 bar, F_3: 2000 mm/min, L_1: 1 mm, D_3: 12 mm)."

Conclusions

"Water pressure was the most effective drilling parameter on roundness error in the AWJ drilling of CFRP composites with three different fiber orientation angles with/without pilot holes."

"Lower kerf angle and roundness error values were achieved in the AWJ drilling of the CFRP composites with the $[0°/45°/90°/-45°]_s$ fiber orientation angle. The results obtained in the multi-objective optimization using the Taguchi-based GRA method to determine the optimum drilling parameter levels for both minimum kerf angle and minimum roundness error: since the pilot hole drilling was found to provide lower kerf angle and roundness error values in the single-objective optimization process, the multi-objective optimization process was conducted only for the CFRP composites with drilled pilot holes."

Parametric Optimization of μEDM Drilling on Titanium Using Principal Component Analysis [115]

This is a machine-generated summary of:

Kar, Siddhartha; Sarmah, Pallab; Baroi, Binoy Kumar; Patowari, Promod Kumar: Parametric optimization of μEDM drilling on titanium using principal component analysis [115].

Published in: Journal of the Brazilian Society of Mechanical Sciences and Engineering (2021).

Link to original: https://doi.org/10.1007/s40430-021-03249-1.

If you want to cite the papers, please refer to the original.

For technical reasons we could not place the page where the original quote is coming from.

Abstract-Summary "The parametric effect of voltage, capacitance and feed rate (FR) on response measures such as material removal rate, linear tool wear rate, overcut and taper is investigated after performing full factorial 27 experimental runs."

"Capacitance and voltage are found to be the most significant factors as compared to FR for all response measures."

"Micrographic analysis reveals minimum burr formation, recast layer and circularity error to occur at lower level of capacitance and voltage."

"An online monitoring is performed by plotting the data of machining time and work feed of the tool movement."

"Significant short circuiting is observed at lower levels of voltage and capacitance, which results in higher machining time due to lower discharge energy."

"Microhole with minimum burr formation, recast layer and circularity error is achieved at highest overall principal component index of 0.263, performed at 120 V, 100 pF and 10 μm/s."

Introduction

"The holes and slots made on the brass material were smoother as compared to Ti-6Al-4 V, as brass is softer material than Ti-6Al-4 V. During fabrication of micro-slots, the shape of the slot for Ti-6Al-4 V became taper, but the shape of the brass slot remained same throughout due to the occurrence of tool wear on machining of Ti-6Al-4 V. Optimization of multiple process parameters was studied by Azad and Puri [116] in microdischarge hole drilling of Ti-6Al-4 V. They observed that both voltage and current (I_p) were significant for single optimization quality characteristics."

"The process parameters were optimized using GRA, wherein higher MRR and lower linear wear rate, recast layer thickness and taper angle were found at T_{on} of 15 μs, I_p of 0.5 A, T_{off} of 10 μs and voltage of 40 V. Mwangi et al. [117] characterized the arcing phenomenon in μEDM and its effect on the key mechanical properties of Nitinol (composed of Ni and Ti)."

Materials and Methods

"Hollow copper (Cu) tool of outer diameter 495 μm is selected as the tool electrode due to its higher wear and corrosion resistance as well as higher electrical and thermal conductivity [118, 119]."

"The physical and mechanical properties of Ti grade 2 collected from some published research articles are found as density = 4.51 g/cc, melting point = 1665 °C, electrical resistivity = 0.000052 Ω cm, thermal conductivity = 16.4 W/m K and specific heat capacity = 0.523 J/g °C, respectively [120, 121]."

"From pilot experiment, it was observed that positive polarity of workpiece actuated more removal of material from it, so polarity of the workpiece (positive) and tool (negative) is fixed during the experiments."

"Linear wear of tool is found by electrical contact of the tool electrode to a fixed point before and after experiment."

Results and Discussion

"A rise in capacitance and voltage increases the discharge energy of the process, which leads to higher spark gap, deeper and larger craters and aids in material removal."

"This can be attributed to higher discharge energy of the process resulting from increment in voltage and capacitance, which ultimately removes material at a faster rate due to the higher intensity of discharges occurring between the two electrodes [122]."

"At capacitance of 1000 pF also, similar trend of LTWR is observed with the increase in voltage from 120 to 160 V, although the values of LTWR are higher as compared to 100 pF. But, a steep rise in LTWR is observed from 160 to 200 V. In case of higher capacitance, at 10,000 pF, the rise in LTWR is higher with the increase in voltage from 120 to 200 V. Elsewhere, LTWR increases with the increase in FR from 10 to 30 μm/s due to faster movement of tool electrode after meeting any short circuiting."

Conclusions

"Highest MRR of 368,261 μm^3/s and lowest LTWR of 0.00335 μm/s were achieved at highest level (10,000 pF, 200 V and 30 μm/s) and lowest level (100 pF, 120 V and 10 μm/s) of parameters considered in this study, respectively. (2) OC and taper increased with the increase in voltage and capacitance due to expansion of plasma channel with increased discharge energy."

"Lowest OC of 11 μm and taper of 0.00929° were achieved at 100 pF, 120 V and 10 μm/s, and 1000 pF, 120 V and 20 μm/s, respectively. (3) With the increase in capacitance and voltage, significant deterioration of workpiece surface takes place at the rim of the microholes in the form of higher circularity error, OC, burr formation and recast layer. (4) By employing PCA, all the response measures were optimized simultaneously, wherein highest OPCI of 0.263 was attained at parametric combination of 100 pF, 120 V and FR at 10 μm/s."

Selection of Parameters in Cost-Tolerance Functions: Review and Approach [123]

This is a machine-generated summary of:

Armillotta, Antonio: Selection of parameters in cost-tolerance functions: review and approach [123].

Published in: The International Journal of Advanced Manufacturing Technology (2020).

Link to original: https://doi.org/10.1007/s00170-020-05400-z.

If you want to cite the papers, please refer to the original.

For technical reasons we could not place the page where the original quote is coming from.

Abstract-Summary "The consistency of the cost-tolerance functions associated with different dimensions of a tolerance chain is a necessary condition for a correct optimization of tolerance values."

"This requires a careful selection of function parameters, in order to take into account all the variables influencing the cost, and thus to compare the costs of the different dimensions in a common scale."

"The cost-tolerance functions proposed in the literature are reviewed, revealing that the parameters are often selected without explicit reference to empirical data or objective rules."

"An original approach is proposed to build cost-tolerance functions for combined processes, for which fewer results or methods are available."

Introduction

"The optimal allocation of tolerances is usually associated with the minimum manufacturing cost."

"The objective function of the allocation problem is the sum of the cost-tolerance functions for the dimensions of a tolerance chain."

"Several types of cost-tolerance functions have been used in solving allocation problems."

"Extensive discussions on cost-tolerance functions are reported in some reviews on tolerance allocation [124–129], and in studies comparing different types of functions [67–71]."

"Relatively less studied is the selection of parameters, which is often treated as a secondary issue with no expected relevance on allocation results; as a consequence, parameter values may not be fully consistent with some requirements of tolerance allocation, such as the need to estimate costs in a common scale for all the dimensions of a tolerance chain."

"The review is limited to the basic formulation of the allocation problem, where tolerances are optimized at product design stage according to general knowledge of available manufacturing options."

General Properties

"A cost-tolerance function is continuous and strictly decreasing, i.e., a dimension can have any tolerance in a given range, and shrinking the tolerance always involves additional manufacturing costs."

"The easiest way to solve the problem analytically is the method of Lagrange multipliers [128], which requires the cost-tolerance functions to be differentiable, of simple form, and of the same type for all the dimensions."

"The tolerance ranges for the optimization have to be preselected, but a relatively simple cost-tolerance function can be accurate enough as it does not have to account for any discontinuity at process changes."

"Under this assumption, the parameters of the cost-tolerance function should change discontinuously at process changes, making the problem impossible to treat analytically."

"All the effort to improve the accuracy of a cost-tolerance function should be made on the variable cost."

Cost-Tolerance Functions

"The comparison of the functions leaves out some aspects already addressed in detail elsewhere, such as the accuracy with respect to published cost data and the suitability to the different optimization methods used for tolerance allocation."

"The function allows an analytical solution of the allocation problem by the Lagrange multiplier method: under the assumption of statistical stackup, optimal tolerances T_i are proportional to parameters b_i."

"The choice between the two most popular functions is not generally based on explicit reasons, although it is often mentioned that the reciprocal power and the exponential functions are respectively more accurate for tight (<0.1 mm) and narrow tolerances; such a comparison could be relevant for allocation problems with given manufacturing processes, while the cost-tolerance functions for combined processes should cover the whole range of tolerance values with similar accuracy."

Cost Data

"In [130], a graph provides relative costs to satisfy flatness tolerances with different processes (milling, planing, surface grinding)."

"After such normalization, some common patterns can be recognized in the different datasets:For an individual process, the cost varies within narrow limits (about 1:2 ratio between extreme values) as long as the tolerance is so large to cause negligible scrap rates; over all the processes applicable to the same type of feature, the cost range is much wider (ratios between 6:1 to 10:1);For each feature type, the cost increases as the tolerance grade decreases (about 1:6 ratio between IT13 and IT5); the relationship is nonlinear but much smoother than what would be obtained if the cost were referred to absolute tolerance. For the same tolerance grade, the cost strongly depends on both part type (prismatic parts cost more to machine than rotational ones) and feature type (external cylinders, internal cylinders, and planes in increasing order of cost)."

Methods for Parameter Selection

"In [131], the parameters of the main cost-tolerance functions are estimated for many manufacturing processes, and the accuracy in approximating cost data is compared for the different function types."

"In [132], a neural network for combined processes is designed to estimate the cost as a function of tolerance and additional influencing factors (machine tools, machining methods, process orders, process time, cutting tools, inspection instruments, skilled operators); these are evaluated by weighted scores to get a single influence coefficient."

"In [133], the parameters of cost-tolerance functions for single processes are estimated from the manufacturing difficulty of the feature with respect to the process, calculated using a method based on fuzzy logic."

"The method proposed in [134] builds cost-tolerance functions for alternative process sequences from elementary functions related to low-level operations (roughing, semi-finishing, finishing); the minimum-cost alternative for each tolerance is then selected as a point of the combined cost-tolerance curve."

Proposed Procedure

"The above review suggests the need to ensure consistency of cost-tolerance functions for the dimensions involved in an allocation problem."

"The cost will be expressed in a relative scale, common to all the features of a tolerance chain."

"Only machining costs will be considered in the combined cost-tolerance function, while quality costs would obviously be relevant when optimizing tolerances with given manufacturing processes."

"The machining cost corresponding to a given tolerance depends on the nominal dimension."

"This can be approximated by a reciprocal power law: where C_0 is the average relative cost for a given nominal dimension, and g is the tolerance grade corresponding to the tolerance for that dimension."

"From [135], the standard tolerance factor depends on the nominal dimension X in mm: The second term, related to measurement uncertainty, can be neglected for $X < 500$ mm."

Conclusions

"The paper reviews the cost-tolerance functions for tolerance allocation problems."

"To shed light on this issue, information is also provided on related problems, such as the collection of cost data and the estimation of function parameters."

"Tolerance allocation problems can be solved with many different cost-tolerance functions."

"Even more than the choice of the function type, the selection of parameters is critical for a correct use of cost-tolerance functions."

"Of how such a calculation procedure could be designed, the paper has proposed a novel formulation of the reciprocal power cost-tolerance function for combined processes."

"The parameters of the function are estimated from currently available cost data, cost estimation models, and basic concepts of tolerance standards."

"Two- or three-dimensional tolerance chains may require different approaches, as the more complex procedures needed for stackup analysis would make the choice of the cost-tolerance function insensitive to such properties as simplicity and analytical tractability."

Multi-Objective Optimization of Process Variables for MWCNT-Added Electro-Discharge Machining of 316 L Steel [136]

This is a machine-generated summary of:

Al-Amin, Md; Abdul-Rani, Ahmad Majdi; Ahmed, Rasel; Shahid, Muhammad Umair; Zohura, Fatema Tuj; Rani, Muhammad Danial Bin Abd: Multi-objective optimization of process variables for MWCNT-added electro-discharge machining of 316 L steel [136].

Published in: The International Journal of Advanced Manufacturing Technology (2021).

Link to original: https://doi.org/10.1007/s00170-021-07169-1.

If you want to cite the papers, please refer to the original.

For technical reasons we could not place the page where the original quote is coming from.

Abstract-Summary "Fabrication of 316 L steel with the desired surface integrity and thin modified recast layer for manufacturing various devices is very challenging using the traditional techniques and requires post-processing."

"Electro-discharge machining (EDM) is an evolving candidate among the non-traditional processes offering concurrent machining and surface alteration for the instrument manufacturing industry."

"To explore its full potential, this research work provides a thorough investigation of process variables on the machining performances and surface features primarily required for processing 316 L steel in the industry."

"Nano multi-walled carbon nanotubes (MWCNT) are utilized to improve the machining and surface responses."

"To this, the parametric optimization is conducted through a Taguchi-based design which assists to obtain the highest material removal rate (MRR) of 42.25 mg/min corresponding to a 10-A peak current, 16-μs pulse-on time, 1-g/l MWCNT amount, and 45% duty cycle while the lowest surface roughness (SR) and recast layer thickness (RLT) of 1.58 μm and 5.243 μm respectively are attained at a 5-A peak current, 8-μs pulse-on time, 0.7-g/l MWCNT amount, and 45% duty cycle."

Introduction

"The stochastic nature and low machining performances including low MRR (material removal rate), high SR (surface roughness), and thick recast layer formation limit the EDM technique applications which lead to high machining cost with poor surface integrity [137–141]."

"The machining performances of EDM method including MRR, recast layer thickness (RLT), and SR depend on its process variables especially peak current, duty cycle, pulse-on time, applied voltage, and tool polarity."

"This research work has used a superior multi-objective optimization technique called NSGA-II to predict the optimal combinations of process variables for the best solution sets as there are conflicting behaviors between the machining responses."

"This research offers an advanced and cost-effective manufacturing method as well as suggests the optimal solution sets to the process engineers for processing 316 L steel with the desired machining and surface responses."

Experimental Procedure

"Multiple-objective optimization of the process factors was conducted through the NSGA-II technique to obtain the Pareto frontiers showing the optimal solution sets for the machining responses."

"The process variables including peak current (A), pulse-on time (μs), MWCNT amount (g/l), and duty cycle (%) were chosen to inspect their influences on the machining and modified surface responses such as MRR, RLT, and SR through extensive literature studies and adequate trial experimentations."

"Surface morphology was carried out by the SEM technique (Model: Zeiss Evo LS15; Origin: Germany) as well applying EHT powder supply of 5 KV to investigate the microcracks, pores, and craters on the modified 316 L steel surface."

"The multi-objective such as maximizing MRR and minimizing both SR and RLT optimization of the process variables was performed through the NSGA-II technique which is widely used as the multi-objective optimization tool."

Results and Discussion

"Based on the LINMAP approach, the optimal solution set for a combination of MRR and SR of 35.363 mg/min and 2.50 μm respectively is predicted with a corresponding peak current of 7.98 A, pulse-on time of 15.84 μs, MWCNT amount of 0.99 g/l, and duty cycle of 54.65%."

"The smallest MRR and SR of 4.22 mg/min and 1.36 μm respectively were estimated corresponding to a 5-A peak current, pulse-on time of 8 μs, MWCNT amount of 0.4 g/l, and duty cycle of 41%."

"According to the LINMAP technique, the best solution set for MRR and RLT of 29.67 mg/min and 7.48 µm respectively is predicted at a 6.74 peak current, 15.74-µs pulse-on time, 1-g/l MWCNT amount, and 54.76% duty cycle."

Conclusion and Suggestions

"The smallest both SR and RLT of 1.58 µm and 5.24 µm respectively are estimated corresponding to a 5-A peak current, 8-µs pulse-on time, 0.7-g/l MWCNT amount, and 45% duty cycle."

"The best-predicted solution set for acquiring large MRR and small SR is measured at a peak current of 6 A, pule-on time of 16 µs, MWCNT amount of 1 g/l, and duty cycle of 50%."

"The required large MRR and low RLT have been predicted at a peak current of 6 A, pule-on time of 16 µs, MWCNT amount of 1 g/l, and duty cycle of 55%."

"To this, the optimal low both RLT and SR are achieved at a 5-A peak current, 8-µs pulse-on time, 0.4-g/l MWCNT amount, and 40% duty cycle."

Gene Expression Programming for Parametric Optimization of an Electrochemical Machining Process [142]

This is a machine-generated summary of:

Mandal, Kishal; Kalita, Kanak; Chakraborty, Shankar: Gene expression programming for parametric optimization of an electrochemical machining process [142].

Published in: International Journal on Interactive Design and Manufacturing (IJIDeM) (2022).

Link to original: https://doi.org/10.1007/s12008-022-00989-9.

If you want to cite the papers, please refer to the original.

For technical reasons we could not place the page where the original quote is coming from.

Abstract-Summary "Like other machining processes, the performance of an ECM process with respect to various product quality characteristics is also influenced by its different input parameters."

"This paper proposes the novel application of an almost unexplored evolutionary algorithm in the form of gene expression programming for parametric optimization of an ECM process while treating electrolyte concentration, electrolyte flow rate, applied voltage and tool feed rate as the input parameters, and material removal rate (MRR) and average surface roughness (Ra) as the responses."

"It is noticed that an optimal parametric intermix as electrolyte concentration = 15.1417 g/l, electrolyte flow rate = 6.1846 l/min, applied voltage = 15.94709 V and tool feed rate = 0.99995 mm/min would lead to simultaneous optimization of both the responses."

"At the derived optimal combination, MRR is increased by 26.14, 3.64, 12.50 and 12.55%, and Ra is reduced by 1.28, 21.97, 21.03 and 7.86% against GA, PSO, ACO and ABC techniques."

Introduction

"Despite the above-cited advantageous features of ECM process, its widespread acceptance and effective utilization is significantly constrained due to high machining cost, difficulty in electrolyte and sludge removal, corrosive action of the electrolyte, high specific energy consumption etc. Like all other machining processes, it has also been observed that the machining performance of an ECM process with respect to material removal rate (MRR), surface roughness (SR) (in the form of average surface roughness (Ra)), overcut (OC), cylindricity error (CE), dimensional deviation etc. is significantly influenced by its various input parameters, like applied voltage, tool feed rate, inter-electrode gap, electrolyte flow rate, electrolyte concentration, type of the electrolyte etc. There also exists complex dynamic relationship between those ECM process parameters and responses (process outputs) which is quite difficult to establish."

Gene Expression Programming

"In GEP, these computer programs as represented by linear strings with fixed length (chromosomes) are subsequently mapped into expression trees (ETs) of varying complexities for fitness function evaluation."

"In spite of having a fixed length, each GEP chromosome forms different ETs with different shapes and sizes."

"Ferreira [143] developed the K-expressions (from the Karva notation) to represent the genes, which are subsequently converted to ETs and sub-ETs to evaluate the fitness value of an objective function."

"It has already been mentioned that GEP chromosomes have fixed lengths, composed of one or more genes."

"Multi-gene chromosomes are more capable of arriving at the global optimal solutions for complex multi-modal functions."

"An example of mutation with single gene is exhibited as below: Hence, it can be observed that after mutation, the corresponding ET changes significantly, leading to a completely new ET with different size and solution."

GEP-Based Parametric Optimization of an ECM Process

"While solving this parametric optimization problem for the ECM process using GEP, the following procedural steps are employed: Step 1: Consider the number of variables in the developed RSM-based equation for MRR or Ra."

"For multi-objective optimization of the considered ECM process, both the responses are optimized simultaneously while solving the following objective function: where Y(Ra) and Y(MRR) are the second-order equations developed for Ra and MRR respectively, and Ra_{min} and MRR_{max} are the minimum Ra and maximum MRR values derived during their single-objective optimization respectively, and w_1 and w_2 are the weights (relative importance) allotted to Ra and MRR respectively."

"Pareto front, two extreme solutions can be derived: (a) For maximum MRR (0.166814 g/min) and maximum Ra (21.74912 μm), the corresponding parametric settings are electrolyte concentration = 29.4545 g/l, electrolyte flow rate = 5.423688 l/min, applied voltage = 15.80966 V and tool feed rate = 0.998214 mm/min."

Results and Discussion

"In ECM process, with increasing values of electrolyte concentration and flow rate, speed of the electrochemical reaction increases due to higher ionic mobility with enhanced electrolyte circulation, which finally results in increased value of MRR."

"Low values of electrolyte concentration and flow rate are responsible for lower MRR mainly due to lack of ionic particles present in the electrolyte solution."

"Higher values of both applied voltage and tool feed rate are responsible for increasing the current density (machining current) in the machining gap, causing rapid dissolution of metals from the workpiece, resulting in higher MRR."

"It has been noticed that increasing values of voltage and tool feed rate are responsible for higher electrolyzing current as well as current density, which finally improves SR of the machined components [144, 145]."

Conclusions

"Derived from both single- and multi-objective optimization of the said ECM process, the following conclusions can be drawn: (a) It is observed that for both single- and multi-objective optimization of the ECM process, GEP outperforms GA, PSO, ACO and ABC techniques with respect to accuracy and variability of the derived solutions, and computational speed and time."

"(b) For single-objective optimization, an optimal parametric combination of electrolyte concentration = 29.4745 g/l, electrolyte flow rate = 5.2339 l/min, applied voltage = 15.76245 V and tool feed rate = 0.99979 mm/min would lead to attainment of the maximum MRR value as 0.1685 g/min."

"(c) In multi-objective optimization, when equal importance is assigned to both the responses, an optimal parametric intermix as electrolyte concentration = 15.1417 g/l, electrolyte flow rate = 6.1846 l/min, applied voltage = 15.94709 V and tool feed rate = 0.99995 mm/min would result in simultaneous achievement of the most desired values for MRR and Ra."

Multi-Cohort Intelligence Algorithm for Solving Advanced Manufacturing Process Problems [146]

This is a machine-generated summary of:

Shastri, Apoorva S.; Nargundkar, Aniket; Kulkarni, Anand J.; Sharma, Kamal Kumar: Multi-cohort intelligence algorithm for solving advanced manufacturing process problems [146].

Published in: Neural Computing and Applications (2020).

Link to original: https://doi.org/10.1007/s00521-020-04858-y.

If you want to cite the papers, please refer to the original.

For technical reasons we could not place the page where the original quote is coming from.

Abstract-Summary "The applicability of the recently developed socio-inspired optimization method referred to as multi-cohort intelligence (Multi-CI) is validated by solving real-world problems from manufacturing processes domain, viz."

"The problems are minimization of surface roughness for abrasive water jet machining (AWJM), electro-discharge machining (EDM), micro-turning and micro-milling processes."

"The taper angle for the AWJM, relative electrode wear rate for EDM, burr height and burr thickness for micro-drilling, flank wear for micro-turning process, machining time for micro-milling processes were minimized."

"In AWJM, Multi-CI achieved 5–8% and 8–23% minimization for surface roughness and taper angle, respectively."

"For EDM, 47–80% maximization of material removal rate; 2–13% and 92–98% minimization of surface roughness and relative electrode wear rate, respectively, have been attained."

"For micro-turning 2% minimization of flank wear and for micro-milling, 2–6% minimization of machining time were attained."

"For micro-drilling, 24% and 16–34% minimization of burr height and burr thickness were attained."

"The contributions in this paper have opened up several avenues for further applicability of the Multi-CI algorithm for solving real-world problems."

Introduction and Motivation

"There are several process parameters associated with the AWJM which need to be optimized in order to minimize surface roughness R_a and taper angle kerf generated due to tool operations [147]."

"Conventional tool-based micro-machining techniques include micro-turning, micro-drilling and micro-milling processes."

"For micro-milling process, predefined cutter diameter, cutting speed and feed are required to be optimized to minimize surface roughness and machining time."

"This paper focuses on optimization of the process parameters associated with the abrasive water jet machining (AWJM), EDM and conventional micro-turning, micro-drilling and micro-milling processes in order to achieve the above-discussed responses."

"The algorithms of variations of CI and Multi-CI exhibited excellent performance against GA, SA, FA, PSO, BPNN, experimental results and RSM solving the problems for AWJM, EDM and micro-machining processes."

"An illustration of Multi-CI algorithm has been elaborated specifically for advanced manufacturing process problems."

Problems and Formulations

"Durairaj and Gowri [148] experimented micro-turning of Inconel alloy with cutting speed and feed as process parameters and further developed regression model for optimizing surface finish and tool wear."

"Kumar [149] experimentally investigated micro-turning of copper alloy with cutting speed and feed rate as process variables with surface finish R_a and MRR being the responses."

"The critical process parameters for tool-based mechanical micro-machining processes are cutting speed and tool feed rate."

"For micro-turning, flank wear and surface roughness R_a are performance responses, whereas for micro-milling process, surface roughness R_a and machining time are responses."

"For micro-turning model, cutting speed (c_1 in m/min), feed (c_2 in μ/rev) and depth of cut (c_3 in μm) are considered as process parameters and flank wear (f_b) and surface roughness (R_a) are performance responses [148]."

"Cutting speed d_1 (in rpm) and feed d_2 (in mm/min) are considered as process parameters, and surface roughness R_a and machining time m_t are process responses."

Multi-Cohort Intelligence Algorithm

"Multi-CI algorithm [150] is a modified version of CI in which intra- and inter-cohort learning mechanism among different candidates is modeled."

"The process begins with the initialization of number of cohorts and candidates in each cohort, sampling interval of every process parameter such as workpiece thickness, nozzle diameter, standoff distance and traverse speed convergence parameter and sampling interval reduction factor."

"Step 1: Every candidate in each cohort generates qualities within its associated sampling interval."

"Step 3: Pool of best behaviors (kerf values) is formed."

"Step 4: The probabilities are calculated related to each candidate excluding the pool of the best candidates."

"Step 5: Using roulette wheel selection approach, every candidate chooses a kerf value from its cohort excluding the best kerf values."

"Step 6: Probabilities are calculated for the pool of the best kerf values."

"Step 8: Every single candidate associated with each cohort picks the best kerf value, viz."

Results and Discussion

"For every following learning attempt, the candidates follow the behavior of the best candidate and at the same time best candidate also explores the better solution due to which candidates do not get trapped in local minimum and gives better mix of exploration and exploitation."

"In Multi-CI exploration and exploitation mechanism is strong as every candidate searches for the best solution from its own cohort as well as other cohorts."

"For micro-turning and micro-drilling processes, multi-CI yielded comparable results with GA, SA and variations of CI."

"For micro-milling with 0.7 mm and with 1 mm tool diameter, the performance of multi-CI exceeded SA and follow best and better variations of CI."

"In solving the micro-machining problems, every candidate in the multi-CI algorithm competed with its own local best behavior as well as the best behavior chosen from the other cohorts."

Conclusion and Future Directions

"Follow better, follow best, alienation and roulette wheel and multi-CI algorithm, have been compared with experimental results from the literature as well as contemporary algorithms such as GA, SA, FA, PSO, BPNN and RSM."

"Multi-CI successfully outperformed experimental results, solutions using regression, FA, SA and PSO for solving kerf and R_a of AWJM problem in terms of solution quality."

"In case of EDM and micro-machining problems, multi-CI results were comparable with GA and PSO."

"Multi-CI outperformed solutions using regression, RSM, FA, SA, BPNN and variations of CI such as roulette wheel, follow better and alienation approaches."

"For most of the problems, the run time for variations of CI is lower as compared to Multi-CI."

Abrasive Water Jet Machining for a High-Quality Green Composite: The Soft Computing Strategy for Modeling and Optimization [151]

This is a machine-generated summary of:

Jagadish, None; Patel, G. C. Manjunath; Sibalija, Tatjana V.; Mumtaz, Jabir; Li, Zhang: Abrasive water jet machining for a high-quality green composite: the soft computing strategy for modeling and optimization [151].

Published in: Journal of the Brazilian Society of Mechanical Sciences and Engineering (2022).

Link to original: https://doi.org/10.1007/s40430-022-03378-1.

If you want to cite the papers, please refer to the original.

For technical reasons we could not place the page where the original quote is coming from.

Abstract-Summary "The use of green composites in various applications has increased rapidly due their superior properties."

"Conventional machining of green composites is extremely difficult due to their inherent properties and high cost."

"Processing of these composites by abrasive water jet machining (AWJM) is a highly relevant topic nowadays, implying a need to minimize the process time and surface roughness simultaneously."

"Experiments were performed to investigate the AWJM parameters (pressure within pumping system, traverse speed, stand-off distance), concerning the two outputs."

"The networks based on genetic algorithm and on back-propagation algorithm were utilized to map the outputs and AWJM parameters, in forward and reverse mode, respectively."

"Three metaheuristics were implemented to optimize the AWJM parameters, including consideration of different output weight fractions: hybrid spider monkey optimization, grey wolf optimization and teaching–learning-based optimization."

Introduction

"In processing Al6082/WC composites, the AWJM parameters were optimized with reference to Ra and MRR using the desirability function approach [152], which has been frequently criticized for providing a sub-optimal solution."

"The metaheuristic algorithms have been also used to address the AWJM optimization, in processing materials other than composites."

"An accurate modeling of the AWJM for green composites has not been broadly studied in the previous works, and the above optimization approaches are not fully suitable and effective for global and simultaneous optimization of multiple responses."

"The intention of this work is twofold: (1) to establish BPNN and GA-NN models as process maps for the forward and reverse AWJM modeling for green composites that could be used for the accurate offline and online process monitoring; (2) to determine the optimal AWJM control variables set that minimizes both the process time (PT) and surface roughness (Ra), considering different weight fractions assigned to the two outputs."

Experimental Procedure

"Green composite made from Sundi wood fiber (as reinforcement) with epoxy resin (as matrix material) was used for machining."

"All linear terms (except PwPS) have significant impact both on the surface roughness and process time."

"TS showed a higher contribution than stand-off distance and pressure within pumping system for the process time in machining green composites."

"Stand-off-distance showed a smaller contribution, which might be due to the fact that reduction in striking force, as the water jet takes away from work material with increased SoD, causes a higher process time."

"Although there are insignificant terms ($PwPS^2$, SoD^2, TS^2, PwPS × SoD, PwPS × TS, TS × SoD) determined for both models developed for PT and surface roughness, they should not be removed from the models and regression equations."

Process Modeling

"The GA and BP algorithms are employed to tune NN parameters during the training (with the set of 1000 input–output data pairs), followed by the testing (using ten random experimental conditions)."

"The other parameters of BPNN and of GA-NN are optimized."

"The BP is utilized for NNs training, with the set of 1000 points in batch mode, aiming to optimize the connection weights and NN parameters to reach a minimum MSE."

"The optimal number of hidden neurons determined through the BPNN parameter study is adopted for the GA-NN."

"The MAPE value for process parameter predictions by the BPNN and GA-NN is estimated by computing the average value of ten random experimental conditions."

"The GA-NN has an inherent adaptability to adjust parameters and modify weights to obtain a minimal training error (MSE: 0.00942 for GA-NN and 0.03412 for BPNN)."

Process Optimization

"This section depicts the three metaheuristic algorithms (HSMO, TLBO and GWO) that are employed to optimize the AWJM parameters, as well as their implementation and benchmark of their results considering the solution accuracy and the convergence rate."

"For case 2, the HSMO algorithm found slightly better solution than TLBO and GWO, converging to the optimum at 6–12th iteration in five runs; TLBO converged to the optimum at the second iteration in all five runs, while GWO needed two to five iterations to locate the optimum."

"Since the solution quality is the most important requests, HSMO algorithm showed the overall best performance due to a slightly better solutions accuracy in case 2, with the convergence rate and computation time comparable to the other two algorithms."

"Since HSMO found a more beneficial solution (lower fitness value) than the other two algorithms, its solution is adopted as the optimal AWJM parameters setting."

Conclusions

"Two neural models (BPNN and GA-NN) are employed for predicting two outputs (PT and Ra) in forward modeling and three machining variables (PwPs, SoD, TS) in reverse modeling."

"The deviation in predicting the PT and Ra values equals 6.61% and 5.32% for the BPNN model, and 4.78% and 4.61% for the GA-NN model."

"The machining variable predictions provided by the GA-NN model are more accurate than those obtained by the BPNN, mainly due to an optimal search mechanism."

"The GA-NN model demonstrates a favorable performance in predicting both the outputs (process responses) and input variables (machining parameters)."

"Three algorithms are employed to identify a set of the AWJM parameters that minimizes PT and Ra simultaneously: HSMO, TLBO and GWO."

"It is successfully verified in a confirmation run; the resulting experimental values are PT = 0.068 s and Ra = 0.22 μm, showing a good agreement with the HSMO predicted output values."

Machining Characteristics and Optimization of Process Parameters in Die-Sinking EDM of Inconel 625 [153]

This is a machine-generated summary of:

Dikshit, Mithilesh K.; Anand, Jatin; Narayan, Devesh; Jindal, Subham: Machining characteristics and optimization of process parameters in die-sinking EDM of Inconel 625 [153].

Published in: Journal of the Brazilian Society of Mechanical Sciences and Engineering (2019).

Link to original: https://doi.org/10.1007/s40430-019-1809-5.

If you want to cite the papers, please refer to the original.

For technical reasons we could not place the page where the original quote is coming from.

Abstract-Summary "Empirical mathematical model for surface roughness and material removal rate (MRR) has been developed to study the influence of die-sinking electrical discharge machining (EDM) parameters, viz."

"Optimal setting of EDM parameters is obtained using composite desirability."

"ANOVA analysis reveals that peak current is the most dominating factor for MRR followed by pulse on time, while surface roughness is mostly influential by pulse on time followed by pulse off time."

"Surface roughness increases with the increase in pulse on time."

"MRR firstly increases with an increase in pulse on time and start decreasing after mid value (0), i.e., 15 μs whereas surface roughness continuously increases with an increase in pulse on time."

"Optimal setting of EDM parameters is obtained by composite desirability function."

"The goal is to minimize surface roughness and maximize MRR simultaneously."

"Results are validated by the confirmation experiment using the optimal set of EDM parameters."

"The confirmation experiments reveal that the experimental and predicted results are very close with error amounting of 2.19% and 2.58% for MRR and surface roughness respectively."

Introduction

"It had been reported that with copper electrode MRR, surface roughness and tool wear rate were mostly influenced by peak current and pulse on time."

"Mohanty et al. [154] had used copper tool electrode to analyze and optimize the influence of EDM process parameters, namely peak current, duty factor and pulse on duration on various performance characteristics like MRR, surface roughness and surface crack density during die-sinking EDM of Inconel 825 using a copper tool electrode."

"Die-sinking EDM was performed by Liu et al. [155] using water in oil emulsion as dielectric to study the effect of peak current, pulse on time and pulse off time with varying water content on MRR, electrode wear rate and surface roughness."

"Gap voltage, peak current, pulse on time, duty factor and flushing pressure were the independent input variables, and the machining performance was measured in terms of MRR, electrode wear rate, surface roughness and surface crack density."

Methodology

"Experiments are planned using response surface methodology (RSM)."

"RSM design is proved to be effective to establish the relationship among measured responses and the independent input variables."

"RSM consists of a dependent variable Y named as the response variable and various independent input parameters such as x_1, x_2, x_3... [156]."

"The responses are material removal rate (MRR) and surface roughness (R_a)."

"To explain nonlinearity in the model a full quadratic model (second-order model) may be given by: The quadratic model considers the influence of linear terms, quadratic terms and interaction terms on the response variables."

"pulse on time (T_{on}), pulse of time (T_{off}) and peak current (I), have been considered and response variables are material removal rate (MRR) and surface roughness (R_a)."

Experimentation

"The objective of the present research is to investigate the influence of major EDM parameters on different performance responses (MRR and R_a) and to find out the optimal setting of EMD parameters."

"Nickel-based superalloy Inconel 625 is used to study the influence of EDM process parameters on MRR and surface roughness."

"Surface irregularities are one of the parameters which reduce the value of the machined product."

"It may be quantified in terms of surface roughness parameters as it is concerned with the size and shape of the surface."

"The surface roughness is the most extensively used indicator for product quality."

"Precise inspection of surface roughness is very important to estimate the quality of a machined component."

"Surface roughness profile average surface roughness (R_a) value is considered."

Results and Discussion

"pulse on time, pulse off time and peak current on surface roughness (R_a) and material removal rate (MRR)."

"The significance of individual parameters and their interactions on MRR and surface roughness are obtained through analysis of variance (ANOVA)."

"It may be observed that the linear terms of the model are significant except pulse off time (T_{off}), whereas square term T_{off} and two interaction terms, $I \times T_{off}$ and $T_{on} \times T_{off}$, are insignificant."

"From the developed quadratic model, it is clearly seen that MRR increases with an increase in all the three EDM factors peak current, pulse on time and pulse off time."

"MRR increases with the increase in peak current and shows a downfall with the further increase in peak current at different hold values of pulse on time."

"Higher surface roughness has been observed at the lower value of peak current and higher pulse off time."

Conclusions

"Central composite design based on the design of experiment has been employed to obtain empirical relationship among EDM parameters and responses are MRR and surface roughness in die-sink EDM of Inconel 625."

"ANOVA analysis of the individual parameters reveals that peak current is the most dominating factor for MRR followed by pulse on time while surface roughness is mostly influential by pulse on time followed by pulse off time."

"The perturbation plots show that both MRR and surface roughness increases with the increase in pulse on time."

"Surface roughness also increases with an increase in pulse of time, while MRR is very little influenced pulse off time."

"Optimal setting of EDM parameters is obtained by response surface optimization to achieve minimum surface roughness and maximum MRR at the same time."

References

1. Ong P, Chong CH, Rahim B, Zulafif M, Lee WK, Sia CK, Ahmad B, Haikal MA (2018) Intelligent approach for process modelling and optimization on electrical discharge machining of polycrystalline diamond. J Intell Manuf. https://doi.org/10.1007/s10845-018-1443-6
2. Maity K, Mishra H (2016) ANN modelling and Elitist teaching learning approach for multi-objective optimization of μ-EDM. J Intell Manuf. https://doi.org/10.1007/s10845-016-1193-2
3. Mohanty CP, Mahapatra SS, Singh MR (2017) An intelligent approach to optimize the EDM process parameters using utility concept and QPSO algorithm. Eng Sci Technol Int J 20(2):552–562. https://doi.org/10.1016/j.jestch.2016.07.003
4. Tsai H, Yan B, Huang F (2003) EDM performance of Cr/Cu-based composite electrodes. Int J Mach Tools Manuf 43(3):245–252
5. McGeough JA (1988) Advanced methods of machining. Springer, Berlin
6. Lin Y-C, Lee H-S (2008) Machining characteristics of magnetic force-assisted EDM. Int J Mach Tools Manuf 48(11):1179–1186. https://doi.org/10.1016/j.ijmachtools.2008.04.004
7. Marini D, Corney JR (2020) Concurrent optimization of process parameters and product design variables for near net shape manufacturing processes. J Intell Manuf. https://doi.org/10.1007/s10845-020-01593-y
8. Zitzler E, Deb K, Thiele L (2000) Comparison of multiobjective evolutionary algorithms: empirical results. Evol Comput 8(2):173–195
9. Fonseca CM, Fleming PJ (1995) An overview of evolutionary algorithms in multiobjective optimization. Evol Comput 3(1):1–16
10. Srinivas N, Deb K (1994) Muiltiobjective optimization using nondominated sorting in genetic algorithms. Evol Comput 2(3):221–248
11. Bäck T, Schwefel HP (1993) An overview of evolutionary algorithms for parameter optimization. Evol Comput 1(1):1–23
12. Pierreval H, Tautou L (1997) Using evolutionary algorithms and simulation for the optimization of manufacturing systems. IIE Trans 29(3):181–189
13. Fonseca CM, Fleming PJ (1998) Multiobjective optimization and multiple constraint handling with evolutionary algorithms. I. A unified formulation. IEEE Trans Syst Man Cybernet Part A Syst Humans 28(1):26–37
14. Aydoğan E, Delice Y, Özcan U, Gencer C, Bali Ö (2019) Balancing stochastic u-lines using particle swarm optimization. J Intell Manuf 30(1):97–111
15. Minghai Y, Hongyan Y, Jinting H, Aimin J (2019) Reconfigurable assembly line balancing for cloud manufacturing. J Intell Manuf 30(6):2391–2405
16. Saif U, Guan Z, Zhang L, Zhang F, Wang B, Mirza J (2019) Multi-objective artificial bee colony algorithm for order oriented simultaneous sequencing and balancing of multi-mixed model assembly line. J Intell Manuf 30(3):1195–1220
17. Sardinas RQ, Santana MR, Brindis EA (2006) Genetic algorithm-based multi-objective optimization of cutting parameters in turning processes. Eng Appl Artif Intell 19(2):127–133

18. Cus F, Balic J (2003) Optimization of cutting process by GA approach. Robot Comput Integr Manuf 19(1–2):113–121
19. Saravanan R, Asokan P, Vijayakumar K (2003) Machining parameters optimisation for turning cylindrical stock into a continuous finished profile using genetic algorithm (GA) and simulated annealing (SA). Int J Adv Manuf Technol 21(1):1–9
20. D'Addona DM, Teti R (2013) Genetic algorithm-based optimization of cutting parameters in turning processes. Proc CIRP 7:323–328
21. Caporalli A, Gileno LA, Button ST (1998) Expert system for hot forging design. J Mater Process Technol 80-81:131–135
22. Denkena B, Behrens BA, Charlin F, Dannenberg M (2011) Integrative process chain optimization using a genetic algorithm. Prod Eng 6(1):29–37
23. Bi ZM, Zhang WJ (2001) Concurrent optimal design of modular robotic configuration. J Robot Syst 18(2):77–87
24. Li Q, Zhang WJ, Chen L (2001) Design for control-a concurrent engineering approach for mechatronic systems design. IEEE/ASME Trans Mechatron 6(2):161–169
25. Rampersad HK (1995) Concurrent design of product, process and robotic assembly system. Assem Autom 15(1):21–28
26. Ouyang PR, Li Q, Zhang WJ, Guo LS (2004) Design, modeling and control of a hybrid machine system. Mechatronics 14(10):1197–1217
27. Marini D (2020) An application protocol for near net shape technologies implementation. Ph.D. Thesis, University of Strathclyde
28. Dixit SR, Das SR, Dhupal D (2018) Parametric optimization of Nd:YAG laser microgrooving on aluminum oxide using integrated RSM-ANN-GA approach. J Ind Eng Int. https://doi.org/10.1007/s40092-018-0295-1
29. Kalita K, Shivakoti I, Ghadai RK (2017) Optimizing process parameters for laser beam micro-marking using genetic algorithm and particle swarm optimization. Mater Manuf Process 32:1101–1108
30. Kibria G, Doloi B, Bhattacharyya B (2014) Modelling and optimization of Nd:YAG laser micro-turning process during machining of aluminum oxide (Al_2O_3) ceramics using response surface methodology and artificial neural network. Manuf Rev 1:1–8
31. Krimpenis AA, Fountas NA, Ntalianis I, Vaxevanidis NM (2014) CNC micromilling properties and optimization using genetic algorithm. Int J Adv Manuf Technol 70:157–171
32. Lian G, Xiao S, Zhang Y, Jiang J, Zhan Y (2021) Multi-objective optimization of coating properties and cladding efficiency in 316L/WC composite laser cladding based on grey relational analysis. Int J Adv Manuf Technol. https://doi.org/10.1007/s00170-020-06486-1
33. Wu Z, Li T, Li Q, Shi B, Li X, Wang X, Lu H, Zhang H-C (2019) Process optimization of laser cladding Ni60A alloy coating in remanufacturing. Opt Laser Technol 120:105718
34. Lian G, Zhang H, Zhang Y, Yao M, Huang X, Chen C (2019) Computational and experimental investigation of micro-hardness and wear resistance of Ni-Based alloy and TiC composite coating obtained by laser cladding. Materials 12(5):793
35. Lian G, Zhao C, Zhang Y, Huang X, Chen C, Jiang J (2019) Investigation of micro-hardness, wear resistance, and defects of 316L stainless steel and TiC composite coating fabricated by laser engineered net shaping. CoatingsTech 9(8):498
36. Lian G, Zhang H, Zhang Y, Tanaka ML, Chen C, Jiang J (2019) Optimizing processing parameters for multi-track laser cladding utilizing multi-response Grey relational analysis. CoatingsTech 9(6):356
37. Sabarish K, Pratheeba P (2020) An experimental analysis on structural beam with Taguchi orthogonal array. Mater Today Proc 22:874–878
38. Chen K-T, Kao J-Y, Hsu C-Y, Hong P-D (2019) Multi-response optimization of mechanical properties for ZrWN films grown using grey Taguchi approach. Ceram Int 45(1):327–333
39. Lian G, Zhao C, Zhang Y, Feng M, Jiang J (2020) Investigation into micro-hardness and wear resistance of 316L/SiC composite coating in laser cladding. Appl Sci 10(9):3167

40. Moghazi E, Sandy N, Wolfe T, Ivey DG, Henein H (2020) Plasma transfer arc additive manufacturing of 17-4 PH: assessment of defects. Int J Adv Manuf Technol. https://doi.org/10.1007/s00170-020-05540-2
41. Yen SK (2006) Determination of the critical temperature for forming a chromium-rich oxide on AISI 430 stainless steel and its corrosion resistance. J Electrochem Soc 143:2493. https://doi.org/10.1149/1.1837036
42. Fayazfar H, Salarian M, Rogalsky A, Sarker D, Russo P, Paserin V, Toyserkani E (2018) A critical review of powder-based additive manufacturing of ferrous alloys: process parameters, microstructure and mechanical properties. Mater Des 144:98–128. https://doi.org/10.1016/j.matdes.2018.02.018
43. Onsoien M, Peters R, Olson DL, Liu S (1995) Effect of hydrogen in an argon GTAW shielding gas: arc characteristics and bead morphology. Weld J 74:10
44. Durgutlu A (2004) Experimental investigation of the effect of hydrogen in argon as a shielding gas on TIG welding of austenitic stainless steel. Mater Des 25:19–23. https://doi.org/10.1016/j.matdes.2003.07.004
45. Tusek J, Suban M (2000) Experimental research of the effect of hydrogen in argon as a shielding gas in arc welding of high-alloy stainless steel. Int J Hydrog Energy 25:369–376. https://doi.org/10.1016/S0360-3199(99)00033-6
46. Wang L, Pratt P, Felicelli S, Kadiri H, Wang P (2009) Experimental analysis of porosity formation in laser-assisted powder deposition process. Min Metals Mater Soc 1:389–396
47. Kalita K, Madhu S, Ramachandran M, Chakraborty S, Ghadai RK (2022) Experimental investigation and parametric optimization of a milling process using multi-criteria decision making methods: a comparative analysis. Int J Interact Des Manuf. https://doi.org/10.1007/s12008-022-00973-3
48. MacCrimmon KR (1968) Decision making among multiple-attribute alternatives: a survey and consolidated approach. No. RM-4823-ARPA. RAND Corporation, Santa Monica
49. Miller D, Starr M (1969) Executive decisions and operations research. Prentice-Hall, Englewood Cliffs
50. Zavadskas EK, Antucheviciene J, Saparauskas J, Turskis Z (2013) MCDM methods WASPAS and MULTIMOORA: verification of robustness of methods when assessing alternative solutions. Econ Comput Econ Cybernet Stud 47(2):5–20
51. Chakraborty S (2011) Applications of the MOORA method for decision making in manufacturing environment. Int J Adv Manuf Technol 54:1155–1166
52. Ghorabaee MK, Zavadskas EK, Olfat L, Turskis Z (2015) Multi-criteria inventory classification using a new method of evaluation based on distance from average solution (EDAS). Informatica 26:435–451
53. Hwang C-L, Yoon K (1981) Methods for multiple attribute decision making. In: Multiple attribute decision making. Lecture notes in economics and mathematical systems. Springer, Berlin, pp 58–191. https://doi.org/10.1007/978-3-642-48318-9_3
54. Yang L, DeVor RE, Kapoor SG (2005) Analysis of force shape characteristics and detection of depth-of-cut variations in end milling. J Manuf Sci Eng 127(3):454–462
55. Kuo Y, Yang T, Huang G-W (2008) The use of grey relational analysis in solving multiple attribute decision-making problems. Comput Ind Eng 55:80–93
56. Chakraborty S, Chakraborty S (2022) A scoping review on the applications of MCDM techniques for parametric optimization of machining processes. Arch Comput Methods Eng. https://doi.org/10.1007/s11831-022-09731-w
57. Trung DD (2021) Application of EDAS, MARCOS, TOPSIS, MOORA and PIV methods for multi-criteria decision making in milling process. J Mech Eng 71(2):69–84
58. Hafezalkotob A, Hafezalkotob A (2016) Extended MULTIMOORA method based on Shannon entropy weight for materials selection. J Ind Eng Int 12:1–13

59. Dilip DG, Panda S, Mathew J (2020) Characterization and parametric optimization of micro-hole surfaces in micro-EDM drilling on inconel 718 superalloy using genetic algorithm. Arab J Sci Eng. https://doi.org/10.1007/s13369-019-04325-4
60. Somashekhar KP, Ramachandran N, Mathew J (2010) Optimization of material removal rate in micro-EDM using artificial neural network and genetic algorithms. Mater Manuf Process 25(6):467–475
61. Mandal D, Pal SK, Saha P (2007) Modeling of electrical discharge machining process using back propagation neural network and multi-objective optimization using non-dominating sorting genetic algorithm-II. J Mater Process Technol 186(1-3):154–162
62. Kuriakose S, Shunmugam MS (2005) Multi-objective optimization of wire-electro discharge machining process by non-dominated sorting genetic algorithm. J Mater Process Technol 170(1-2):133–141
63. Khullar VR, Sharma N, Kishore S, Sharma R (2017) RSM- and NSGA-II-based multiple performance characteristics optimization of EDM parameters for AISI 5160. Arab J Sci Eng 42(5):1917–1928
64. Pandey AK, Gautam GD (2018) Grey relational analysis-based genetic algorithm optimization of electrical discharge drilling of Nimonic-90 superalloy. J Braz Soc Mech Sci 40:117. https://doi.org/10.1007/s40430-018-1045-4
65. Kolli M, Ram Prasad AVS, Naresh DS (2021) Multi-objective optimization of AAJM process parameters for cutting of B_4C/Gr particles reinforced Al 7075 composites using RSM-TOPSIS approach. SN Appl Sci. https://doi.org/10.1007/s42452-021-04699-x
66. Armillotta A (2022) An extended form of the reciprocal-power function for tolerance allocation. Int J Adv Manuf Technol. https://doi.org/10.1007/s00170-021-08579-x
67. Chase KW, Parkinson AR (1991) A survey of research in the application of tolerance analysis to the design of mechanical assemblies. Res Eng Des 3:23–37
68. Ashiagbor A, Liu HC, Nnaji BO (1998) Tolerance control and propagation for the product assembly modeler. Int J Prod Res 36:75–93
69. Jefferson TR, Scott CH (2001) Quality tolerancing and conjugate duality. Ann Oper Res 105:185–200
70. Ghie W (2009) Functional requirement cost for product using Jacobian-Torsor model. CIRP, Annecy
71. Sahani AK, Jain PK, Sharma SC, Bajpai JK (2014) Design verification through tolerance stack up analysis of mechanical assembly and least cost tolerance allocation. Procedia Mater Sci 6:284–295
72. Bjørke Ø (1989) Computer-aided tolerancing. ASME Press, New York
73. Niazi A, Dai JS, Balabani S, Seneviratne L (2006) Product cost estimation: technique classification and methodology review. ASME J Manuf Sci Eng 128:563–575
74. Garcia-Crespo A, Ruiz-Mezcun B, Lopez-Cuadrado JL, Gonzalez-Carrasco I (2011) A review of conventional and knowledge-based systems for machining price quotation. J Intell Manuf 22:823–841
75. Armillotta A (2020) Selection of parameters in cost-tolerance functions: review and approach. Int J Adv Manuf Technol 108:167–182
76. Das PP, Chakraborty S (2020) Lexicographic method-based parametric optimization of non-traditional machining processes for ceramic materials. OPSEARCH. https://doi.org/10.1007/s12597-020-00439-8
77. Demarbaix A, Rivière-Lorphèvre E, Ducobu F, Filippi E, Petit F (2017) The hybrid machining of ceramic: the choice of production stage. In: Proceedings of the 17th international conference of the European Society for Precision Engineering and Nanotechnology. EUSPEN, Cranfield, pp 111–112
78. Bhattacharyya B, Doloi BN, Sorkhel SK (1999) Experimental investigations into electrochemical discharge machining (ECDM) of non-conductive ceramic materials. J Mater Process Technol 95:145–154

79. Kibria G, Doloi B, Bhattacharyya B (2013) Predictive model and process parameters optimization of Nd:YAG laser micro-turning of ceramics. Int J Adv Manuf Technol 65:213–229
80. Goswami D, Chakraborty S (2017) Multi-objective optimization of electrochemical discharge machining processes: a posteriori approach based on bird mating optimizer. OPSEARCH 54(2):306–335
81. Leone C, Genna S, Tagliaferri V (2021) An integrated approach for the modelling of silicon carbide components laser milling process. Int J Adv Manuf Technol. https://doi.org/10.1007/s00170-021-07516-2
82. Tuersley IP, Jawaid A, Pashby IR (1994) Various methods of machining advanced ceramic materials. J Mater Process Technol 42(4):377–390
83. Ferraris E, Vleugels J, Guo Y, Bourell D, Kruth JP, Lauwers B (2016) Shaping of engineering ceramics by electro, chemical and physical processes. CIRP Ann 65:761–784. https://doi.org/10.1016/j.cirp.2016.06.001
84. Calignano F, Denti L, Bassoli E, Gatto A, Iuliano L (2013) Studies on electrodischarge drilling of an Al_2O_3-TiC composite. Int J Adv Manuf Technol 66:1757–1768. https://doi.org/10.1007/s00170-012-4455-y
85. Sabur A, Ali MY, Maleque MA, Khan AA (2013) Investigation of material removal characteristics in EDM of nonconductive ZrO2 ceramic. Proc Eng 56:696–701. https://doi.org/10.1016/j.proeng.2013.03.180
86. Monteverde F, Bellosi A, Scatteia L (2008) Processing and properties of ultra-high temperature ceramics for space applications. Mater Sci Eng A 485:415–421. https://doi.org/10.1016/j.msea.2007.08.054
87. Umer U, Mohammed MK, Al-Ahmari A (2017) Multi-response optimization of machining parameters in micro milling of alumina ceramics using Nd:YAG laser. Measurement 95:181–192
88. Leone C, Bassoli E, Genna S, Gatto A (2018) Experimental investigation and optimisation of laser direct part marking of Inconel 718. Opt Lasers Eng 111:154–166. https://doi.org/10.1016/j.optlaseng.2018.08.004
89. Genna S, Tagliaferri F, Papa I, Leone C, Palumbo B (2017) Multi-response optimization of CFRP laser milling process based on response surface methodology. Polym Eng Sci 57:595–605. https://doi.org/10.1002/pen.24560
90. Abdo BM, El-Tamimi AM, Anwar S et al (2018) Experimental investigation and multi-objective optimization of Nd:YAG laser micro-channeling process of zirconia dental ceramic. Int J Adv Manuf Technol 98:2213–2230. https://doi.org/10.1007/s00170-018-2374-2
91. Kasman Ş (2013) Impact of parameters on the process response: a Taguchi orthogonal analysis for laser engraving. Measurement 46:2577–2584. https://doi.org/10.1016/j.measurement.2013.04.022
92. Wang X, Duan J, Jiang M, Zhang F, Ke S, Wu B, Zeng X (2017) Investigation of processing parameters for three-dimensional laser ablation based on Taguchi method. Int J Adv Manuf Technol 93:2963–2974. https://doi.org/10.1007/s00170-017-0749-4
93. Campanelli SL, Ludovico AD, Bonserio C, Cavalluzzi P, Cinquepalmi M (2007) Experimental analysis of the laser milling process parameters. J Mater Process Technol 191:220–223. https://doi.org/10.1016/j.jmatprotec.2007.03.005
94. Alahmari AM, Darwish S, Ahmed N (2016) Laser beam micro-milling (LBMM) of selected aerospace alloys. Int J Adv Manuf Technol 86:2411–2431. https://doi.org/10.1007/s00170-015-8318-1
95. Pham DT, Dimov SS, Petkov PV (2007) Laser milling of ceramic components. Int J Mach Tools Manuf 47:618–626. https://doi.org/10.1016/j.ijmachtools.2006.05.002
96. Pham DT, Dimov SS, Ji C, Petkov PV, Dobrev T (2004) Laser milling as a "rapid" micromanufacturing process. Proc Inst Mech Eng B J Eng Manuf 2018:1–7. https://doi.org/10.1243/095440504772830156

97. Guarino S, Ponticelli GS, Giannini O, Genna S, Trovalusci F (2018) Laser milling of yttria-stabilized zirconia by using a Q-switched Yb:YAG fiber laser: experimental analysis. Int J Adv Manuf Technol 94:1373–1385. https://doi.org/10.1007/s00170-017-1020-8
98. Xu ZM, Hong ZH (2014) Experimental investigation on laser milling of Al_2O_3 ceramic. Appl Mech Mater 457-458:160–163. https://doi.org/10.4028/www.scientific.net/AMM.457-458.160
99. Knowles MRH, Rutterford G, Karnakis D, Ferguson A (2007) Micro-machining of metals, ceramics and polymers using nanosecond lasers. Int J Adv Manuf Technol 33:95–102. https://doi.org/10.1007/s00170-007-0967-2
100. Heyl P, Olschewski T, Wijnaendts RW (2001) Manufacturing of 3D structures for micro-tools using laser ablation. Microelectron Eng 57-58:775–780. https://doi.org/10.1016/S0167-9317(01)00485-3
101. Leone C, Genna S, Tagliaferri F, Palumbo B, Dix M (2016) Experimental investigation on laser milling of aluminium oxide using a 30 W Q-switched Yb:YAG fiber laser. Opt Laser Technol 76:127–137. https://doi.org/10.1016/j.optlastec.2015.08.005
102. Tsai CH, Chen HW (2003) Laser milling of cavity in ceramic substrate by fracture-machining element technique. J Mater Process Technol 136:158–165. https://doi.org/10.1016/S0924-0136(03)00133-X
103. Tsai CH, Ou CH (2004) Machining a smooth surface of ceramic material by laser fracture machining technique. J Mater Process Technol 155-156:1797–1804. https://doi.org/10.1016/j.jmatprotec.2004.04.363
104. Montgomery DC (2019) Design and analysis of experiments, 10th edn. Wiley, New York
105. Coleman DE, Montgomery DC (1993) A systematic approach to planning for a designed industrial experiment. Technometrics 53:1–12. https://doi.org/10.1080/00401706.1993.10484984
106. Del Re F, Contaldi V, Astarita A et al (2018) Statistical approach for assessing the effect of powder reuse on the final quality of AlSi10Mg parts produced by laser powder bed fusion additive manufacturing. Int J Adv Manuf Technol 97:2231–2240. https://doi.org/10.1007/s00170-018-2090-y
107. Buscema PM, Massini G, Breda M et al (2018) Artificial neural networks. Stud Syst Decis Control 131:11–35. https://doi.org/10.1007/978-3-319-75049-1_2
108. Leone C, Matarazzo D, Genna S, D'Addona DM (2019) A cognitive approach for laser milled PMMA surface characteristics forecasting. Opt Laser Technol 113:225–233. https://doi.org/10.1016/j.optlastec.2018.12.025
109. Ceylan H, Gopalakrishnan K, Bayrak MB, Guclu A (2013) Noise-tolerant inverse analysis models for nondestructive evaluation of transportation infrastructure systems using neural networks. Nondestruct Test Eval 28:233–251. https://doi.org/10.1080/10589759.2012.742084
110. Natarajan E, Kaviarasan V, Lim WH, Tiang SS, Parasuraman S, Elango S (2019) Non-dominated sorting modified teaching–learning-based optimization for multi-objective machining of polytetrafluoroethylene (PTFE). J Intell Manuf. https://doi.org/10.1007/s10845-019-01486-9
111. Suresh P, Venkatesan R, Sekar T, Elango N, Sathiyamoorthy V (2014) Optimization of intervening variables in MicroEDM of SS 316L using a genetic algorithm and response-surface methodology. Stroj Vestn-J Mech E 60(10):9–10. https://doi.org/10.5545/sv-jme.2014.1665
112. Sathiyamoorthy V, Sekar T, Elango N (2015) Optimization of processing parameters in ECM of die tool steel using nanofluid by multiobjective genetic algorithm. Sci World J 2015:6. https://doi.org/10.1155/2015/895696
113. Rao RV, Rai DP, Balic J (2019) Multi-objective optimization of abrasive waterjet machining process using Jaya algorithm and PROMETHEE method. J Intell Manuf 30(5):2101–2127. https://doi.org/10.1007/s10845-017-1373-8
114. Altin Karatas M, Motorcu AR, Gökkaya H (2020) Optimization of machining parameters for kerf angle and roundness error in abrasive water jet drilling of CFRP composites

with different fiber orientation angles. J Braz Soc Mech Sci Eng. https://doi.org/10.1007/s40430-020-2261-2
115. Kar S, Sarmah P, Baroi BK, Patowari PK (2021) Parametric optimization of μEDM drilling on titanium using principal component analysis. J Braz Soc Mech Sci Eng. https://doi.org/10.1007/s40430-021-03249-1
116. Azad MS, Puri AB (2012) Simultaneous optimisation of multiple performance characteristics in micro-EDM drilling of titanium alloy. Int J Adv Manuf Technol 61:1231–1239. https://doi.org/10.1007/s00170-012-4099-y
117. Mwangi JW, Bui VD, Thüsing K et al (2020) Characterization of the arcing phenomenon in micro-EDM and its effect on key mechanical properties of medical-grade Nitinol. J Mater Process Technol 275:116334. https://doi.org/10.1016/j.jmatprotec.2019.116334
118. Rahul DS, Masanta M (2018) Surface integrity and metallurgical characteristics of the EDMed work surfaces of A2 tool steel (SAE 304SS), inconel 601 and Ti-6Al-4V: a comparative analysis. SILICON 10:1557–1572. https://doi.org/10.1007/s12633-017-9640-z
119. Czelusniak T, Higa CF, Torres RD et al (2019) Materials used for sinking EDM electrodes: a review. J Braz Soc Mech Sci Eng 41:14
120. Gisario A, Mehrpouya M, Venettacci S, Barletta M (2017) Laser-assisted bending of Titanium Grade-2 sheets: experimental analysis and numerical simulation. Opt Lasers Eng 92:110–119. https://doi.org/10.1016/j.optlaseng.2016.09.004
121. Kar S, Patowari PK (2019) Experimental investigation of machinability and surface characteristics in microelectrical discharge milling of titanium, stainless steel and copper. Arab J Sci Eng 44:7843–7858. https://doi.org/10.1007/s13369-019-03918-3
122. Mondol K, Azad MS, Puri AB (2015) Analysis of micro-electrical discharge drilling characteristics in a thin plate of Ti-6Al-4 V. Int J Adv Manuf Technol 76:141–150. https://doi.org/10.1007/s00170-013-5414-y
123. Armillotta A (2020) Selection of parameters in cost-tolerance functions: review and approach. Int J Adv Manuf Technol. https://doi.org/10.1007/s00170-020-05400-z
124. Singh PK, Jain PK, Jain SC (2009) Important issues in tolerance design of mechanical assemblies, part 2: tolerance synthesis. Proc Inst Mech Eng B J Eng Manuf 223:1249–1287
125. Karmakar S, Maiti J (2012) A review on dimensional tolerance synthesis: paradigm shift from product to process. Assem Autom 32:373–388
126. Chase KW, Greenwood WH (1988) Design issues in mechanical tolerance analysis. Manuf Rev 1:50–59
127. Wu Z, ElMaraghy WH, El Maraghy HA (1998) Evaluation of cost-tolerance algorithms for design tolerance analysis and synthesis. Manuf Rev 1:168–179
128. Chase KW, Greenwood WH, Loosli BG, Hauglund LF (1990) Least cost tolerance allocation for mechanical assemblies with automated process selection. Manuf Rev 3:49–59
129. Chase KW (1999) Minimum-cost tolerance allocation. In: Drake PJ (ed) Dimensioning and tolerancing handbook. McGraw-Hill, New York
130. Kanai S, Onozuka M, Takahashi H (1995) Optimal tolerance synthesis by genetic algorithm under the machining and assembling constraints. CIRP, Tokyo, pp 235–250
131. Dong Z, Hu W, Xue D (1994) New production cost-tolerance models for tolerance synthesis. Trans Am Soc Mech Eng 116:199–206
132. Cao Y, Zhang H, Mao J, Yang J (2010) Novel cost-tolerance model based on fuzzy neural networks. Proc Inst Mech Eng B J Eng Manuf 224:1757–1765
133. Wang Y, Zhai W, Yang L, Wu W, Ji S, Ma Y (2007) Study on the tolerance allocation optimization by fuzzy-set weight-center evaluation method. Int J Adv Manuf Technol 33:317–322
134. Dong Z, Wang GG (1998) Automated cost modeling for tolerance synthesis using manufacturing process data, knowledge reasoning and optimization. In: El Maraghy HA (ed) Geometric design tolerancing: theories, standards and applications. Chapman & Hall, London, pp 282–293

135. ISO 286-1 (2010) Geometrical product specifications (GPS), ISO code system for tolerances on linear sizes, part 1: basis of tolerances, deviations and fits. International Organization for Standardization, Geneva
136. Al-Amin M, Abdul-Rani AM, Ahmed R, Shahid MU, Zohura FT, Rani MDBA (2021) Multi-objective optimization of process variables for MWCNT-added electro-discharge machining of 316L steel. Int J Adv Manuf Technol. https://doi.org/10.1007/s00170-021-07169-1
137. Al-Amin M, Majdi AA-R, Rasel A, Rao TV (2021) Multiple-objective optimization of hydroxyapatite-added EDM technique for processing of 316L-steel. Mater Manuf Process 36(4):1–14
138. Al-Amin M, Majdi AA, Danish M, Harvey MT, Azeez AA, Hastuty S, Fatema TZ, Bryant MG, Rubaiee S, Rao TV (2020) Assessment of PM-EDM cycle factors influence on machining responses and surface properties of biomaterials: a comprehensive review. Precis Eng 66:531–549
139. Hanif M, Wasim A, Hakim AS, Sahar N, Sajid M, Nasir M (2019) Optimization of process parameters using graphene-based dielectric in electrical discharge machining of AISI D2 steel. Int J Adv Manuf Technol 103:3735–3749
140. Sachin M, Harmesh K (2017) Parametric optimization of multiwalled carbon nanotubes-assisted electrical discharge machining of Al-10%SiCp metal matrix composite by response surface methodology. Mater Manuf Process 32:263–273
141. Rajeswari R, Shunmugam MS (2020) Finishing performance of die-sink EDM with ultrasonic vibration and powder addition through pulse train studies. Mach Sci Technol 24:245–273
142. Mandal K, Kalita K, Chakraborty S (2022) Gene expression programming for parametric optimization of an electrochemical machining process. Int J Interact Des Manuf. https://doi.org/10.1007/s12008-022-00989-9
143. Ferreira C (2001) Gene expression programming: a new adaptive algorithm for solving problems. Complex Syst 13(2):87–129
144. Babar PD, Jadhav BR (2013) Experimental study on parametric optimization of titanium based alloy (Ti-6Al-4V) in electrochemical machining process. Int J Innov Technol Exploring Eng 2(1):171–175
145. Ngo LP, Nguyen TP, Tran TT (2020) Experimental investigation of effect of process parameters on surface roughness in electrochemical machining. In: Parinov IA, Chang S-H, Long BT (eds) Advanced materials. Springer International Publishing, Cham, pp 255–270. https://doi.org/10.1007/978-3-030-45120-2_22
146. Shastri AS, Nargundkar A, Kulkarni AJ, Sharma KK (2020) Multi-cohort intelligence algorithm for solving advanced manufacturing process problems. Neural Comput & Applic. https://doi.org/10.1007/s00521-020-04858-y
147. Armağan M, Arici AA (2017) Cutting performance of glass-vinyl ester composite by abrasive water jet. Mater Manuf Process 32(15):1715–1722
148. Durairaj M, Gowri S (2013) Parametric optimization for improved tool life and surface finish in micro turning using genetic algorithm. Proc Eng 64:878–887
149. Kumar SL (2019) Measurement and uncertainty analysis of surface roughness and material removal rate in micro turning operation and process parameters optimization. Measurement 140:538–547
150. Shastri AS, Kulkarni AJ (2018) Multi-cohort intelligence algorithm: an intra-and inter-group learning behavior based socio-inspired optimisation methodology. Int J Parallel Emerg Distrib Syst 33(6):675–715
151. Jagadish N, Patel GC, Sibalija TV, Mumtaz J, Li Z (2022) Abrasive water jet machining for a high-quality green composite: the soft computing strategy for modeling and optimization. J Braz Soc Mech Sci Eng. https://doi.org/10.1007/s40430-022-03378-1
152. Kumar KR, Sreebalaji VS, Pridhar T (2018) Characterization and optimization of abrasive water jet machining parameters of aluminium/tungsten carbide composites. Measurement 117:57–66

153. Dikshit MK, Anand J, Narayan D, Jindal S (2019) Machining characteristics and optimization of process parameters in die-sinking EDM of Inconel 625. J Braz Soc Mech Sci Eng. https://doi.org/10.1007/s40430-019-1809-5
154. Mohanty A, Talla G, Gangopadhyay S (2014) Experimental investigation and analysis of EDM characteristics of Inconel 825. Mater Manuf Process 29(5):540–549. https://doi.org/10.1080/10426914.2014.901536
155. Liu Y, Zhang Y, Ji R, Cai B, Wang F, Tian X, Dong X (2013) Experimental characterization of sinking electrical discharge machining using water in oil emulsion as dielectric. Mater Manuf Process 28(4):355–363. https://doi.org/10.1080/10426914.2012.700162
156. Hicks CR (1993) Fundamental concepts in the design of experiments. Saunders College Publishing, Philadelphia

Chapter 3
Optimization Methods in Advanced Machining Processes

Introduction by the Editor

Machining processes are one of the very essential processes in manufacturing systems. The processes are not only associated with material removal, but also associated with the finishing of the surface where the shape and size of the object does not change. The chapter covers mainly the experimental and analytical methods and solutions solving the problems of grinding wheel dressing, cladding morphology prediction, material removal rate associated with the spherical error of the lapping process, optimal warehouse distribution plan, high-speed laser surface hardening process. In addition, the complex different finite element approaches for temperature distribution of laser surface hardening have also been discussed. Moreover, several machine learning models such as regression analysis, artificial neural networks, support vector machines, gaussian regression models employed to the grinding machine parameter optimization are reviewed along with certain test cases and real-world case studies. The testing of applicability of the soft computing methods for optimization of rolling parameter prediction and subsequent schedule setting have also been elaborated. The linear programming, TOPSIS and fuzzy logic approaches addressing the consideration of manufacturability in early design stages are discussed with case studies. Also, the monarch butterfly algorithm and differential evolution approaches were discussed in accordance with the efficient facility/ store layout.

The work such as remanufacturing oriented multilayer cladding morphology prediction has been reviewed. The work gives the conclusion that the material difference between the substrate and the cladding layer in laser remanufacturing, and the characteristics of the melting pool, more optimized morphology can be tried in future studies to obtain more accurate prediction models. In addition, the review of the varying demand patterns and life-cycles associated with each product class highlights that the different planning decisions possibly with different objectives;

A. J. Kulkarni (ed.), *Optimization Methods in Manufacturing Processes*,
https://doi.org/10.1007/978-981-96-5257-0_3

cost-efficiency for basic products and time-responsiveness for fashion products. Another study highlights the attempts to employ a novel hybrid welding technique to overcome root defects during the joining of aluminium pipes using friction stir welding (FSW) process. The further reviews discuss about the validation of optimized parameters, materials and methods. It concludes that the empirical-statistical modeling and optimization showed as being a valuable tool with low computational cost to estimated deposition parameters to obtain coatings with desirable geometric characteristics.

The work on the roundness prediction in centreless grinding using physics-enhanced machine learning techniques is critically reviewed which proposes a model for suggesting optimal process configuration in plunge centreless grinding operations. It concludes that the results from attempting to combine first principle and machine learning techniques into hybrid models to forecast performance of a centreless grinding process in terms of workpiece final roundness. It also suggests to improve roundness predicted by the numerical models based on first principles. The review of optimization of metal rolling control using soft computing approaches is also provided in detail. It includes computational methods in metal rolling, direction in rolling schedule, etc. It concludes that the soft computing method has achieved remarkable results in rolling parameter prediction and rolling schedule setting. The reviews of the principle of lapping balls with variable-radius groove plate (VRGP) and associated material removal model have been discussed. It concludes that the results of kinematic analysis and the single material removal were combined to obtain the dynamic material removal model used to describe the lapping process of ceramic balls.

A modelling framework to support integrated design of production systems is also provided. The case studies worked upon and studied conclude that the SSMM is a key artefact in the proposed framework and embodies three aspects of a CEPS: system architecture, decisions regarding the allocation of design options to design requirements for configuring a design alternative, and dynamic behaviour of an alternative. A unique monarch butterfly optimization for an unequal area facility layout problem is critically reviewed, it concludes that the better solutions with MBO can be generated. Some researchers have taken specific operators from other metaheuristic algorithms. This kind of effort could enable the algorithm to generate better solutions in reasonable amounts of time. The algorithm generates powerful solutions within reasonable amounts of computation time, but the solutions of each iteration tend to fluctuate. It also underscores the worthwhile research challenge to adopt new types of metaheuristic algorithms to achieve better performance in terms of solution quality.

Furthermore, the review underscores an extreme learning machine and knowledge base-based dynamic scheduling method consisting of a knowledge base-based scheduling rule selection and an extreme learning machine (ELM)-based product mix matching method. The study provides a dynamic scheduling problem of material handling for mixed-model assembly lines based on line-integrated supermarkets that considers both the effect of changing market environment (the variable product mix and weights of scheduling criteria) and changing system state (the random failure of the equipment and the instability of the cycle time) in order to better meet the real-world manufacturing process. In addition, the advantages of applying AI in product lifecycle management (PLM) have been critically discussed. It discusses the advantages and challenges. It includes the smart customization, computer vision, smart robots, and autonomous vehicles, complexity, reliability, accuracy, and timeliness in industrial scenarios, etc. These highlights certainly provides the research directions involving hybridization.

Machine Generated Summaries

Disclaimer: The summaries in this chapter were generated from Springer Nature publications using extractive AI auto-summarization: An extraction-based summarizer aims to identify the most important sentences of a text using an algorithm and uses those original sentences to create the auto-summary (unlike generative AI). As the constituted sentences are machine selected, they may not fully reflect the body of the work, so we strongly advise that the original content is read and cited. The auto generated summaries were curated by the editor to meet Springer Nature publication standards. To cite this content, please refer to the original papers.

Machine generated keywords: laser, surface, steel, grind, profile, speed, ceramic, layer, powder, remanufacture, contact, diameter, removal, material removal, angle.

Laser Deflection Tangential Dressing Complex-Concave Surface Bronze-Bonded Diamond Wheels [1]

This is a machine-generated summary of:

Zhou, Wei; Chen, Genyu; Pan, Huajun; Luo, Fengrong; Wei, Yi; Li, Mingquan: Laser deflection tangential dressing complex-concave surface bronze-bonded diamond wheels [1].

Published in: The International Journal of Advanced Manufacturing Technology (2022).

Link to original: https://doi.org/10.1007/s00170-022-09647-6.

If you want to cite the papers, please refer to the original.

For technical reasons we could not place the page where the original quote is coming from.

Abstract-Summary "A novel method named laser deflection tangential dressing was proposed for profiling concave complex-concave surface diamond wheels."

"The profile error after the tangential dressing was analyzed, and the compensation method was developed."

"The effects of the laser deflection angle and energy on the accuracy of dressing bevel edge profiles were investigated, and the impacts of the deflection angle and the compensation radius on dressing arc profiles were studied."

"The experimental results illustrated that the method of the laser deflection tangential dressing effectively solved errors in the dressing process."

"The PV value of the bevel edge profile dressing accuracy could reach about 8 μm at a deflection angle of 1–1.5°."

"With a suitable deflection angle, the radius deviation of the circular dressing profile could reach 10 μm with the compensation radius of 0.04 mm (for convex arc radius of 0.25 mm), 0.03 mm (for concave arc radius of 0.2 mm), and 0.06 mm (0.5 mm for concave arc), respectively."

Introduction

"Laser dressing is capable of non-selective removal of diamond abrasives and bond material for shaping purposes, as well as selective removal of bond material without removing abrasive grains to sharpen grinding wheels [2, 3]."

"Dold et al. [4] used a Yb:YAG pulsed picosecond laser to perform tangential incidence dressing experiments on electroplated diamond wheels, compared the surface morphology and grinding performance with the mechanical dressing method, and demonstrated the possibility of tangential dressing with high accuracy and high reliability."

"Deng and Hu [5, 6] finished the laser tangential shaping experiment on the convex-arc resin-bonded diamond grinding wheel, proposed a new sharpening method, and analyzed the material change and surface quality after dressing."

"A new laser deflection tangential dressing method for concave surface diamond wheels was proposed, which could be used to process concave surface diamond wheels of various shapes with high profile accuracy."

Experimental Procedures

"In order to observe the real-time effect of laser dressing, a backlight is used to project the opposite side of the grinding wheel into the camera view."

"The designed wheel profile is imported into the computer and located in the field of camera vision, and thus, the actual dressing profile can be compared with the designed profile in real-time."

"It is unreasonable to simply detect the contour shape of a certain section of the grinding wheel to evaluate the dressing accuracy of the grinding wheel."

"The dressed wheel profile was on-machine replicated on the graphite board firstly, and then, the graphite board can be moved by a displacement mechanism to ensure that it enters the proper vision field of the camera."

"The surface morphology of the diamond wheel after the dressing is observed by KEYENCE Ultra Depth of Field VH6000."

Results and Discussions

"When the laser dresses a diamond wheel with a complex concave surface, the depth h of the profiled groove has a significant influence on the dressing accuracy."

"The laser energy is lost due to SLE and SLEE, and a residual area is generated during the dressing process of the grinding wheel."

"The results indicated that the bevel edge error PV of the complex-concave profile wheel was greatly influenced by the parameters of laser deflection dressing."

"This was because the increase of deflection angle would lead to more concentrated laser energy, and it improved the accuracy of the laser dressing contour."

"The SLE and SLEE in the dressing process would be reduced because of laser deflection, so the dressed angle of the bevel edge profile could meet the design requirements at last."

"The laser deflection dressing can effectively process the bevel edges with any angle of inclination on the concave surface."

Conclusion

"A novel method for accurately dressing concave-shaped diamond wheels with deflected laser was proposed, which can effectively dress and prepare diamond wheels with complex contours."

"When the laser dressing concave and convex arc contours, the compensation radius needs to be considered in the programmed path."

"The laser deflection angle and laser power have dramatic influence on the dressing accuracy of bevel edge profiles."

"For arc contours, in addition to laser deflection, the proper radius compensation is also indispensable."

"Based on the laser deflection dressing technology, the overall PV value of the dressed grinding wheel could reach 25 μm."

Remanufacturing Oriented Multilayer Cladding Morphology Prediction Using a New Second Order Fitting Method [7]

This is a machine-generated summary of:

Wang, Zisheng; Jiang, Xingyu; Yang, Guozhe; Liu, Weijun; Song, Boxue: Remanufacturing oriented multilayer cladding morphology prediction using a new second order fitting method [7].

Published in: Environmental Science and Pollution Research (2022).
Link to original: https://doi.org/10.1007/s11356-021-16303-1.

If you want to cite the papers, please refer to the original.
For technical reasons we could not place the page where the original quote is coming from.

Abstract-Summary "The coupling law of laser power, scanning speed, and powder feeding speed on the geometry of cladding layer is deeply analyzed, and the process-dimension model of single-track melting layers is established."

"Based on the characteristics of the cross-sectional morphology of the cladding layer, a parabolic fitting model of the cross-sectional profile of a single-track cladding layer and an analytical model of the influence of the planar lap of multiple cladding trajectories on the melting width are proposed."

"The cross-sectional profile curves of the multilayer cladding were derived and the prediction of the multilayer cladding geometry was achieved."

"Validation experimental data for laser cladding with 304 powder showed that the average relative errors of melting height and width between the predicted results and the experimental samples were 5.18% and 1.53%, respectively, indicating that the proposed model can accurately predict the cross-sectional shape of the multilayered laser cladding."

Introduction

"Although the existing research on the prediction of laser cladding layer morphology can initially meet the basic requirements for the prediction of melt height and melt width of single-track single-layer cladding layers, it still cannot meet the requirements for the quality control of large-area and high-precision multilayer laser cladding layer morphology [8, 9]."

"Scholars have investigated the dimensional change and morphology forming law of the cladding layer from multiple perspectives, such as process parameters, laser deposition, melt pool heat accumulation, and energy compensation system, and have basically achieved the prediction of the single-track cladding layer morphology."

"This paper proposes a process-dimension model for single-track single-layer cladding layer, establishes the correspondence between laser cladding process parameters (laser power, scanning speed, powder feeding speed) and single-track cladding height and width."

Process-Dimensional Model for Single-Track, Single-Layer Laser Cladding Layer

"In order to obtain the influence law, it is necessary to design single-track laser cladding process experiments, analyze the process and dimensional data quantitatively, construct a process-dimension model for single-track single-layer laser cladding layer, and realize the prediction of melt height and width of single-track cladding layer."

"In the figure, the three axes indicate the laser power, scanning speed, and powder feeding speed, and the melt height and width data are indicated by color."

"According to the laser melting energy\mass transfer law, the change of laser power, powder feeding speed, and scanning speed directly affects the mass of material involved in the melting with the laser energy size per unit time, resulting in the variation law of the melted layer size in the figure."

"Multiple linear regression analysis was used to model the relationship between cladding process parameters and dimensions, so that the melt width and melt height can be predicted separately."

Cross-Sectional Morphology Analysis and Morphology Prediction Model for Cladding Layers

"The correspondence between the cross-sectional morphology of multilayer cladding and single-layer cladding is derived based on the process-dimension model, and a multichannel multilayer cladding morphology prediction model is established to realize the prediction of the overall morphology of the cladding layer."

"The cross-sectional morphology of 25 sets of experimental samples from single-track experiments was observed using an optical microscope, and the surface morphological characteristics of the molten cladding layer were analyzed under the condition of achieving a good metallurgical bond between the molten cladding layer and the base material."

"In the figure, the height H_0 and the width W_0 are defined with the bonding interface between the cladding layer and the substrate as the horizontal coordinate x-axis and the geometric symmetry center of the cross-sectional morphology as the vertical coordinate y-axis."

"The value of the overlap rate is calculated from the melt width of a single clad layer."

Verification of Accuracy of Multi-Track Multilayer Cladding Morphology Prediction Model

"The experimental results show that when the scanning speed and powder feeding speed remains unchanged and the laser power increases from 600 W to 800 W, the melt height of the clad layer increases by 174.92 μm, and the average prediction error of the model is 3.58%; the melt width of the clad layer increases by 981.69 μm, and the average prediction error of the model is 1.25%."

"When the laser power and scanning speed remain unchanged and the powder feeding speed increases from 8 to 12 g/min, the melt height of the clad layer increases by 1300.02 μm, and the average prediction error of the model is 6.30%; the melt width of the clad layer decreases by 316.42 μm, and the average prediction error of the model is 1.51%."

Conclusion

"This paper addressed the morphological quality of multilayer cladding, and the main work can be concluded as follows: The interaction principle of laser power, scanning speed and powder feeding speed on the melting height and width was analyzed, and the process-dimension model of single-track single-layer cladding was established."

"Under the assumption of the same content of cladding material between layers, an ensemble morphology prediction model based on quadratic algebraic equations for multilayered cladding was further derived to achieve the prediction of molding morphology and melting width and height dimensions for m-tracks and n-layers with lap ratio λ."

"Considering the material difference between the substrate and the cladding layer in laser remanufacturing, and the characteristics of the melting pool, more optimized morphology can be tried in future studies to obtain more accurate prediction models."

Theoretical and Experimental Study of Ceramic Balls Lapping with Eccentric Plates and Eccentric V-groove [10]

This is a machine-generated summary of:

Darmiyani, Mohammad Reza; Amirabadi, Hossein; Khosravi, Mohammad: Theoretical and experimental study of ceramic balls lapping with eccentric plates and eccentric V-groove [10].

Published in: The International Journal of Advanced Manufacturing Technology (2021).

Link to original: https://doi.org/10.1007/s00170-021-07585-3.

If you want to cite the papers, please refer to the original.

For technical reasons we could not place the page where the original quote is coming from.

Abstract-Summary "Ceramic balls are produced via powder metallurgy and compared to steel balls are preferred due to their special physical and mechanical properties."

"Ceramic balls can reach the desired surface finishing, sphericity, and diameter by finishing processes such as grinding, lapping, and polishing."

"By considering the length of time and high costs of the finishing process, creating an economic mechanism is of great importance in the use of ceramic balls."

"The present paper was conducted to investigate the eccentric lapping plates mechanism with eccentric V-grooves for ceramic balls lapping."

"This mechanism is proposed to increase the material removal rate (MRR) and to reduce the sphericity error of the ceramic balls."

"The effects of parameters including lower and upper lapping plate rotation speeds, the eccentricity of lapping plates, and lapping force were investigated on MRR, and the reduction of sphericity error."

Introduction

"Kang [11] proposed the lapping mechanism with an eccentric V-groove for ceramic balls lapping."

"Feng et al. [12] proposed a spiral V-groove mechanism for ceramic balls lapping and performed the finishing process of ceramic balls with a higher MRR and better surface roughness."

"Pasichnyi [13] proposed a V-groove of variable curvature, which has provided analytically and experimentally good results for ceramic balls lapping."

"Various mechanisms have been developed for ceramic ball lapping to increase the MRR and surface quality."

"The rate of reduction of sphericity error is very important in the lapping process, since increasing the MRR alone cannot be a desirable goal for the ball finishing process."

"This paper aimed to provide eccentric lapping plates with eccentric V-groove to increase the ball rolling, achieve a higher MRR, and improve the reduction of sphericity error."

Eccentric Lapping Plate Mechanism

"The concentric V-groove mechanism was proposed in 1996 by Kurobe and others In this mechanism, the upper lapping plate is flat and the lower lapping plate has a concentric V-groove."

"Different lapping mechanisms could be compared to each other by examining the balls rolling."

"In a rotation of the lapping plate, any mechanism that gives the ball further uniform rolling works better."

"In the kinematic analysis of the lapping mechanism, the ball has three points of contact."

"In the eccentric V-groove mechanism, the velocity of the two contact points of the ball is variable while in the concentric V-groove mechanism, the velocity of the three contact points is constant."

"The mechanism of the eccentric lapping plates for establishing three contact points with variable velocities was proposed."

Kinematic Analysis of the Proposed Mechanism

"We determined the trajectory of the ball contact points; the trajectory length and uniformity of distribution could be used for a quantitative assessment of the proposed mechanism."

"E shows the distance between the axis of rotation of the upper and lower lapping plates and r_g is the distance between the center of rotation of the lower lapping plate and the point of contact of the ball with the upper lapping plate."

"The trajectory of the contact points and trajectory length on the surface of the ball are calculated in the next section."

"The effect of increasing the rotation speed of the upper lapping plate was less than that of the lower lapping plate because the ball had one point of contact with the upper lapping plate."

Lapping Experiment

"The number of points and their distribution on the surface of the ball is important."

"The surface of the ball was measured with different number of points, being 6, 9, 12, and 15 in three replicates."

"Measuring the ball with 15 points indicated better repeatability."

"15 contact points were used to measure the dimensions of the ball."

"For the evaluation of the performance of the proposed mechanism experimentally, a set of experiments was performed to measure the effect of input parameters on MRR and reduction of sphericity error."

"The balls were measured before and after the lapping process to determine MRR and the reduction of sphericity error."

"The balls were lapped for 20 min and eventually measured."

Discussion

"The reduction of sphericity error reduced down to 12.56 rad/s and provided higher values at 9.42 and 15.70 rad/s. Unlike MRR, the ball sphericity was primarily on a decreasing trend and then on an increasing trend as the rotation speed of the lower lapping plate increased."

"At low forces, as the speed of rotation of the lower lapping plate increased, the ball's rolling within the V-groove increased in both the groove direction and in other directions."

"The rotation of the upper lapping plate increased the ball rolling inside the V-groove in different directions and improved the ball sphericity."

"The increase in the rotation speed of the upper lapping plate at low forces reduced the ball rolling and MRR."

"At high forces, the decreasing effect of the rotation speed of the upper lapping plate was eliminated and increased the rolling of the ball to some extent."

Conclusion

"The MRR was directly related to the rotation speed of the lower lapping plate, and the reduction of spherical error at 12.56 rad/s was minimal and improved at higher speeds."

"As the lapping force increased, the effect of lower lapping plate speed on the MRR decreased."

"The upper lapping plate speed improved the reduction of spherical error."

"With the increase in lapping force, the decreasing effect of lower lapping speed on the MRR disappeared and increased slightly."

"The lapping plates eccentricity had a little effect on the MRR, which increased at higher forces."

"The reduction of sphericity error decreased with the increase in eccentricity of the lapping plates."

"The MRR increased with the increase lapping force and the reduction of sphericity error increased up to 20 N and then decreased."

Distribution Planning for Products with Varying Life Cycles [14]

This is a machine-generated summary of:

Sainathuni, Bhanuteja; Guthrie, Bradley; Parikh, Pratik J.; Kong, Nan: Distribution planning for products with varying life cycles [14].

Published in: Flexible Services and Manufacturing Journal (2018).

Link to original: https://doi.org/10.1007/s10696-018-9314-1.

If you want to cite the papers, please refer to the original.

For technical reasons we could not place the page where the original quote is coming from.

Abstract-Summary "Distribution planning becomes increasingly complex when warehouses handle products with varying life cycles; long (basic) and short (fashion)."

"The varying demand patterns and life-cycles associated with each product class require different planning decisions possibly with different objectives; cost-efficiency for basic products and time-responsiveness for fashion products."

"To address this challenge, we introduce a novel problem to the literature, the two-product class, multi-product quantities, multi-echelon, multi-period, distribution planning problem."

"The uniqueness of this problem is in jointly deriving optimal distribution plans for both product classes, combined with optimal warehouse design decisions (e.g., technology and workforce), in order to minimize the total cost."

"Results suggest that with an increase in the proportion of fashion products, there is a substantial increase in warehouse workload variation, and subsequently total cost."

Introduction

"The novelty of this problem is in its ability to capture, in a single model, various tactical (e.g., warehouse technology, permanent workforce, and fashion product window) and operational decisions (e.g., temporary workforce during each time period, inbound and outbound transportation, and inventory of products at warehouse and stores)."

"The key contributions of our research include (1) a novel mixed-integer bilinear mathematical model for TPC-WITP that jointly determines distribution plans (warehouse, inventory, and transportation decisions) for both basic and fashion products, (2) a neighborhood search heuristic that can be used independently for computationally efficient optimization (upwards of 1000 products and 200 stores), (3) analysis of the impact of joint handling of both products on warehousing workload, workforce planning (permanent and temporary), and costs, and (4) sensitivity of the distribution plans to fashion window, warehouse worker cost, and product mix."

Relevant Literature

"Several conceptual models have been presented in the literature that discuss the right approach for companies to match their functional and innovative products with physically efficient and market responsive supply chains, respectively [15]."

"According to Khan et al. [16], the unprecedented shift in the supply chain strategies in the fashion industry over the last decade from product-centric to customer-centric has had a major impact on the changing risk profile and responsiveness of fashion retailers."

"Khan and Creazza [17] present a roadmap for a design centric business to help facilitate better integration between product design and their supply chain."

"While a supply chain/distribution network for fashion products should not only be responsive, but also accurate in meeting the demand."

"Weng [18] develops a framework to address the problem of coordinating operating decisions of the manufacturing and distribution supply chains to meet price-sensitive random demand for products with short life cycles."

The Two-Product Class Warehouse-Inventory-Transportation Problem (TPC-WITP).

"In order to jointly identify an optimal distribution plan for two product classes (basic and fashion) such that the total distribution cost (comprised of warehousing, inventory, and transportation) is minimized, the TPC-WITP focuses on the following decisions: Warehouse technology selection, number of permanent workers, and number of periods that fashion products are held at the warehouse—these are usually part of overall tactical decision making."

"Number of temporary workers at the warehouse during each period, for each basic product, the (1) inventory at warehouse and store, (2) inbound shipment times and quantities from the vendor (i.e., inbound schedule), and (3) outbound to store (i.e., outbound schedule), and for each fashion product the (1) timing of inbound shipment to the warehouse and (2) timing of outbound shipment for each product to store—these are usually part of overall operational decision making."

An Iterative Local Search (ILS) Based Heuristic for the TPC-WITP

"The selected putaway technology then determines the worker productivity, and thus the number of putaway workers required in each time-period. (b) Improve basic inbound A solution is improved by implementing several neighborhood moves; e.g., advance partial, delay partial, advance complete and delay complete moves only on the basic products that are inbound from basic vendors to warehouse."

"The selected picking technology then determines the picker productivity, and thus the number of pickers required in each time-period. (d) Improve basic outbound For the best inbound solution found thus far (s*), the outbound solution (which consists of outbound basic and fashion product quantities, outbound shipment schedules, store inventory, and warehouse technology and workforce for picking) is iteratively improved using ideas similar to those in the inbound phase."

Numerical Study with Industry-Sized Problem Instances

"Focusing on the technology selection, as the total number of products (basic and fashion) flowing through the warehouse increases, so does the sophistication of the selected technology."

"We evaluated the effect of product mix (ratio of basic and fashion products), the fashion window, and warehouse worker cost in the heuristic solutions for large problem instances."

"The instances were run on a dataset with 1:1 product mix (i.e., 500 basic and 500 fashion products), the permanent (temporary) cost of workers at the warehouse as \$40 (\$30) per hour, and the selection of technology being a decision."

"One way to limit this increase in worker cost is through the selection of appropriate technology to be employed at the warehouse and also by changing the number and/or timing (either delaying or advancing) the inbound and outbound shipments."

Conclusions and Future Research

"The main objective of the TPC-WITP is to identify an optimal distribution plan jointly for both basic and fashion products, along with optimal warehouse technology and workforce decisions, in order to minimize the total cost (which includes warehouse, inventory, and transportation)."

"If the length of the fashion window could be increased, then it should be pursued as it allows for more time and options for the warehouse manager to handle the inbound and outbound of these fashion products, thus reducing the workload variation and total cost."

"This has several advantages: (1) it helps consolidate outbound shipments, (2) holding these products at the warehouse is typically less expensive than at the stores, and (3) although this induces substantial workload variation over those few time-periods, it can be handled by the addition of temporary workers."

"Our approach is expected to help warehouse managers identify factors (e.g., proportion of fashion products vs. labor cost) that drive workload variation in both putaway and picking areas, and manage it much better."

Experimental Investigation on Joining Process of Aluminium Alloy 6063-T6 Pipes Using Hybrid Friction Stir Welding [19]

This is a machine-generated summary of:

Senthil, S. M.; Nathan, S. Ragu; Parameshwaran, R.: Experimental investigation on joining process of aluminium alloy 6063-T6 pipes using hybrid friction stir welding [19].

Published in: Journal of the Brazilian Society of Mechanical Sciences and Engineering (2021).

Link to original: https://doi.org/10.1007/s40430-021-03054-w.

If you want to cite the papers, please refer to the original.

For technical reasons we could not place the page where the original quote is coming from.

Abstract-Summary "This study has attempted to employ a novel hybrid welding technique to overcome root defects during the joining of aluminium pipes using friction stir welding (FSW) process."

"Aluminium 6063-T6 pipes, with outer diameter 500 mm and 5 mm thickness, are first welded on the root side using TIG welding to a calculated depth of 1 mm, which is followed by FSW on the face side to the remaining depth of 4 mm."

"Hybrid welded pipes are compared with pure friction stir welded pipes for mechanical properties."

"Taper cylindrical tool at a rotational speed of 2000 rpm and welding speed of 0.6 rpm exhibited a superior weld strength of 176 MPa, which is 73% of base metal strength and 87.2% of pure friction stir welded pipes."

Introduction

"Pipe joints are fabricated mostly and usually using fusion welding techniques."

"Recent techniques such as friction stir welding (FSW) and friction welding are being considered as beneficial techniques to fabricate pipe joints requiring higher mechanical strength, especially for aluminium alloys."

"When using a friction welding process for joining pipes, tight concentricity requirements (zero eccentricity between the two pipes during welding) are of major concern, which is not in the case of FSW."

"Implementation of FSW technique in pipes could extend its use to the petroleum, petrochemical, and natural gas industries where high weld volume would justify the upfront costs of FSW [20]."

"To address this problem, root welding could be one of the successful techniques."

"This work aims to explore the feasibility of performing root closure using TIG welding process for AA pipes to avoid most prone root defects."

Materials and Methods

"As most of the root defects occur after the pin length area (since usually FSW tool pin is ~1 mm less than the material thickness), it is decided to root weld the pipes to a depth of 1 mm."

"The root welded pipes are mounted on the indigenously developed rotary fixture for face welding using FSW process."

"The pin length of the FSW tool is modified accordingly to facilitate the welding process, i.e. for the depth of 4 mm thickness."

"Pure FSW successful samples are also fabricated using tool rotational speed of 2000 rpm and weld speed of 0.6 rpm for comparison reasons using a tool and fixture consideration, which resulted after extensive optimization process [21]."

"After welding process, the successful pipe joints are subjected to radiography tests in order to check for flaws."

"Longitudinal samples are cut from the weld, and tensile test is carried out using universal testing machine as per ASTM E8 standards."

Results and Discussion

"The weld zone a, b, and c clearly depicts the extrusion of material beneath the probe to the middle region (b) and it forms a hook-like structure in the swirl zone which in turn improves the stir zone quality."

"Sound stir zone formation without defect yields good mechanical properties, metallurgical characteristics decide the sustainability of the weld quality irrespective of environment."

"There is clear boundary interaction between stir zone and TMAZ which might increase the tensile behaviour of the weld joint."

"The fracture surface of the joint welded with 2000 rpm and 0.6 rpm shows coarse and populated dimples, whereas fine and deep dimples were found across the transgranular fracture region."

"For the root side of hybrid weld, an average value of 50 VHN is obtained in the weld zone."

Conclusions

"An attempt has been made to avoid root defects by incorporating TIG welding into FSW, i.e. hybrid welding."

"Samples are made out of pure FSW and hybrid welding and results are compared."

"Pure FSW sample exhibited a tensile strength of 201.81 MPa which is 83.7% of the base metal strength [22]."

"Around, 87.2% of pure FSW sample strength has been retained by the hybrid weld sample."

Finite Element Analysis and Experimental Validation of High-Speed Laser Surface Hardening Process [23]

This is a machine-generated summary of:

Anusha, Epparla; Kumar, Adepu; Shariff, Sulthan Mohiddin: Finite element analysis and experimental validation of high-speed laser surface hardening process [23].

Published in: The International Journal of Advanced Manufacturing Technology (2021).

Link to original: https://doi.org/10.1007/s00170-021-07303-z.

If you want to cite the papers, please refer to the original.

For technical reasons we could not place the page where the original quote is coming from.

Abstract-Summary "A finite element method is presented with an aim to predict the temperature distribution for optimizing laser surface hardening process on a cylindrical steel solid rod."

"The proposed quasi-stationary laser beam processing technique utilizing high-speed rotation was employed to obtain a uniform hardened layer along the entire workpiece surface with complete elimination of deleterious inter-pass softening effects in the overlapped regions and melting."

"Two principal quality-determining factors of laser surface hardened layer, namely, total hardened case depth and hardness distribution, across and along its depth and length of the rod are correlated and compared."

"The temperature distribution profiles simulated along the treated layer longitudinally and through thickness, processed with high rotary axis speed and appropriate linear speed, exhibited uniform treated layer with complete elimination of inter-pass tracking softening effects with surface temperature being above critical temperature throughout the processed length."

Introduction

"As laser surface hardening employing high-speed rotation entails removal of deleterious back-tempering effects with effective overlapping of laser tracks, evaluating thermal history of the processed part helps envisage in assessing and analyzing case profile variation and hardness distribution along its processed length."

"In another recently reported study, Barath and co-workers could simulate the surface temperature distribution profile effectively by adopting a transient thermal model correlate with on-line temperature monitoring using a high-speed camera during laser hardening of En-19 steel [24]."

"Although many studies attempted to optimize the laser surface hardening process utilizing high-speed processing, none of them used the FEM technique to correlate the temperature distribution with the critical responses such as hardness distribution behavior along the treated layer to minimize deleterious softening effects."

Laser Processing Setup and Simulation

"In the case of high-speed rotational process, the laser spot rotates multiple times within the circumferential periphery resulting in virtual ring-shaped laser beam termed as a quasi-stationary beam spot that envelops the diameter of the cylindrical rod."

"The first task undertaken was to simulate the process of determining laser dwell time (duration of the laser beam exposed without any linear traverse along the length of the cylindrical rod) required to produce a uniform quasi-stationary ring of 12-mm diameter."

"Further to the optimum LDT of 310 ms required for single-spot stationary beam processing (minimum hardened case depth of 300 μm), similar residual time for laser beam movement through the lateral traverse direction (along the length of the rod) has been considered."

"The hardened case profile that evolves due to quasi-stationary beam process adopting high-speed rotation is analogous to the processing of the steel rod with a ring-shaped laser beam (equivalent to the diameter of the rod)."

Results and Discussion

"Simulated contour profiles corroborated that maximum austenitization temperature through the depth of the layer increased from 900 to 1045 °C with increase in laser power from 2500 to 3500 W. Similar effects of case depth and hardness distribution through its case depth with heat input were observed in a reported study involving high-speed laser surface hardening of steel rod involving apparent spot technique [25]."

"In similar studies involving correlation of simulated thermal history (maximum austenitization and surface temperatures) with actual hardness distribution results of steel, increase in laser power directly enhanced hardness with minimized variation and attributed to enhanced effective thermal diffusion time and cooling rate [26, 27]."

"To validate the proposed model, quantitative prediction of surface temperature distribution along the length of treated rod and thermal diffusion length governed by A_{C1} boundary with variation in RPM and linear speed was simulated with thermal history and correlated with their respective experimental values of TCD and DHL."

Conclusions

"The effects of three critical processing parameters, namely, laser power, linear speed, and rotary axis speed on the quality of laser-hardened layer in terms of total case depth (TCD) and hardness distribution, namely, mean hardness-cross-sectional (MHC) and differential hardness-longitudinal (DHL), were investigated and assessed."

"There exists an optimum processing parameter range over the investigated range of processing conditions (2.5–3.5-kW laser power, 8–12-mm/s linear speed, and 1600–2000-RPM rotary axis speed) determined by the cylindrical steel rod diameter and laser spot dimensions, over which a uniform hardened layer could be achieved with high-speed rotation with complete elimination of softening effect associated with overlapping of circumferential tracks."

"The simulated temperature distribution as per the FEM model facilitated in quantitatively assessing its case depth with hardness distribution influenced by the critical processing parameters of laser power, linear speed, and rotary axis speed."

Laser Cladding of Stellite-6 on AISI 316 L Austenitic Stainless Steel: Empirical-Statistical Modeling and Parameter Optimization [28]

This is a machine-generated summary of:

Pacheco, Jeferson Trevizan; da Silva, Leandro João; Barbetta, Luiz Delagnelo; Ferreira, Henrique Santos; Veiga, Marcelo Taveira; Forni, Renato; Teixeira, Moisés Felipe: Laser Cladding of Stellite-6 on AISI 316 L Austenitic Stainless Steel: Empirical-Statistical Modeling and Parameter Optimization [28].

Published in: Lasers in Manufacturing and Materials Processing (2020).

Link to original: https://doi.org/10.1007/s40516-020-00132-0.

If you want to cite the papers, please refer to the original.

For technical reasons we could not place the page where the original quote is coming from.

Abstract-Summary "This work aimed to develop an empirical statistical model to determine the optimized parameters for depositing multi-bead single layer Stellite-6 coating on AISI 316 L austenitic stainless steel by laser cladding."

"A full factorial experimental design was carried out to establish a correlation between processing parameters (laser power, scanning speed and powder feed rate) and geometric characteristics of the coatings (dilution, peak height and surface waviness) through multiple regression."

"The microstructure and microhardness were assessed as a mean of estimate the performance of optimized coating."

"The empirical-statistical model and optimization method proposed in this work were demonstrated to be valuable tools for process optimization with low computational cost."

Introduction

"Among several processing parameters of laser cladding process, those with the most significant impact on the coating characteristics are laser power [W], scanning speed [mm/min] and powder feed rate [g/min]."

"The present work aimed to develop an empirical statistical model to determine the optimized parameters for deposition multi-bead single layer Stellite-6 coating on AISI 316 L austenitic stainless steel by laser cladding."

"A full factorial experimental design was carried out to establish a correlation between deposition parameters (laser power, scanning speed and powder feed rate) and geometric characteristics of the coatings (dilution, peak height and surface waviness) through multiple regression."

"The main contributions of the present work were to present a parameter selection procedure to deposit coatings with desirable geometric characteristics by laser cladding, as well as, to confirm trend presented in the current literature."

Materials and Methods

"Stellite-6 gas atomized powder was deposited by laser cladding on a substrate of AISI 316 L steel with dimensions of 15 × 63 × 250 mm."

"The laser source and powder feeder are connected to a coaxial deposition head (Fraunhofer Coax 8)."

"The coatings were obtained by depositing areas composed of 10 overlapped beads with length of 60 mm."

"The coatings were randomly deposited to mitigate systematic errors and the substrates alternated between depositions to keep a similar interpass temperature for all depositions."

"The selected dependent variables (responses) were dilution, peak height and waviness."

"Dilution is one of most important characteristics of the laser cladding coatings."

"A small height difference between peak and valley (surface waviness) is desirable to avoid excessive machining to obtain a flat surface."

"The limits were established based on the literature and experiments previously carried out, and sensitivity coefficients were adjusted according to the desired response [29–31]."

Results

"An interesting characteristic observed was the increasing of the dilution during multiple beads depositions (covering an area), which could be linked with the accumulation of energy caused by sequential depositions [32]."

"The correlation equations for dilution [%], peak height [mm] and coating surface waviness [mm] as a function of laser power [W], scanning speed [mm/min] and powder feed rate [g/min] through multiple regression were: where P is power [W], V is scanning speed [mm/min] and H is hatch distance [mm]."

Discussions

"The laser power had a greater influence on dilution than scanning speed."

"The dilution level tends be higher as the higher the hatch distance, which can be related with a higher interaction between laser beam and substrate surface."

"It can be observed that the scanning speed had greater impact on the peak heigh than the laser power."

"As the higher the scanning speed, the lower the peak height due to the lower the deposition rate by length unit."

"The coating peak height tends to be higher as lower the hatch distance, which could be related with a higher overlapping between the beads."

Validation of Optimized Parameters

"The coating deposited using the optimized parameters estimated by desirability function presented sound visual aspect and to be free of superficial defects."

"It can be noted that the microstructure of the coating is composed of dendrites rich in cobalt, and the interdendritic region consists mainly of chromium carbides, corroborating the literature [33]."

"The microhardness of the coating was measured through the cross section, from the top of the coating towards the substrate."

"These results suggesting that the performance was not compromised in favor of the coating geometry."

Conclusions

"The present work aimed to develop an empirical statistical model to determine optimized parameters for deposition multi-bead single layer Stellite-6 coating on AISI 316 L austenitic stainless steel by laser cladding."

"Based on the presented results, the following conclusions can be drawn: The empirical-statistical modeling and optimization showed as being a valuable tool with low computational cost to estimated deposition parameters to obtain coatings with desirable geometric characteristics."

"The geometric characteristics of the coating obtained in the validation experiment reached the limits imposed on the desirability function, showing small differences in relation to the values estimated by the empirical-statistical model."

"The dilution showed greater dependence of laser power followed by to scanning speed."

"It is worth mentioning that the proposed model is valid for the tested conditions (laser cladding system and alloy)."

Roundness Prediction in Centreless Grinding Using Physics-Enhanced Machine Learning Techniques [34]

This is a machine-generated summary of:

Safarzadeh, Hossein; Leonesio, Marco; Bianchi, Giacomo; Monno, Michele: Roundness prediction in centreless grinding using physics-enhanced machine learning techniques [34].

Published in: The International Journal of Advanced Manufacturing Technology (2020).

Link to original: https://doi.org/10.1007/s00170-020-06407-2.

If you want to cite the papers, please refer to the original.

For technical reasons we could not place the page where the original quote is coming from.

Abstract-Summary "This work proposes a model for suggesting optimal process configuration in plunge centreless grinding operations."

"Seven different approaches were implemented and compared: first principles model, neural network model with one hidden layer, support vector regression model with polynomial kernel function, Gaussian process regression model and hybrid versions of those three models."

"Complexity in centreless grinding models arises from phenomena like contact length dependency on local compliance, contact force and grinding wheel roughness, unpredicted material properties of the grinding wheel and workpiece, precision of the manual setup done by the operator, wheel wear and nature of wheel wear."

"In order to improve the overall model prediction accuracy and allow automated continuous learning, several machine learning techniques have been investigated: a Bayesian regularized neural network, an SVR model and a GPR model."

"To exploit the a priori knowledge embedded in physical models, hybrid models are proposed, where neural network, SVR and GPR models are fed by the nominal process parameters enriched with the roundness predicted by the first principle model."

"Those hybrid models result in an improved prediction capability."

Introduction

"Lizarralde [35] has applied similar approaches to guide setup and optimization of centreless plunge grinding processes, in order to reduce setup time and avoid geometric instabilities as a function of WP height and blade angle, taking into account machine-WP dynamic interaction."

"To the contact length, variability in material properties of the grinding wheel and workpiece, systematic precision/error of the manual setup done by the operator as a parameter, should be tackled in overall calculations with a robust approach, in order to reach accurate results."

"Zakharov et al. [36] showed that the setup of centreless superfinishing machine tools entails the creation of geometric, kinematic and mechanical models of the shaping process and the use of formal optimization methods."

"Barrenetxea [37] developed an assistant tool for the setup and optimization of the centreless grinding process, optimizing productivity considering process data and machine characteristics (dynamics included)."

Process Model

"As Rowe [38] describes, considering a plunge centreless grinding, the shape of the cylindrical WP is defined by the radial reduction r(θ) occurring at the grinding point (i.e. at the contact between wheel and WP), where θ defines WP rotation, starting from zero at the beginning of grinding."

"Given its physical meaning, I (θ) cannot become negative, as the process either subtracts material, or leaves the surface unvaried when the wheel detaches: thus, the actual depth of cut must be "clipped" to zero if negative: where K_1 and K_2 are the well-known coefficients relating WP displacement at the grinding contact to radius variation at work rest and rubbing wheel contacts, respectively: $K_1 \triangleq \sin \beta/\sin(\alpha + \beta)$ and $K_2 \triangleq \sin \alpha/\sin(\alpha + \beta)$."

Machine Learning Techniques

"The neural network model is defined via few hyperparameters, as the number of hidden layers and the number of neurons in each layer, etc. They must be selected before the training stage but there is no unique or predefined way of selecting them."

"The performance of the networks is measured by the achieved minimum mean square error (MSE) and maximum correlation value R. A network with one hidden layer has been considered since the number of effective parameters and overall performance of model in terms of its prediction precision did not change significantly by adding more hidden layers."

"Parameters "x" are a subset of "z", as wheel dullness indicator (that is considered by the ML models) does not play any role in the adopted physical model: For doing so, an NN model is considered as before, with one hidden layer and number of neurons selected, based on previously explained method."

Discussion

"This paper presents results from attempting to combine first principle and machine learning techniques into hybrid models to forecast performance of a centreless grinding process in terms of workpiece final roundness."

"For obtaining the benefits of both methods, they have been combined into a hybrid modelling approach."

"It was also noticed that, using the SVR and GPR approaches, the improvement from normal model to the hybrid model was more significant."

"Comparing these seven models, the best test results was obtained by hybrid GPR model that, in addition, takes considerably less time for training with respect to an NN hybrid with a similar precision."

Conclusion

"A hybrid approach has been suggested to improve roundness predicted by the numerical models based on first principles."

"Data-driven models, based on nominal process parameters augmented by additional outputs provided by a Physics-based model, deliver an optimal estimation that alleviates the effect of the unavoidable uncertainties and epistemic errors."

"Future activities will be aimed at improving the first principle model."

Study of the Surface Roughness of a Remanufactured Bimetallic AISI 1045 and 316 L SS Part Obtained by Hybrid Manufacturing (DED/HSM) [39]

This is a machine-generated summary of:

Barragan De Los Rios, German Alberto; Ferreira, Rodrigo; Mariani, Fabio Edson; da Silva, Eraldo Jannone; Coelho, Reginaldo Teixeira: Study of the surface roughness of a remanufactured bimetallic AISI 1045 and 316 L SS part obtained by hybrid manufacturing (DED/HSM) [39].

Published in: The International Journal of Advanced Manufacturing Technology (2022).

Link to original: https://doi.org/10.1007/s00170-022-09179-z.

If you want to cite the papers, please refer to the original.

For technical reasons we could not place the page where the original quote is coming from.

Abstract-Summary "Directed energy deposition (DED) is an additive manufacturing process that produces remanufacturers and repairs metal parts and its unique characteristics enable the combination of materials with different properties and surface functionalization."

"DED alone does not always produce surfaces and dimensional quality suitable for most intended applications, thus requiring post-processing steps."

"Research on post-processing materials obtained by metal additive manufacturing has focused on non-conventional processes, primarily over flat surfaces."

"An injection mold part, originally in AISI 1045, was remanufactured with AISI 316 L stainless steel with the use of laser-based directed energy deposition (DED-LB) and high-speed machining (HSM) towards improving surface finishing and dimensional accuracy."

"The parameters and strategy employed by DED-LB obtained a material with no significant porosity, cracks, or microstructure defects."

"The HSM process reduced the Sa value by approximately 90% with relatively short times in comparison to other surface finishing processes."

"The remanufacturing of components by DED-LB deposition of stainless steel onto carbon steels has opened a wide range of new applications, especially for complex 3D surfaces belonging to high-value components."

Introduction

"Remanufacturing components using the DED-LB process employing different materials may widen the range for its applications."

"Some studies covering the post-processing operations tending to surface improvements of additive manufactured 316 L material were found in the literature."

"Yang et al. [40] studied the combination of DED process and milling in the same machine (hybrid process) in 316 L stainless steel, evaluating both end and side milling concerning Sa."

"Yang et al. [41] proposed a new method, called hybrid directed energy deposition and thermal milling process, which combines DED deposition with machining each set of 316 L stainless steel layers deposited when they reach certain temperatures."

"The milling of 316 L stainless steels manufactured by the DED additive manufacturing process has been little explored, with studies focusing primarily on flat surfaces."

"It allows the use of materials with different mechanical and thermophysical properties that influence the characteristics of the final component, such as the roughness values and surface quality generated during the process."

Materials and Methods

"An initial pre-machining operation is performed on the original material (AISI 1045), followed by AM construction by DED-LB for replacing the worn part with AISI 316 L stainless steel."

"The study intends to test the restoration sequence of a remanufacturing process, analyzing the feasibility and performance of the component with particular attention to the surface roughness when a combination of DED-LB and HSM is employed."

"The manufacture or remanufacture of components by DED with such material combinations opens a wide range of applications, especially when components corrosion or mechanical behavior must be improved [42]."

"The restoration step of a general remanufacturing sequence encompasses three phases, namely pre-machining, deposition, and post-processing (finishing machining)."

"After pre-machining, the second phase begins with the deposition of the new material in the same machine by the DED-LB head positioned next to the spindle."

Results and Discussion

"Several partially melted grains and adhered ones were detected on the surface of the deposited AISI 316 L and later removed by the milling operation."

"No significant change was detected in the substrate related to the DED-LB process after deposition."

"The values of the surface roughness of the top region were approximately 1.0 μm, due to the contact of the bottom of the ball nose cutter at lower speeds, but with a significant plowing effect."

"In both cases, the reduction was above 90%, indicating milling is an essential step for reducing the roughness values of geometries obtained by DED with shorter processing time in comparison to electrochemical methods [43] or magnetic abrasive finishing [44]."

"Another aspect related to the DED/Milling process should be improved edge preparation for materials with refined grain and crystallographic orientation and textures during the remanufacturing obtained by the DED process."

Conclusions

"This study addressed a remanufacturing that used HM through the association of DED-LB and HSM in a sequence of processes in a same machine."

"The sequence has proven feasible and able to reproduce the original feature using AISI 316 L, creating a new piece of improved corrosion resistance in comparison to the original AISI 1045."

"The AISI 316 L microstructure obtained was primarily dendritic, and some increase in its hardness was observed, which might contribute to improving wear resistance, hence, its service life."

"The material (AISI 1045) was initially removed, and a new and more suitable one was deposited (AISI 316 L); the same complex shape was re-machined according to the HSM technique."

"The AISI 316 L added showed superior performance in corrosion resistance compared with the original material, and the AISI 316 L hardness was higher than the conventional one and closer to the values found for AISI 1045, which may also improve wear resistance."

Optimization of Metal Rolling Control Using Soft Computing Approaches: A Review [45]

This is a machine-generated summary of:

Hu, Ziyu; Wei, Zhihui; Sun, Hao; Yang, Jingming; Wei, Lixin: Optimization of Metal Rolling Control Using Soft Computing Approaches: A Review [45].

Published in: Archives of Computational Methods in Engineering (2019).

Link to original: https://doi.org/10.1007/s11831-019-09380-6.

If you want to cite the papers, please refer to the original.

For technical reasons we could not place the page where the original quote is coming from.

Abstract-Summary "Rolling exerts an essential impact on material properties and product quality as an important part of the production in steel industry."

"With an increasing rolling speed, more difficulties like process information monitoring, behavior characteristics modeling, and controlling of high speed operating are manifested in high-speed continuous rolling mills."

"The existing control system of rolling process is difficult to cope with the condition changes of high-speed rolling and the specification changes of complex products."

"In order to achieve precise control of large-scale and high-speed rolling, the analysis of rolling process rules based on industrial big data should be considered to establish the dynamic process model, and multi-objective real-time computational method of rolling schedules should also be introduced."

"Through a summary of steel industry and a review of the history of rolling optimization, the purpose is to explore the relationship between the optimization objectives of the rolling schedule and the process parameters of the rolling process, reveal the rules of how rolling conditions affecting rolling process in high speed rolling and provide theoretical basis and technical support to the production of steel industry."

Introduction

"Rolling is the main method of metal plastic forming with its products accounting for more than 90% of metal plastic processing."

"Steel, non-ferrous metals and composite materials can be processed by rolling."

"Rolling control system has evolved from traditional rolling process research, the optimization of rolling model to intelligent and high precision control."

"In order to further solve the problem of parameter prediction and intelligent control in rolling process, the computational methods designed for engineering are introduced into the control of metal rolling process."

"Steel industry belongs to large-scale complex process industry, and has not yet formed a process-integrated control and overall coordination optimization."

"This paper will review the rolling technology, calculation methods and procedures optimization objectives used in the steel industry, taking the important part of steel processing as an example, and propose a prospect for the future development of steel industry, including the whole process management, information physics system, multidisciplinary intersection, etc".

Computational Methods in Metal Rolling

"Soft computing techniques have been introduced to predict process data, optimize control parameter and build global model."

"The improvement of the rolling process systems by the computational methods is mainly reflected in the solution to complex problems."

"The application of software calculation in steel rolling is mainly reflected in three aspects. (1) On the basis of mathematical model, the finite element method is used to simulate the rolling deformation region; (2) The rolling deformation zone is regarded as a black box, and uses neural network method to predict the rolling force energy data; (3) The rolling procedure is regarded as a multi-objective optimization problem, and uses optimization algorithm to solve the load distribution coefficient."

"To deal with the increasing computation complexity, computing method based on Graphic Processing Unit(GPU) has been applied to solve finite element simulation, and rolling schedule optimization [46]."

Optimization of Rolling Schedule

"Reasonable load distribution method can not only improve the shape quality and thickness accuracy, but also have an important impact on reducing rolling energy consumption and reducing roll loss."

"The load distribution coefficient method is balancing the load of each rack, avoiding the high load operation of individual racks, and making close the ratio of the production load to the maximum load of each rack, including the reduction mode, the rolling force mode and power modes."

"There are many inherent defects in the traditional rolling schedule optimization method based on the distribution of consume curve. (2) The traditional load distribution coefficient method requires a large number of operations on the matrix, and the final result is very sensitive to the initial value, which is difficult to meet the requirements of the field application."

"By properly distributing the reduction amount of each rack, the rolling mill load and the rolling energy consumption can be reduced while meeting the requirements of equipment and process."

Future Research Direction in Rolling Schedule

"The rolling process parameter prediction model is constructed and its parameters are determined to improve the prediction accuracy of the rolling force model under different rolling conditions. (2) For the problem of parameter setting optimization under rolling schedule, historical information and environmental information are employed to construct a multi-objective evolutionary algorithm with balanced convergence performance and decomposition performance, and make autonomous also dynamic solution of load distribution parameters in line with decision makers' preferences. (3) Through multi-objective optimization algorithm and deep learning modeling, the horizontal coordination of raw material parameters, inspection information, and finished products data and the vertical perforation of production planning, rolling control, and information feedback are realized."

"It is necessary to establish a process quality big data platform based on the whole process of steel production, form process technique quality data integration technology from metallurgical composition, slab quality to rolling full table, and conduct online monitoring also retrospective analysis evaluation and number quality online rating on rolling process parameters and rolling stock quality combined with steel surface quality defects, trapping and internal grain structure performance on-line detection technology."

Conclusion

"The soft computing method has achieved remarkable results in rolling parameter prediction and rolling schedule setting."

"The soft computing method for metal rolling develops from parameter awareness and data mining in vertical direction to the whole process industry in horizontal direction."

"In the further, raw material parameters, inspection information, production data will be cooperate with each other horizontally and reduction planning, rolling control, information feedback will be cooperated vertically."

"Hopefully, this review will provide the practitioner an overview of soft computing methods for metal rolling and enable them to select the most appropriate approach for optimization."

Establishment of Material Removal Model for Lapping Ceramic Balls with Variable-Radius Groove Plate [47]

This is a machine-generated summary of:

Zhou, Fenfen; Yao, Weifeng; Yuan, Julong; Li, Fan; Lyu, Binghai; Zhao, Ping: Establishment of material removal model for lapping ceramic balls with variable-radius groove plate [47].

Published in: The International Journal of Advanced Manufacturing Technology (2020).

Link to original: https://doi.org/10.1007/s00170-020-06259-w.

If you want to cite the papers, please refer to the original.

For technical reasons we could not place the page where the original quote is coming from.

Abstract-Summary "The material removal model of a single ball was established based on Hertz contact theory."

"Combining the previous research works based on kinematic analysis and the material removal model, the forming process of single ball was simulated and evaluated by the deviation from spherical form (Δ_{Sph})."

"The validity of material removal model of lapping ceramic balls was verified experimentally."

Introduction

"The theoretical research on these lapping methods mainly focuses on two aspects: lapping trajectory analysis based on kinematics and material removal mechanism."

"Wang et al. [48] described the distribution of lapping trajectory in DRP method, then meshed the ball surface according to the direction of longitude and latitude, and then calculated the standard deviation value of the lapping trajectory point to evaluate the distribution uniformity."

"In order to calculate the standard deviation value more accurately, Zhao et al. [49] divided the ball surface by triangle to evaluate the lapping trajectory in DRP method."

"Zhou et al. [50] set the removal amount of contact points between the ball and lapping plates as a constant value at each moment, then obtained the material removal distribution on the ball surface based on kinematics."

"The material removal model of ceramic ball was established by combining the lapping trajectory analysis in kinematics with single material removal amount."

The Principle of Lapping Balls with VRGP

"The equipment is mainly composed of a flat plate (lower plate), variable-radius groove plate (upper plate), cycling feeding transporter, pressure device, and driving device."

"During lapping, balls gradually outward continuously spin and around the rotating spindle revolving along the variable-radius groove which are driven by the rotating lower plate."

"A single processing cycle is the time of a ball from the inlet through the whole variable-radius groove to the outlet."

"Balls are carried in sequence from the outlet along the plate edge to the inlet in the center of groove by cycling feeding transporter to realize cycle lapping."

Material Removal Model

"The ball surface is meshed, and the initial value of each mesh is assigned by Matlab to form the height distribution matrix D. Then the coordinates of the contact points at each moment are calculated."

"The contact deformation d_{A0} will form at the contact point A. Referring to spherical-spherical Hertz contact empirical formula [51], contact deformation d_{j0} can be expressed aswhere v_1 is the Poisson ratio of ceramic ball and v_2 is the Poisson ratio of lapping plate; E_1 is the elasticity modulus of ceramic ball and E_2 is the elasticity modulus of lapping plate; E_3 is the equivalent elastic modulus; R is the radius of curvature of grooved raceway; r is the equivalent radius of ball and ball Hertz contact."

"The experiment was repeated three times; then, the average removal mass was calculated, and the height removal amount of workpiece can be obtained, finally by the Preston's formula to calculate K_0, and then, K_0 can be expressed aswhere MRR represents the material removal height at unit time, p is the pressure of contact area, v is the relative speed between the center of workpiece and lapping plate."

Results and Discussion

"The experiment for lapping ceramic ball with VRGP was carried out to observe the change of deviation from spherical form Δ_{Sph}."

"The roundness of each ball in three perpendicular directions was measured and the maximum value of roundness was the actual value of Δ_{Sph}."

"After lapping of every 2 h, five balls were randomly selected to measure the Δ_{Sph}."

"Simulation results show that the Δ_{Sph} gradually decreases with the increase of lapping time."

"The value of Δ_{Sph} has been dropped from about 1 μm to about 0.3484 μm, and the Δ_{Sph} decreases quickly within 2–8 h. The experimental results show that the Δ_{Sph} also decreases with the increase of processing time."

"The variation trend of experimental results is consistent with the simulation results, and the maximum deviation between experimental value and simulation value is 17.7%."

Conclusions

"The method of lapping ceramic balls with variable-radius groove plate (VRGP) is developed in recent years."

"The single material removal model of a ball was established based on Hertz contact."

"The results of kinematic analysis and the single material removal were combined to obtain the dynamic material removal model used to describe the lapping process of ceramic balls."

"The effectiveness of the material removal model of lapping ceramic balls with VRGP has been verified."

Evaluation of Dimensional Accuracy and Surface Integrity of Cylindrical Array Microelectrodes and Cylindrical Array Microholes Machined by EDM [52]

This is a machine-generated summary of:

Sun, Yao; Gong, Yadong; Wen, Xuelong; Xin, Bo; Yin, Guoqiang; Meng, Fantao; Tang, Benjia: Evaluation of dimensional accuracy and surface integrity of cylindrical array microelectrodes and cylindrical array microholes machined by EDM [52].

Published in: Archives of Civil and Mechanical Engineering (2022).

Link to original: https://doi.org/10.1007/s43452-021-00366-5.

If you want to cite the papers, please refer to the original.

For technical reasons we could not place the page where the original quote is coming from.

Abstract-Summary "8 × 8 cylindrical array microelectrodes with diameter of about 185.18 μm and length of 2168.79 μm were fabricated by the loss self-finishing method for the first time."

"Besides, the size accuracy, surface morphology and surface roughness parameters of cylindrical array microholes machined using array microelectrodes prepared by loss self-finishing method were evaluated."

"Experimental results disclosed that the average consistent errors of inlet and outlet diameter for cylindrical array microholes, respectively, were 1.495 μm and 3.13 μm, indicating that the cylindrical array microelectrodes obtained by loss self-finishing method are capable to manufacture cylindrical array microholes with good surface quality and high dimensional accuracy."

Introduction

"Takahata et al. [53] adopted the LIGA technology to prepare 20 × 20 array microelectrodes with diameter of 20 μm and then fabricated array microholes on stainless steel plate by micro-EDM machine."

"Zhang et al. [54] used reverse EDM process to fabricate 3 × 3 tungsten electrode arrays and 5 × 5 tungsten electrode arrays, and array microholes were manufactured by array microelectrodes machining method on the tool steel."

"The literature has proven that the parameter optimization and special shape microelectrodes can improve surface integrity and size accuracy of microholes fabricated by EDM methods, but these studies mainly are focused only on adopting single microelectrode method to fabricate microhole instead of array microelectrodes method."

Materials and Methods

"The helix microelectrode is selected to fabricate microhole due to its superiority in improving flushing condition and machining stability compared to solid cylindrical microelectrode [55]."

"Cylindrical array microholes were machined by array microelectrodes."

"To reduce the prepared complexity and save prepared time of cylindrical array microelectrodes, the loss self-finishing method was proposed in this study, which can change the square array microelectrodes into cylindrical array microelectrodes without the fabrication of middle electrode template made of microholes and single helix microelectrode."

"The fabrication of helix microelectrodes was conducted on the low speed wire electrical discharge machine (CA20, Beijing Agie Charmilles Industrial Electronics Co., Ltd.)."

Results and Discussion

"Inspired by this phenomenon, the loss self-finishing method is proposed, which can take the cylindrical array microholes machined by circular wear portion of square microelectrodes as middle electrode to conduct the EDM reverse copy process; and after repeating self-finishing process, the cylindrical array microelectrodes can be directly obtained from square array microelectrodes."

"The loss self-finishing method proposed in this study was capable to fabricate cylindrical array microelectrodes due to it not only can maintain good dimensional precision and surface quality of self-finishing method, but also can significantly improve prepared efficiency due to it does not need to fabricate middle reversely copy electrodes and single microelectrodes."

"The 3 × 3 array microholes were fabricated with cylindrical array microelectrodes obtained by loss self-finishing method."

"It can be revealed that cylindrical array microelectrodes fabricated by loss self-finishing are capable to obtain cylindrical array microholes with high dimensional accuracy and good surface quality."

Conclusions

"The EDM loss self-finishing method is first proposed to prepare cylindrical array microelectrodes without intermediate electrode plate and the secondary clamp, and the 8 × 8 cylindrical array microelectrodes with the average diameter of 185.18 μm and aspect ratio of 11.71 were first obtained by EDM loss self-finishing method, which provides a new way to fabricate cylindrical array microelectrodes."

"The array microholes were fabricated with cylindrical array microelectrodes obtained by loss self-finishing method, and the average consistent errors of inlet diameter and surface roughness for 3 × 3 cylindrical array microholes, respectively, can be confined to 1.495 μm and 1.07 μm, which indicated that the cylindrical array microelectrodes fabricated by the loss self-finishing were capable to obtain cylindrical array microholes with high dimensional accuracy and good surface quality."

A Modelling Framework to Support Integrated Design of Production Systems at Early Design Stages [56]

This is a machine-generated summary of:

Abdoli, Shiva: A modelling framework to support integrated design of production systems at early design stages [56].

Published in: International Journal on Interactive Design and Manufacturing (IJIDeM) (2022).

Link to original: https://doi.org/10.1007/s12008-022-00987-x.

If you want to cite the papers, please refer to the original.

For technical reasons we could not place the page where the original quote is coming from.

Abstract-Summary "This does not address the impact of single design decisions on overall system performance, which may lead to inconsistencies between different disciplines or failures."

"This model is constructed in Finite-State-Machine formalism to achieve an executable architecture and integrated with optimization models to allow simulation of alternatives and to observe the impact of design decisions on system behaviour."

"Supportive algorithms are introduced for refinements of design alternatives according to the simulation results."

"The framework integrates simulation and fuzzy assessment results and performs a multi-criteria assessment to select an alternative for the detailed design."

"The framework can stand as a decision support framework at early design stages, giving insights to designers about the impact of single design decisions on system overall performance and satisfaction of various objectives."

Introduction

"2–2 Assessment against eighter qualitative or quantitative criteria and not both concurrently 3-Domain specific (not generic) approaches Four overall phases can be defined in the design of CEPSs, requirements definition, conceptual design, configuration design, and detail design."

"Sequential approaches (not integrated) do not completely reflect the impact of single decisions on overall performance due to not having a holistic approach in addressing all types of design requirements simultaneously."

"It is necessary to use methods, such as simulation, that can capture the dynamic aspect of a CEPS to assess the impact of design decisions on its performance."

"Integrated design in this paper means assessing the impact of design decisions on a CEPS's performance as a function of the dynamic interactions between its resources."

"Enabling multi-aspect design assessment by supporting the decision-making regarding the selection of better alternatives by considering different types of performance criteria."

Literature Review

"In CEPS design, modelling can serve three purposes; developing system architecture, design assessment, and validation [57]."

"Thiers [58] introduced a design methodology to support the analysis of logistics systems by using model-based systems engineering."

"Lack of an integrated generic modelling framework that: (1) addresses different aspects of modelling at early design stages, (2) provides a systematic interconnection between different models to interchange information."

"An integrated modelling approach can overcome the design methodologies' shortcomings that mainly guide the design process without generating solutions."

"Different modelling approaches answer narrowly defined design questions."

"There is a lack of a generic modelling approach providing prescriptive guidelines for system architecting, multi-aspect assessment, and validation, which can assist in the integrated design of CEPSs [59, 60]."

"Two designers may use one modelling approach and resultant models can be different with different capabilities."

Framework Structure and Applied Methods

"The framework develops an executable architecture for observing the behaviour (dynamic performance) of alternatives by simulation."

"The framework uses the OO method and architecting guidelines are developed for capturing the complexity of holistically modelling CEPSs."

"The framework uses optimization/quantification models to develop a numerical model for alternatives, which are later simulated for validation."

"The framework uses Linear-Programming (LP) and Automated Layout Design Program (ALDeP)."

"The framework includes a fuzzy assessment approach to assess the alternatives against qualitative objectives."

"This framework uses simulation results as one and fuzzy assessment as another source of information in the trade study step."

"This framework uses TOPSIS to rank the alternatives and identify the final alternative that demonstrates better performance by considering simulation and fuzzy assessment results."

"Examples of a warehouse design are given in the following sections to explain the methods' applications and the connection between different artefacts in the framework."

Developing an Executable System Architecture

"A CEPS is firstly modelled as a parent state in FSM and then is decomposed into sub-states to embody the CEPS's processes, called process-state."

"Each process-state is decomposed into further sub-states that represent its needed enablers, called enabler-state."

"Further, each enabler-state is decomposed into sub-states, which address its specific DRs, called DR-state."

"The possible design options/genres for each DR-state are modelled as its sub-states, called object-state."

"'SKU-Allocation' is defined as a DR-state, which includes two sub-states showing two possible operational policy options: 'Class-Based Storage' (CBS) and 'Random'."

"FSM allows defining functions for a state to model its behaviour."

"A function is associated with each object-state to model its dynamic behaviour, called object-functions."

"In SSMM, a design alternative is configured by activating one object-state for a DR-state."

Trade Study Process

"This framework utilizes a novel fuzzy framework for assessment of alternatives against qualitative objectives by modelling the domain expert's opinion [61]."

"The fuzzy framework has a hybrid-hierarchal structure and comprises fuzzy models in three levels: design-object, process, and system."

"The 'Object-fuzzy-model' assesses the goodness of a design-object to be employed in an alternative."

"Process-fuzzy-model' assesses the goodness of a process as the configuration of the employed design-object that interact with each other."

"The first input set to 'Process-fuzzy-model' is the goodness of design-objects (output of 'Object-fuzzy-model')."

"The expert's opinion about the interaction efficiency between every two design-objects is modelled as the second fuzzy input for the 'Process-fuzzy-model'."

"The introduced framework provides two sources of information to analyse/assess the performance of alternatives: the simulation results for the quantitative objectives and the fuzzy assessment for the qualitative ones."

Case Studies

"The warehouse managers defined DRs for process enablers and their possible design-objects."

"The CBS policy allocates the high-demand SKUs in modules that require less travel-time (operation time) in storing/picking processes, while this increases the travel-time for low-demand SKUs."

"These alternatives differ in their employed design-objects in storing/picking processes."

"If a sequential design approach was followed without having an integrated design approach in addressing all design requirements at the same time including the operational policies, the results could have been different in cost minimization."

"If two subsequent processes do not share resources, if the first process operates properly such that the second can start its operation on time, the employed design-objects in the first process might not impact the operation of the latter process."

"This paper defines operational policies as another enabler to show the importance of having an integrated design approach."

Conclusion and Future Work

"The simulation takes a long time when the number of variables is large, while the number of design variables is large at early design stages."

"SSMM is a key artefact in the proposed framework and embodies three aspects of a CEPS: system architecture, decisions regarding the allocation of design options to design requirements for configuring a design alternative, and dynamic behaviour of an alternative."

"Such a rich embodiment is one of the key contributions of this research and is the fruit of a novel application of proper modelling approach (OO), innovative application of a modelling formalism (FSM), and seamless incorporation of the framework artefacts such as SSMM, fuzzy assessment framework, TOPSIS, quantification models, and supporting algorithms."

"The introduced approach can stand as a decision support framework for the integrated design of complex engineering processed based systems at their early design stages."

A Monarch Butterfly Optimization for an Unequal Area Facility Layout Problem [62]

This is a machine-generated summary of:

Kim, Minhee; Chae, Junjae: A monarch butterfly optimization for an unequal area facility layout problem [62].

Published in: Soft Computing (2021).

Link to original: https://doi.org/10.1007/s00500-021-06076-7.

If you want to cite the papers, please refer to the original.

For technical reasons we could not place the page where the original quote is coming from.

Abstract-Summary “Unequal area facility layout problems deal with the placement of departments in a particular area.”

“Monarch butterfly optimization, a recently developed algorithm, is presented to solve an unequal area facility layout problem.”

“A slicing tree representation is used to form a layout structure as well as greedy acceptance, which accelerates the monarch butterfly optimization’s search performance.”

“A set of well-known instances from existing studies is tested to evaluate the algorithm’s effectiveness.”

“The proposed algorithm generates solutions that match the best results from previous research, and it provides such solutions within a comparable amount of time.”

Introduction

“Normal facility layout problems (FLPs) deal with the arrangement of rectangular blocks (which are called departments in this paper) within a planar area without overlapping [63].”

“UA-FLPs with a large number of departments are NP-hard problems, and an optimal solution might not be found in a reasonable amount of computation time [64].”

“A new approach with a recently announced metaheuristic algorithm called monarch butterfly optimization (MBO) is adopted to solve a UA-FLP to gain better solutions in a reasonable time.”

“To solve a UA-FLP with particular metaheuristic algorithms, a layout representation method is needed to compose the structure and location of each department.”

“These layout representations can be used to solve a UA-FLP when the departmental area is given rather than specified by fixed dimensions.”

“To extend the range of layout representations, we adopt the STS with MBO to solve the UA-FLP.”

Background and Literature Review

"The UA-FLP can be described as a resolution of possible layouts in a given area using relative locations of the departments for the distance measurement."

"Kulturel-Konak and Konak [65] proposed a new relaxed FBS representation by allowing dead spaces in a bay for greater flexibility in large department problems."

"Friedrich et al. [66] introduced an integrated STS-based parallel tempering heuristic to solve the UA-FLP, and their heuristic was able to determine the P/D points of every department."

"The empty spaces can be simply located at the end of the bays in the FBS, but the empty spaces in the STS are regarded as additional areas for each department, and this might degrade the solution quality of the STS."

"Employing these advantages, this study proposes a UA-FLP approach that applies MBO with an STS to demonstrate its effectiveness."

General UA-FLP

"We describe the basic formulations to locate unequal blocks in a floor space."

"Meller et al. [63] then modified the basic MIP model."

"The model has subsequently been modified by many researchers [67, 68]."

"N blocks must be located in a fixed floor space size with the limitation being the aspect ratio of the width and height."

"The optimal solutions of large-sized problems are not found by MIP models, but they furnish the UA-FLP with an excellent support base."

MBO Method in a UA-FLP

"The department sequence (1–5–3–2–4) is horizontally separated into two branches (1–5 and 3–2–4) because the cut orientation is 2 with cut code 0."

"The population generation concept in MBO centers on the migration behavior of the butterfly between the two lands, and the existing butterfly's solution can be preserved by restricting the component changing range."

"The department sequence of a selected butterfly is used in the swapping process."

"As shown in other studies that have modified the MBO via different operators and algorithm processes, the mutation process, which changes the solution representation, can be applied differently depending on the characteristics of the problem being studied [69]."

"For the updating process, the original MBO updates all newly generated butterfly individuals."

"The component changing process in the migration operator and the swapping process in the butterfly adjusting operator are conducted to find a neighborhood solution while maintaining the characteristics of the initial random generation of the solution."

Computational Results

"Instances with a compactness of 1.0 (Category 1), compactness of less than 1.0 (Category 2) and open space (Category 3) are tested to evaluate the performance of the proposed algorithm."

"The experiment is validated through the comparison of known optimal solutions using the instances in Category 1 because the optimal solutions for most of the example problems in Category 1 are known."

"Most researchers using MBO in their studies have followed the original parameter settings, and this has been verified to be meaningful via tests with many optimization problems [70–72]."

"According to previous research, the instances of O7, O8, O9, FO7, FO8, FO9, vC10s and vC10a have been solved using several different MIP models, and their optimal solutions can also be found via our MBO algorithm."

"The proposed MBO does identify the best solution compared to other methods for the 62-department problem."

Conclusion

"The proposed algorithm tended to generate favorable solutions with a compactness value of 1.0."

"To generate better solutions with MBO, some researchers have taken specific operators from other metaheuristic algorithms."

"This kind of effort could enable the algorithm to generate better solutions in reasonable amounts of time."

"Additional efforts to reduce solution times would be advantageous for MBO, which is an evolutionary algorithm and has an iterative process."

"MBO generates powerful solutions within reasonable amounts of computation time, but the solutions of each iteration tend to fluctuate."

"Hybridizing the heuristic by applying local search methods to individual representations to generate better solutions in reasonable amounts of time could serve as a fruitful future research direction."

"Another worthwhile research challenge would be to adopt new types of metaheuristic algorithms to achieve better performance in terms of solution quality."

A Dynamic Material Handling Scheduling Method Based on Elite Opposition Learning Self-Adaptive Differential Evolution-Based Extreme Learning Machine (EOADE-ELM) and Knowledge Base (KB) for Line-Integrated Supermarkets [73]

This is a machine-generated summary of:

Zhou, Binghai; Zha, Wenfei; Ye, Lyujiangnan; He, Zhaoxu: A dynamic material handling scheduling method based on elite opposition learning self-adaptive differential evolution-based extreme learning machine (EOADE-ELM) and knowledge base (KB) for line-integrated supermarkets [73].

Published in: Soft Computing (2021).

Link to original: https://doi.org/10.1007/s00500-021-06385-x.

If you want to cite the papers, please refer to the original.

For technical reasons we could not place the page where the original quote is coming from.

Abstract-Summary "Since the highly diversified consumer demands and changing market environment pose a great challenge for the automobile industry, this paper presents an extreme learning machine and knowledge base-based dynamic scheduling method to solve the dynamic scheduling problem of material handling for mixed-model assembly lines based on line-integrated supermarkets."

"An extreme learning machine and knowledge base-based dynamic scheduling method is constructed, consisting of a knowledge base-based scheduling rule selection and an extreme learning machine (ELM)-based product mix matching method."

"Considering the defects of extreme learning machine, an elite opposition learning self-adaptive differential evolution-based extreme learning machine (EOADE-ELM) is proposed to optimize the parameters of ELM."

Introduction

"Compared to the traditional line stocking and kitting strategies, the material delivery strategy based on the line-integrated supermarket has the advantages of the stable supply, saving labors and less inventory at the line."

"We propose a dynamic material handling scheduling problem (DMSP) for MMALs based on line-integrated supermarket considering the change of enterprise demands and market environment."

"An extreme learning machine and knowledge base-based dynamic scheduling method (EKDSM) is proposed to maximize the weight sum of the throughput of the assembly line and the number of logistic workers under the condition of the variable product mix and weights of scheduling criteria, considering the random failure of the equipment and the instability of cycle time."

"This is the first study of the DMSP for MMALs based on the concept of line-integrated supermarket in a variable manufacturing environment in this paper."

Literature Review

"In order to solve the problem of the intensive computational effort and the limited performance, the machine learning approach has become a popular method of solving the scheduling problem of material handling in dynamic manufacturing environments."

"Machine learning approach is utilized to predict the performance of the current system state and select the optimal scheduling rule."

"Liu et al. [74] utilized neural network trained by simulation results to select the most efficient scheduling rules for job shop scheduling problem and used it at real-time scheduling process."

"Chen et al. [75] studied the dynamic scheduling problem of dispatching tow trains for MMALs with the constraint of last-in-first-out (LIFO) and proposed a multiple-criteria real-time scheduling approach (MCRS) employing general regression neural network (GRNN) to select the optimal scheduling rule."

"An elite opposition learning self-adaptive differential evolution-based extreme learning machine (EOADE-ELM) is developed to predict the optimal scheduling rule for current manufacturing system."

Problem Description and Mathematical Model

"The scheduling system determines when the logistic workers are dispatched to perform the material handling."

"Let x represent the minimum batch size performed by logistic workers at a time, so the minimum batch size of x is denoted as MBS-x. If the number of refill events for assembly line is less than x, logistic workers will wait until the number of refill events reaches x. Then, the scheduling system decides to dispatch a logistic worker to replenish materials."

"Some studies present that the dispatching decision problem is adopted to determine how to choose logistic workers to perform the material handling."

"After determining the dispatch of logistic worker to perform material handling, the selection problem of refill event is first solved when the number of unexecuted refill events is greater than the maximum number of refill events that logistic workers can perform at a time."

EKDSM

"The DMSP based on the line-integrated supermarket consists of four sub-decision problems, and each decision problem can adopt corresponding scheduling rules."

"The optimal rule combination can be selected from the KB at any given product mix and weights of scheduling criteria."

"If the given product mix and rule combination are not stored in the KB, the product mix in the KB which is closest to the given product mix can be found by the product mix matching method, and the knowledge in the KB can be used to determine the optimal rule combination for any given product mix and weights of scheduling criteria."

"An extreme learning machine and knowledge base-based dynamic scheduling method (EKDSM) is proposed to solve the DMSP based on the line-integrated supermarket."

EOADE-ELM

"ELM developed from SLFN consists of the input layer, the hidden layer and the output layer."

"Due to the random generation of the input weights and the hidden layer biases, the learning speed and the generalization ability of ELM are excellent."

"Since SLFNs needs to train a large number of parameters in the iteration process, which results in slow convergence speed and easy to fall into local optimum."

"Opposition-based learning (OBL) was proposed by Rahnamayan et al. [76], and Wang et al. [77] introduced an enhanced DE algorithm with OBL to improve the convergence speed and the quality of solutions."

"In the iteration process, the mutation operator decreases, which can accelerate the convergence speed and enhance the local optimization ability."

"The crossover operator increases gradually as the iteration, and this can avoid the algorithm falling into the local optimal and increase the diversity of population."

Computational Experiments

"It shows the average number of logistic workers required by the assembly line under the same product mix group or scheduling rule combination."

"It can be concluded that since no scheduling rule combination can be adopted to obtain the maximum value of objective function under all the product mix groups, it is necessary to carry out dynamic scheduling for the assembly line."

"The selection and sequencing rules of refill events SS and SD can obtain more throughput by considering the shortest slack time, so the value of objective function is larger, while scheduling rule combinations DD and DS consider the distribution distance of logistic workers."

"It can be seen that when the weights of scheduling criteria take different values, the importance of the throughput of assembly line and the number of logistic workers is different, and the optimal scheduling rule combination is also different."

Conclusions

"We have studied a dynamic scheduling problem of material handling for mixed-model assembly lines based on line-integrated supermarkets that considers both the effect of changing market environment (the variable product mix and weights of scheduling criteria) and changing system state (the random failure of the equipment and the instability of the cycle time) in order to better meet the real-world manufacturing process."

"The objective is trade-off between the throughput of assembly line and the number of logistic workers to maximize the objective."

"In order to solve the proposed problem, an extreme learning machine and knowledge base-based dynamic scheduling method (EKDSM) is proposed."

"The EKDSM method includes the offline phase and the online phase to dispatch logistic workers for material handling with the optimal scheduling rule combination."

"The objective function in this paper only considers the throughput of assembly line and the number of logistic workers."

Artificial Intelligence in Product Lifecycle Management [78]

This is a machine-generated summary of:

Wang, Lei; Liu, Zhengchao; Liu, Ang; Tao, Fei: Artificial intelligence in product lifecycle management [78].

Published in: The International Journal of Advanced Manufacturing Technology (2021).

Link to original: https://doi.org/10.1007/s00170-021-06882-1.

If you want to cite the papers, please refer to the original.

For technical reasons we could not place the page where the original quote is coming from.

Abstract-Summary "Artificial intelligence (AI) technology receives extensive attention in the manufacturing field."

"In the context of smart manufacturing, this paper reviews various theories, algorithms, and technologies of AI to different stages of PLM (i.e., product design, manufacturing, and service)."

"This paper also discusses the opportunities and challenges of applying AI for PLM."

Introduction

"The rapid advancement of emerging information technologies, such as cloud computing, the Internet of Things (IoT), big data, and artificial intelligence (AI), is transforming the paradigm of advanced manufacturing through impacts on all aspects of PLM [79]."

"Compared with other fields, AI applications for PLM in the context of advanced manufacturing are relatively limited [80]."

"Due to the high requirements for quality, reliability, accuracy, and cost-effectiveness in manufacturing, manufacturers are naturally hesitant about adopting AI in PLM, which is especially true for small and medium-sized manufacturers (SMM) [81]."

"Many AI algorithms are compatible with industrial big data, which drives the manufacturing decision-making towards a more data-driven fashion."

"The technical dimensions of AI are mapped to the stages of PLM to make visible the opportunities of leveraging AI to enhance the efficiency, effectiveness, and intelligence of PLM."

Developing Trend of Product Lifecycle Management (PLM)

"The pursuit of higher product quality, innovativeness, and cost-effectiveness is imposing new requirements on PLM. (1) To quickly respond to market changes, manufacturers must adjust their strategies to shorten a product life cycle, especially the cycling pace of the product design and manufacturing stages. (2) The increasing market demands for customized and personalized products require manufacturers to develop a more modularized, flexible, and responsive design, manufacturing, and service processes. (3) To design, manufacture, and service customized products globally, manufacturers, must collaborate with relevant stakeholders, share available resources, and develop viable manufacturing networks towards a more sustainable manufacturing ecosystem. (4) The intensified competition for lower cost, higher quality, and shorter lead time requires manufacturers to enhance the intelligence of their manufacturing processes and systems, which put forward higher requirements for process automation and equipment intelligence. (5) High customer expectation on service quality requires manufacturers to improve information monitoring, tracking, and management system and establish a more responsive online service platform to provide remote product repair, maintenance, and upgrade. (6) The emergence of a massive volume of industrial big data requires manufacturers to process, mine, and integrate a variety of high-dimensional, unlabeled, unstructured, and non-standardized data."

AI and Its Applications

"In light of the existing definitions, the authors regard AI as the theories, methodologies, technologies, and tools that are intended to understand human intelligence, develop artificial systems with intelligence, empower artefacts to perform intellectual tasks, and leverage computational means (both software and hardware) to simulate intelligent behaviors."

"Starting from the late 1970s, the emergence of expert systems, which simulated how human experts solve domain-specific problems using knowledge, represented AI's reorientation from theoretical investigations to practical applications."

"AGI represents a more advanced state of AI, where AI is regarded, more or less, equivalent to human intelligence in terms of its ability to perform generic intellectual tasks."

"Various statistical methods in probability models can support AI algorithms to identify overfitting probability models (i.e., expectation-maximization, Kalman filter, Bayesian model, Markov model, etc) ML studies how computers learn progressively by discovering implicit patterns and performing an intellectual task without explicit instructions in advance [82]."

AI in PLM

"The popularity of AI technology is spawned by supporting the product design from two mainly aspects: analyze market trends by using AI algorithms PLM data mining; provide decision-making aid for designers to achieve a fast and personalized product design."

"Virtual technologies (e.g., virtual reality, augmented reality, and virtual prototype) have the function of modifying the product design continuously in a virtual environment and obtain the simulation data of a large sample size [83]."

"Based on the sales big data, the combined application of deep learning and recommendation algorithms (e.g., DNN + collaborative filtering) can enhance the personalized decision level of recommendation by analyzing the demand associated information of user set and product set [84]."

"Intelligent recycling technologies encapsulate recycled product resources as a service, including the optimization of recycling resources by heuristic algorithms and the evaluation of re-manufacturability/reusability towards the highest recycling rate [85]."

Advantages of Applying AI in PLM

"Smart customization is most applicable to the design and manufacturing of consumer products, where there are high demands for differentiated, customized, and personalized products and services."

"Many mature AI applications can be implemented to improve the automation and intelligentization of the product manufacturing processes, such as computer vision, smart robots, and autonomous vehicles."

"Smart manufacturing systems can logically plan, coordinate, and adjust a complex production process with minimum human intervention, improving the collaborative configuration of manufacturing resources on a shop-floor."

"When a production process is interrupted by uncertainties (e.g., equipment failure, order change, and temporary insertion), manufacturing systems can respond to the change in real-time and fulfill differentiated production tasks."

"The product manufacturing system is substantially complex since the reliability is affected by numerous factors such as materials, process, environment, personnel, equipment, etc. AI can improve the system reliability from many aspects."

Challenges

"As the unstructured data size increases in PLM, it is increasingly difficult to verify data consistency, and the complexity of AI algorithms increases proportionately."

"Complex PLM requires more advanced AI algorithms to improve the reliability, accuracy, and timeliness in industrial scenarios."

"The design parameter recommendation based on expert systems requires the support of a large amount of data in PLM."

"There is a need to improve AI algorithms' transparency with protecting data security and business intelligence."

"Unlike a laboratory environment where most AI algorithms are developed and validated, PLM's actual environment is full of uncertainties, affecting data quality and algorithm reliability."

"Many AI algorithms are designed to learn new rules based on big data, while they cannot cope with accidental low probability emergencies."

"A significant challenge for AI&CSP is the mapping between application requirements and AI algorithms."

Conclusions and Future Works

"Recommendations for the future are summarized as follows: (1) Specific application of AI technologies in different industrial scenes."

"This paper discusses the existing and potential applications of AI technologies in each stage of PLM."

"The successful implementation of AI technologies depends on data, communication technology, and hardware resources."

"Industries, systems, devices, and algorithms have different standards concerning data formats and communication protocols, making it difficult for AI technologies to integrate with other technologies."

"Building a unified standard system or studying communication among different standards will help integrate and develop AI systems."

Flexible Automated Warehouse: A Literature Review and An Innovative Framework [86]

This is a machine-generated summary of:

Custodio, Larissa; Machado, Ricardo: Flexible automated warehouse: a literature review and an innovative framework [86].

Published in: The International Journal of Advanced Manufacturing Technology (2019).

Link to original: https://doi.org/10.1007/s00170-019-04588-z.

If you want to cite the papers, please refer to the original.

For technical reasons we could not place the page where the original quote is coming from.

Abstract-Summary "This paper aims to review published papers in the last ten years related to flexible automation in warehouses and to construct a framework that could guide future researchers in the construction of an innovative conceptual model that may be applied at warehouses in the future."

"A total of 113 papers published between January 2008 and December 2018 have been selected, reviewed, and categorized to construct a useful foundation of past research."

"Results showed the key point to achieve a flexible automated warehouse is the combination of automated equipment, data collection technologies, and management solutions."

"Based on the reviewed papers, an innovative framework of a flexible automated warehouse is proposed."

Introduction

"Just-in-time philosophy, mass customization, e-commerce, and omni-channel distributions influence the supply chain causing the need to automate logistics processes and to achieve an improvement in performance in a shorter period of time [87, 88]."

"Automated warehouses are evolving into more intelligent storage systems without the need of installing fixed structures that complicate and reduce the scalability and flexibility to future changes."

"Although the importance of automation systems is notable, challenges as flexibility against market fluctuation and the management risk owning to high investment costs need to be discussed and solved."

"Based on the need to automate warehouse process and the bottlenecks about the lack of flexibility, it is necessary to research the concept of flexible automation applied to warehouses to help upcoming research focus on future directions."

"The purpose of this research is to perform a comprehensive literature review of flexible automation in warehouses in order to construct a framework that could help to develop and design a flexible automated warehouse."

Evolution of Automation

"Fixed automation, also known as hard automation, has little or no flexibility to accommodate product variety."

"In programmable automation or soft automation, the equipment is designed with the capability to change the program of instructions to allow production of different parts or products."

"This adaptable feature results in lower production rates and more flexibility, when compared to fixed automation."

"Although programmable automation offers more flexibility allowing to produce a variety of goods, the required time spent to change the equipment configuration, known as setup time, affects production."

"Flexible automation allows any required change in the program of instructions enabling continuous production of different parts or product styles [89]."

"The fourth industrial revolution, also known as Industry 4.0, introduced some emerging technologies that have been integrated into automation systems in order to control the whole system in real time and to allow rapid adaptability according to the dynamic market changes."

Classification of Order Picking Systems in Warehouses

"According to [90], (1) warehouse is a storage place for products and its main activities include receipt of product, storage, shipment, and order picking; (2) distribution warehouse (DW) is a warehouse that stores finished goods and from which customer orders are assembled; and (3) distribution center (DC) is a warehouse facility which holds inventory from manufacturing pending distribution to the appropriate stores."

"As this article focuses on order picking and material handling activities from inventory, the term warehouse will be used throughout the paper."

"An order picking system (OPS) is the process which retrieves products from storage according to a specific customer request."

"Dallari et al. [91] developed a methodology to support warehouse designers in choosing the most suitable OPS and proposed a classification in five main classes, which are picker-to-parts, pick-to-box, pick-and-sort, parts-to-picker, and automated picking."

"In a picker-to-parts system, pickers walk along the aisles to pick items according to the customer order."

Research Methodology

"Most of the irrelevant papers excluded were related to data warehouse, energy generation for automated systems or did not add to the selected theme."

"The main contribution of this article is to build a framework that should help upcoming research and specialists in warehouses to construct and design a flexible automated warehouse that overcomes issues mentioned before, as the increase of e-commerce, just-in-time philosophy and products customization."

"Based on the warehouse's activities listed, three constructs were selected and analyzed in the papers."

"The reviewed articles were categorized according to these constructs in order to build a framework that should give future directions in flexible automation in warehouses."

"The problem solved, and technology analyzed/proposed/implemented will be analyzed in reviewed papers in order to identify possible technologies that can help in the construction of a flexible warehouse framework."

Detailed Analyses of the Literature

"Problems discussed in data collections technologies' category papers include the necessity to (1) reduce the time from picking activities, (2) control inventory on real time, and (3) estimate material positions integrating it with a management system in real time to allow faster decision-making."

"In order to solve issues from the current logistics industry, a layout of a flexible automated warehouse joining RFID and QR code information to provide robot-positioning data was proposed by Xiao-Long et al. [92]."

"Petković et al. [88] presented an algorithm for human intention recognition in flexible robotized warehouses in order to operate efficiently and safely and demonstrated that the proposed framework predicts human warehouse worker's desires in an intuitive manner."

Flexible Automated Warehouse Framework

"The reviewed literature points out that the critical point to achieve a flexible automated warehouse is the combination of automated equipment, data collection technologies, and management solutions."

"The review shows that the use of unique technologies cannot characterize a flexible automated warehouse."

"The combination of technologies mentioned is necessary to represent a flexible automated warehouse."

"The purpose of the framework is to describe technologies that can be applied in a warehouse to obtain a flexible automated warehouse."

"The framework supports managers to assertively decide about technologies that may be applied in the warehouse to enhance flexibility in warehouses, based on the real or expected automation level in the warehouse."

"The framework can be used to help upcoming research in the construction of a conceptual model for a flexible automated warehouse capable of adapting according to dynamic market changes."

Conclusion

"This paper presents a comprehensive literature review of recent papers about flexible automation applied in warehouses."

"One hundred thirteen published papers between January 2008 and September 2018 were selected, reviewed, categorized, and analyzed to find the future directions and opportunities of research in flexible automated warehouses."

"Based on the reviewed literature, the main contribution of this study is an innovative framework of a flexible automated warehouse to guide future researches on the construction of a conceptual model."

"The proposed framework exposes the idea that the critical point to achieve flexibility in an automated warehouse is the combination of automated equipment, data collection technologies, and management solutions."

References

1. Zhou W, Chen G, Pan H, Luo F, Wei Y, Li M (2022) Laser deflection tangential dressing complex-concave surface bronze-bonded diamond wheels. Int J Adv Manuf Technol. https://doi.org/10.1007/s00170-022-09647-6

2. Chen GY, Deng H, Zhou X, Zhou XC, Zhou C, He J, Cai S (2015) Online tangential laser profiling of coarse-grained bronze-bonded diamond wheels. Int J Adv Manuf Technol 79:1477–1482. https://doi.org/10.1007/s00170-015-6963-z
3. Deng H, Zhou X, Zhu PC, Ying HQ (2019) Optimization of efficiency and uniformity of bond removal during laser sharpening. Int J Adv Manuf Technol 103:5–8. https://doi.org/10.1007/s00170-019-03759-2
4. Dold C, Transchel R, Rabiey M, Langenstein P, Jaeger C, Pude F, Kuster F, Wegener K (2011) A study on laser touch dressing of electroplated diamond wheels using pulsed picosecond laser sources. CIRP Ann Manuf Technol 60:363–366. https://doi.org/10.1016/j.cirp.2011.03.117
5. Deng H, Xu Z (2020) Laser dressing of arc-shaped resin-bonded diamond grinding wheels. J Mater Process Technol. https://doi.org/10.1016/j.jmatprotec.2020.116884
6. Deng H, Xu Z (2021) Laser-dressing topography and quality of resin-bonded diamond grinding wheels. Opt Lasers Eng. https://doi.org/10.1016/j.optlaseng.2020.106322
7. Wang Z, Jiang X, Yang G, Liu W, Song B (2022) Remanufacturing oriented multilayer cladding morphology prediction using a new second order fitting method. Environ Sci Pollut Res. https://doi.org/10.1007/s11356-021-16303-1
8. Shu F, Tian Y, Zhou J, He W, Zhao H, Xu B (2019) Microstructure and mechanical properties of two-layer laser-cladded Co-based coatings. J Therm Spray Technol 28:1330–1338
9. Zheng H, Cong M, Liu D, Dong H, Liu Y (2018) Optimization of STL model and layer shape for laser cladding forming. Int J Adv Manuf Technol 100:599–608
10. Darmiyani MR, Amirabadi H, Khosravi M (2021) Theoretical and experimental study of ceramic balls lapping with eccentric plates and eccentric V-groove. Int J Adv Manuf Technol. https://doi.org/10.1007/s00170-021-07585-3
11. Kang J, Hadfield M (1970) A study on the lapping of ceramic balls. WIT Trans Eng Sci 25:1
12. Feng M, Wu Y, Yuan J, Ping Z (2017) Processing of high-precision ceramic balls with a spiral V-groove plate. Front Mech Eng 12(1):132–142
13. Pasichnyi O (2019) Technology precision machining of ceramic balls in the V-grooves of variable curvature. In: Simulation and experiments of material-oriented ultra-precision machining. Springer, Singapore, pp 237–264
14. Sainathuni B, Guthrie B, Parikh PJ, Kong N (2018) Distribution planning for products with varying life cycles. Flex Serv Manuf J. https://doi.org/10.1007/s10696-018-9314-1
15. Fisher M (1997) What is the right supply chain for your product? Harv Bus Rev 75(2):105–117
16. Khan O, Christopher M, Burnes B (2008) The impact of product design on supply chain risk: a case study. Int J Phys Distrib Logist Manag 38(5):412–432
17. Khan O, Creazza A (2009) Managing the product design-supply chain interface: towards a roadmap to the design centric business. Int J Phys Distrib Logist Manag 39(4):301–319
18. Weng ZK (1999) The power of coordinated decisions for short-life-cycle products in a manufacturing and distribution supply chain. IIE Trans 31(11):1037–1049
19. Senthil SM, Nathan S, Parameshwaran R (2021) Experimental investigation on joining process of aluminium alloy 6063-T6 pipes using hybrid friction stir welding. J Braz Soc Mech Sci Eng. https://doi.org/10.1007/s40430-021-03054-w
20. Lammlein D, Gibson B, DeLapp D, Cox C, Strauss A, Cook G (2012) The friction stir welding of small-diameter pipe: an experimental and numerical proof of concept for automation and manufacturing. Proc Inst Mech Eng B J Eng Manuf 226(3):383–398
21. Senthil SM, Parameshwaran R, Ragu Nathan S, Bhuvanesh Kumar M, Deepandurai K (2020) A multi-objective optimization of the friction stir welding process using RSM-based-desirability function approach for joining aluminum alloy 6063–T6 pipes. Struct Multidiscip Optim. https://doi.org/10.1007/s00158-020-02542-2
22. Sashank JS, Sampath P, Krishna PS, Sagar R, Venukumar S, Muthukumaran S (2018) Effects of friction stir welding on microstructure and mechanical properties of 6063 aluminium alloy. Mater Today Proc 5(2):8348–8353

23. Anusha E, Kumar A, Shariff SM (2021) Finite element analysis and experimental validation of high-speed laser surface hardening process. Int J Adv Manuf Technol. https://doi.org/10.1007/s00170-021-07303-z
24. Barath VR, Manish T, Padmanaban R, Padmanabham G (2019) Adaptive process control for uniform laser hardening of complex geometries using iterative numerical simulation. Mater Performance Character 8(6):1178–1191
25. Giorleo L, Ceretti E, Giardini C (2013) Fe modeling of the apparent spot technique in circular laser hardening. Int J Adv Manuf Technol 69:1961–1969
26. Fakir R, Barka N, Brousseau J (2018) Case study of laser hardening process applied to 4340 steel cylindrical specimens using simulation and experimental validation. Case Stud Thermal Eng 11:15–25
27. Sehyeok O, Ki H (2017) Prediction of hardness and deformation using a 3-D thermal analysis in laser hardening of AISI H13 tool steel. Appl Therm Eng 121:951–962
28. Pacheco JT, da Silva LJ, Barbetta LD, Ferreira HS, Veiga MT, Forni R, Teixeira MF (2020) Laser cladding of stellite-6 on AISI 316 L austenitic stainless steel: empirical-statistical modeling and parameter optimization. Lasers Manuf Mater Process. https://doi.org/10.1007/s40516-020-00132-0
29. de Oliveira U, Ocelík V, De Hosson JTM (2005) Analysis of coaxial laser cladding processing conditions. Surf Coat Technol 197(2–3):127–136. https://doi.org/10.1016/j.surfcoat.2004.06.029
30. Javid Y (2020) Multi-response optimization in laser cladding process of WC powder on Inconel 718. CIRP J Manuf Sci Technol. https://doi.org/10.1016/j.cirpj.2020.07.003
31. Zhao Y, Guan C, Chen L, Sun J, Yu T (2020) Effect of process parameters on the cladding track geometry fabricated by laser cladding. Optik 223:165447. https://doi.org/10.1016/j.ijleo.2020.165447
32. Lian G, Yao M, Liu Z, Yang S, Chen C, Wang H, Xiang Y, Cong W (2019) Near-net shaping control of triangular stacking in laser cladding process. Proc Manuf 34:233–238. https://doi.org/10.1016/j.promfg.2019.06.144
33. Xavier MGC, Souza TMG, Cheung N, Garcia A, Spinelli JE (2020) Effects of cobalt and solidification cooling rate on intermetallic phases and tensile properties of a -cu, -Zn, -Fe containing Al-Si alloy. Int J Adv Manuf Technol 107(1–2):717–730. https://doi.org/10.1007/s00170-020-05077-4
34. Safarzadeh H, Leonesio M, Bianchi G, Monno M (2020) Roundness prediction in centreless grinding using physics-enhanced machine learning techniques. Int J Adv Manuf Technol. https://doi.org/10.1007/s00170-020-06407-2
35. Lizarralde R, Barrenetxea D, Gallego I, Marquinez JI, Bueno R (2005) Practical application of new simulation methods for the elimination of geometric instabilities in centerless grinding. CIRP Ann 54(1):273–276. https://doi.org/10.1016/S0007-8506(07)60101-2
36. Zakharov OV, Datskovskaya EA (2010) Setup of centerless superfinishing machine tools. Russ Eng Res 30(12):1263–1267. https://doi.org/10.3103/S1068798X10120191
37. Barrenetxea D, Marquinez JI, Álvarez J, Fernández R, Gallego I, Madariaga J, Garitaonaindia I (2012) Model-based assistant tool for the setting-up and optimization of centerless grinding process. Mach Sci Technol 16(4):501–523. https://doi.org/10.1080/10910344.2012.729480
38. Rowe WB (1979) Research into the mechanics of Centreless grinding. Precis Eng 1(2):75–84. https://doi.org/10.1016/0141-6359(79)90137-5
39. De Los B, Rios GA, Ferreira R, Mariani FE, da Silva EJ, Coelho RT (2022) Study of the surface roughness of a remanufactured bimetallic AISI 1045 and 316L SS part obtained by hybrid manufacturing (DED/HSM). Int J Adv Manuf Technol. https://doi.org/10.1007/s00170-022-09179-z
40. Yang Y, Gong Y, Qu S et al (2020) Densification, mechanical behaviors, and machining characteristics of 316L stainless steel in hybrid additive/subtractive manufacturing. Int J Adv Manuf Technol 107:177–189. https://doi.org/10.1007/s00170-020-05033-2

41. Yang Y, Gong Y, Li C et al (2021) Mechanical performance of 316 L stainless steel by hybrid directed energy deposition and thermal milling process. J Mater Process Technol 291:117023. https://doi.org/10.1016/j.jmatprotec.2020.117023
42. Ahsan MRU, Tanvir ANM, Ross T et al (2020) Fabrication of bimetallic additively manufactured structure (BAMS) of low carbon steel and 316L austenitic stainless steel with wire + arc additive manufacturing. Rapid Prototyp J 26:519–530. https://doi.org/10.1108/RPJ-09-2018-0235
43. Tyagi P, Goulet T, Riso C et al (2019) Reducing the roughness of internal surface of an additive manufacturing produced 316 steel component by chempolishing and electropolishing. Addit Manuf 25:32–38. https://doi.org/10.1016/j.addma.2018.11.001
44. Yamaguchi H, Fergani O, Wu PY (2017) Modification using magnetic field-assisted finishing of the surface roughness and residual stress of additively manufactured components. CIRP Ann 66:305–308. https://doi.org/10.1016/j.cirp.2017.04.084
45. Hu Z, Wei Z, Sun H, Yang J, Wei L (2019) Optimization of metal rolling control using soft computing approaches: a review. Arch Comput Methods Eng. https://doi.org/10.1007/s11831-019-09380-6
46. Otsuka T, Sakamoto M, Takamachi Y, Higashida Y, Segawa Y, Takeshima S (2017) An online rolling model for plate mill using parallel computation. ISIJ Int 57(11):2042–2048. https://doi.org/10.2355/isijinternational.ISIJINT-2017-299
47. Zhou F, Yao W, Yuan J, Li F, Lyu B, Zhao P (2020) Establishment of material removal model for lapping ceramic balls with variable-radius groove plate. Int J Adv Manuf Technol. https://doi.org/10.1007/s00170-020-06259-w
48. Wang Z, Lv B, Yuan J, Yang F (2009) On the evaluation of lapping uniformity for precision balls. Key Eng Mater 416:558–562
49. Zhao P, Yu W, Tang K, Lv B, Yuan J (2010) The quantitative evaluation on the lapping uniformity of rotated dual-plates lapping mode. Appl Mech Mater 37-38:1534–1539
50. Zhou Z, Zheng J, Yuan J, Lv B, Wang Z (2007) Simulation of a ceramic ball lapping process under dual rotating plates lapping mode. Int J Comput Appl Technol 29:247–251
51. Wen S, Huang P (2008) Principles of tribology, 3rd edn. Tsinghua University Press, Beijing, pp 49–56
52. Sun Y, Gong Y, Wen X, Xin B, Yin G, Meng F, Tang B (2022) Evaluation of dimensional accuracy and surface integrity of cylindrical array microelectrodes and cylindrical array microholes machined by EDM. Arch Civil Mech Eng. https://doi.org/10.1007/s43452-021-00366-5
53. Takahata K, Gianchandani YB (2002) Batch mode micro-electro-discharge machining. J Microelectromech Syst 11:102–111
54. Zhang Y, Xie B (2021) Investigation on hole diameter non-uniformity of hole arrays by ultrasonic vibration-assisted EDM. Int J Adv Manuf Technol 112:3083–3091
55. Gong S, Sun Y, Jin L, Su Z (2020) Experimental study on fabricating micro- holes in DD5 single-crystal nickel-based superalloy using electrical discharge drilling. Arch Civil Mech Eng 20(3):1–16
56. Abdoli S (2022) A modelling framework to support integrated design of production systems at early design stages. Int J Interact Des Manuf. https://doi.org/10.1007/s12008-022-00987-x
57. Abdoli S, Kara S (2017) A review of modelling approaches for conceptual design of complex engineering systems (CESs). In: 2017 IEEE international conference on industrial engineering and engineering management (IEEM). IEEE, pp 1266–1270
58. Abdoli S, Kara S, Kornfeld B (2017) Application of dynamic value stream mapping in warehousing context. Mod Appl Sci 11(1):1913–1852
59. Thiers G (2014) A model-based systems engineering methodology to make engineering analysis of discrete-event logistics systems more cost-accessible. Georgia Institute of Technology
60. Tomiyama T, Gu P, Jin Y, Lutters D, Kind C, Kimura F (2009) Design methodologies: industrial and educational applications. CIRP Ann 58(2):543–565
61. Abdoli S (2022) Application of fuzzy-logic for design assessment of complex engineering systems in the early design stages. J Eng Des 33(3):234–258

62. Kim M, Chae J (2021) A monarch butterfly optimization for an unequal area facility layout problem. Soft Comput. https://doi.org/10.1007/s00500-021-06076-7
63. Meller RD, Narayanan V, Vance PH (1998) Optimal facility layout design. Oper Res Lett 23(3–5):117–127. https://doi.org/10.1016/S0167-6377(98)00024-8
64. Meller RD, Gau K-Y (1996) The facility layout problem: recent and emerging trends and perspectives. J Manuf Syst 15(5):351–366. https://doi.org/10.1016/0278-6125(96)84198-7
65. Kulturel-Konak S, Konak A (2011a) A new relaxed flexible bay structure representation and particle swarm optimization for the unequal area facility layout problem. Eng Optim 43(12):1263–1287. https://doi.org/10.1080/0305215X.2010.548864
66. Friedrich C, Klausnitzer A, Lasch R (2018) Integrated slicing tree approach for solving the facility layout problem with input and output locations based on contour distance. Eur J Oper Res 270(3):837–851. https://doi.org/10.1016/j.ejor.2018.01.001
67. Sherali HD, Fraticelli BMP, Meller RD (2003) Enhanced model formulations for optimal facility layout. Oper Res 51(4):629–644. https://doi.org/10.1287/opre.51.4.629.16096
68. Bukchin Y, Tzur M (2014) A new MILP approach for the facility process-layout design problem with rectangular and L/T shape departments. Int J Prod Res 52(24):7339–7359. https://doi.org/10.1080/00207543.2014.930534
69. Ghetas M, Yong CH, Sumari P (2016) Harmony-based monarch butterfly optimization algorithm. In: Proceedings—5th IEEE international conference on control system, computing and engineering, ICCSCE. IEEE, pp 156–161. https://doi.org/10.1109/ICCSCE.2015.7482176
70. Chen S, Chen R, Gao J (2017) A monarch butterfly optimization for the dynamic vehicle routing problem. Algorithms 10(3):107. https://doi.org/10.3390/a10030107
71. Ghanem WAHM, Jantan A (2018) Hybridizing artificial bee colony with monarch butterfly optimization for numerical optimization problems. Neural Comput & Applic 30:163–181. https://doi.org/10.1007/s00521-016-2665-1
72. Wang GG, Hao GS, Cheng S, Qin Q (2016) A discrete monarch butterfly optimization for Chinese TSP problem. In: Lecture notes in computer science (including subseries lecture notes in artificial intelligence and lecture notes in bioinformatics), vol 9712. LNCS, pp 165–173. https://doi.org/10.1007/978-3-319-41000-5_16
73. Zhou B, Zha W, Ye L, He Z (2021) A dynamic material handling scheduling method based on elite opposition learning self-adaptive differential evolution-based extreme learning machine (EOADE-ELM) and knowledge base (KB) for line-integrated supermarkets. Soft Comput. https://doi.org/10.1007/s00500-021-06385-x
74. Liu H, Dong J (1996) Dispatching rule selection using artificial neural networks for dynamic planning and scheduling. J Intell Manuf 7(3):243–250. https://doi.org/10.1007/bf00118083
75. Chen C, Xi LF, Zhou BH, Zhou SS (2011) A multiple-criteria real-time scheduling approach for multiple-load carriers subject to LIFO loading constraints. Int J Prod Res 49(16):4787–4806. https://doi.org/10.1080/00207543.2010.510486
76. Rahnamayan S, Tizhoosh HR, Salama MMA (2008) Opposition-based differential evolution. Adv Differ Evol 143:155–171. https://doi.org/10.1109/sde.2011.5952059
77. Wang H, Wu Z, Rahnamayan S (2011) Enhanced opposition-based differential evolution for solving high-dimensional continuous optimization problems. Soft Comput 15(11):2127–2140. https://doi.org/10.1007/s00500-010-0642-7
78. Wang L, Liu Z, Liu A, Tao F (2021) Artificial intelligence in product lifecycle management. Int J Adv Manuf Technol. https://doi.org/10.1007/s00170-021-06882-1
79. Tao F, Qi Q (2017) New IT driven service-oriented smart manufacturing: framework and characteristics. IEEE Trans Syst Man Cybern Syst 49(1):81–91
80. Zhong RY, Xu X, Klotz E, Newman ST (2017) Intelligent manufacturing in the context of industry 4.0: a review. Engineering 3(5):616–630
81. Tao F, Zhang M, Liu Y, Nee AYC (2018) Digital twin driven prognostics and health management for complex equipment. CIRP Ann Manuf Technol 67(1):169–172
82. Hutter F, Kotthoff L, Vanschoren J (2019) Automated machine learning. Springer, New York

83. Zhang Y, Jing QIU, Guanjun LIU, Peng YANG (2012) A fault sample simulation approach for virtual testability demonstration test. Chin J Aeronaut 25(4):598–604
84. Betru BT, Onana CA, Batchakui B (2017) Deep learning methods on recommender system: a survey of state-of-the-art. Int J Comput Appl 162(10):17–22
85. Wang XV, Wang L (2019) Digital twin-based WEEE recycling, recovery and remanufacturing in the background of Industry 4.0. Int J Prod Res 57(12):3892–3902
86. Custodio L, Machado R (2019) Flexible automated warehouse: a literature review and an innovative framework. Int J Adv Manuf Technol. https://doi.org/10.1007/s00170-019-04588-z
87. Casado F, Losada DP, Santana-Alonso A et al (2017) Pose estimation and object tracking using 2d images. Proc Manuf 11:63–71
88. Petković T, Marković I, Petrović I (2017) Human intention recognition in flexible robotized warehouses based on markov decision processes. In: Iberian robotics conference. Springer, Cham, pp 629–640
89. MacDuffie JP, Pil FK (1997) From fixed to flexible: automation and work organization trends from the INternational Assembly Plant. Transforming Automobile Assembly, pp 238–254
90. CSCMP (2013) Supply chain management. TERMS and GLOSSARY. Supply chain management terms and glossary (August): 1222
91. Dallari F, Marchet G, Melacini M (2009) Design of order picking system. Int J Adv Manuf Technol 42(1–2):1–12
92. Xiao-Long W, Chun-Fu W, Li G-D, Qing-Xie C (2017) A robot navigation method based on rfid and QR code in the warehouse. In: 2017 Chinese Automation Congress (CAC). IEEE, pp 7837–7840

Chapter 4
Optimization Methods in Production Planning and Scheduling

Introduction by the Editor

The production planning and scheduling include all the kinds of administrative processes necessary for manufacturing systems to have sufficient and raw material, human resource, etc. so that the supply chain will not be disturbed. This significantly helps in efficient planning of the operations at the shop floor as well as every segment of the supply chain. The processes are associated with estimation of the product demand, alternatives of the resources, continuous monitoring and adjustments and revisions in the production schedule. Optimization methods and tools play a significant part in this process which is very much required for the small and medium enterprises. This chapter discusses optimization methodologies solving the associated single and multi-objective problems. The methods such as Krill Heard based algorithm solving the scheduling problem, multi-level capacitated lot-sizing and scheduling problem using particle swarm optimization, Petri nets for the production floor processes problems, stochastic mixed-integer program, CPLEX, Jaya algorithm and simulated annealing algorithms, reinforcement learning, for maintenance scheduling and programming have been discussed. In addition, mixed-integer linear programming model for addressing disassembly scheduling model, ML Methods for transportation scheduling, reverse logistics network for product recovery using best worst method, contemporary MCDM techniques, NSGA for job-shop scheduling, etc. have also been discussed in very details with several test and real-world cases.

The chapter in detail reviews the algorithmic framework that uses process mining tools to extract the underlying industrial processes. The review underscores that the use of GA procedure which initializes from a randomly generated population of solutions and evolves good local solutions by mimicking the process of natural selection using mechanisms such as mutation to generate variants and crossover to improve combinations. The associated cases studies conclude that an increase in the

A. J. Kulkarni (ed.), *Optimization Methods in Manufacturing Processes*,
https://doi.org/10.1007/978-981-96-5257-0_4

population size in the algorithm improves the quality of the solutions only up to a certain point, after which only the solution time increases, although for small-sized instances of the problem we recommend the use of commercial/exact solvers. The review highlights the problem formulation as a multi-objective optimization problem for lean and resource efficient manufacturing systems. The analysis focuses on the approaches that use discrete event simulation for modeling and improving manufacturing systems. It also reveals that a suitable combination of real implementable measures for the execution of the long-term target state is required to be chosen along with a suitable optimization method. The review reveals that the optimal maintenance scheduling of transmission assets is very necessary. The problem discussed involved maintenance scheduling of a system subject to electrical, seasonal, legal, security, logistical, among other constraints. A deep reinforcement learning assisted simulated annealing algorithm for a maintenance planning problem is also reviewed. The study underscored the importance of the Machine Learning (ML) algorithms in which different properties of the spare parts such as failure rates, holding costs, repair rates, etc. are used as clustering features for the ML algorithm.

The chapter in detail review the benefits of the robust multi-objective optimization for the important case studies involving flexible automotive assembly line balancing. The review highlights that the robust multi-objective optimization and simulation techniques provide managers with a set of robust and equally preferred solutions for assembly line balancing. This also added with the conclusion that the adverse managerial effects of not having flexible line configuration when demand changes can be alleviated. The application of the mixed-integer linear programming (MILP) model is also discussed in the detail in connection with capacitated disassembly scheduling problem. The review highlights numerical analysis, managerial insights. The insight concludes that one of promising avenues such as investigating more realistic framework in which the quality of returned items and its impact on the disassembly techniques can be studied. The chapter discusses that the numerical experiments of the robust berth scheduling problem. It reveals that the handling times can be forecasted using ML techniques and the influence on the robustness of the resulting berthing schedules is to be studied.

A fuzzy hybrid multi-criteria decision-making approach is discussed in detail for its application to reverse logistics optimization of an industrial air conditioner manufacturing company for designing sustainable supply chain. The model is original and is proposed first time to have discussions of this approach. This model reflects the basic characteristics of the reverse supply chain. The results of the study applied to an industrial air conditioner manufacturer working in practice are expected to have a significant impact on sustainability. The study concludes that the efficient and appropriate design of the logistics network leads to meeting economic advantages of using returned products, improving customer satisfaction, increasing market share, reducing costs and adding value to logistics network. A critical review of the machine learning and optimization for production rescheduling in Industry 4.0 is also elaborated in detail. The reviews are associated with the Problem definition and modeling phase, scheduling optimization phase, machine learning–based classification phase along with the analysis of the numerical experiments.

Machine Generated Summaries

Disclaimer: The summaries in this chapter were generated from Springer Nature publications using extractive AI auto-summarization: An extraction-based summarizer aims to identify the most important sentences of a text using an algorithm and uses those original sentences to create the auto-summary (unlike generative AI). As the constituted sentences are machine selected, they may not fully reflect the body of the work, so we strongly advise that the original content is read and cited. The auto generated summaries were curated by the editor to meet Springer Nature publication standards. To cite this content, please refer to the original papers.

Machine generated keywords: scheduling, planning, schedule, job, maintenance, scheduling problem, route, flow, logistic, shop, oil, activity, framework, uncertainty, unit.

Multi-Product Scheduling Through Process Mining: Bridging Optimization and Machine Process Intelligence [1]

This is a machine-generated summary of:

Choueiri, Alexandre Checoli; Portela Santos, Eduardo Alves: Multi-product scheduling through process mining: bridging optimization and machine process intelligence [1].

Published in: Journal of Intelligent Manufacturing (2021).

Link to original: https://doi.org/10.1007/s10845-021-01767-2

If you want to cite the papers, please refer to the original.

For technical reasons we could not place the page where the original quote is coming from.

Abstract-Summary "Small and medium enterprises (SMEs) may not have the maturity to put forward and unfold all the benefits from an ERP based system, a vital tool for production planning."

"We propose an algorithmic framework that uses process mining tools to extract the underlying industrial process via Petri nets, and then retrieve all product tree necessary information to perform the multi-level scheduling."

"A faster solution decoding is proposed, for algorithms that uses random-keys."

Introduction

"Since scheduling functionalities are closely related to information systems (such as ERP or even the MRP), it becomes a challenge for SMEs to achieve efficient planning and control activities in their production floor, unveiling a gap from this companies regarding a more ready to use framework."

"We thus propose a framework that uses the information generated by machines in the form of event logs, extracts the underlying process to a Petri net, and from the net and the data log reconstruct the product tree (dependency graph)."

"This would enable to structure a production planning and control for SMEs, even without the use of major ERP or MRP-based systems, using the already collected data from machines."

"The reminder of the paper is organized as follows: in "Literature review" section we provide some basic concepts of process mining, as well as some related works conducted on multi-level product structures and the use of random keys for scheduling."

Literature Review

"On the second part we expose previous work on multi-level scheduling, and the use of random keys for coding solutions."

"Mohammadi et al. [2] has provided a mathematical model, and a relax and fix heuristic for the multi-product multi-level capacitated lotsizing problem with sequence-dependent setups."

"Yan et al. [3] solve the multi-level capacitated lot-sizing and scheduling problem using particle swarm optimization."

"The work of Chen et al. [4] is the first, to our knowledge, to conduct a case study for the multi-level product scheduling encountered in a light source manufacturer."

"Puongyeam et al. [5] designed a Krill Heard (KH) based algorithm to solve the scheduling problem, the authors have also used real data to validate their approach, from the capital goods industry."

"The first author to propose a mathematical model for the multi-level scheduling integrated with preventive production maintenance is Chansombat et al. [6]."

Proposed Framework

"The final information of the product trees regards their processing times; in this phase (III), we cope with the stochastic behavior of the process by collecting 3 different processing times for each machine, which will enable the different scheduling scenarios evaluation."

"That the Petri net provide the dependency graph, along with the permissible machine sets, but it is not possible to derive processing times and quantities only with the net, that would require the new discovered graph and the log (that is explained on further sections)."

"Pessimistic As expected, in a pessimistic scenario, all machines process their respective products using as long a time as possible, that is, the upper value of the intervals In this way, industrial managers are better equipped to deal with different types of clients and suppliers; when bargaining a new client due date, managers could use the pessimistic scenario of scheduling, which would hardly be extrapolated, hence providing the client with a secure delivery date."

Computational Results

"We chose the GA as it has many parameters that influences the total number of decoded solutions, and an overview of time differences in this complex search space may provide a good estimate of time savings for other meta-heuristic approaches."

"Since we are interested in quantify the speed performance of algorithms, we did not provide a comparison of scheduling objective functions, as we use the same GA for both random keys decoding schemes; which implies that the scheduling outputs were the same, differing only by their required computational time."

"Besides the GA parameters, one particularly interesting parameter is the due date of the demand; as the bisection decoding algorithm speeds up the process when the first decoding is not feasible."

"Our decoding has a time increase practically constant while due dates approach 0."

Conclusion

"With the Petri nets describing the production floor processes, along with the event log, we propose an algorithm that extract the multi-level product tree (dependency graph) for each product, as well as the relevant information to perform the scheduling."

"In order to perform 3 different schedule outputs for each manager input, the optimization engine should be fast enough as to not compromise time restrictions; we then provide an acceleration method to decode solutions based on random-keys."

"As an implication of our work, SMEs that does not yet posses a scheduling engine are now able to build one from scratch, with little efforts regarding systems and databases integration; by only using the log generated by machines, they are able, through the use of process mining, to perform scheduling with multiple scenarios, giving production managers a response answer for different customer demands and supplier restrictions."

Integrated Preventive Maintenance and Flow Shop Scheduling Under Uncertainty [7]

This is a machine-generated summary of:

Seif, Javad; Dehghanimohammadabadi, Mohammad; Yu, Andrew Junfang: Integrated preventive maintenance and flow shop scheduling under uncertainty [7].

Published in: Flexible Services and Manufacturing Journal (2019).

Link to original: https://doi.org/10.1007/s10696-019-09357-4

If you want to cite the papers, please refer to the original.

For technical reasons we could not place the page where the original quote is coming from.

Abstract-Summary "This paper is concerned with stochastic scheduling of production and maintenance activities in a permutation flow shop setting."

"A two-stage stochastic mixed-integer program (SMIP) that adapts the conventional permutation flow shop scheduling problem for incorporating multiple preventive maintenance activities with various meter-based intervals."

"The model handles uncertainties in both processing times and the duration of maintenance activities."

"The concept of combining maintenance activities in scheduling problems is introduced and formulated, along with other practical considerations."

Introduction

"Although incorporating maintenance decisions into production scheduling problems increases the practicality of existing models and solution methods, it also increases their complexity, and considering uncertainty makes the problem even more complex."

"This paper introduces an integrated preventive maintenance and flow shop scheduling problem under uncertainty as a stochastic mixed integer program (SMIP)."

"Performing maintenance in a time window has been modeled in different types of scheduling problems, yet they are not precise in timing of MAs."

"We model the concept of combined maintenance activities in scheduling problems for the first time."

"The contributions and features of this paper are: the conventional permutation flow shop scheduling problem is extended to a stochastic flow shop scheduling integrated with maintenance planning, the concepts of combining MAs and prolonged (maintenance-dependent) processing times are introduced and incorporated to the model in order to increase its practicality, two methodologies (simulation–optimization and stochastic programming) are developed and contrasted for validating the results and evaluating their performances."

Problem Definition and Mathematical Formulation

"To do this, first, we define and formulate the concepts of combining maintenance activities and discuss the prolonged processing times."

"None, one, or more than one of the variables on the right-hand side can take the value 1 (we can decide to perform none, one, or more than one type of maintenance activity before processing a certain job on a certain machine), but only one of the variables on the left-hand side must take the value 1 (only one combination can represent the chosen set of maintenance activities)."

"An time-based maintenance activity allows a machine to operate only for certain number of hours, called maintenance interval."

"This example involves a single machine (for simplicity), two types of maintenance activities (MA_1 and MA_2, with time intervals R_1 and R_2), three production jobs (J_1, J_2, and J_3), and two probabilistic scenarios."

Simulation Optimization

"Since the model is stochastic, in each iteration of GA, the generated solutions are evaluated in a simulated environment to calculate the expected value of the objective function (the cost)."

"The GA procedure initializes from a randomly generated population of solutions, and evolves good local solutions by mimicking the process of natural selection using mechanisms such as mutation to generate variants and crossover to improve combinations [8]."

"Yu and Seif [9] employed a factorial design of computational experiments for studying the parameters of a GA that was designed for solving a similar problem: a deterministic flow shop scheduling problem with multiple maintenance activities."

"This solution representation determines (1) the sequence of jobs in the flow shop system, and (2) the number of MAs needed to be performed on each machine prior to each job by choosing one of the MA combinations."

Computational Experiments

"We test the quality of solutions by comparing the results of SO with those of CPLEX on small-size test problems, in which the global optimal is known."

"We evaluate the performance of the SO method for larger problem sizes under a limited solution time."

"One observation is that, as the number of jobs (problem size) increases, the quality of CPLEX solutions decreases under a limited solution time."

"It can be concluded that under a limited solution time, the proposed SO algorithm outperforms a commercial solver as the problem size increases."

"As the last part of the experiments, we want to make sure that the quality of solutions of the algorithm (as measured by the gap in OFV), as well as the time to find a solution, are not dependent on how we generate the input data of the test problems."

Case Study

"Because the equipment operate in series to complete specific operations (such as the earth-moving) in each location, finding the optimal sequence for routing the equipment between the sites can be treated as a flow shop scheduling problem in which the locations are equivalent to production jobs."

"The focus of this case study is on stochastic optimization of a realistic problem applicable to operations and maintenance scheduling in construction projects."

"The input data and description of this case study are adapted from the case study by Yu and Seif [9] that is designed for a deterministic flow shop scheduling with multiple maintenance activities (MAs) in which combining the MAs was not considered and processing times were fixed regardless of the machines' health state."

"This solution provides a schedule for routing of the machines between the construction locations and a maintenance station, as well as the maintenance plan."

Conclusion

"We incorporated the concept of combined maintenance activities in the permutation flow shop and considered the impact of the health of machines on the processing times of jobs."

"We concluded that: an increase in the population size in the algorithm improves the quality of the solutions only up to a certain point, after which only the solution time increases, although for small-sized instances of the problem we recommend the use of commercial/exact solvers, for medium to large-scale instances and under a limited time frame, the proposed solution method outperforms these computationally and financially expensive solvers, and the quality of the solutions and solution time of the presented simulation–optimization method is not sensitive to the input data under a limited solution time, which alludes to the robustness of the method."

"The concept of combined maintenance activities can be incorporated in other production settings such as flexible flow shop and job shop scheduling."

Multi-Objective Optimization of Lean and Resource Efficient Manufacturing Systems [10]

This is a machine-generated summary of:

Greinacher, Sebastian; Overbeck, Leonard; Kuhnle, Andreas; Krahe, Carmen; Lanza, Gisela: Multi-objective optimization of lean and resource efficient manufacturing systems [10].

Published in: Production Engineering (2020).

Link to original: https://doi.org/10.1007/s11740-019-00945-9

If you want to cite the papers, please refer to the original.

For technical reasons we could not place the page where the original quote is coming from.

Abstract-Summary "In the manufacturing industry, target-oriented and efficient use of resources is gaining importance, alongside economic optimization."

"The economic and organizational optimization of manufacturing systems according to the lean principles is only partly compatible with the goals of resource-efficient manufacturing."

"This paper proposes an approach for the multi-objective optimization of lean and resource-efficient manufacturing systems."

"To predict the dynamic effects of several configurations of manufacturing systems, material, energy, and information flows of a discrete event simulation are coupled with an assessment model, based on objectives of lean and resource-efficient manufacturing."

Introduction

"For manufacturing companies, the target-oriented and efficient use of resources, such as material and energy, is gaining importance, alongside a pure economic optimization."

"A fundamental compatibility and synergy between the principles of lean based on Toyota Production Systems (TPS) and resource-efficient production can be found [11–14]."

"Such assessment and optimizations are necessary to tailor the use of lean and resource-efficient improvement measures specifically to an individual manufacturing system."

"Many businesses are lacking the appropriate approaches to systematically and holistically assess and optimize their manufacturing procedures, while simultaneously taking profitability, energy and material efficiency into account [15, 16]."

"This paper addresses gaps in previous research by introducing an approach for multi-objective optimization of lean and resource-efficient manufacturing systems."

"The approach includes: (1) a simulation model for the assessment of different manufacturing system configurations, taking dynamic material, energy, and information flows into account, and (2) a meta-model for the multi-objective optimization of manufacturing systems."

State of the Art

"Section, assessment and optimization approaches for lean and resource-efficient manufacturing systems in discrete production environments are compared."

"The analysis focuses on the approaches that use discrete event simulation for modeling and improving manufacturing systems."

"To the design and evaluation of lean and resource-efficient manufacturing systems, several approaches pursue an optimization of the modeled system."

"Both approaches rely on a meta-model of the analyzed manufacturing system, which is adapted to the response data of the simulation model, based on DoE, but focus exclusively on selected lean criteria and do not consider relevant aspects in regards to resource efficiency."

"It can be derived from the literature review that so far, an approach that fully meets the requirements for multi-objective optimization of lean and resource-efficient manufacturing systems does not exist."

"No approaches can be identified that integrate resource requirements into simulation-based modeling of manufacturing."

Simulation Framework

"The modules interact through material, energy and information interfaces."

"Subsequent to the transformation in the process module and the buffering in the inventory module, the material flow exit the system in terms of finished products or waste."

"Modules request energy and material from each other."

"The behavior of the process module is modeled for all six operating states."

"The module's behavior can be described through the three operating states, "operational", "malfunction", and "off"."

"The module behavior can be described through the two operating states "operational" and "off"."

"The module behavior can be described by the operating states of "operation" and "off"."

"The production planning and control module (PPC module) contains a higher-level logic for controlling all material, energy, and information flows."

"The total energy consumption per variant corresponds to the product of the time-based energy consumption rate, and operating duration for the module per operating state and variant summed over all energy carriers."

Process Model for the Multi-objective Optimization of Manufacturing Systems and Industrial Application

"For the multiple objectives CPP, TPT, SE, and DR, Gaussian-process meta-models are computed based on the training data."

"For the multiple objectives CPP, TPT, SE, and DR, Gaussian-process meta-models are computed based on the training data using Kriging."

"Within the experiments, meta-model predictions of the objectives CPP, TPT, SE, and DR are generated for 1.25 million randomly distributed factor combination and transformed into utility values."

"The results of the meta-model-based multi-objective optimization are compared with the results of an established simulation-based PSO using the external parameter optimization software optiSLang v4.2 [17]."

"Based on the simulation model, the same multi-objective optimization studies as in the meta-models are carried out and the determined optimization results are compared."

"This confirms that the meta-model-based approach for multi-objective optimization of manufacturing systems achieves similar results when compared to traditional simulation-based optimization algorithms."

Summary

"An approach for meta-model-based multi-objective optimization of manufacturing systems was presented."

"The use of meta-models allows the optimization of manufacturing systems to be completed in substantially less time than it would be if using simulations."

"A suitable combination of real implementable measures for the execution of the long-term target state should be chosen."

Optimal Maintenance Scheduling of Transmission Assets in the Brazilian Electric System [18]

This is a machine-generated summary of:

Evangelista Araujo Neto, José; Castro, Carlos A.: Optimal Maintenance Scheduling of Transmission Assets in the Brazilian Electric System [18].

Published in: Journal of Control, Automation and Electrical Systems (2021).

Link to original: https://doi.org/10.1007/s40313-020-00678-6

If you want to cite the papers, please refer to the original.

For technical reasons we could not place the page where the original quote is coming from.

Abstract-Summary "In order to enforce an increase in the quality of service, the ANEEL (Portuguese for The Brazilian National Electric Energy Agency) introduced penalties in contracts with public electricity transmission utilities in case transmission facilities are taken out of service due to accidents, failures or scheduled maintenance."

"Traditional maintenance approaches to electric power transmission systems are based on actions taken periodically."

"This paper proposes an approach for addressing this problem based on reliability/cost to find the best strategies for performing maintenance on electric power transmission equipment (assets), with a mathematical model that provides a feasible maintenance calendar while respecting the constraints of the electrical system operation."

Introduction

"The main objective of this work is to fill the gap in the literature concerning the predictive maintenance of transmission assets, by presenting an innovative solution to the problem of maintenance scheduling in the current scenario of the Brazilian electric sector, where the maintenance policy of electrical assets has become a fundamental part in obtaining better financial results for companies."

"The main contributions of this paper are: Bibliographic review on the maintenance scheduling of transmission assets; Elaboration and implementation of mathematical modeling for the problem; Development of a new and efficient solution method (combining a constructive heuristic with dynamic programming) for the problem; Evaluation of the proposed methodology through simulations with real data associated with a substation of the Brazilian National Interconnected System; Presentation of tools and technical inferences that can contribute to improving efficiency and increasing the reliability of the electricity industry."

The Maintenance Schedule Problem

"The proposed problem can be defined as a maintenance scheduling of a system subject to electrical, seasonal, legal, security, logistical, among other constraints."

"The problem of determining the best times to turn off electrical power equipment in order to carry out the respective maintenance is a multidisciplinary, complex, and challenging engineering problem, especially because it involves binary and real variables, as well as coupled restrictions in time and in space."

"Marwali and Shahidehpour [19] proposed an approach based on the duality theory to solve the problem of scheduling maintenance of transmission lines with systemic restrictions."

"Araujo Neto et al. [20] proposed a strategy that consisted of minimizing maintenance costs and the risk of failure (based on Markov chains) for application in making a feasible timetable (that respects systemic restrictions) with logistical optimization."

The Maintenance Programming Model

"According to Araujo Neto et al. [20], the development of mathematical models that represent more accurately the operation of electrical equipment and maintenance procedures adopted by utilities is one of the greatest challenges for engineers and researchers."

"The development of the model is based on the following steps: Development of Markovian model (determination of the failure rate for each equipment); Determination of a commitment interval between maintenance (harmonize the cost–benefit ratio of each maintenance); Making the optimal maintenance schedule (elaboration of the feasible calendar with rational logistics)."

"These models can be used to represent, for example, the process of equipment deterioration, preventive maintenance, random failures, among others."

"According Araujo Neto et al. [20], the relevance of scheduled maintenance is due to the fact that all equipment fail, and these failures have major consequences for production processes."

Solution Method

"Due to the specific inability of the exact methods to handle certain programming problems, heuristic methods have been proposed to solve several complex problems in the power systems area."

"Heuristics are procedures that are supported by an intuitive approach, in which the particular structure of the problem can be explored to obtain an adequate, good-quality solution."

"Although the definition may seem discouraging, many current heuristic procedures are able to achieve very good solutions to complex combinatorial problems at an adequate computational cost."

"Dynamic programming is used to obtain the optimal solution of the reduced problem."

"The constructive heuristic described by Algorithm 1 aims to reduce the search space to make the further resolution process by dynamic programming more efficient."

"The adapted DP lends itself to obtain the optimal solution for the new search space determined by the constructive heuristics."

Case Study

"The tests were used to compare the actual maintenance schedule taken into effect with the scheduling suggested by the proposed method, considering the time horizon from January 1, 2018, to December 31, 2018."

"Only one type of maintenance activity was considered per category."

"The data provided by the utility were used to compute the quantities defined in the proposed method, for the sake of a fair comparison."

"The computing times for the monthly solution corresponding to the maintenance types WS, PE, and LL were, respectively, 0.0902, 0.1409, and 0.0031 s."

"The VPU provided by the proposed method was much lower (approximately 50%, average due to better distribution of maintenance activities in the programming window) than the actual case, for all types of maintenance."

Conclusions

"An efficient method was proposed for the maintenance scheduling of transmission assets."

"The simulation results presented showed that the proposed method resulted in significant improvement in all cases and aspects evaluated."

"In all instances assessed, the combination of constructive heuristics with dynamic programming proved to be extremely efficient, presenting significant quality improvements in comparison with the solutions adopted in practice (actual case)."

"The solutions provided by the proposed method consumed the necessary and sufficient resources for a rational and concatenated maintenance schedule, thus presenting the best cost/reliability ratio of the studied transmission assets."

Benefits of Robust Multiobjective Optimization for Flexible Automotive Assembly Line Balancing [21]

This is a machine-generated summary of:

Chica, Manuel; Bautista, Joaquín; de Armas, Jesica: Benefits of robust multiobjective optimization for flexible automotive assembly line balancing [21].

Published in: Flexible Services and Manufacturing Journal (2018).

Link to original: https://doi.org/10.1007/s10696-018-9309-y

If you want to cite the papers, please refer to the original.

For technical reasons we could not place the page where the original quote is coming from.

Abstract-Summary "Automotive manufacturers must adapt their decisions when balancing the assembly line by considering different flexible solutions."

"Our proposal is using robust multiobjective optimization and simulation techniques to provide managers with a set of robust and equally-preferred solutions for assembly line balancing."

"The problem is addressed by considering a robust multiobjective model for assembly line balancing based on a high number of production plans."

"After the selection of six different assembly line configurations, we study the implications of robustness metrics based on workstations' overload."

"We show that the adverse managerial effects of not having flexible line configuration when demand changes are alleviated."

"For the real Nissan automotive case, our analysis and conclusions show the managerial and industrial benefits of using robust assembly lines."

Introduction

"These engines' demands are not usually fixed and certain, and when the assembly line produces mixed products in a given sequence [22], the model cannot only consider the operation time of the tasks as the averaged times of the different products and their demand."

"The model links robustness with the flexibility of an assembly line configuration when demand changes according to a set of real production plans."

"The model identifies and measures how robust a assembly line configuration is for a set of production plans according to both operation time and the linear area (i.e., one dimensional length) needed for workers and tools to perform the tasks of the assembly line."

"Three temporal non-robustness metrics, set as optimization constraints, are proposed to measure the robustness of the assembly line configurations."

Background and Research Methodology

"Chica et al. [23] also defined a set of scenarios and proposed a visual representation of the optimal solutions to quantitatively measure and represent how robust the assembly line configuration is on the set of scenarios or production plans."

"The work of Papakostas et al. [24] also evaluated a posteriori the solutions of a model for minimizing time and cost through a set of demand profiles but they used single-objective particle swarm optimization."

"This is a multiobjective optimization problem which models uncertainty in the demand by setting production plans as an input."

"The goal of this model is to identify the most flexible assembly line configurations for a set of production plans by searching for the most robust solutions based on three non-robustness temporal metrics."

"In the next two sub-sections we first define the r-TSALBP non-robustness metrics and second, how to use them as constraints within an optimization method to find flexible assembly line solutions."

The Automotive Case Study

"This fact means that, although the assembly line maintains a daily production of 270 units, it should be able to adapt to different production plans based on the partial demands of each type of engine."

"We have considered all the 23 demand plans to search for the most robust assembly line configurations."

"In order to generate more representative scenarios with the available demand information, we extract statistical information of the demand for each engine (such as minimum, maximum, average, or median) from the production plans."

"A set of different engines' demand values are built once we have obtained the probability distributions which represent the processing times of each engine."

"Monte Carlo simulation was chosen to perform this process because of its simplicity and appropriateness for evaluating a set of artificially generated demand plans."

Experimental Results

"We show how a simulation technique is used as a tool to intensively use thousands of plans and compare the robustness metrics obtained for the six assembly line configurations."

"In these two cases (see supplemental material file), the standard EMO algorithm provides assembly line solutions where the stations are overloaded more frequently than the configuration found by the robust multiobjective method."

"The solution obtained with the robust EMO algorithm is less robust than the corresponding 18-stations assembly line, but still those metric values lead to confirm that the robustness of the solution is higher than the obtained by the standard algorithm."

"Monte Carlo simulation can help us to extend the evaluation of the robustness of the assembly line configurations by using a higher number of demand plans instead of using the reduced number of them."

Final Discussion and Concluding Remarks

"There are also possible negative effects in the workstations and production line when global demand does not change but there are changes in the production mix with respect to the one considered when balancing the assembly line (reference plan)."

"Our case study from the Nissan assembly line showed how our robust TSALBP model, EMO algorithms, and Monte Carlo simulation techniques can help the decision maker to find flexible assembly line configurations which have less risk when the demand of the products changes."

"This is done through robustness metrics applied to an assembly line configuration with respect to a set of demand plans (E)."

"We understand these robustness metrics as the capacity of the assembly line configuration to absorb the possible demand variations in the set of products to be assembled in the same assembly line (I)."

An Optimization Model for Management of Empty Containers in Distribution Network of a Logistics Company Under Uncertainty [25]

This is a machine-generated summary of:

Hosseini, Ahmad; Sahlin, Tobias: An optimization model for management of empty containers in distribution network of a logistics company under uncertainty [25].

Published in: Journal of Industrial Engineering International (2018).

Link to original: https://doi.org/10.1007/s40092-018-0286-2

If you want to cite the papers, please refer to the original.

For technical reasons we could not place the page where the original quote is coming from.

Abstract-Summary "This study addresses the problem of empty container repositioning (ECR) in the distribution network of a European logistics company, where some restrictions impose decision making in an uncertain environment."

"A multi-period optimization model is developed to help make tactical decisions under uncertainty and data shortage for flow management of empty containers over a predetermined planning horizon."

"Employing the operational law of uncertainty programming, a new auxiliary chance-constrained programming is established for the ECR problem, and we prove the existence of an equivalence relation between the ECR plans in the uncertain network and those in an auxiliary deterministic network."

"Exploiting this new problem, we give the uncertainty distribution of the overall optimal ECR operational cost."

Introduction

"Minimizing the logistics and distribution costs arising from the container flow management across different locations has emerged as a major problem that companies and affiliated third-party logistics firms face in everyday routine [26, 27]."

"As a class of multi-period distribution planning problems [28–30], this work focuses on the ECR problem in a dynamic environment that occurs in the context of intermodal distributions and transportation operations."

"To respond to such a need for the analysis of ECR in dynamic uncertain environment, we illustrate a modeling process to take the uncertain nature of ECR parameters into account using the operational law of uncertainty programming."

"We develop an uncertain ECR model that can accommodate uncertainty to ensure an efficient repositioning policy at minimal operational cost, while meeting the time-dependent demand, supply, and capacity requirements."

"We then employ the uncertainty operational law to originally model, solve, and analyze the uncertain ECR problem in the framework of uncertainty programming."

Literature Review

"Ji et al. [31] study an ECR problem in short sea liner service and develop a dynamic optimization model of ECR for each stage in short sea liner service minimizing the overall repositioning cost during a decision-making horizon."

"Olivo et al. [32] develop an integer program for the ECR between depots and ports through an inland transportation network and solve the model by a linearization technique."

"Lam et al. [33] develop a dynamic stochastic model for a two-port two-voyages relocation of empty containers to solve the ECR problems."

"Shintani et al. [34] model the ECR problem as an integer program to optimize the repositioning in the hinterland."

"Long et al. [35] formulate a two-stage stochastic programming model with random demand, ship weight capacity, and ship space capacity in order to incorporate uncertainty in the ECR problem and minimize the expected operational cost for container repositioning."

Uncertain ECR Model and Description

"The network consists of several locations (nodes) with different functionalities: depots, cleaning station, railway terminals, port terminals, rail ship terminals, supply customers (unloading sites), demand customers (loading sites), and various transportation links between locations."

"It is assumed that a dirty empty container is always sent to a cleaning station for cleaning before reloading at demand customer sites."

"A typical supplier s may be linked to depot j, cleaning station c, railway terminal r, port terminal p and rail ship terminal v. Assumptions necessary to postulate are listed below, all of which are standard in majority of container management models."

"Transportation capacities between locations, transportation costs over links, and storage costs in locations are assumed to be independent of the type and the condition of the container."

"The model takes several aspects of the problem into account: uncertainty, multiple traveling modes, multiple containers, multiple ports and terminals, and time-varying storage and transportation mode capacities."

Computational Study and Experiments

"To account for variability in demand and to investigate the model's performance, we examine the proposed model under different planning horizons using one-year actual historical shipping patterns performed by the company, with key practical considerations such as delivery time periods and intermodal transports."

"To highlight certain features of the model and to reveal how the distribution cost depends on the model's parameters, we present a sensitivity analysis and we investigate the effect of the parameters on the container management."

"The data obtained from the company include time-dependent demands over one year for all customers, transportation costs over links for all transportation modes, cleaning times, storage times, storage costs, transit capacities for all modes, and storage capacities at intermodal terminals."

"We have observed that although the 7-day, and 21-day models consider longer planning horizons, the optimal solutions only suggest a higher cost during the early periods."

Conclusion

"Exploiting the operational law of independent uncertain variables, the uncertain ECR problem is originally modeled and analyzed in the framework of uncertainty programming."

"By exploiting this relation, the original uncertain ECR problem is solved and the uncertainty distribution of the optimal logistics cost is obtained."

"The proposed model captures many time-varying parameters, which makes it possible to quantify the trade-offs between different criteria of the ECR problem."

"The model includes many practical considerations needed for ECR distribution networks: meeting demand and supply requirements, multimodal transports, cleaning processes, transport and storage capacity, internal and external locations, various types of ports, terminals, and depots, etc. However, there are several possibilities for extending the model proposed in this study."

"We note that proposed model is concerned with one of the main objectives of the ECR distribution networks, viz."

A Capacitated Disassembly Scheduling Problem Considering Processing Technology Selection and Parts Commonality [36]

This is a machine-generated summary of:

Darghouth, M. N.; Abdel-Aal, M.: A capacitated disassembly scheduling problem considering processing technology selection and parts commonality [36].

Published in: Journal of Remanufacturing (2021).

Link to original: https://doi.org/10.1007/s13243-021-00104-3

If you want to cite the papers, please refer to the original.

For technical reasons we could not place the page where the original quote is coming from.

Abstract-Summary "A multi-product multi-period capacitated disassembly scheduling problem with parts commonality, along with start-up and setup costs, and processing technology selection is formulated as a mixed-integer linear programming (MILP) model."

"Energy consumption cost is associated with each disassembly processing technique."

"The model aims at quantifying energy costs by considering the expense incurred during the disassembly process to illustrate realistic remanufacturing concerns under several restrictions."

"The objective is to determine the optimal decisions on the quantity, the timing, and the processing technology of disassembling root items to satisfy leaf item demand over the planning horizon."

"The periodic capacity of disassembling root items is considered to be limited."

"A comprehensive sensitivity analysis is conducted to demonstrate the significance of the proposed mathematical model."

Introduction

"The main goal is to investigate the disassembly scheduling problem with energy efficiency consideration."

"No prior research work on disassembly scheduling problems with energy efficiency consideration was reported in the literature."

"The multi-product, multi-period, capacitated disassembly scheduling problem with parts commonality, setup and start-up costs, and processing technology selection under different energy consumption and dependent yield ratios is formulated and investigated as a comprehensive MILP model."

"The developed model is the first that integrates energy efficiency with dependent yield ratios into the capacitated disassembly scheduling problem."

"The performed computational experiments on problem instances with different sizes and input parameters showed that increasing the number of disassembly techniques helps to dramatically reduce the total cost by offering the best disassembly schedule considering the tradeoff between the energy cost and the disassembly yield ratio."

Literature Review

"Kim et al. [37] considered the disassembly scheduling problem for multiple product types with parts commonality."

"Kim and Xirouchakis [38] considered a variant of the capacitated multi-product disassembly scheduling problem with stochastic demand for leaf items for a two-level product structure."

"Tian and Zhang [39] examined the capacitated disassembly scheduling problem considering multi-item and multi-period pricing of returned products with price-dependent yields."

"Tan et al. [40] established a multi-objective production scheduling optimization model for the economic dispatch problem considering time-varying electricity costs for hot rolling processing."

"This paper complement existing studies in disassembly scheduling especially the work performed by [41] where the authors formulated the problem as a MILP considering simultaneously capacity restrictions, parts commonality and start-up cost by introducing the following two ideas: The cost of energy characterized by the selection of the disassembly technique to be used to satisfy the time-varying demand initially investigated by [42] to study the capacitated production planning and scheduling problem (CPPS) with sequence-dependent setups is considered."

"The multi-product, multi-period, capacitated disassembly scheduling problem with parts commonality, setup and start-up costs, and processing technology selection under different energy consumption and dependent yield ratios is formulated and investigated as a comprehensive MILP model."

Problem Statement and Model Formulation

"A yield ratio is defined by the parameter a_{oji} and it is associated with the disassembly technique o of the root item i to the leaf item j. A time-varying cost e_{ot} is associated with each disassembly processing technique."

"The total disassembly cost includes the setup costs, the start-up costs, the energy costs in disassembly, the procurement costs of the root items, and the inventory and backlogs costs."

"Each disassembly technique is associated with a specific yield ratio of leaf items and charge different energy costs."

"The second term indicates the generated quantity of leaf item j using the disassembly technique o to disassemble the root item i in period t. We assume that a_{oji} is independent of the time period t. Terms three and four present the demand and the backlogged amount of leaf item j at period t, respectively."

Numerical Analysis

"Several computational experiments focusing on the impact of the number of disassembly techniques with different characteristics on the objective function components, including the costs of inventory, energy, procurement, setup, start-up, and backlogging were conducted on various test instances including different problem sizes and different inputs parameters."

"Test instances with different values of the total number of root items (N), the total number of leaf items (M), and the sizes of the available disassembly techniques set (O) have been investigated for different time periods varying from 5 to 15."

"We can see that the CPLEX solver can reach the optimal solution within a reasonable computing time, especially for low and medium-size problem test instances up to a planning horizon consisting of 10 time periods, except for some test instances with large-sized disassembly techniques."

Managerial Insights

"Increasing both the number of disassembly techniques and the capacity restriction has a beneficial impact on the total disassembly cost."

"It is worth pointing out the negative correlation between the total cost, relative Gap, and CPU time with the available disassembly capacity."

"Managers should weigh and analyze the trade-off between larger disassembly capacity investment and lower operational disassembly costs, and improved relative Gap and CPU time [41, 43]. have demonstrated similar results."

"Like the number of disassembly techniques, increasing the capacity constraint has a limited beneficial impact on the total disassembly cost."

"It is essential to mention that considering a high number of root items may have a negative effect on the total disassembly cost."

"In a case in point, managers are advised to consider more disassembly techniques to improve cost reduction performance."

Conclusions and Possible Extensions

"To address these challenging issues in the context of the operational disassembly of EOL products, we establish a multi-product, multi-period capacitated disassembly scheduling problem with parts commonality, along with start-up and setup costs, considering disassembly technology selection as a MILP model."

"Increasing the number of disassembly techniques and the capacity constraint beyond a certain level has a limited beneficial impact on the total disassembly cost."

"Cutting down the energy cost with a greater number of root items, especially when the number of disassembly techniques is limited, is difficult to achieve."

"Managers are advised to consider more disassembly techniques to achieve an improved cost reduction performance."

"One of promising avenues for future research is to investigate a more realistic framework in which the quality of returned items and its impact on the disassembly techniques and the corresponding incurred cost is considered with a certain level of uncertainty."

Robust Berth Scheduling Using Machine Learning for Vessel Arrival Time Prediction [44]

This is a machine-generated summary of:

Kolley, Lorenz; Rückert, Nicolas; Kastner, Marvin; Jahn, Carlos; Fischer, Kathrin: Robust berth scheduling using machine learning for vessel arrival time prediction [44].

Published in: Flexible Services and Manufacturing Journal (2022).

Link to original: https://doi.org/10.1007/s10696-022-09462-x

If you want to cite the papers, please refer to the original.

For technical reasons we could not place the page where the original quote is coming from.

Abstract-Summary "The aim of robust berth scheduling is to derive conflict-free vessel assignments at the quay of a terminal, taking into account uncertainty regarding the actual vessel arrival times which may result from external influences as, e.g., cross wind and sea current."

"The calculation and use of so-called dynamic time buffers (DTBs), which are derived from the different AIS-based forecasts and whose length depends on the estimated forecast reliability, in the berth scheduling model enhance the robustness of the resulting schedules considerably, as is shown in an extensive numerical study."

"The optimization model does not only lead to more robust solutions, but also to less actual waiting times for the vessels and hence to an enhanced service quality, as can be shown by studying the resulting schedules for real vessel data."

"It turns out that the accuracy of the resulting berthing schedules, measured as the deviation of planned and actually realisable schedules, exceeds the accuracy of all forecasts which underlines the usefulness of the DTB approach."

Introduction

"This work is aimed at filling this gap with respect to a specific problem by developing and presenting a new machine-learning-based approach for forecasting vessel arrival times."

"Through the use of real AIS data from the past on which the forecasts are based, but for which the actual vessel arrival times are already known as well, the approach taken in this work allows an evaluation of the forecast quality of the different forecasting methods."

"The approach chosen in this work-the combination of AIS data exploration for forecasting, the subsequent use of an optimization approach which is based on the respective forecasts and the judgement of the quality of the results based on actual vessel arrivals-is, to the authors' knowledge, unique, even if some publications use AIS data for forecasting."

Literature Review

"With respect to the topic of this work, it should be mentioned that De Leon et al. [45] as well as Cheimanoff et al. [46] and many other authors in the field use randomly generated vessel arrival times in their computational studies, i.e., no real data are used and no arrival time prediction is employed, even in the most recent publications."

"It can be concluded from this literature review that the approaches taken so far in the literature are in many cases based on trajectory or lane prediction from which then an arrival time can be deduced, with Parolas [47] and Grida and Lee [48] being notable exceptions who base their predictions on AIS data."

"Building up on their work, the approach suggested in this paper is the application of different ML methods to AIS data in order to directly derive arrival time predictions from them."

Data Pre-processing

"In terms of data selection, it is recommendable to consider only recent entries from the available historic AIS data, as the overall world fleet's characteristics may change over time, e.g., due to a larger proportion of vessels with higher transport capacities."

"All AIS messages broadcasted with a current position inside the quay area and moreover with the status "5", corresponding to a vessel moored at the quay, are considered to be vessels arriving at the designated container terminal, i.e., a "port event.""

"The first AIS message of each trajectory from inside the quay area ending at the designated container terminal is used to determine the arrival time of that vessel at the port."

"Knowing the actual arrival times of all calling vessels from the past data, the actual remaining travel time can be calculated for each AIS message from the vessels' port approach."

Robust Berth Scheduling

"The robust model with DTB for the BAP with continuous quay (ro-DTB-BAPc) which is presented below is based on the "Model 1" by Liu et al. [49] who, as stated above, also aim at a robust berth allocation, taking uncertainty regarding vessel arrival and handling times into account through a scenario-based approach."

"Three aspects of the service quality provided are considered: (1) the vessel waiting time, (2) the deviation from the pre-scheduled berthing position and (3) the non-robust assignments and postponement of vessels, i.e., a high service level."

"The model formulation of the ro-DTB-BAPc can be presented using the notation introduced above:subject to The objective (1) is to minimize all penalty costs resulting from the three aspects of service quality: The potential waiting time, the spatial deviation from the pre-scheduled berthing position and the non-robust assignments or postponement of vessels."

Numerical Experiments

"The number of vessels that are assigned to a robust berthing position by the ro-DTB-BAPc, i.e., that are scheduled according to their DTB, then is used as the desired minimum service level for the benchmark model by Liu et al. [49]."

"The ro-DTB-BAPc reaches an average actual waiting time of about 566 min per berthing schedule considering the robustly scheduled vessels, whereas the "Model 1" leads to an actual waiting time of about 902 min on average according to the vessels' actual arrival times."

"Overall the "Model 1" leads to significantly more conflicts per case than the ro-DTB-BAPc with respect to the actual arrival times, providing an average true service level of 65.49% only considering conflicts between the scheduled vessels."

"The actual waiting time is much higher than for the ro-DTB-BAPc, indicating that the vessels often are assigned too late in planning horizon such that the "Model 1" results in a more conservative berthing schedule."

Conclusion and Outlook

"To the best of the authors' knowledge, the approach presented in this work-to develop an AIS data based forecast using ML methods for vessel arrival time prediction at a designated container terminal, which is subsequently used in an optimization model for the robust berth allocation problem, the ro-DTB-BAPc-is unique and fills part of the research gap on data-driven decision making for container terminals as identified by Heilig et al. [50]."

"It is to be analysed whether the handling times can be forecasted using ML techniques and the influence on the robustness of the resulting berthing schedules is to be studied."

"It can be evaluated if the accuracy of the vessel arrival time prediction for past data is a good predictor of the future accuracy and hence if more specific weights--i.e., the more successful the method, the larger the weight of the respective forecast-lead to even better results."

Reverse Logistics Optimization of an Industrial Air Conditioner Manufacturing Company for Designing Sustainable Supply Chain: A Fuzzy Hybrid Multi-Criteria Decision-Making Approach [51]

This is a machine-generated summary of:

Ali, Sadia Samar; Paksoy, Turan; Torğul, Belkız; Kaur, Rajbir: Reverse logistics optimization of an industrial air conditioner manufacturing company for designing sustainable supply chain: a fuzzy hybrid multi-criteria decision-making approach [51].

Published in: Wireless Networks (2020).

Link to original: https://doi.org/10.1007/s11276-019-02246-6

If you want to cite the papers, please refer to the original.

For technical reasons we could not place the page where the original quote is coming from.

Abstract-Summary "Product recovery involves reuse, repair, refurbishing, remanufacturing and materials recycling, requires an efficient network design known as reverse logistic network and offers economical benefits in terms of fewer procurement of raw material, inventory management and less disposal."

"A mixed integer linear programming model designed on a multi-stage reverse logistics network for product recovery is proposed which considers different recovery options-product remanufacturing, component reprocessing and material recycling for sustainable outcomes."

"The model is designed to find optimal solutions for fulfilling demand and revenue needs by focusing on strategic locations for collection centers, reprocessing centers, remanufacturing plants and transportation options and simultaneously achieving sustainability goals."

"The model is converted into a multi-objective programming model in accordance with the importance of each objective suiting the business needs."

"The model applied has originality and uniqueness for applications to solve multi-objective problems under uncertain environment and tends to strike a balance between economic and environmental objectives."

Introduction

"Crumbling under environmental, social, legal and political pressures companies are bypassing the profit margin angles and are increasingly and intrinsically aligning reverse logistic in their supply chains for reduced energy consumption and lower emissions [52, 53]."

"Reverse logistics is known as the ability of firms to use their product values effectively and efficiently and to reuse them with recycling activities by considering both environmental and social factors."

"The objectives of any business integrating reverse logistics in their supply chains are simple: maximization of asset recovery, supply chain efficiencies and reduced operational and environmental cost."

"With the help of given study, we wish to explore the following research objectives; (1) To understand processes in a reverse supply chains for environmental benefits. (2) To identify suitability of these processes for cost reduction. (3) To formulate model for cost effectiveness in reverse logistics."

Literature

"Researchers have propounded several optimization methods and postulated models to deal with these complexities of reverse logistics process."

"Supply chain network designs are deemed to incorporate multi dimensions, multi objective and multiple processes so mathematical modelling helps to experiment with these multi-combinations for reaching optimal decisions involving new configurations with specific structure of SC network [54, 55]."

"Modelling techniques used by previous researchers has majorily been applied in the forward supply chain scenario, and reverse logistics area is relatively unexplored."

"In a research conducted by Agrawal et al. [56], it has been pointed out that research area is lacking in the field of secondary market, key performers indicators for forecasting product returns, modelling of Reverse logistics network."

"This optimization based mathematical model is proposed to be designed strategically based on data collected from the industry identified to have strong emphasis for the research based on carbon footprint index, product carbon footprint index—overall Supply chain greenhouse gas to reduce carbon emission."

The Case Study

"Pac Ltd. (name changed) was in business of sale and purchase of second-hand AC for almost 20 years having established a strong network of buyers and sellers in four districts covering a wide customers base."

"The company had its main production center in Saudi Arabia and expanded the other units in China and India."

"The company's main production setup was in the northern zone of India, where weather remained hot and humid most of the time, and the demand of AC was definite and continuous."

"Their business model had 4 collection centers, one in each district where the old products remained lodged for 2 weeks."

"They had one reprocessing center and were catering to limited dealers of second- hand AC market for resale of their products."

The Reverse Logistics Network

"Goods collected in each CC transported to the reprocessing centers as soon as possible so that they do not incur any holding costs. (5) Dismantling operations are carried out in the RPC, where the components are disassembled, inspected and sorted for reuse, remanufacture, spare and recycle. (6) Remanufacturing requires used as well new products, which were readily available and RMP accounts only for inventory cost for used products and for newer products, the new supplier operates on Just in Time inventory management principles. (7) The cost of transportation is assumed to be full truck load and is calculated keeping distance and overhead cost in account. (8) The CC, RMP, ROC have fixed monthly cost and new setup for RPC will have fixed cost on rental sharing basis. (9) The cost of new products for remanufacturing purpose is inclusive of the transportation cost. (10) The RMP will store the excess of components till their demand is made. (11) The percentage of disposal of returned products is 13%; remaining returned modules are considered for remanufacturing in a factory, or to spare market."

Methodologies

"The process should be followed for j different objective function for j times."

"Best Worst method (BWM), latest MCDM technique, proposed by Rezaei [57] is applied to determine the weights of objective functions."

"Determine the preference of the best criterion over all the other criteria, using a number between 1 and 9."

"The resulting best-to-others (BO) vector would be: $A_B = (a_{B1}, a_{B2}, \ldots, a_{Bn})$, where a_{Bj} indicates the preference of the best criterion B over criterion j. It is clear that $a_{BB} = 1$."

"Determine the preference of all the criteria over the worst criterion, using a number between 1 and 9."

"The resulting others-to-worst (OW) vector would be: $A_W = (a_{1W}, a_{2W}, \ldots, a_{nW})^T$ where a_{jW} indicates the preference of the criterion j over the worst criterion W. It is clear that $a_{WW} = 1$."

Mathematical Model and Data Set

"Transportation cost (TTC) Transportation in model is carried out from customers to collection centers; from collection centers to the reprocessing centers; from the reprocessing centers to remanufacturing plants, spare markets, recycling center and disposal site."

"Constraint 27 ensures that the total amount of product coming to reprocessing centers is equal to the total amount of part going from reprocessing centers to remanufacturing plants, spare markets, recycling center and disposal site."

"Constraint 28 ensures that the sum of remanufactured product sending from remanufacturing plants to secondary market and end of period stock in remanufacturing plants are equal to the sum of used part sending from reprocessing centers to remanufacturing plants, new part purchasing from new module suppliers to remanufacturing plants and previous period stock in remanufacturing plants for each period."

"Constraint 30 provides that a certain percentage of the parts in reprocessing centers to be sent to remanufacturing plants."

"Constraints 34–40 are capacity constraints of collection centers, reprocessing centers, remanufacturing plants, spare markets, recycling center and disposal site respectively."

Computational Analysis

"It can be said that the model tries to maintain the balance between cost reduction and environmental protection by using different technology levels and transportation options at each stage of the network."

"To analyze the effect of CO_2 emissions on the model CO_2 emissions in the example (released from RPC and RMP plants and vehicles) were increased and decreased by 10%, 20%, 30%, 40%, 50%, and the change in the costs was examined and the information obtained was reported."

"Because as the CO_2 emission amount increases, the model considers the environmental aspect, and plants with technology-level 1 will be used."

"As the CO_2 emission amount increases, the model tends to use more environment-friendly vehicles because it considers the environmental aspects."

"Transportation Cost (TTC), which is fixed up to first 5 ratios, increases as a result of the increased use of environment-friendly vehicles."

Discussion and Managerial Implications

"This is achieved by using three different methods: AHP, FAHP and BWM and weights obtained are further integrated into linear programing model to have three different sets of results."

"According to the solution results, the weights for the objective functions are close to each other by the three methods."

"The inconsistency increases in AHP, because the number of pairwise comparisons is higher in AHP compared to BWM."

"FAHP, on the other hand, satisfaction levels approach the performance of BWM, since allows comparison in a wider range scale in pairwise comparisons."

"Importance weights can be found by any methods and the appropriate solution can be obtained for the problem."

"Fuzzy AHP was previously used by Kumar et al. [58] for weighing various factors to develop a linear programing model so this study became a base for our research."

Uniqueness of the Model and Solution

"The model is original and is proposed first time to have discussions of this approach."

"This model reflects the basic characteristics of the reverse supply chain, and so, the results of our study applied to an industrial air conditioner manufacturer working in practice are expected to have a significant impact on the concept of 'sustainability'."

"The proposed method is a useful decision-making tool for multi-objective problems in an uncertain environment."

"The system can easily be implemented by practitioners and academics, thus making a significant contribution to improving the efficiency of supply chains."

Conclusions

"Efficient and appropriate design of the logistics network leads to meeting economic advantages of using returned products, improving customer satisfaction, increasing market share, reducing costs [59], and adding value to logistics network."

"We focused on the design of a multi-stage reverse logistics network for product recovery."

"Manufacturer recovers each returned product in appropriate ways and provides new parts from outside when needed to meet product demand and optimize total cost."

"Future studies may contain uncertainty about the amount and quality of return products, sources and constraints, too, therefore they can be modeled as stochastic parameters and some constraints such as capacity and demand can be considered fuzzy as well as objective functions."

"It can be strengthened in a closed network model, which is the integration of both forward and reverse logistics and customized for different problems."

Agent-Based, Hybrid Control Architecture for Optimized and Flexible Production Scheduling and Control in Remanufacturing [60]

This is a machine-generated summary of:

Groß, Sebastian; Gerke, Wolfgang; Plapper, Peter: Agent-based, hybrid control architecture for optimized and flexible production scheduling and control in remanufacturing [60].

Published in: Journal of Remanufacturing (2020).

Link to original: https://doi.org/10.1007/s13243-020-00081-z

If you want to cite the papers, please refer to the original.

For technical reasons we could not place the page where the original quote is coming from.

Abstract-Summary "To handle these uncertainties an agent-based hybrid control architecture comprising centralized and decentralized components is presented."

"In the former, the scheduling takes place including the consideration of the use of automated guided vehicles (AGV) to realize flexible material handling within the RS."

"The scheduling of machines and AGVs is thereby considered simultaneously and not separately, as it is the case in currently available control systems."

"For the optimization of the simultaneous scheduling Constraint Programming (CP) is used."

"In the decentralized component, all participants within the RS will be networked as a cyber-physical system and controlled by respective agents."

"Simulation results, using benchmark instances, show that simultaneous scheduling results in a 19.7% reduction of the makespan."

"The CP-based approach delivers the best results, compared to other approaches for simultaneous scheduling, which are also achieved in a significantly shorter computing time."

Introduction

"This paper is structured as follows: in the second part the state-of-the-art regarding the PPC in the domain of remanufacturing, MAS in PPC, fleet management for AGVs and SSMA is presented; in the third part the architecture and methodology of the proposed agent-based, hybrid control architecture as well as the used CP approach regarding the SSMA will be represented; in the fourth part simulation results are presented to show the superiority of the proposed approach and in the last part a conclusion is established and future work will be described."

"Some of these challenges are: Unknown product condition of the used-products due to varying stresses during their usage unknown inspection results of the used-products unknown condition of the disassembled parts varying processing times balance between customer demand and return of used-products stochastic routings for materials and products unknown arrival time and quantity of used-products these characteristics, unique to the domain of remanufacturing, require a different approach for PPC, and then in the field of traditional manufacturing systems [61]."

Methodology

"The following constraints are used to define the required precedence constraints of operations and transport tasks as well as the resource constraints of machines and AGVs:endBeforeStart(s_{pi-1}, s_{pi}): Constraint regarding the sequence of operations for each job p.endBeforeStart(a_{fi-1}, a_{fi}): Constraint regarding the transport sequence for each AGV f.endBeforeStart(a_{fi}, s_{fi}): Transport of the job to the according machine must be completed before processing on the machine can start.endBeforeStart(s_{pi-1}, a_{fi}): Previous operation must be completed before transport of the job to the next machine can start.interval t_{fi}, optional: AGV only needs to execute transport order if no other AGV executes it.alternativ($a_{fi,}t_{fi}$): Each transport order can alternatively be executed by any available AGV, but only one will be selected to carry out the transport order.noOverlap(M_i): A machine can only process one operation at a time. noOverlap(F_j, L, true): Each AGV can only execute one transport order at a time and, if the current position of the AGV and the starting point of the transport order are not identical, it must also perform a corresponding empty trip to the starting point of the transport order."

Experiments

"It can be seen that the optimization potential regarding the makespan reduction through simultaneous scheduling, in comparison to sequential scheduling, is higher for the benchmark instances with a $t/p > 0.25$ than for the instances with a $t/p < 0.25$."

"The results of the approach presented in this paper were achieved as follows: For each of the 82 instances 10 runs were performed and the average value is presented for both makespan and computational time."

"Regarding the instances with $t/p < 0.25$, the presented approach achieves an average makespan of 1.3 s, as well as a median of 0.03 s. For all instances where the computational time was more than 1 s, it can be seen that the approach is able to find solutions close to the optimum within 1 s."

Conclusions

"Because these both scheduling problems depend and influence each other, the simultaneous scheduling approach generates better results regarding the makespan optimization then the sequential approach."

"Simulation studies using benchmark instances show that the simultaneous scheduling approach generates on average results with a 19.7% reduced makespan compared to sequential scheduling."

"The proposed optimization approach for the SSMA using Constraint Programming was compared to other state-of-the-art works in the field of simultaneous scheduling."

"The presented approach will be implemented and used for scheduling as well as the control of the individual resources, the products and the AGV within the model factory."

"This model factory has already been implemented in parts, including the SSMA, as well as the control of different robots and an AGV through the presented approach."

Machine Learning and Optimization for Production Rescheduling in Industry 4.0 [62]

This is a machine-generated summary of:

Li, Yuanyuan; Carabelli, Stefano; Fadda, Edoardo; Manerba, Daniele; Tadei, Roberto; Terzo, Olivier: Machine learning and optimization for production rescheduling in Industry 4.0 [62].

Published in: The International Journal of Advanced Manufacturing Technology (2020).

Link to original: https://doi.org/10.1007/s00170-020-05850-5

If you want to cite the papers, please refer to the original.

For technical reasons we could not place the page where the original quote is coming from.

Abstract-Summary "To find a tradeoff between rescheduling frequency and the growing accumulation of delays, we propose a rescheduling framework, which integrates machine learning (ML) techniques and optimization algorithms."

"To prove the effectiveness, we first model a flexible job-shop scheduling problem with sequence-dependent setup and limited dual resources (FJSP) inspired by an industrial application."

"We train the ML classification model for identifying rescheduling patterns."

"Through observing the simulation results, we find the integration of these techniques can provide a good compromise between rescheduling frequency and scheduling delays."

"The main contributions of the work are the formalization of the FJSP problem, the development of ad hoc solution methods, and the proposal/validation of an innovative ML and optimization-based framework for supporting rescheduling decisions."

Introduction

"In the scheduling sector, new data sources help to trigger improvements in several aspects: shortening wasted time, improving reliability, shortening setup times, reducing waste, handling exceptions in real-time, and controlling fixing times [63]."

"We focus on the real-time exception management and, in particular, on the problem of determining when rescheduling is worthwhile during the ongoing production."

"The reasons for rescheduling come from several factors such as accumulation of delays in production, the unexpected arrival of urgent orders, machine faults, or absence of the operator."

"To enforce a rescheduling, it is necessary to compute a new schedule balancing the possible time savings and efforts to implement the changes."

"Given an optimization technique to create a new schedule, determining the best rescheduling time remains the main problem."

"In the view of improving the rescheduling strategy, our paper proposes a new rescheduling framework by combining metaheuristic optimization algorithms and machine learning (ML) techniques."

Literature Review

"In [64], the authors develop mathematical models for the job-shop scheduling problem with sequence-dependent setup times and solve them through several search methods."

"In [65], the authors describe why the production process in a Belgian rubber company is a job-shop scheduling problem."

"We formalize the scheduling problem in P&R4.0, namely the flexible job-shop scheduling problem with sequence-dependent setup time and limited dual resources (FJSP), where dual resources mean general-purpose machines and setup workers."

"Another similar work has been proposed [66], which considers optimization scheduling and rescheduling under I4.0 and introduces a new decision-making scheme by using Tolerance Scheduling (identifying scenarios where a given schedule remains acceptable) to mitigate the rescheduling changes in the dynamic environment."

"We validate our rescheduling framework through a scheduling problem and numerical experiments."

Methodology

"To enable the automation of rescheduling, meaning that the system knows when to reschedule without or with minimal human intervention, we use ML classification algorithms to learn from the historical data and create a model for future prediction."

"Demonstrate the potential effectiveness, run the subsequent numerical experiments: (a) Implement and test a heuristic approach capable of finding good schedules in a reasonable amount of time; (b) Create data to simulate the information from technologies provided in the I4.0 framework; (c) Train ML model to learn the rescheduling patterns for deciding when to reschedule (i.e., when to trigger the heuristic for getting a new schedule and then to update the production schedule); (d) Compare the performance on the same scenarios followed by the proposed rescheduling framework and the commonly used periodical rescheduling that do not align with ML and real-time data analysis."

Problem Definition and Modeling Phase

"The optimization problem being considered is the flexible job-shop problem with sequence-dependent setup time and limited dual resources (FJSP)."

"Based on conventional JSP, our FJSP introduces: The flexibility in selecting machines as there may be more than one machine capable of the same operations; The limited resources of setup workers and machines; The sequence-dependent setup, which is under the control of both machines and setup workers."

"The key assumptions of the model are: No preemption is allowed for each operation, operations between different jobs are independent; One machine and one worker can only work on one operation at a time; All jobs, machines, and workers are known at the start."

Scheduling Optimization Phase

"GA then uses improved solutions from TS to start a new evolution."

"This means that the i th presence of the job number j is the i th operation of job j. In the decoding phase, the assignment of machines and the calculation of start and end time are performed."

"The availability of both a machine and a setup worker is considered to calculate the start and end time of each operation."

"TS records and encodes the best solution found in a chromosome, and then returns it to the GA population."

"Although it is likely to transform into better solutions by running for more generations, there should be a tradeoff between the running time of TS and that of GA in HA."

Machine Learning–Based Classification Phase

"From the dataset, we extract a set of features for each state, considering information on processing time variations (PTV), planned scheduling, and plant information (including the available resources for each operation and customer orders to be managed)."

"Following the approach of [67, 68], we do not consider automatic feature extraction but we exploit the experience of the involved company to define the following features: t: the time step, at t, the features related to operation o: OPT_o: remaining processing time, PTV_o: processing time variation, ρ_o: ratio of the available machines able to perform o. Since considering all operations can lead to overfitting (the number of operations is higher than the number of scenarios), with the ratio specified as a measure, only the operations with a high ratio, meaning those with more flexibility in changing machines, are considered."

Numerical Experiments

"When comparing under the same running time, there is only one row of positive values (1 out of 20 instances) available in the first three gap columns, demonstrating that, in most cases, all the three heuristics perform significantly better than ES."

"The column Approach indicates the rescheduling mechanism, N represents the rescheduling times, and avgI indicates the average improvement of the makespan (i.e., the average saving of the production time, calculated by averaging the improvements of all the rescheduling occurrences)."

"ML performed best in terms of both average value and standard deviation by rescheduling a few times."

"Considering the proposed approaches—ML and periodical ones—we can see that by recommending to reschedule less frequently and at the right time, ML got satisfactory outcomes not only in saving overall production time but also in the rescheduling effectiveness, which avoided wasting resources in managing machine and worker changes."

Conclusions and Future Research

"We have proposed a new framework for coping with rescheduling under the context of I4.0."

"This work represents the preliminary approach to use ML and optimization together in the rescheduling field by assuming the availability of real-time data analysis."

"The average waiting time of patients, the number of patients, and the number of available nurses can be exciting features to be included within the ML approach."

"The work has shown that the performance of a heuristic is possible to be learned for a machine learning technique."

"This general aspect could be applied in several other contexts and opens several general research lines as, for example, the possibility to use the ML techniques not just to classify the application of a heuristic but to guide and calibrate it on the ongoing setting."

Joint Optimization of Production Scheduling and Group Preventive Maintenance Planning in Multi-Machine Systems [69]

This is a machine-generated summary of:

Mishra, Aseem K.; Shrivastava, Divya; Tarasia, Devesh; Rahim, Abdur: Joint optimization of production scheduling and group preventive maintenance planning in multi-machine systems [69].

Published in: Annals of Operations Research (2021).

Link to original: https://doi.org/10.1007/s10479-021-04362-z

If you want to cite the papers, please refer to the original.

For technical reasons we could not place the page where the original quote is coming from.

Abstract-Summary "This paper addresses the joint optimization model of production scheduling, work-in-process (WIP) inventory control and group preventive maintenance (PM) planning in a multi-machine system with multi-components."

"The objective is to obtain optimum production sequence, PM intervals and grouping of components, which minimize the total expected cost per unit time of the system."

"The optimum group of components is obtained based on the integer multiples of individual PM intervals."

"Computational results reveal that the proposed integrated model along with group PM yields up to 25% cost reductions when compared to the integrated model with individual maintenance as well as 37% savings while no integration is performed."

"The obtained results show that Jaya and SA yielded comparable results, however, PSO is least productive."

"The proposed approach yields better economic performance and brings more improvised solutions as compared to the conventional methods of integrated scheduling and maintenance optimization problems."

Introduction

"Joint optimization of PM and production scheduling in a multi-machine system is more complex than in a single-machine system."

"Imperative, the joint optimization of shop floor performance measures of multi-machine systems brings in computational complexity that poses another challenge in terms of receptiveness in the value chain."

"Although the researchers are constantly aiming to improve operations planning by answering these challenges, the interrelationships and practicality of the problem collectively make the integrated optimization of the multi-machine system to be more complex and interesting."

"To alleviate the limitations of the existing literature and to further improve the economic performance of the shop floor system, the present paper proposes the problem of joint optimization of production scheduling, WIP inventory control and group maintenance planning in multi-machine systems with multi-components."

Literature Review

"Mishra and Shrivastava [70] addressed the permutation flow shop scheduling problem to minimize WIP inventory and tardiness costs considering machine set-up times."

"This section focuses on the integrated approaches of production scheduling and group maintenance planning considering type II (single-machine multi-component) or type III systems (multi-machine single-component)."

"Chen [71] proposed an integrated model of production scheduling and imperfect PM in flow shop system to maximize the total profit by minimizing the production and maintenance costs."

"While considering the group PM approach, Li et al. [72] proposed the integrated approach of opportunistic maintenance and production scheduling in the flow shop system to minimize the total weighted maintenance cost."

"Boufellouh and Belkaid [73] presented an integrated scheduling and maintenance model in a flow shop system with a bi-objective of minimizing the makespan and production costs."

Problem Description

"The periodic imperfect PM is considered based on the restoration factor method."

"For any maintenance activity (PM or CM) the maintenance and downtime costs are associated with each component."

"The total completion time of the job, as well as PM actions, are minimized leading to the reduced total expected cost of the system including scheduling, inventory and maintenance costs."

"According to them, there are eight methods for modeling imperfect PM action."

"In the present study, we apply the restoration factor method to model the time-based imperfect preventive maintenance actions [74]."

Mathematical Models

"The objective of the group PM approach is to identify the grouping of components on which simultaneous PM is performed that minimize the expected total maintenance cost of the system."

"The grouped PM is performed on the jth machine after the ith job is processed."

"The Weibull distribution for time-to-failure, the reliability of a component k in machine j while the ith job is processed can be determined as follows: Since the components in the machine are assumed to be arranged in series, therefore the reliability of the jth machine while the ith job is processed can be defined as: Hence, the probability that the machine j fails while the ith job is processed will be: Therefore, the expression for the completion time of batch i on first machine can be formulated as:where G_k represents the group of components on which PM is performed simultaneously."

Solution Procedure

"In the group maintenance model, while optimum PM intervals can be easily obtained by the classical methods the implications of integer multiples lead to a large number of possible solutions."

"The steps involved in the solution of the proposed problem by Jaya algorithm are given as follows."

"Step 2: Random population vector For each population, an initial random set of PM intervals and integer multiples are generated same as in Jaya algorithm."

"The integrated cost incorporating group PM intervals is calculated and the solution is updated again as per the probability acceptance criterion given in the previous section (Steps 4–8)."

"It can also be noticed that the machine running hours for M1 and M2 are 167 and 158 respectively whereas their optimum group PM intervals (obtained from the group maintenance model) are 182 and 134 h. Thus it is intuitive for the PM locations in the retrieved optimum solution."

Computational Experiments

"It is seen that the optimum PM intervals and minimum maintenance costs of machines 1, 2 and 5 are similar for the mono group and optimum group model."

"The overall maintenance cost per unit time of the system considering mono and optimum group models are 2757 Rs./h and 2481 Rs./h respectively."

"The optimum grouped PM (GPM) locations of components obtained from the group maintenance model are integrated with production sequences, yielding a large number of solutions."

"While comparing with the results of the no integration model, the integrated results yield substantial cost reductions of 37% and 43% with respective optimum and mono groups and 53% with no group policy."

"All, it can be concluded that the proposed integrated model provides the optimum production sequence and grouped PM intervals with minimum expected cost as compared to the isolated models."

Sensitivity Analysis

"Both integrated and isolated models are less sensitive towards the shift in penalty cost, which shows the minimum and a maximum variation of 0.05% and 0.29% respectively."

"Variation in PM, CM and labor cost results in respective maximum variations of 3.5%, 3.1% and 2.6% in the expected total cost of the system which appears to be moderately sensitive."

"This suggests that inaccuracies in the estimation of cost parameters do not bring substantial change in the solution obtained from the base model."

"Variation in restoration factor leads to a major change in the objective function value in the range 10–15% for various integrated and non-integrated models under grouping and no grouping methodologies."

"This is intuitively appearing as, in an integrated model, the effect of scheduling parameters also affects the maintenance cost of the system and vice-a-versa."

Conclusions and Future Extensions

"This paper develops a new approach for the integration of production scheduling and group maintenance planning in a multi-machine system with multi-components."

"The integrated model allows the joint optimization of job sequence, PM intervals, and an optimum group of components, which minimizes the expected total cost of the system."

"The outcome of this research is a dynamic, systematic, and practical as well as manufacturer-oriented approach for joint optimization of scheduling, inventory and maintenance decisions in multi-machine manufacturing systems."

"Due to the complexity of the system, certain assumptions have been made in the model such as the grouping of components is within the machines, the minimal repair is performed for failures, and some constant input parameters, which would be variable in the true sense."

"The other configuration of machines/components such as job shop, parallel shop along with other objectives including machine availability, improving quality can be modeled which would be more practical but rather more complex."

Constructing and Evaluating a Master Surgery Schedule Using a Service-Level Approach [75]

This is a machine-generated summary of:

Deklerck, Loïc; Akbarzadeh, Babak; Maenhout, Broos: Constructing and evaluating a master surgery schedule using a service-level approach [75].

Published in: Operational Research (2021).

Link to original: https://doi.org/10.1007/s12351-021-00677-8

If you want to cite the papers, please refer to the original.

For technical reasons we could not place the page where the original quote is coming from.

Abstract-Summary "We consider the problem to build a cyclic master surgery schedule for a case study of a small-to-medium-sized Belgian hospital."

"The problem considers the objectives of different stakeholders and aims to maximise the adjacency of operating room blocks assigned to an individual surgeon (group) and to optimise the bed capacity usage in the connected downstream units."

"To account for different sources of uncertainty surrounding the patient admissions and the required bed capacity in the downstream units, we formulate and solve a deterministic optimisation model relying on the service level concept to define higher-than-expected values for the patient demand."

"The yielded master surgery schedule establishes suitable bed capacity buffers in the downstream ward units, which improve the robustness to variations in the stochastic variables avoiding bed capacity overruns."

"The proposed method increased both the adjacency between OR blocks and the bed capacity usage significantly as all yielded solutions outperformed the master surgery schedule currently in use in the visited hospital."

Introduction

"Crucial in the organisation of the OR department is the decision-making on the tactical level with the purpose of steering the allocation of resources efficiently and effectively via the so-called master surgery schedule (MSS), which defines a cyclic timetable, managing the operating theatre by organising the OR time in blocks."

"We optimise the MSS to improve the assignment of surgeons to OR blocks while focusing on the effect of the MSS on the bed usage in the different involved hospital wards."

"Via this study, the visited hospital wanted to increase their insight in the linkage between the MSS and the usage of bed resources in related downstream ward units, taking into account the hospital's patient-related data and the uncertainty surrounding the bed capacity usage."

Literature Review

"Different constraints have been considered related to the surgical demand, the preferences of surgical groups, avoiding parallel sessions, downstream bed and nursing resources, etc. The considerations related to the surgical demand refer to demand requested by the different surgical groups and is usually modelled via a lower/upper bound on the number of OR blocks that can be assigned to a surgical group and/or via a priority-based objective such that more capacity is allocated to higher prioritised surgery types (see [76–80])."

"Since many stakeholders (i.e. surgeons, nurses and the hospital's management) are involved [81, 82], multiple, often conflicting, objectives have been proposed related to the master surgery scheduling problem, which can be grouped in different categories."

"Cappanera et al. [83] use simulation and optimisation to compare alternative objective functions related to number of surgeries scheduled, workload balancing among OR resources and robustness involving overruns of OR and bed utilisation."

Problem Description and Formulation

"The MSS is composed by the hospital's planning manager and embodies a cyclic schedule that defines the allocation of OR time blocks in different ORs to a specific surgeon or group of surgeons."

"We make use of the following notation to describe the problem under study: Sets and indices Parameters Stochastic input variables Decision variables Objective function values In this section, we discuss relevant types of constraints involving (1) the internal organisation of the OR department and the assignment of surgeons to blocks and (2) the bed usage in the downstream ward units."

"Calculation of bed usage for a particular ward and day In order to estimate the patient's demand for care in a particular ward w and day i, we need to take the patients' LOS into account, which is defined as the average number of days patients spend in a hospital."

Solution Methodology

"In a second stage, the two objectives related to the usage of bed capacity in the downstream ward units, i.e. the optimisation and levelling of the bed capacity usage, are considered to build a MSS without degrading the objective value for the objective of the first stage."

"Based upon this deterministic patient demand, we formulate a deterministic optimisation model for the problem under study by calculating the associated bed capacity usage, which is a function of the allocations of surgeons to OR blocks."

"We decided to consider this hierarchy in the objectives based on the following considerations, i.e. (1) the large priority given to this objective by the different stakeholders; (2) the simplicity and comprehension of this approach for practitioners, improving their decision-making ability; and (3) the large number of alternative solution points as a result of the objective function structure."

"This has been taken into account in the solution methodology by constructing MSSs following a deterministic optimisation model relying on a service level approach."

Computational Experiments

"According to Penn et al. [84], there are five essential categories of input factors that should be obtained from the hospital under discussion to establish relevant MSSs, i.e. (1) the amount of available theatre time, (2) the case-mix, (3) the number of available hospital beds, (4) the availability of surgeons, and (5) the availability of other resources."

"Other relevant benchmarks, based on the trace-driven simulation experiment, are the obtained values for the bed capacity usage and levelling objectives, which are for the current schedule equal to 4382.72 and 577.84 respectively."

"By defining the postulated objective to maximise the adjacency of blocks, the hospital wants to take into account efficiency considerations when constructing a MSS to allow the scheduling of surgeons to consecutive OR blocks."

"This justifies why the MSS focusing on the levelling objective has the highest level of robustness, i.e. the lowest number of bed capacity overruns."

Conclusion

"The optimisation model is based on the requirements of a private non-for-profit hospital and considers three optimisation criteria, i.e. (1) to optimise the adjacency of OR blocks for a particular surgeon in the same OR as much as possible, (2) to maximise bed capacity usage and (3) to level capacity-related workload in the downstream units."

"The problem is inherently a stochastic optimisation problem due to three different sources of uncertainty, i.e. patient admissions, the patient transfer between hospital wards, and the patient LOS, impacting the bed capacity usage in the downstream units."

"In order to improve the robustness of the yielded solutions regarding the (stochastic) bed usage, we embed adequate bed capacity buffers in the downstream units over the planning horizon."

"These proxy values are determined based on the accounted service level, reflected in the higher-level strategic capacity decisions of the visited hospital."

Optimization of Discrete Cavities with Guide Vanes in A Centrifugal Compressor Based on A Comparative Analysis of Optimization Techniques [85]

This is a machine-generated summary of:

Ma, Sang-Bum; Roh, Min-Su; Kim, Kwang-Yong: Optimization of Discrete Cavities with Guide Vanes in A Centrifugal Compressor based on A Comparative Analysis of Optimization Techniques [85].

Published in: International Journal of Aeronautical and Space Sciences (2021).

Link to original: https://doi.org/10.1007/s42405-020-00341-z

If you want to cite the papers, please refer to the original.

For technical reasons we could not place the page where the original quote is coming from.

Abstract-Summary "Discrete cavities with guide vanes were developed and optimized to improve the operating stability of a centrifugal compressor."

"Various combinations of search algorithms and surrogate models were tested to find the best optimization methods."

"The yaw and pitch angles of the guide vanes and axial distance between cavities were selected as design variables."

"The stall margin was used as an objective function for the design optimization."

"The best combination was found by testing four surrogate models (response surface approximation, Kriging, radial basis neural network, and deep neural network models) and three searching algorithms (a genetic algorithm, particle swarm optimization, and hybrid PSO-GA)."

"The optimum design showed increments of 13.36% and 3.78% in the stall margin compared to compressors with a smooth casing and the reference cavity design, respectively."

Introduction

"This flow recirculation induces a positive pre-swirl, which improves the operating stability of the compressor in the same way as VIGV."

"Surrogate modeling has been widely used to reduce the computing time required for design optimization [86–88]."

"There are several ways to search for the optimal solution on a hyper-surface generated by a surrogate model."

"The optimal solution varies depending on the surrogate model and search algorithm, even for the same problem."

"The combination of surrogate models and search algorithms has not been investigated extensively in any engineering optimization problem so far."

"Guide vanes are proposed for the discrete cavities to enhance the operating stability of a centrifugal compressor further by effectively controlling the pre-swirl just upstream of the impeller."

"The design of the discrete cavities with guide vanes (DCGV) was optimized to maximize the operating stability of the compressor."

Centrifugal Compressor

"The inlet angles of the impeller blades at the hub and tip are 54.12° and 62.96°, and the outlet angles are 28.98° and 32.14°, respectively."

"The reference shapes of discrete cavities and guide vanes were an optimal shape obtained in a previous work [89] and a reference shape from Tamaki [90], respectively."

"Three geometric parameters related to the shape of the DCGV were considered: the yaw (α_Y) and pitch (α_P) angles of the guide vanes and the axial distance between the cavities (P)."

"The reference DCGV has the following parameter values: $\alpha_Y = 65°$, $\alpha_P = 0°$, P/D = 0.093, and L/D = 0.202, where D and L are the diameter of the impeller and the axial length of the guide vanes, respectively."

Numerical Methods

"The frozen-rotor method [91] was applied at the interface between the rotating and stationary domains to generate a steady-state solution."

"Atmospheric conditions were applied at the inlet of the compressor with a total pressure of 101,325 Pa and total temperature of 295.15 K. A mass flow rate was assigned at the outlet of the compressor."

"The solid surfaces in the computational domain were considered to be aerodynamically smooth with adiabatic and no-slip conditions."

"In the impeller domain, O-type grids were constructed around the blades, and H/C/L/J-type grids were constructed in the other regions."

"ANSYS-Turbo Grid® and ICEM® [91] were used to generate the computational grids in the rotating and stationary domains, respectively."

"A grid-dependency test was performed to ensure the reliability of the numerical analysis."

"Under the design conditions, the test was performed for the adiabatic efficiency and total pressure ratio of the compressor."

Optimization Techniques

"A RANS analysis is used to calculate the objective function values at the sample points, which are used to construct the surrogate model."

"The search algorithm is used to find an optimum solution in the fitness function (hyper-surface) generated by the surrogate model."

"The KRG model is applied with an interpolation method using a Gaussian random function and can be expressed as follows [92]:where μ is a constant global model, and $Z(x)$ represents a local deviation from the global model."

"The following termination criteria were used: function tolerance = 10^{-8}, maximum iterations = 2000, number of particles = 1000, and initial velocity of particles = 2.5."

"The following termination criteria were used: function tolerance = 10^{-8}, maximum iterations = 2000, population size (i.e., number of particles) = 1000, crossover fraction = 0.35, mutation = 0.75, and initial velocity of particles = 2.5."

Results and Discussion

"A total of 12 combinations using four surrogate models (RSA, KRG, RBNN, and DNN) and three search algorithms (PSO, GA, and hybrid PSO-GA), were tested for optimization."

"DNN predicts the best objective function (the largest stall margin), followed by RBNN, while the RSA prediction is the worst except in the case with PSO, irrespective of the search algorithm."

"In the cases of the GA and hybrid PSO-GA, the optimization results are similar regardless of the surrogate model."

"Relative errors exceeding 10% are observed for RSA and KRG models, but those of the neural network-based models are less than 2% in all cases of search algorithms, and the DNN model shows the least error in any combination."

"The combinations of hybrid PSO-GA with neural network models showed the best performance."

"The hybrid PSO-GA with DNN model predicted the objective function value most economically and efficiently with the highest accuracy."

Conclusion

"The average computation time for search algorithm was shortest for PSO, followed by hybrid PSO-GA, and GA, and also for RSA, followed by KRG, RBNN, and DNN, regardless of the combination."

"The DNN model predicted the largest stall margin in the optimum design, followed by the RBNN model, regardless of the search algorithm."

"The GA and hybrid PSO-GA predicted similar optimization results that were better than those with PSO, regardless of the surrogate model."

"Those of the neural network-based models were less than 2% for all search algorithms, and the DNN model showed the best prediction accuracy."

"Hybrid PSO-GA with reasonable computation time is expected to help solve huge optimization problems."

"Although the DNN model showed relatively long computation time, it predicted the best objective function value in the optimum design with the least prediction error when coupled with hybrid PSO-GA."

Optimization of Fixture Locating Layout Design Using Comprehensive Optimized Machine Learning [93]

This is a machine-generated summary of:

Qazani, Mohammad Reza Chalak; Parvaz, Hadi; Pedrammehr, Siamak: Optimization of fixture locating layout design using comprehensive optimized machine learning [93].

Published in: The International Journal of Advanced Manufacturing Technology (2022).

Link to original: https://doi.org/10.1007/s00170-022-10061-1

If you want to cite the papers, please refer to the original.

For technical reasons we could not place the page where the original quote is coming from.

Abstract-Summary "The workpiece's deformation, which affects machining precision, is greatly influenced by the positioning of fixture elements around the workpiece."

"Long short-term memory (LSTM), multilayer perception (MLP), and adaptive neuro-fuzzy inference system (ANFIS) are three machine-learning approaches employed to model the connection between locator and clamp positions and maximum workpiece deformation throughout end milling."

"Among developed methods, MLP optimized using BOA (BOA-MLP) reached the highest accuracy among all developed models in predicting the response surface."

"The developed model had a lower computational load than the final element model in calculating the response surface during the machining process."

"At the final step, the prementioned five evolutionary algorithms were implemented in the developed BOA-MLP to extract the optimal parameters of the fixture to decrease the deflection of the workpiece throughout the machining."

"The outcomes showed that the mentioned model was efficient enough compared with the previous method, such as optimized response surface methodology in the point view of 0.0441 μm lower workpiece deflection."

Introduction

"Optimized machining fixture layout refers to properly selecting the position and orientation of locators and clamps around the workpiece to reduce its deformation throughout machining."

"The usage of machine-learning methods and evolutionary algorithms to extract the optimal fixture layout has been focused by lots of researchers in recent years to extract the optimal fixture layout."

"Wu et al. [94] used the hybrid model of FEM and GA to extract the optimal layout of the fixture in end-milling process of the fan blade."

"The proposed optimized machine-learning methods are compared with each other, and the best candidate is selected as a proper representer for calculating the workpiece elastic deformation during the machining process."

"The selected model is employed in evolutionary algorithms to extract the optimal 3–2-1 locator's position of the machining fixture to achieve the minimum workpiece deformation."

Fixture Layout and Finite Element Simulation

"The position of fixture elements has a substantial impact on workpiece deformation throughout machining and clamp actuation."

"It is critical to identify the proper position for the locators and clamps to reduce the workpiece's maximum deformation."

"In the mentioned research, a 2D simulation was performed using finite element analysis software to calculate the machining forces acting on a deformable part."

"At least two clamping forces must also act on the opposite edges of the locators so that the workpiece stays stable during machining."

"The objective function is to minimize the maximum amount of elastic deformation of the workpiece under the clamping and machining forces."

"Locators with positive reaction forces keep the workpiece connected with all the locators throughout machining, while those with opposing reaction forces induce the workpiece to be decoupled from the locators."

Methodology

"Three machine-learning methods, as the most common methods in this field (ANFIS, MLP and LSTM), are chosen to be trained based on the calculated data on FEM software."

"The evolutionary algorithms are used to extract the optimal hyperparameters of the developed machine-learning methods and extract the optimal fixture locating and clamping positions using developed machine-learning methods."

"Three models, including ANFIS, MLP, and LSTM, were utilized in this paper to estimate the elastic workpiece deformation during machining based on the locating and clamping positions of the fixture."

"The hyperparameters of the three investigated machine-learning methods, namely, ANFIS, MLP, and LSTM, are chosen based on five evolutionary algorithms in the last five subsections of this section."

"Another notable distinction between BOA and typical evolutionary algorithms is that BOA computes the value of the fitness function for creating original solutions."

Results and Discussions

"Three models optimized by five evolutionary methods have been investigated in this study to estimate the elastic deformation of the workpiece throughout machining based on the different positions of the locating and clamping points."

"The investigated methods are shown by GA, PSO, GWO, BOA, and WOA for optimized method and ANFIS, MLP, and LSTM for machine-learning methods."

"BOA is the best evolutionary-based optimization method among MLP models in extracting the optimal hyperparameters with the lowest deformation of the workpiece during the machining process."

"WOA is the best evolutionary-based optimization method in calculating the workpiece deformation during the machining process."

"The five explained evolutionary-based optimization methods used this extracted model to calculate the best locating and clamping position of the fixture layout to reach the lowest elastic deformation of the workpiece."

Conclusion

"Researchers used optimization methods, machine learning, and FEM models to extract the optimal fixture layout."

"This study presented a comprehensive evaluation for optimal designing of the machining fixture using three standard machine-learning techniques, including ANFIS, MLP, and LSTM, which were optimized using five evolutionary algorithms including GA, PSO, GWO, BOA, and WOA."

"Among the investigated methods, BOA-MLP reached the most accurate result based on the regenerated dataset in ANSYS software for the workpiece elastic deformation desiring the machining inside the fixture."

"The results showed the 0.18% lower elastic deformation of the workpiece inside the 3–2-1 fixture during the machining using our proposed method compared to the previous GA- and PSO-RSM method."

Surrogate-Based Design Optimisation Tool for Dual-Phase Fluid Driving Jet Pump Apparatus [95]

This is a machine-generated summary of:

Mifsud, D.; Verdin, P. G.: Surrogate-Based Design Optimisation Tool for Dual-Phase Fluid Driving Jet Pump Apparatus [95].

Published in: Archives of Computational Methods in Engineering (2019).

Link to original: https://doi.org/10.1007/s11831-019-09373-5

If you want to cite the papers, please refer to the original.

For technical reasons we could not place the page where the original quote is coming from.

Abstract-Summary "A comparative study of four well established surrogate models used to predict the non-linear entrainment performance of a dual-phase fluid driving jet pump (JP) apparatus is performed."

"The stacked/ensemble models methodology was integrated in this study, to improve the predictive model results, thus providing accurate GA that facilitate the multi-variable non-linear response design optimisation."

"The RSM, Kriging, RBFANN and stacked models formed part of the surrogate-based optimisation, having the entrainment ratio as the main objective function."

"A total of 60 optimisation problems were formulated and solved with all three approximation models."

"Results showed that the hybrid formulation having the level-2 ensemble Kriging model performed best, predicting the experimental performance results for all JP models within an error margin of less than 10% in 90% of the cases."

Introduction

"Considering the potential of using a unique experimental data-set (an unpublished data-set which forms part of an ongoing investigatory work related to the performance of JPs under dual-phase driving fluids), motivated the authors to use this data to build surrogate models to perform surrogate-based optimisation."

"Perform a comparison between level-1 and level-2 (ensemble models) and select the surrogate models showing acceptable accuracy in the non-linear optimisation model (consider multiple optimisation algorithms) to identify the best design parameters which optimise entraintment performance under various motive fluid gas volume fractions (GVFs)."

"This research shows novelty as (1) different surrogate modelling methodologies were applied and compared in the design of a dual-phase (water and air) driving gas jet pump apparatus, and (2) it is shown that the ensemble surrogate modelling approach/stacked-generalisation approach having Kriging as a level-2 learning model, is suitable for predicting the optimised design parameters of a highly non-linear design problem attributed to the complex design of a dual-phase JP apparatus."

Background: Surrogate Models

"The response surface methodology is the simplest and most common method applied for analysing the results of physical experiments and to generate empirically based models for response values [96, 97]."

"Ensemble methods are algorithms which amalgamate several machine learning methodologies into one predictive model, to either decrease the variance, or improve predictions."

"Four of the most common types of ensemble methods are: (a) model averaging, (b) bagging, (c) stacking, and (d) boosting [98, 99]."

"Bagging uses a subset of samples to train each model to train the base learners."

"If the RBFANN method is used, multiple models are generated to include all subsets of sample data."

"Stacking, better known as meta-ensembling, is a model ensemble technique which combines response data from a plurality of predictive models, to generate a new and improved model."

"N trained models are used to predict output of the new sample."

Applied Methodology: JP Device Apparatus

"The sample data for all learning models was obtained from a unique experimental data-set comprising tests performed on a dual-phase (water and air) facility located in the Process Systems Engineering Lab at Cranfield University, UK."

"The first set (training set) included three quarters of the whole sub-set samples which was used to build and train the level-1 models: RSM, Kriging, RBFANN, while the other set (validation set) included the remaining quarter of samples, which were used for model validation and verification of the same level-1 models."

"Study, including a single JP model M, a sub-set comprising an array of 144 samples and responses was used for processing each learning model."

"For each Kriging model, the testing of models included a total of 26 data points; thereby including the samples which were not considered for training purposes during the build-up of the models."

"The input and response samples data sets were randomly selected by the syntax 'dividerand' and divided into three main categories: training, validation and testing."

Results and Discussions for Global Approximations

"As a final comparison of the accuracy of both individual and ensemble learning models, 6 pairs of optimisation problems were first formulated and then solved for each JP model."

"To proceed with the three types of optimisation algorithms, four separate predictive functions (one for each learning model) were created."

"To check the accuracy of the predicted optima and prediction errors, the optimum design values X, GVF and At, for the level-1 Kriging model and the level-2 ensemble-Kriging model were again used as inputs to the predictor function of their respective model."

"It was noted that the three algorithms, including: (a) the 'interior-point' algorithm, (b) the Augmented Lagrangian genetic algorithm (ALGA) and (c) a hybrid formulation, produced similar results for the same optimisation algorithms, but varied for the four types of applied global approximation methods."

"Higher accuracy and consistency were obtained for the optimisation algorithms based on the predicted data from both RBFANN and ensemble models."

Extended Genetic Algorithm for Solving Open-shop Scheduling Problem [100]

This is a machine-generated summary of:

Rahmani Hosseinabadi, Ali Asghar; Vahidi, Javad; Saemi, Behzad; Sangaiah, Arun Kumar; Elhoseny, Mohamed: Extended Genetic Algorithm for solving open-shop scheduling problem [100].

Published in: Soft Computing (2018).

Link to original: https://doi.org/10.1007/s00500-018-3177-y

If you want to cite the papers, please refer to the original.

For technical reasons we could not place the page where the original quote is coming from.

Abstract-Summary "Due to this fact, this problem has attracted many researchers over the past decades and numerous algorithms have been proposed for that."

"This paper investigates the effects of crossover and mutation operator selection in Genetic Algorithms (GA) for solving OSSP."

"Computational results show that selection of genetic operation type has a great influence on the quality of solutions, and the proposed algorithm could generate better solutions compared to other developed algorithms in terms of computational times and objective values."

Introduction

"In open-shop scheduling problem (OSSP), n jobs are processed by m machines."

"If each job is processed in a specific route by m machines, the problem will be a job-shop type problem; and if the routes of job processing are not deterministic, the problem will be classified as open-shop type [101–103]."

"Each job includes n operations that must be processed in a deterministic processing time interval by a machine [104]."

"At a specified time, each machine is able to process only one operation of a job."

"Operations of each job can be processed only after a time called release time for each job."

"Jobs and operations are as inputs, and the minimization of the finish time of all jobs is the output of the proposed algorithm."

Related Work

"Low and Yeh [105] first defined the OSSP problem as a binary programming model and then introduced a Hybrid Genetic Algorithm (HGA) considering various restrictions including independent setup and dependent removal time for solving the problem with the purpose of makespan reduction."

"An efficient HGA called Hybrid Non-dominated Sorting Genetic Algorithm (HNSGA) was developed to solve the multi-objective open-shop problem in Noori-Darvish and Tavakkoli-Moghaddam [106]."

"In Ciro et al. [107], two multi-objective methods named Non-dominated Sorting Genetic Algorithm (NSGA-II) and NSGA-III were proposed and compared."

"The asymptotic optimality of the General Dense Scheduling (GDS) algorithm was investigated for the static and dynamic versions of the Flexible Open-Shop (FOS) makespan problem in Baia et al. [108]."

"Harmanani and Ghosn [109] proposed a Simulated Annealing Algorithm (SA) for solving the non-preemptive open-shop scheduling problem in order to minimize the makespan by determining a schedule for operations on the machines."

Problem Description

"At any moment, each machine can only process one operation related to the job."

"Two or more operations of a job cannot be processed on different machines at the same time."

"It indicates that the total processing time of jobs must at least be equal to their defined required time to process their operations."

"The processing time of each machine must at least be equal to the sum of all times required for operating its assigned jobs."

The Proposed Algorithm

"Four different crossover operations are used in the EGA_OS to study the effects on the type of the selected crossover operation on solving the problem."

"Once the parents are selected, the Homologous Crossover Genetic Operation (C.S) first passed the homologous genes in the two parents on to the child, and then the remaining of the genes are stochastically selected from parent 1 or 2 and passed on to the child."

"The Two-Point Crossover Operation (C.D) first selects two random numbers in the interval from 0 to n (length of the chromosome)."

"After the selection of the parents, the Displacement Mutation Operation (M.A) randomly selects two genes and displaces them."

"The EGA_OS pseudo-code algorithm is a combination of the two crossover operations (the One-point and Two-point crossover operations), with the C.O having a 70% chance, and the C.D having a 30% chance, of being selected for creating children."

Evaluation of the Proposed Algorithm

"The C.O crossover operation generated more desirable chromosomes, and better solutions compared to the other crossover operations were obtained because it merged information that resulted from the two parents better."

"By applying the M.A mutation operation, the generated children are very similar to the parents."

"If the M.S. mutation operation was used, children very different from the parents could be created."

"This would cause greater diversity in chromosomes, but, compared to the M.A mutation operation, the solution would be found much more slowly."

Computational Results for the Designed Test Problems

"In approximation algorithms, population convergence had a direct impact on ending and finding the optimal solution."

"Some approximation algorithms were converged in a lower number of repetitions and this convergence may occur in an inappropriate space of the problem, so that obtained solution was not optimal."

"An algorithm is useful if it could investigate more space of solutions in a shorter time and also convergence occurs in a short time."

"If an algorithm investigates more space of problem in a shorter time, and it gets more time to converge, the chance of achieving optimum solution increases."

Computational Results for the Benchmarks

"This section provides some scenarios containing different jobs and machines for solving the problem considered for this study using the EGA_OS."

"The EGA_OS was tested on Taillard Benchmarks [110]."

"The EGA_OS could obtain optimum solutions in most benchmark instances and compete with compared algorithms."

Conclusion

"The effects of selected crossover and mutation operators on the proposed algorithm (EGA_OS) for solving the Open-Shop Scheduling Problem are investigated and it has been proved that these operators have greatly influenced the efficiency of the Genetic Algorithm."

"Applying the proposed Genetic Algorithm (EGA_OS) for solving the Open-Shop Scheduling Problem was largely dependent on selecting the type of crossover and mutation operators."

"Employment of suitable crossover and mutation operators in the EGA_OS led to a goal-oriented dispersion of the chromosomes in the problem space and to find better solutions."

"Results show that hybrid selection in the Genetic Algorithm worked well for solving the Open-Shop Scheduling Problem and that the use of the One-point Crossover Operator together with the Displacement Mutation Operator resulted in finding solutions better and faster."

A Solution Framework for the Integrated Periodic Supply Vessel Planning and Port Scheduling in Oil and Gas Supply Logistics [111]

This is a machine-generated summary of:

Seimetz Chagas, Rennan Danilo; Soares, Juliana Beatriz Carvalho de Oliveira; Longhi, Rafael Pedro; Vieira, Bruno Ferreira; de Arruda, Edilson Fernandes; Leite, Laura Silvia Bahiense da Silva; Ferreira Filho, Virgílio José Martins: A solution framework for the integrated periodic supply vessel planning and port scheduling in oil and gas supply logistics [111].

Published in: Optimization and Engineering (2022).

Link to original: https://doi.org/10.1007/s11081-022-09723-6

If you want to cite the papers, please refer to the original.

For technical reasons we could not place the page where the original quote is coming from.

Abstract-Summary "We deal with a real-world problem on oil and gas upstream logistics, comprehending the transport of goods from ports to maritime units, through vessels called Platform Supply Vessels (PSVs)."

"An integrated methodology to define the routes of these vessels and port schedules in a three-phase framework."

"In the first phase, we decompose the problem using a clustering heuristic and then solve periodic supply vessel routing problems for each cluster."

"In the third phase, given port departure times, the routes are re-sequenced to respect opening time constraints at installations, aiming to reduce waiting times and to balance the intervals between successive services."

Introduction

"Due to the combinatorial nature of both problems, an integrated model comprising port operations and PSV routing becomes impractical since a comprehensive analysis of the entire supply chain network may demand extensive computational time, making impossible to incorporate the tool on the periodic planning."

"To circumvent this, we developed an innovative methodology through which a solution is found collaboratively within a three-phase framework: PSV routing, port scheduling, and maritime units re-sequencing for time windows adequacy."

"Of the framework, we divide the problem into smaller ones using a clustering heuristic, and then we solve a routing problem, considering capacity, maximum port berth time and maximum service time constraints."

"Knowing each route's sequence of visits, one can find the loading time on port and in second step of the framework we schedule the activities at the onshore base though a berth allocation problem (BAP)."

Problem Definition

"PSV routing consists of defining which maritime units should be visited on each route, as well as the order of visits and the amount of supplies to be loaded on each vessel, with the objective of minimizing distance traveled."

"The available inputs of the problem are expected demands for each maritime unit (for both load and backload cargoes), quantity of crane movements performed on each maritime unit, the capacity of vessels and maximum service, travel and berthing times."

"The planning objective is to minimize the deviation between interval of departures of a route, ensuring visits to the same maritime units with uniform time spacing throughout the week."

"Compared to PSV routing, there are additional constraints related to the maximum and minimum times between services to ensure that visits to maritime units are well spaced."

Literature Review

"Mardaneh et al. [112] dealt with the problem of scheduling cargo supplies to maritime units with time windows at base and clients, with a solution found in a two-step approach: first a heuristic was used to obtain an initial feasible solution, which is them use initialize solver to obtain final optimal schedule."

"In the work of Halvorsen-Weare et al. [113], simulation was combined with a recourse optimization procedure to address uncertainties in sailing times and daily production rates within a routing and scheduling problem with time windows."

"Stanzani et al. [114] studied the problem of transporting crude oil from maritime units, proposing a multi-start heuristic combined with a local search procedure to solve larger instances of the multiship routing and scheduling problem with inventory constraints and pickup-delivery operations."

Framework

"Phase 3 uses the vessel departure times from Phase 2 to reformulate the routes when necessary, to minimize costs while respecting the actual operating time constraints (time windows) of each maritime unit."

"After assigning maritime units to clusters, we solve a capacitated vehicle routing problem (CVRP) with total travel time constraints for each cluster to create routes that respect the capacity constraint of the demand flow, and the total travel and berthing time."

"These times would only be available a posteriori because, to define the port schedule, it is necessary to know the loading times of the vessel, which depends on its occupancy according to the maritime units served on each route."

"In this phase, the model only deals with routes with at least one maritime unit with opening time constraints."

Realistic Case Study

"For clustering, routing, and maritime unit re-sequencing, the computer used in the experiments had the following configuration: Intel(M) Core i7 processor running at 3.0 GHz, with 8 GB of RAM and Windows 8 operating system."

"To define the number of groups that maritime units should be divided (NG) into, we evaluated the routing solution and computational time considering different numbers of clusters, considering scenarios with 1–4 maritime unit clusters."

"That the large gap in the instance A-2 even if the installations are divided into four clusters happens due to the bigger size of the instance with respect to the others, and because of the presence of maritime units with opening time constraints."

Conclusions

"Another goal of this study was to introduce a new type of berth allocation problem."

"Traditional literature has primarily focused on methods to solve dynamic berth allocation problems for commercial port operators."

"In relation to the berth allocation model, there are plans to enhance the objective function, approximate solutions to operational goals, and accelerate optimization."

"These numbers were defined arbitrarily, according to specialists' experience, but studying the operation's stochasticity can help in selecting better operation parameters."

"Although the intermediate solutions of each model are not proven optimal, they are obtained in reasonable times and provide satisfactory results with improvement over the company's practice."

"Due to the large number of operational constraints that need to be considered, one of the hardest challenges was to obtain a feasible solution, something that was rarely achieved manually by the decision makers."

Scheduling of a Class of Partial Routing FMS in Uncertain Environments with Beam Search [115]

This is a machine-generated summary of:

Cherif, G.; Leclercq, E.; Lefebvre, D.: Scheduling of a class of partial routing FMS in uncertain environments with beam search [115].

Published in: Journal of Intelligent Manufacturing (2021).

Link to original: https://doi.org/10.1007/s10845-021-01801-3

If you want to cite the papers, please refer to the original.

For technical reasons we could not place the page where the original quote is coming from.

Abstract-Summary "A particular class of flexible manufacturing systems (FMSs) with partial precedence constraints is proposed where some operations are performed with total precedence constraints and others with full routing flexibility (namely partial routing FMSs)."

"Interruptions may occur due to unavailability of resources and interruption of operations."

"Such interruptions may deviate the trajectory from the planed schedule."

"For the modeling of the partial routing FMS, a systematic multi-level formalism based on the hierarchical structuration of the operations is introduced."

"The new algorithm is based on a new filtering mechanism that uses the cost function to selectively explore the state space of Petri net model in order to find a control sequence from an initial state to a reference one with a trade-off between performance and robustness."

"Examples are used to illustrate the efficiency of the proposed scheduling approach."

Introduction

"Another MILP model is presented by Birgin et al. [116], but their algorithm requires very long times to generates a scheduling solution."

"The advantage of exact methods is to guarantee optimality but the drawback is that the computational time needed in order to obtain the optimal schedule, grows exponentially with respect to the problem size."

"In that sense, adding a time margin to some operations (the ones that require the manipulator) will decrease the risk of interruption and increase the robustness of the schedule."

"Based on the resulting model and on the new cost function, a variant of beam search method that we refer to as Generation Filtered Beam Search (GFBS) is detailed to compute sequences with minimal cost by exploring selectively the PN state space."

State of the Art

"In order to model the job shop scheduling problem, System of Simple Sequential Processes with Resources (S3PR) has been defined as a subclass of ordinary and conservative PN [117–119]."

"The method allows us to formulate scheduling problems in workshops with precedence constraints on some operations and total flexibility on others."

"It is true that exploring the whole state space of the problem [120, 121] leads to the optimal solution, but the problem with these methods is that they are NP-hard which makes the scheduling method applicable to small systems only."

"In order to overcome these limitations, graph search methods have been widely used in the scheduling literature: the A* search algorithm [122–124], which only explores the most promising branches of the tree, presents a well-known technique used to overcome the problem of complexity in using global methods."

Timed Petri Nets

"A T-TPN is a 6-tuple $N = < P, T, W_{PR}, W_{PO}, D, M_0 >$ where $< P, T, W_{PR}, W_{PO}, M_0 >$ is a marked PN, and $D: T \rightarrow R^+$ is a firing time function that assigns a positive real delay D(t) to each transition t of the net."

"Controllable transitions fire according to earliest firing policy with time specifications similar to the one used for T-TPNs [125] where each transition $t \in T_C$ fires as soon as its time constraint D(t) is satisfied."

"The enabling memory policy [126] is used for controllable transitions where after firing a transition t and passing from a marking M to another M', elapsed time for still enabled transitions are kept and decremented and elapsed time associated with disabled transitions are forgotten."

Complex FMS Modeling

"Open (J_1, J_2,…,J_h) is the function used to model, with full routing flexibility, a set of h jobs (or eventually operations) given as inputs."

"The two basic jobs of the first level {J_{111}, J_{211}} are composed of two static operations with full routing flexibility: {o_1, o_2} for J_{111} and {o_3, o_4} for J_{211}."

"The basic job J_{121} of the second level is composed of a sequence of two dynamic operations {o_{11}, o_{21}} representing the two basic jobs of the first level."

"The basic job of the first level J_{112} is composed of two static operations with full routing flexibility {o_5, o_6}."

"The basic job of the second level J_{122} is a sequence of three operations: one dynamic operation o_{112} and two static operations {o_7, o_8}."

"In order to get the whole model of J_2, the dynamic operations o_{112} will be replaced by the basic job J_{112}."

Beam Search for Scheduling Control

"The combined use of dynamic transitions and places embeds the models at levels 1,…, l − 1 into the dynamic transitions and dynamic places of level l: the markings, residual durations and residual risks of each basic job at level 1,…, l − 1 are propagated at level l to change respectively the marking of the dynamic places, the firing time and risk of the dynamic transitions."

"The objective is to explore selectively the PN state space according to successive generations of candidates in order to reach the reference marking M_{ref} starting from an initial marking M_0 with a marking trajectory that minimizes both the duration of the jobs and the risk of failures during the execution of the trajectory."

Computational Experiments and Results

"The simulation shows that the mean duration of the simulations obtained when taking into account the failures risk (GFBS-DR algorithm with a mean total cost equal to 51.82 TU) is improved by 3% compared to that based only on the duration (GFBS-D algorithm with a mean total cost equal to 53.55 TU)."

"The results are also improved when simulating the schedule obtained with GDFBS algorithm (with a mean total cost equal to 49.72 TU) compared to the simulations of the obtained schedule with GFBS-DR algorithm (with a mean total cost equal to 51.82 TU) and the gain is 5%."

"Comparing the GDFBS algorithm with other algorithms that takes into account the duration and the risk when selecting the candidates, the gain in space complexity for GDFBS algorithm, with 2830 explored candidates, is improved by 7% compared to GFBS-DR algorithm (3030 candidates) and by 17% compared to FBS-DR algorithm (3381 candidates)."

Conclusion

"A new scheduling method—Generation Double Filtered Beam Search—was presented in order to improve the scheduling performance and to accelerate the convergence of the method to the reference."

"The method incrementally computes schedules for complex FMSs with respect to uncertainties."

"The proposed cost function, which takes into account not only the duration but also the risk, is used in order to find the next operation with the objective of getting a schedule with a compromise between duration and risk."

"The idea is to avoid such markings in order to improve the robustness of the schedule."

"The robustness will be quantified in order to evaluate a given scheduling in a more general perspective."

A Deep Reinforcement Learning Assisted Simulated Annealing Algorithm for a Maintenance Planning Problem [127]

This is a machine-generated summary of:

Kosanoglu, Fuat; Atmis, Mahir; Turan, Hasan Hüseyin: A deep reinforcement learning assisted simulated annealing algorithm for a maintenance planning problem [127].

Published in: Annals of Operations Research (2022).

Link to original: https://doi.org/10.1007/s10479-022-04612-8

If you want to cite the papers, please refer to the original.

For technical reasons we could not place the page where the original quote is coming from.

Abstract-Summary "Maintenance planning aims to improve the reliability of assets, prevent the occurrence of asset failures, and reduce maintenance costs associated with downtime of assets and maintenance resources (such as spare parts and workforce)."

"To find such optimal planning is a nontrivial task due to the (1) complex and usually nonlinear inter-relationship between different planning decisions (e.g., inventory level and workforce capacity), and (2) stochastic nature of the system (e.g., random failures of parts installed in assets)."

"To alleviate these challenges, we study a joint maintenance planning problem by considering several decisions simultaneously, including workforce planning, workforce training, and spare parts inventory management."

"Further, we also develop a Machine Learning (ML) algorithm (i.e., K-Median) as another benchmark in which different properties of spare parts (e.g., failure rates, holding costs, and repair rates) are used as clustering features for the ML algorithm."

Introduction

"A high number of skilled workforce would lead to faster repairs for failed parts and reduce throughput times in the maintenance facility, which would result in achieving the same availability level for assets with a less number of spare stocks."

"Solving the modeled joint problem and finding the (near) optimal maintenance plans are nontrivial tasks due to the challenges associated with (1) the size of the decision space to search for (i.e., the number of feasible plans including, amount of spare stocks to keep on inventory for each part type, the number of maintenance workers to utilize in the maintenance facility and the number of possible combinations of assignments of skills to workers), and (2) the stochastic nature of the system (e.g., random failures of parts installed in assets and service rate of maintenance)."

Literature Review

"Reinforcement Learning (RL) and Deep Reinforcement Learning (DRL) have gained popularity in solving hard combinatorial optimization problems."

"The maintenance planning (e.g., frequencies of corrective and preventive maintenance and condition-based maintenance policies) of spare parts inventory management models is well-studied in the literature with the various optimization models."

"Some recent studies employ RL to solve maintenance planning problems in various sectors."

"Huang et al. [128] propose a preventive maintenance (PM) model in the Markov Decision Process (MDP) framework and solve the resulting problem employing the DRL algorithm."

"Mahmoodzadeh et al. [129] study an RL-based CBM planning model for corrosion-related maintenance management of dry gas pipelines."

"The literature review in RL-DRL in maintenance optimization reveals the vacancy of research on the DRL-based solution approaches to corrective maintenance planning."

"This study contributes to maintenance planning optimization by developing a hybrid solution algorithm that combines Deep Reinforcement Learning (DRL) and Simulated Annealing (SA) algorithm."

Problem Description and Formulation

"The studied maintenance planning problem consists of (1) a set of assets that includes components/parts subject to fail, (2) a maintenance facility divided into repair cells, (3) repair cells with a different number of cross-trained maintenance workforce that has different skill sets, and (4) a spare parts stock point where multiple types of repairable spare parts; i.e., stock keeping units (SKUs), are kept as inventory."

"To the base-level salary, a cross-trained workforce (that has the ability to repair one or more types of failed parts) can get the compensation payment depending on the number of skills the worker has."

"When the number of repair cells is equal to one, it is said that the maintenance facility is fully flexible/fully cross-trained, and in this type of facility each worker can fix all types of failures."

Solution Approach

"The approach basically contains three subroutines: a Simulated Annealing (SA) meta-heuristic, a Double Deep Q-Network-based Deep Reinforcement Learning (DRL) algorithm, and a simple greedy heuristic that uses a queuing approximation."

"RL exhibits several limitations in practice when deployed in high-dimensional and complex stochastic domains, mainly manifesting algorithmic instabilities with solutions that significantly diverge from optimal regions, or exhibiting slow value updates at infrequently visited states [130]."

"Q-learning aims at learning the optimal action-value functions (also known as the Q-value functions or Q-functions) to derive the optimal maintenance management policy [129]."

"Traditional Q-learning uses a tabular structure(Q-table) to learn how to map action-state pairs to Q-values."

"Because of the complexity of the action-state space in our problem, using a Q-table-based solution wouldn't be feasible."

"This type of architecture is called Double Deep Q-Network (DDQN) and leads to less overestimation of the Q-learning values, as well as improved stability, and hence improved performance [131]."

Benchmark Methods

"The first set of benchmarks includes three well-known meta-heuristic algorithms, namely a Simulated Annealing (SA), a Genetic Algorithm (GA), and a Variable Neighborhood Search (VNS)."

"Further, the details of benchmark GA and VNS algorithms together with their parameter settings can be found in Turan et al. [132]."

"In the second set of benchmarks, we consider a machine learning-based (ML-based) clustering algorithm with four variants (K-Median1, K-Median2, K-Median3, K-Median4)."

Computational Study

"We observe that the optimal number of repair cells never reaches the number of SKUs N (such that dedicated design)."

"We also observe that fully flexible design achieves the lowest total cost in 11 of 128 cases, and dedicated designs never achieve the minimum cost."

"The cost reduction amounts are sensitive to problem factors in comparison to both dedicated and fully flexible design."

"When workforce cost is higher, the cost of dedicated design results in a larger cost due to the extensive number of workers."

"The cost reduction is more sensitive to cross-training costs in comparison to fully flexible design due to an increase in training cost (compensation payment) of workers for each SKU."

"This section investigates how repair cell formation and cross-training depend on problem parameters and solution algorithms."

"The number of SKUs N is another essential problem factor that affects the average cross-training percentage."

Conclusions and Future Research

"We develop a heuristic algorithm inspired by DRL to solve a joint maintenance planning problem by taking into account several decisions simultaneously, including workforce planning, workforce training, and spare parts inventory management."

"This work is an initial attempt to improve solution algorithms using DRL for corrective maintenance problems."

"The computational results show that the proposed solution algorithm is promising in solving the defined problem."

"Machine learning-enforced optimization algorithms may provide relatively better solutions."

"We note that the performance of algorithms is sensitive to problem parameters."

"The model is analyzed under static policies (e.g. routing of failed parts in the maintenance facility) to ensure the computational tractability of the queuing approximations."

A Data-Driven Hierarchical MILP Approach for Scheduling Clinical Pathways: A Real-world Case Study from a German University Hospital [133]

This is a machine-generated summary of:

Schwarz, Karsten; Römer, Michael; Mellouli, Taïeb: A data-driven hierarchical MILP approach for scheduling clinical pathways: a real-world case study from a German university hospital [133].

Published in: Business Research (2019).

Link to original: https://doi.org/10.1007/s40685-019-00102-z

If you want to cite the papers, please refer to the original.

For technical reasons we could not place the page where the original quote is coming from.

Abstract-Summary "To support this endeavor, this paper proposes a flexible hierarchical mixed-integer linear programming (MILP)-based approach for the day-level scheduling of clinical pathways (CP)."

"CP form sequences of ward stays and treatments to be performed during a patient's hospitalization under consideration of all relevant resources such as beds, operating rooms and clinical staff."

"Since in most hospitals CP-related information needed for planning is not readily available, we propose a data-driven approach in which the structure of the CP to be scheduled including all CP-related constraints is automatically extracted from standardized hospital billing data available in every German hospital."

"The approach uses a flexible multi-criteria objective function considering several patient- and hospital-related aspects which makes our approach applicable in various scenarios."

"In contrast to other approaches, it considers several practically relevant aspects ensuring the implementability of the scheduling results such as multiple ward stays per hospitalization and gender-separated room assignments."

"We evaluate our approach in a case study with real-world data from a German university hospital showing that our approach is able to solve instances with a planning horizon of 1 month exhibiting 1088 treatments and 302 ward stays of 286 patients."

Introduction

"Motivated from a case study with a German university hospital, this paper addresses most of the issues discussed above and proposes a tactical pathway-scheduling approach intended to support hospital case managers responsible for the coordination and scheduling of treatments and ward stays."

"We propose a day-level pathway-scheduling approach considering all pathway constraints and hospital resources on an adequate level of detail which is flexible enough to consider multiple scheduling criteria: it considers multiple ward stays during hospitalization and gender-separated room assignments."

"To support scheduling pathways for a broad range of possible objective structures, the objective function in our approach involves a weighted combination of multiple hospital- and patient-related criteria including fairness and workload-balancing criteria."

"Like the other authors dealing with clinical pathway planning (see e.g., [134, 135]), we assume that additional uncertainties such as absence of medical staff, resource breakdowns or stochastic patient arrivals are addressed by embedding our pathway-scheduling approach in a rolling planning process."

Related Work

"The most promising way to avoid bottlenecks, idle time and overtime during hospital patient flow is to schedule not only surgeries, possibly augmented with induced bed resources, but each patient's full CP."

"Vissers [136] proposes a master surgery scheduling approach for CP involving cardiothoracic surgeries based on computing the patient mix with a MILP model."

"Saadani et al. [137] propose a MILP-based CP scheduling approach in which treatments using different resources are scheduled to minimize the patients' length of stay."

"Scheduling CP involves various complex aspects to be considered by a CP scheduling approach to achieve practically useful and implementable schedules: Given the multitude of relevant objectives, the approach should allow to include multiple objective criteria related to both patients and resources: Important patient-related objectives encompass the consideration of LOS (the main driver of schedule-related costs in a DRG system), admission day preferences and hospitalization without delays."

Constraint-Based CP Representation

"While the formal representation of a CP is of secondary importance if it serves merely descriptive purposes, it is crucial when CP form the basis of scheduling patient flow."

"Conforti et al. [138], for example, consider a CP as a set of treatments to be scheduled without imposing inter-treatment dependencies."

"Gartner and Kolisch [134] represent the possible schedules for each CP on a directed graph in which the arcs form minimum time lags between the clinical activities."

"The CP scheduling approach proposed in this work relies on the constraint-based CP representation introduced in Helbig et al. [139]."

"On the second level of the CP representation, a treatment t corresponds to a surgery or any other clinical service to be performed to cure a disease."

A Hierarchical MILP Approach for Scheduling Clinical Pathways

"With respect to the level of detail considered, the pathway-scheduling problem addressed in this paper is situated between the problems considered in Gartner and Kolisch [134] and in Burdett and Kozan [135]: both the problems considered in Gartner and Kolisch [134] and in this paper schedule treatments on the time-aggregated level of days, but in contrast to the problem addressed here, Gartner and Kolisch [134] do not consider multiple ward stays, bed assignment to gender-separated rooms and resource eligibility constraints."

"That all other decisions considered in the following model such as the start and end days of the ward stays, the aggregated gender-separated room assignments and the scheduled dates of the non-complex treatment groups are not transferred to the second-stage model as instructions but merely serve the purpose of anticipation."

"Besides the fact that in the second-stage model, certain decisions are considered as fixed, the main differences between the first-stage and the second-stage model are the level of detail on which the bed resources are modeled and the fact that in the second-stage model, all treatment groups are scheduled under consideration of their resource requirements."

Real-World Case Study from a German University Hospital

"To obtain a realistic set of parameters, we discussed all 259 different OPS-Codes that had occurred during our planning horizon (March 2011) with the case manager of the department for urology who provided us with information regarding the required resources and resource times for each code."

"This approximation is based on the assumption provided by the case manager that all required resources need to be available for the full surgery duration."

"Observation, we assume that 60% of the total available time of all treatment resources owned by the urology department can be scheduled for elective patients."

"By the case manager at our partner hospital department, a smooth resource allocation is favored by the staff members and establishes a similar level of flexibility for reacting to emergency cases on each day of the planning horizon."

Experimental Results

"The objective function gaps between the monolithic and the hierarchical approach can mainly be attributed to the fact that the first stage model determining the cornerstones of the CP schedule is based on an aggregate representation of room assignments and only considers the resource requirements for complex treatments."

"As will become clear in the next section, in contrast to the monolithic model, the hierarchical approach allows obtaining high-quality feasible solutions with a fully flexible admission date within a time limit of 2 h. Compared to other pathway-scheduling approaches, our model operates on a higher level of detail."

"It turns out that even the largest possible instance with completely flexible patient admission days was solved in less than 2 h. We suspect that this considerable decrease in solving time compared to the first scenario results from the fact that in the second scenario, the objective function is less symmetric."

Conclusion

"The CP scheduling approach considers all relevant resources on an adequate level of detail; in particular, it models bed assignment on the level of rooms and allows considering treatment resources individually to avoid disaggregation problems of complex resources such as operating rooms."

"The results presented in this paper offer multiple opportunities for future research: The robustness of the schedules obtained with our approach could be assessed by a simulation study accounting for sources of uncertainty such as such as emergency patients, stochastic treatment times and short-term resource unavailability."

"While we are aware that this goal is presently out of reach mainly due to the present organizational structure in hospitals and due to a lack of integrated information systems, we believe that providing an approach that is data-driven and considers many practically relevant aspects is a step in the right direction."

A Solution Framework for the Integrated Periodic Supply Vessel Planning and Port Scheduling in Oil and Gas Supply Logistics [140]

This is a machine-generated summary of:

Chagas, Rennan Danilo Seimetz; Soares, Juliana Beatriz Carvalho de Oliveira; Longhi, Rafael Pedro; Vieira, Bruno Ferreira; de Arruda, Edilson Fernandes; Leite, Laura Silvia Bahiense da Silva; Filho, Virgílio José Martins Ferreira: A solution framework for the integrated periodic supply vessel planning and port scheduling in oil and gas supply logistics [140].

Published in: Optimization and Engineering (2022).

Link to original: https://doi.org/10.1007/s11081-022-09742-3

If you want to cite the papers, please refer to the original.

For technical reasons we could not place the page where the original quote is coming from.

Abstract-Summary "We deal with a real-world problem on oil & gas upstream logistics, comprehending the transport of goods from ports to maritime units, through vessels called Platform Supply Vessels (PSVs)."

"An integrated methodology to define the routes of these vessels and port schedules in a three-phase framework."

"In the first phase, we decompose the problem using a clustering heuristic and then solve periodic supply vessel routing problems for each cluster."

"In the third phase, given port departure times, the routes are re-sequenced to respect opening time constraints at installations, aiming to reduce waiting times and to balance the intervals between successive services."

Introduction

"Due to the combinatorial nature of both problems, an integrated model comprising port operations and PSV routing becomes impractical since a comprehensive analysis of the entire supply chain network may demand extensive computational time, making impossible to incorporate the tool on the periodic planning."

"To circumvent this, we developed an innovative methodology through which a solution is found collaboratively within a three-phase framework: PSV routing, port scheduling, and maritime units re-sequencing for time windows adequacy."

"Of the framework, we divide the problem into smaller ones using a clustering heuristic, and then we solve a routing problem, considering capacity, maximum port berth time and maximum service time constraints."

"Knowing each route's sequence of visits, one can find the loading time on port and in second step of the framework we schedule the activities at the onshore base though a berth allocation problem (BAP)."

Problem Definition

"PSV routing consists of defining which maritime units should be visited on each route, as well as the order of visits and the amount of supplies to be loaded on each vessel, with the objective of minimizing distance traveled."

"The available inputs of the problem are expected demands for each maritime unit (for both load and backload cargoes), quantity of crane movements performed on each maritime unit, the capacity of vessels and maximum service, travel and berthing times."

"The planning objective is to minimize the deviation between interval of departures of a route, ensuring visits to the same maritime units with uniform time spacing throughout the week."

"Compared to PSV routing, there are additional constraints related to the maximum and minimum times between services to ensure that visits to maritime units are well spaced."

Literature Review

"Mardaneh et al. [112] dealt with the problem of scheduling cargo supplies to maritime units with time windows at base and clients, with a solution found in a two-step approach: first a heuristic was used to obtain an initial feasible solution, which is them use initialize solver to obtain final optimal schedule."

"In the work of Halvorsen-Weare et al. [113], simulation was combined with a recourse optimization procedure to address uncertainties in sailing times and daily production rates within a routing and scheduling problem with time windows."

"Stanzani et al. [114] studied the problem of transporting crude oil from maritime units, proposing a multi-start heuristic combined with a local search procedure to solve larger instances of the multiship routing and scheduling problem with inventory constraints and pickup-delivery operations."

Framework

"Phase 3 uses the vessel departure times from Phase 2 to reformulate the routes when necessary, to minimize costs while respecting the actual operating time constraints (time windows) of each maritime unit."

"After assigning maritime units to clusters, we solve a capacitated vehicle routing problem (CVRP) with total travel time constraints for each cluster to create routes that respect the capacity constraint of the demand flow, and the total travel and berthing time."

"These times would only be available a posteriori because, to define the port schedule, it is necessary to know the loading times of the vessel, which depends on its occupancy according to the maritime units served on each route."

"In this phase, the model only deals with routes with at least one maritime unit with opening time constraints."

Realistic Case Study

"For clustering, routing, and maritime unit re-sequencing, the computer used in the experiments had the following configuration: Intel(M) Core i7 processor running at 3.0 GHz, with 8 GB of RAM and Windows 8 operating system."

"To define the number of groups that maritime units should be divided (NG) into, we evaluated the routing solution and computational time considering different numbers of clusters, considering scenarios with 1–4 maritime unit clusters."

"That the large gap in the instance A-2 even if the installations are divided into four clusters happens due to the bigger size of the instance with respect to the others, and because of the presence of maritime units with opening time constraints."

Conclusions

"Another goal of this study was to introduce a new type of berth allocation problem."

"Traditional literature has primarily focused on methods to solve dynamic berth allocation problems for commercial port operators."

"In relation to the berth allocation model, there are plans to enhance the objective function, approximate solutions to operational goals, and accelerate optimization."

"These numbers were defined arbitrarily, according to specialists' experience, but studying the operation's stochasticity can help in selecting better operation parameters."

"Although the intermediate solutions of each model are not proven optimal, they are obtained in reasonable times and provide satisfactory results with improvement over the company's practice."

"Due to the large number of operational constraints that need to be considered, one of the hardest challenges was to obtain a feasible solution, something that was rarely achieved manually by the decision makers."

A State of the Art Review of Intelligent Scheduling [141]

This is a machine-generated summary of:

Fazel Zarandi, Mohammad Hossein; Sadat Asl, Ali Akbar; Sotudian, Shahabeddin; Castillo, Oscar: A state of the art review of intelligent scheduling [141].

Published in: Artificial Intelligence Review (2018).

Link to original: https://doi.org/10.1007/s10462-018-9667-6

If you want to cite the papers, please refer to the original.

For technical reasons we could not place the page where the original quote is coming from.

Abstract-Summary "Intelligent scheduling covers various tools and techniques for successfully and efficiently solving the scheduling problems."

Introduction

"Owing to the operational diversity and intrinsic difficulty of most scheduling problems, artificial intelligence techniques have a great place in the toolkit of the scheduling researchers."

"Because of these barriers, many researchers in scheduling have turned to artificial intelligence approaches."

"There are several studies in the literature which look at the problems of scheduling using artificial intelligence approaches."

"Rajpathak categorized the various techniques developed for the intelligent scheduling into seven categories which are heuristic scheduling approach, constraint-based scheduling, fuzzy approaches, neural network approach, iterative improvement techniques, distributed scheduling, and meta-scheduling [142]."

"Brown and others provided a survey of intelligent scheduling systems based on artificial and computational intelligence techniques."

"Considering all these categories, in the current study, we investigate five major artificial intelligence techniques for scheduling problems."

Fuzzy Scheduling

"Huang et al. [143] proposed an improved algorithm to search optimal solutions to the flow shop scheduling problems with fuzzy processing times and fuzzy due dates."

"Sakawa and Mori [144] by considering the imprecise or fuzzy nature of the data in real-world problems, formulated job shop scheduling problems with fuzzy processing time and fuzzy due date and proposed a genetic algorithm which is suitable for solving the formulated problems."

"Li et al. [145] proposed a hybrid particle swarm optimization (PSO) algorithm for solving the job shop scheduling problem with fuzzy processing times."

"Wang et al. [146] proposed an effective estimation of distribution algorithm (EDA) to solve the flexible job shop scheduling problem with fuzzy processing time."

"Li and Pan [147] proposed a hybrid algorithm combining particle swarm optimization (PSO) and tabu search (TS) to solve the job shop scheduling problem with fuzzy processing time."

Expert System in Scheduling

"The intelligent solutions, based on expert systems, to solve problems in scheduling are becoming more and more widespread nowadays."

"Metaxiotis et al. [148] proposed the rule-based expert system, GENESYS (generic expert system for scheduling), for the solution of the production scheduling problem in manufacturing systems."

"The goal of this study is to develop a knowledge-based production scheduling software with expert's technical knowledge."

"This paper formulates a scheduling problem by using the result of analysis, and expresses a solution methods based on expert's empirical selection rule and regards that the expert's empirical and the present state is modeled by using actual production information, particularly interview in the printing process [149]."

"Xing et al. [150] presented a knowledge-based ant colony optimization (KBACO) algorithm for the flexible job shop scheduling problem."

"Wang and Wang [151] proposed an effective knowledge-based multi-agent evolutionary algorithm (KMEA) for solving the semiconductor final testing scheduling problem (SFTSP)."

Machine Learning in Scheduling

"Tang et al. [152] proposed a neural network model and algorithm to solve the dynamic hybrid flow shop scheduling problem."

"Azadeh et al. [153] proposed an algorithm based on computer simulation and artificial neural networks to select the optimal dispatching rule for each machine from a set of rules in order to minimize the makespan in stochastic job shop scheduling problems (SJSSPs)."

"In another paper, Azadeh et al. [153] presented an integrated computer simulation and Artificial Neural Network (ANN) algorithm for a stochastic Two-Stage Assembly Flow shop Scheduling Problem (TSAFSP) with setup times under a weighted sum of makespan and mean completion time (MCT) criteria, known as bi-criteria."

"The proposed neural network-based scheduling framework combined with the multi-agent intelligence is a unique approach to efficiently deal with the resource selection problem [154]."

Stochastic Local Search Optimization Algorithms in Scheduling

"Ogbu and Smith [155] presented an approach for solving the n-job, m-machine flow shop scheduling problem with minimum completion time using both heuristic and randomly generated solutions as initial seeds to a simulated annealing optimization process."

"An integer linear programming (ILP) model was presented for the problem to minimize the makespan, which considers intercellular moved and non-consecutive multiple processing of parts on a machine A particle swarm optimization algorithm with and without a proposed local search was developed by Bank et al. [156] to determine a job sequence with minimization of the total tardiness criterion."

"Lin et al. [157] presented a new hybrid swarm intelligence algorithm consists of particle swarm optimization, simulated annealing technique, and multi-type individual enhancement scheme to solve the job shop scheduling problem."

"Gupta et al. [158] proposed a heuristic procedure based on genetic algorithms with the potential to address more generalized objective function such as weighted flow time variance for an n-job, single machine scheduling problem with an objective to minimize the flow time variance."

Scheduling with Constraint Programming

"Weil et al. [159] presented the efficiency of constraint programming for solving nurse scheduling problem."

"Yun and Gen [160] considered a preemptive and non-preemptive scheduling model as one of the advanced scheduling problems considered by a constraint programming technique and proposed a new genetic algorithm (GA) considering simultaneously the preemptive and non-preemptive cases of the activities of jobs under single machine job-shop scheduling problems."

"Sevaux et al. [161] developed a genetic algorithm and a constraint programming based branch-and-bound algorithm to solve multiprocessor task scheduling problem in hybrid flow-shop environments."

"Xujun and Zhimin [162] presented a constraint satisfaction model for the steelmaking-continuous casting scheduling problem solved by constraint programming methods."

"They presented two constraint programming models that rely on a disjunctive scheduling representation of the problem."

"Gökgür et al. [163] presented constraint programming models that aim to solve scheduling and tool assignment problems in parallel machine environments."

Conclusion

"The intelligent scheduling is one of the momentous issues in the domain of scheduling research."

"Most of the best performing methods were described and the properties of the various artificial intelligence techniques used in solving the scheduling problems have been surveyed."

"We have investigated about 540 papers on intelligent scheduling that have appeared since 1980."

"It demonstrates the capabilities of artificial intelligence methods in scheduling problems and clearly needs to better understand the potential of artificial intelligence techniques, like machine learning, for performing scheduling operations."

References

1. Choueiri AC, Santos P, Alves E (2021) Multi-product scheduling through process mining: bridging optimization and machine process intelligence. J Intell Manuf. https://doi.org/10.1007/s10845-021-01767-2
2. Mohammadi M, Ghomi SF, Karimi B, Torabi SA (2010) Rolling-horizon and fix-and-relax heuristics for the multi-product multi-level capacitated lotsizing problem with sequence-dependent setups. J Intell Manuf 21(4):501–510
3. Yan J, Li L, Zhao F, Zhang F, Zhao Q (2016) A multi-level optimization approach for energy-efficient flexible flow shop scheduling. J Clean Prod 137:1543–1552
4. Chen K, Ji P, Wang Q (2011) A case study for advanced planning and scheduling (aps). J Syst Sci Syst Eng 20(4):460–474
5. Puongyeam H, Pongcharoen P, Vitayasak S (2014) Application of krill herd (kh) algorithm for production scheduling in capital goods industries. In: International conference on recent challenges in engineering and technology. IEEE, Piscataway, pp 67–72
6. Chansombat S, Pongcharoen P, Hicks C (2019) A mixed-integer linear programming model for integrated production and preventive maintenance scheduling in the capital goods industry. Int J Prod Res 57(1):61–82
7. Seif J, Dehghanimohammadabadi M, Yu AJ (2019) Integrated preventive maintenance and flow shop scheduling under uncertainty. Flex Serv Manuf J. https://doi.org/10.1007/s10696-019-09357-4
8. Trevino V, Falciani F (2006) GALGO: an R package for multivariate variable selection using genetic algorithms. Bioinformatics 22(9):1154–1156. https://doi.org/10.1093/bioinformatics/btl074
9. Yu AJ, Seif J (2016) Minimizing tardiness and maintenance costs in flow shop scheduling by a lower-bound-based GA. Comput Ind Eng 97:26–40
10. Greinacher S, Overbeck L, Kuhnle A, Krahe C, Lanza G (2020) Multi-objective optimization of lean and resource efficient manufacturing systems. Prod Eng. https://doi.org/10.1007/s11740-019-00945-9
11. Dües CM, Tan KH, Lim M (2013) Green as the new lean: how to use lean practices as a catalyst to greening your supply chain. J Clean Prod 40:93–100. https://doi.org/10.1016/j.jclepro.2011.12.023
12. Florida R (1996) Lean and green: the move to environmentally conscious manufacturing. Calif Manag Rev 39(1):80–105. https://doi.org/10.2307/41165877

13. Bergmiller GG, McCright PR (2009) Are lean and green programs synergistic. In: Proceedings of the 2009 industrial engineering research conference. ACM, New York
14. Larson T, Greenwood R (2004) Perfect complements: synergies between lean production and eco-sustainability initiatives. Environ Qual Manag 13(4):27–36. https://doi.org/10.1002/tqem.20013
15. Bey N, Hauschild MZ, McAloone TC (2013) Drivers and barriers for implementation of environmental strategies in manufacturing companies. CIRP Ann 62(1):43–46. https://doi.org/10.1016/j.cirp.2013.03.001
16. Neugebauer R, Westkämper E, Klocke F, Spath D, Schenk M, Michaelis A, ten Hompel M, Weidner E (2008) Untersuchung zur Energieeffizienz in der Produktion. FhG, München. Retrieved from https://www.tib.eu/de/suchen/id/TIBKAT%3A576816582/Untersuchung-zur-Energieeffizienz-in-der-Produktion/
17. Diaz-Elsayed N, Jondral A, Greinacher S, Dornfeld D, Lanza G (2013) Assessment of lean and green strategies by simulation of manufacturing systems in discrete production environments. CIRP Ann Manuf Technol 62(1):475–478. https://doi.org/10.1016/j.cirp.2013.03.066
18. Evangelista Araujo Neto J, Castro CA (2021) Optimal maintenance scheduling of transmission assets in the Brazilian Electric System. J Control Autom Electr Syst. https://doi.org/10.1007/s40313-020-00678-6
19. Marwali MKC, Shahidehpour SM (2000) Short-term transmission line maintenance scheduling in a deregulated system. In: Proceedings of the 21st international conference on power industry computer applications. IEEE, Santa Clara. https://doi.org/10.1109/PICA.1999.779382
20. Araujo Neto JE, Moretti AC, Castro CA (2015) Transmission asset maintenance programming optimization—the Brazilian electric system case. IEEE Lat Am Trans 13:1414–1420
21. Chica M, Bautista J, de Armas J (2018) Benefits of robust multiobjective optimization for flexible automotive assembly line balancing. Flex Serv Manuf J. https://doi.org/10.1007/s10696-018-9309-y
22. Boysen N, Fliedner M, Scholl A (2010) Level scheduling under limited resequencing flexibility. Flex Serv Manuf J 22(3–4):236–257
23. Chica M, Cordón O, Damas S, Bautista J (2013) A robustness information and visualization model for time and space assembly line balancing under uncertain demand. Int J Prod Econ 145:761–772
24. Papakostas N, Pintzos G, Giannoulis C, Nikolakis N, Chryssolouris G (2014) Multi-criteria assembly line design under demand uncertainty. Proc CIRP 25:86–92
25. Hosseini A, Sahlin T (2018) An optimization model for management of empty containers in distribution network of a logistics company under uncertainty. J Ind Eng Int. https://doi.org/10.1007/s40092-018-0286-2
26. Rashidi H, Tsang E (2015) Vehicle scheduling in port automation: advanced algorithms for minimum cost flow problems. CRC Press, Boca Raton
27. Bhattacharya A, Kumar S, Tiwari M, Talluri S (2014) An intermodal freight transport system for optimal supply chain logistics. Transp Res Part C Emerg Technol 38:73–84
28. Ahmad Hosseini S, Şahin G, Ünlüyurt T (2017) A penalty-based scaling algorithm for the multi-period multi-product distribution planning problem. Eng Optim 49(4):583–596
29. Gen M, Syarif A (2005) Hybrid genetic algorithm for multi-time period production/distribution planning. Comput Ind Eng 48(4):799–809
30. Hosseini S (2013) A model-based approach and analysis for multi-period networks. J Optim Theory Appl 157(2):486–512
31. Ji Y, Yang H, Zhu Q, Xing Y (2016) Optimization model of empty container reposition of liner alliance based on mutually renting strategy in short sea service. In: Proceedings of the 6th international Asia conference on industrial engineering and management innovation. Springer, Cham, pp 1055–1064
32. Olivo A, Zuddas P, Di Francesco M, Manca A (2005) An operational model for empty container management. Merit Logist 7(3):199–222

33. Lam SW, Lee LH, Tang LC (2007) An approximate dynamic programming approach for the empty container allocation problem. Transp Res Part C Emerg Technol 15(4):265–277
34. Shintani K, Konings R, Imai A (2010) The impact of foldable containers on container fleet management costs in hinterland transport. Transp Res Part E Logist Transp Rev 46(5):750–763
35. Long Y, Lee LH, Chew EP (2012) The sample average approximation method for empty container repositioning with uncertainties. Eur J Oper Res 222(1):65–75
36. Darghouth MN, Abdel-Aal M (2021) A capacitated disassembly scheduling problem considering processing technology selection and parts commonality. J Remanuf. https://doi.org/10.1007/s13243-021-00104-3
37. Kim H-J, Lee D-H, Xirouchakis P, Züst R (2003) Disassembly scheduling with multiple product types. CIRP Ann 52(1):403–406
38. Kim H-J, Xirouchakis P (2010) Capacitated disassembly scheduling with random demand. Int J Prod Res 48(23):7177–7194
39. Tian X, Zhang Z-H (2019) Capacitated disassembly scheduling and pricing of returned products with price-dependent yield. Omega 84:160–174
40. Tan M, Duan B, Su Y, He F (2015) Optimal hot rolling production scheduling for economic load dispatch under time-of-use electricity pricing. arXiv: 1501.05502
41. Ji X, Zhang Z, Huang S, Li L (2016) Capacitated disassembly scheduling with parts commonality and start-up cost and its industrial application. Int J Prod Res 54(4):1225–1243
42. Liang J, Wang Y, Zhang Z-H, Sun Y (2019) Energy efficient production planning and scheduling problem with processing technology selection. Comput Ind Eng 132:260–270
43. Slama I, Ben-Ammar O, Dolgui A, Masmoudi F (2020) New mixed integer approach to solve a multi-level capacitated disassembly lot-sizing problem with defective items and backlogging. J Manuf Syst 56:50–57
44. Kolley L, Rückert N, Kastner M, Jahn C, Fischer K (2022) Robust Berth scheduling using machine learning for vessel arrival time prediction. Flex Serv Manuf J. https://doi.org/10.1007/s10696-022-09462-x
45. de León AD, Lalla-Ruiz E, Melián-Batista B, Moreno-Vega JM (2017) A machine learning-based system for berth scheduling at bulk terminals. Expert Syst Appl 87:170–182. https://doi.org/10.1016/j.eswa.2017.06.010
46. Cheimanoff N, Fontane F, Kitri MN, Tchernev N (2021) A reduced VNS based approach for the dynamic continuous berth allocation problem in bulk terminals with tidal constraints. Expert Syst Appl 168:114215. https://doi.org/10.1016/j.eswa.2020.114215
47. Parolas I (2016) ETA prediction for containerships at the port of Rotterdam using machine learning techniques. TU Delft, Delft
48. Grida M, Lee C-Y (2018) An empirical model for estimating berth and sailing times of mega container ships. Marit Policy Manag 45:1078–1093. https://doi.org/10.1080/03088839.2018.1452304v
49. Liu C, Xiang X, Zheng L (2017) Two decision models for berth allocation problem under uncertainty considering service level. Flex Serv Manuf J 29:312–344. https://doi.org/10.1007/s10696-017-9295-5
50. Heilig L, Stahlbock R, Voß S (2020) From digitalization to data-driven decision making in container terminals. In: Böse JW (ed) Handbook of terminal planning, 2nd edn. Springer, Cham, pp 125–154. https://doi.org/10.1007/978-3-030-39990-0_6
51. Ali SS, Paksoy T, Torğul B, Kaur R (2020) Reverse logistics optimization of an industrial air conditioner manufacturing company for designing sustainable supply chain: a fuzzy hybrid multi-criteria decision-making approach. Wirel Netw. https://doi.org/10.1007/s11276-019-02246-6
52. Tautenhain CP, Barbosa-Povoa AP, Nascimento MC (2018) A multi-objective matheuristic for designing and planning sustainable supply chains. Comput Ind Eng 135:1203–1223. https://doi.org/10.1016/j.cie.2018.12.062
53. Zhen L, Huang L, Wang W (2019) Green and sustainable closed-loop supply chain network design under uncertainty. J Clean Prod 227:1195–1209

54. Wang F, Lai X, Shi N (2011) A multi-objective optimization for green supply chain network design. Decis Support Syst 51:262–269. https://doi.org/10.1016/j.dss.2010.11.020
55. Jamshidi R, Fatemi Ghomi S, Karimi B (2012) Multi-objective green supply chain optimization with a new hybrid memetic algorithm using the Taguchi method. Sci Iranica 19(6):1876–1886
56. Agrawal S, Singh RK, Murtaza Q (2015) A literature review and perspectives in reverse logistics. Resour Conserv Recycl 97:76–92
57. Rezaei J (2015) Best-worst multi-criteria decision-making method. Omega 53:49–57
58. Kumar D, Rahman Z, Chan F (2017) A fuzzy AHP and fuzzy multi-objective linear programming model for order allocation in a sustainable supply chain: a case study. Int J Comput Integr Manuf 30:535–551. https://doi.org/10.1080/0951192X.2016.1145813
59. Kilic E, Ali SS, Weber GW, Dubey R (2014) A value-adding approach to reliability under preventive maintenance costs and its applications. Optimization 63(12):1805–1816
60. Groß S, Gerke W, Plapper P (2020) Agent-based, hybrid control architecture for optimized and flexible production scheduling and control in remanufacturing. J Remanuf. https://doi.org/10.1007/s13243-020-00081-z
61. Cui Y, Guan Z, He C, Yue L (2017) Research on remanufacturing scheduling problem based on critical chain management. IOP Conf Ser Mater Sci Eng 215:215. https://doi.org/10.1088/1757-899X/215/1/012005
62. Li Y, Carabelli S, Fadda E, Manerba D, Tadei R, Terzo O (2020) Machine learning and optimization for production rescheduling in industry 4.0. Int J Adv Manuf Technol. https://doi.org/10.1007/s00170-020-05850-5
63. McKinsey. (2015). *Industry 4.0 how to navigate digitization of the manufacturing sector*. Retrieved from https://www.mckinsey.com/business-functions/operations/our-insights/industry-four-point-o-how-to-navigae-the-digitization-of-the-manufacturing-sector
64. Tamaki H, Hasegawa Y, Kozasa J, Araki M (1993) Application of search methods to scheduling problem in plastics forming plant: a binary representation approach. In: Proceedings of 32nd IEEE conference on decision and control. IEEE, Piscataway, pp 3845–3850
65. Sels V, Steen F, Vanhoucke M (2011) Applying a hybrid job shop procedure to a Belgian manufacturing company producing industrial wheels and castors in rubber. Comput Ind Eng 61(3):697–708
66. Rossit DA, Tohme F, Frutos M (2018) Industry 4.0: smart scheduling. Int J Prod Res 57(12):3802–3813. https://doi.org/10.1080/00207543.2018.1504248
67. Castrogiovanni P, Fadda E, Perboli G, Rizzo A (2020) Smartphone data classification technique for detecting the usage of public or private transportation modes. IEEE Access 8:58377–58391. https://doi.org/10.1109/ACCESS.2020.2982218
68. Fadda E, Mana D, Perboli G, Vallesio V (2018) Sustainable mobility and user preferences by crowdsourcing data: the open agora project. In: 2018 IEEE 14th international conference on automation science and engineering (CASE). IEEE, Piscataway. https://doi.org/10.1109/COASE.2018.8560512
69. Mishra AK, Shrivastava D, Tarasia D, Rahim A (2021) Joint optimization of production scheduling and group preventive maintenance planning in multi-machine systems. Ann Oper Res. https://doi.org/10.1007/s10479-021-04362-z
70. Mishra A, Shrivastava D (2018) A TLBO and a Jaya heuristics for permutation flow shop scheduling to minimize the sum of inventory holding and batch delay costs. Comput Ind Eng 124(July):509–522. https://doi.org/10.1016/J.CIE.2018.07.049
71. Chen X (2015) An integrated model of production scheduling and maintenance planning under imperfect preventive maintenance. Maintence Reliability 17(1):70–79
72. Li M, Li H, Liu Q (2010) Integrated production scheduling and opportunistic preventive maintenance in the flowshop manufacturing system. In: 2nd international conference on information science and engineering. ICISE, Qui Nhơn, pp 294–298. https://doi.org/10.1109/ICISE.2010.5690889

73. Boufellouh R, Belkaid F (2020) Bi-objective optimization algorithms for joint production and maintenance scheduling under a global resource constraint: application to the permutation flow shop problem. Comput Oper Res 122:104943. https://doi.org/10.1016/j.cor.2020.104943
74. Lim J-H, Park DH (2007) Optimal periodic preventive maintenance schedules with improvement factors depending on number of preventive maintenances. Asia-Pac J Oper Res 24(1):111–124. https://doi.org/10.1142/S0217595907001139
75. Deklerck L, Akbarzadeh B, Maenhout B (2021) Constructing and evaluating a master surgery schedule using a service-level approach. Oper Res. https://doi.org/10.1007/s12351-021-00677-8
76. Santibáñez P, Begen M, Atkins D (2007) Surgical block scheduling in a system of hospitals: an application to resource and wait list management in a British Columbia health authority. Health Care Manag Sci 10(3):269–282. https://doi.org/10.1007/s10729-007-9019-6
77. Kharraja S, Albert P, Chaabane S (2007) Block scheduling: toward a master surgical schedule. Int Conf Serv Syst Serv Manag 1:429–435
78. Samanlioglu F, Ayag Z, Batili B, Evcimen E, Yilmaz G, Atalay O (2010) Determining master schedule of surgical operations by integer programming: a case study. Inst Ind Eng 2010:2010
79. Agnetis A, Coppi A, Corsini M, Dellino G, Meloni C, Pranzo M (2014) A decomposition approach for the combined master surgical schedule and surgical case assignment problems. Health Care Manag Sci 17(1):49–59
80. Fügener A (2015) An integrated strategic and tactical master surgery scheduling approach with stochastic resource demand. J Bus Logist 36(4):374–387. https://doi.org/10.1111/jbl.12105
81. Cardoen B, Demeulemeester E, Beliën J (2010) Operating room planning and scheduling: a literature review. Eur J Oper Res 201(3):921–932
82. Zhu S, Fan W, Yang S, Pei J, Pardalos P (2018) Operating room planning and surgical case scheduling: a review of literature. J Comb Optim 37:757–805
83. Cappanera P, Visintin F, Banditori C (2014) Comparing resource balancing criteria in master surgical scheduling: a combined optimisation-simulation approach. Int J Prod Econ 158:179–196
84. Penn ML, Potts CN, Harper PR (2017) Multiple criteria mixed-integer programming for incorporating multiple factors into the development of master operating theatre timetables. Eur J Oper Res 262(1):194–206. https://doi.org/10.1016/j.ejor.2017.03.065
85. Ma S-B, Roh M-S, Kim K-Y (2021) Optimization of discrete cavities with guide vanes in a centrifugal compressor based on a comparative analysis of optimization techniques. Int J Aeronaut Space Sci. https://doi.org/10.1007/s42405-020-00341-z
86. Kim JH, Ovgor B, Cha KH, Kim JH, Kim KY (2014) Optimization of the aerodynamic and aeroacoustic performance of an axial-flow fan. AIAA J 52:2032–2044. https://doi.org/10.2514/1.j052754
87. Kim KY, Samad A, Benini E (2019) Design optimization of fluid machinery. Wiley, Hoboken
88. Robinson TD, Eldred MS, Willcox KE, Haimes R (2008) Surrogate-based optimization using multifidelity models with variable parameterization and corrected space mapping. AIAA J 46:2814–2822. https://doi.org/10.2514/1.36043
89. Ma SB, Kim KY (2017) Optimization of discrete cavities in a centrifugal compressor to enhance operating stability. Aerosp Sci Technol 68:308–319. https://doi.org/10.1016/j.ast.2017.05.029
90. Hunziker R, Dickmann HP, Emmrich R (2001) Numerical and experimental investigation of a centrifugal compressor with an inducer casing bleed system. Proc Inst Mech Eng Part A 215:783–791. https://doi.org/10.1243/0957650011538910
91. ANSYS CFX-15.0 (2013) ANSYS CFX-solver theory guide. ANSYS, Canonsburg
92. Kleijnen JPC (2009) Kriging metamodeling in simulation: a review. Eur J Oper Res 192:707–716. https://doi.org/10.1016/j.ejor.2007.10.013

93. Chalak QMR, Parvaz H, Pedrammehr S (2022) Optimization of fixture locating layout design using comprehensive optimized machine learning. Int J Adv Manuf Technol. https://doi.org/10.1007/s00170-022-10061-1
94. Wu B, Zheng Z, Wang J, Zhang Z, Zhang Y (2021) Layout optimization of auxiliary support for deflection errors suppression in end milling of flexible blade. Int J Adv Manuf Technol 115(5):1889–1905
95. Mifsud D, Verdin PG (2019) Surrogate-based design optimisation tool for dual-phase fluid driving jet pump apparatus. Arch Comput Methods Eng. https://doi.org/10.1007/s11831-019-09373-5
96. Box GE, Draper NR (1987) Empirical model-building and response surfaces. Wiley, New York
97. Morris MD, Mitchell TJ (1995) Exploratory designs for computational experiments. J Stat Plan Inf 43:381–402
98. Opitz D, Maclin R (1999) Popular ensemble methods: an empirical study. J Artif Intell Res 11:169–198
99. Caruana R, Niculescu-Mizil A, Crew G, Ksikes A (2004) Ensemble selection from libraries of models. In: Proceedings of the twenty-first international conference on machine learning. ACM, New York, p 18
100. Hosseinabadi R, Asghar A, Vahidi J, Saemi B, Sangaiah AK, Elhoseny M (2018) Extended genetic algorithm for solving open-shop scheduling problem. Soft Comput. https://doi.org/10.1007/s00500-018-3177-y
101. Shamshirband S, Shojafar M, Hosseinabadi AR, Kardgar M, Nizam MH, Nasir M, Ahmad R (2015a) OSGA: genetic-based open-shop scheduling with consideration of machine maintenance in small and medium enterprises. Ann Oper Res 229(1):743–758
102. Hosseinabadi AR, Siar H, Shamshirband S, Shojafar M, Nizam MH, Nasir M (2015) Using the gravitational emulation local search algorithm to solve the multi-objective flexible dynamic job shop scheduling problem in Small and Medium Enterprises. Ann Oper Res 229(1):451–474
103. Shojafar M, Kardgar M, Hosseinabadi AR, Shamshirband S, Abraham A (2016) TETS: a genetic-based scheduler in cloud computing to decrease energy and makespan. In: The 15th international conference on hybrid intelligent systems (HIS 2015), chapter advances in intelligent systems and computing, vol 420. Springer, Seoul, pp 103–115
104. Hosseinabadi AR, Farahabadi AB, Rostami MS, Lateran AF (2013) Presentation of a new and beneficial method through problem solving timing of open-shop by random algorithm gravitational emulation local search. Int J Comput Sci Issues 10(1):745–752
105. Low C, Yeh Y (2009) Genetic algorithm-based heuristics for an open-shop scheduling problem with setup, processing, and removal times separated. Robot Comput Integr Manuf 25:314–322
106. Noori-Darvish S, Tavakkoli-Moghaddam R (2011) Solving a bi-objective open-shop scheduling problem with fuzzy parameters. J Appl Oper Res 3:59–74
107. Ciro GC, Dugardin F, Yalaoui F, Kelly R (2016) A NSGA-II and NSGA-III comparison for solving an open-shop scheduling problem with resource constraints. In: 8th IFAC conference on manufacturing modelling, management and control MIM 2016, vol 49. IFAC, Troyes, pp 1272–1277
108. Baia D, Zhang Z, Zhang Q (2016) Flexible open-shop scheduling problem to minimize makespan. Comput Oper Res 67:207–215
109. Harmanani HM, Ghosn SB (2016) An efficient method for the open-shop scheduling problem using simulated annealing. Adv Intell Syst 448:1183–1193
110. Taillard E (1993) Benchmarks for basic scheduling problems. Eur J Oper Res 64:278–285
111. Chagas S, Danilo R, de Oliveira SJBC, Longhi RP, Vieira BF, de Arruda EF, da Silva LLSB, Filho F, Martins VJ (2022) A solution framework for the integrated periodic supply vessel planning and port scheduling in oil and gas supply logistics. Optim Eng. https://doi.org/10.1007/s11081-022-09723-6

112. Mardaneh E, Lin Q, Loxton R, Wilson N (2017) Cargo scheduling decision support for offshore oil and gas production: a case study. Optim Eng 18:991–1008
113. Halvorsen-Weare E, Fagerholt K, Rönnqvist M (2013) Vessel routing and scheduling under uncertainty in the liquefied natural gas business. Comput Ind Eng 64(1):290–301
114. Stanzani AL, Pureza V, Morabito R, Silva BJV, Yamashita D, Ribas PC (2018) Optimizing multiship routing and scheduling with constraints on inventory levels in a Brazilian oil company. Int Trans Oper Res 25:1163–1198
115. Cherif G, Leclercq E, Lefebvre D (2021) Scheduling of a class of partial routing FMS in uncertain environments with beam search. J Intell Manuf. https://doi.org/10.1007/s10845-021-01801-3
116. Birgin EG, Feofiloff P, Fernandes CG, De Melo EL, Oshiro MT, Ronconi DP (2014) A MILP model for an extended version of the flexible job shop problem. Optim Lett 8(4):1417–1431
117. Ezpeleta J, Colom JM, Martinez J (1995) A Petri net based deadlock prevention policy for flexible manufacturing systems. IEEE Trans Robot Autom 11:173–184
118. Xing KY, Xing KL, Li JM (1996) Deadlock avoidance controller for a class of manufacturing systems. In: Proceedings, IEEE international conference on robotics and automation, vol 1. IEEE, Piscataway, pp 200–204
119. Abdul-Hussin M (2015) Design of a Petri net based deadlock prevention policy supervisor for S3PR. In: 6th international conference on intelligent systems, modeling and simulation, Kuala Lumpur, Malaysia. IEEE, Piscataway, pp 46–52
120. Lefebvre D (2018) Approximated timed reachability graphs for performance evaluation and control of DES. IFAC PapersOnLine 51(7):224–229
121. Lefebvre D, Daoui C (2018) Control design for bounded partially controlled TPNs using timed extended reachability graphs and MDP. IEEE Trans Syst Man Cybern Syst 50(6):1–11
122. Hart PE, Nilsson NJ, Raphael B (1968) A formal basis for the heuristic determination of minimum cost paths. IEEE Trans Syst Sci Cybernet 4(2):100–107
123. Lee DY, DiCesare F (1994) Scheduling flexible manufacturing systems using Petri nets and heuristic search. IEEE Trans Robot Autom 10(2):123–132
124. Ow PS, Morton TE (1988) Filtered beam search in scheduling. Int J Prod Res 26(1):35–62
125. Ramchandani C (1974) Analysis of asynchronous concurrent systems by timed Petri nets. Massachusetts Institute of Technology, Cambridge
126. Molloy MK (1982) Performance analysis using stochastic Petri nets. IEEE Trans Comput 31(9):913–917
127. Kosanoglu F, Atmis M, Turan HH (2022) A deep reinforcement learning assisted simulated annealing algorithm for a maintenance planning problem. Ann Oper Res. https://doi.org/10.1007/s10479-022-04612-8
128. Huang J, Chang Q, Arinez J (2020) Deep reinforcement learning based preventive maintenance policy for serial production lines. Expert Syst Appl 160:113701
129. Mahmoodzadeh Z, Wu K-Y, Droguett EL, Mosleh A (2020) Condition-based maintenance with reinforcement learning for dry gas pipeline subject to internal corrosion. Sensors 20:5708
130. Andriotis CP, Papakonstantinou KG (2018) Managing engineering systems with large state and action spaces through deep reinforcement learning. CoRR, arXiv:1811.02052
131. Hu J, Niu H, Carrasco J, Lennox B, Arvin F (2020) Voronoi-based multi-robot autonomous exploration in unknown environments via deep reinforcement learning. IEEE Trans Veh Technol 69:14413–14423. https://doi.org/10.1109/TVT.2020.3034800
132. Turan HH, Kosanoglu F, Atmis M (2020) A multi-skilled workforce optimisation in maintenance logistics networks by multi-thread simulated annealing algorithms. Int J Prod Res. https://doi.org/10.1080/00207543.2020.1735665
133. Schwarz K, Römer M, Mellouli T (2019) A data-driven hierarchical MILP approach for scheduling clinical pathways: a real-world case study from a German university hospital. Bus Res. https://doi.org/10.1007/s40685-019-00102-z
134. Gartner D, Kolisch R (2014) Scheduling the hospital-wide flow of elective patients. Eur J Oper Res 233:689–699. https://doi.org/10.1016/j.ejor.2013.08.026

135. Burdett RL, Kozan E (2018) An integrated approach for scheduling health care activities in a hospital. Eur J Oper Res 264:756–773. https://doi.org/10.1016/j.ejor.2017.06.051
136. Vissers J, Adan I, Bekkers J (2005) Patient mix optimization in tactical cardiothoracic surgery planning? A case study. IMA J Manag Math 16(3):281–304. https://doi.org/10.1093/imaman/dpi023
137. El Saadani N, Houda ZB, Bouras A (2014) A linear mathematical model for patients' activities scheduling on hospital resources. IEEE, Piscataway. https://doi.org/10.1109/codit.2014.6996871
138. Conforti D, Guerriero F, Guido R, Cerinic MM, Conforti ML (2011) An optimal decision making model for supporting week hospital management. Health Care Manag Sci 14:74–88. https://doi.org/10.1007/s10729-010-9144-5
139. Helbig K, Römer M, Mellouli T (2015) A clinical pathway mining approach to enable scheduling of hospital relocations and treatment services. In: Motahari-Nezhad HR, Recker J, Weidlich M (eds) Business process management, vol 9253. Springer, Cham, pp 242–250
140. Chagas RDS, de Oliveira SJBC, Longhi RP, Vieira BF, de Arruda EF, da Silva LLSB, Ferreira FVJM (2022) A solution framework for the integrated periodic supply vessel planning and port scheduling in oil and gas supply logistics. Optim Eng. https://doi.org/10.1007/s11081-022-09742-3
141. Zarandi F, Hossein M, Asl S, Akbar A, Sotudian S, Castillo O (2018) A state of the art review of intelligent scheduling. Artif Intell Rev. https://doi.org/10.1007/s10462-018-9667-6
142. Rajpathak DG (2001) Intelligent scheduling—a literature review. Knowledge Media Institute, Milton Keynes
143. Huang C-S, Huang Y-C, Lai P-J (2012) Modified genetic algorithms for solving fuzzy flow shop scheduling problems and their implementation with CUDA. Expert Syst Appl 39(5):4999–5005
144. Sakawa M, Mori T (1999) An efficient genetic algorithm for job shop scheduling problems with fuzzy processing time and fuzzy due date. Comput Ind Eng 36(2):325–341
145. Li J et al (2012) Solving fuzzy job shop scheduling problem by a hybrid PSO algorithm. In: Swarm and evolutionary computation. Springer, Berlin, pp 275–282
146. Wang S et al (2013) An effective estimation of distribution algorithm for the flexible job shop scheduling problem with fuzzy processing time. Int J Prod Res 51(12):3778–3793
147. Li JQ, Pan YX (2013) A hybrid discrete particle swarm optimization algorithm for solving fuzzy job shop scheduling problem. Int J Adv Manuf Technol 66(1–4):583–596
148. Metaxiotis KS, Psarras JE, Askounis DT (2002) GENESYS: an expert system for production scheduling. Ind Manag Data Syst 102(6):309–317
149. Matsumoto S et al (2009) Design of knowledge-based scheduling solution based on expert's technical knowledge in printing process and proposal of its improvement. Int J Innov Comput Inform Control 5(11):4125–4143
150. Xing L-N, Chen Y-W, Wang P, Zhao Q-S, Xiong J (2010) A knowledge-based ant colony optimization for flexible job shop scheduling problems. Appl Soft Comput 10:888–896. https://doi.org/10.1016/j.asoc.2009.10.006
151. Wang S, Wang L (2015) A knowledge-based multi-agent evolutionary algorithm for semiconductor final testing scheduling problem. Knowl-Based Syst 84:1–9
152. Tang L, Liu W, Liu J (2005) A neural network model and algorithm for the hybrid flow shop scheduling problem in a dynamic environment. J Intell Manuf 16(3):361–370
153. Azadeh A, Negahban A, Moghaddam M (2012) A hybrid computer simulation-artificial neural network algorithm for optimisation of dispatching rule selection in stochastic job shop scheduling problems. Int J Prod Res 50(2):551–566
154. Ezugwu AE et al (2016) Neural network-based multi-agent approach for scheduling in distributed systems. Concurr Comput Pract Exp 29:3887
155. Ogbu FA, Smith DK (1990) The application of the simulated annealing algorithm to the solution of the n/m/C max flow shop problem. Comput Oper Res 17(3):243–253

156. Bank M, Ghomi SF, Jolai F, Behnamian J (2012) Application of particle swarm optimization and simulated annealing algorithms in flow shop scheduling problem under linear deterioration. Adv Eng Softw 47(1):1–6
157. Lin T-L et al (2010) An efficient job shop scheduling algorithm based on particle swarm optimization. Expert Syst Appl 37(3):2629–2636
158. Gupta MC, Gupta YP, Kumar A (1993) Minimizing flow time variance in a single machine system using genetic algorithms. Eur J Oper Res 70(3):289–303
159. Weil G et al (1995) Constraint programming for nurse scheduling. IEEE Eng Med Biol Mag 14(4):417–422
160. Yun Y-S, Gen M (2002) Advanced scheduling problem using constraint programming techniques in SCM environment. Comput Ind Eng 43(1–2):213–229
161. Sevaux M, Jouglet A, Oguz C (2005) Combining constraint programming and memetic algorithm for the hybrid flowshop scheduling problem. In: ORBEL 19th annual conference of the SOGESCI-BVWB, vol 25. ACM, New York
162. Xujun Z, Zhimin L (2009) Model and solution for steelmaking-continuous casting scheduling problem based on constraint programming method. In: International conference on information technology and computer science, ITCS, vol 1. IEEE, Piscataway
163. Gökgür B, Hnich B, Özpeynirci S (2018) Parallel machine scheduling with tool loading: a constraint programming approach. Int J Prod Res 30:1–17

Chapter 5
Optimization Methods in Assembly Line Management

Introduction by the Editor

The assembly line is one of the major systems in any manufacturing industry, where several finished components are brought together and assembled sequentially leading to the customer ready product. This chapter discusses several contributions associated with the complexities and corresponding methods and approaches in finding optimal solutions. The reconfigurable manufacturing systems are very much required to accommodate the flexibility in assembly line. It is very much demanded and indirectly driven by the market trend. The methods such as Linear Programming minimizing the overall cost of the workstation activation and reconfiguration, cycle time are discussed along with the maximization of the process quality of the assembly tasks. Moreover, automated storage/retrieval system optimization using numerical simulations are described. The work associated with the energy consumption minimization with a view point of assembly line balancing and control systems using genetic algorithm as well as particle swarm optimization, artificial immune systems have been discussed. The chapter also discusses the disassembly process efficiency maximization using genetic algorithms, goal programming and artificial bee colony algorithm using several case studies. The chapter elaborates certain test cases of two-sided assembly line balancing using several exact methods. Lastly, the Industry-4.0 approaches such as considering late customization, Resequencing Assembly lines have also been reviewed.

Few of the important contributions reviewed are discussed here. The use of the AI based optimization algorithm referred to as genetic algorithm is reviewed for energy-efficient integration of assembly line balancing and part feeding. The study and experiments underscore the need of further use of constraints, such as incompatible pair tasks or uncertain task times, more types of assembly lines can be taken into consideration, such as U-shaped assembly lines or multiproduct parallel assembly lines, etc. The MOGA and NSGA-II algorithms have been reviewed for the

A. J. Kulkarni (ed.), *Optimization Methods in Manufacturing Processes*,
https://doi.org/10.1007/978-981-96-5257-0_5

bridge structures design applications. More specifically, the study deals with the use of an active suspension system, artificial intelligence-based methods and PID optimization being used to control the vibrations. A novel version of the particle swarm optimization referred to as the Immune particle swarm optimization has been reviewed for the SUV rollover crashworthiness and weight. The immune algorithm simulates the process of antibody recognition in an immune system to find the optimal solution. The particle swarm optimization helps to quickly search a group of non-dominated solutions on the Pareto front. The work associated with the disassembly line balancing optimization method for high efficiency and low carbon emission have been discussed. It involves the case studies such as the disassembly sequences of an automobile engine analysis verifying the effectiveness of the multi-objective optimization models. It demonstrates that the analysis obtained a disassembly sequence having a reduction in disassembly time and carbon emissions. A case study on Artificial Bee Colony Algorithm for multi-objective two-sided disassembly line balancing optimization from the automotive domain is also elaborated. The study aimed to reduce the number of workstations, idle time, a multi-objective two-sided disassembly line balancing model, etc. based on Pareto non-dominated relationship, crossover and mutation operators of GA for disassembly scheme optimization.

A review is provided involving the recent scenario of the utilization of robots in different areas such as packaging, loading/unloading, transportation, and especially assembly lines enhance the productivity of production systems. It includes the discussion on the problem structure and mathematical model, data collection and solution approach. More specifically the real-world assembly line balancing problems such as resource optimization, idle time minimization etc. motivates the author to think about the improvement in this direction have been analysed along with the optimization using Lingo software. A hybrid version of the genetic algorithm and discrete Levy flight algorithm for a robust mixed-model assembly line balancing problem has been reviewed. It investigates the effect of uncertainty parameters on production efficiency. This provides the future directions that the mixed-model assembly line balancing and sequencing with interval uncertainty could be considered simultaneously. The reconfigurable manufacturing systems have also been detailed. The associated illustrative examples include product configuration optimization according to the customer requirements, meaning that the product module instances need to be selected to meet the product functions required by the customer. The findings provided further research direction aiming to investigate this trade-off by comparing this integrated approach to the separated ones.

A two-sided assembly line balancing problem using an exact solution approach such as Lingo has also been discussed. It required precedence constraints since tasks are assigned on both sides of the line as well as different models. The objective of the study is to assign the set of tasks of models to different stations of two-sided assembly line so that the number of single stations, idle time are reduced along with the space and computational time. An important Industry 4.0 approach to assembly line resequencing has been critically reviewed and elaborated. This involves production planning and control as well as resequencing assembly lines in Industry 4.0 environments.

Machine Generated Summaries

Disclaimer: The summaries in this chapter were generated from Springer Nature publications using extractive AI auto-summarization: An extraction-based summarizer aims to identify the most important sentences of a text using an algorithm and uses those original sentences to create the auto-summary (unlike generative AI). As the constituted sentences are machine selected, they may not fully reflect the body of the work, so we strongly advise that the original content is read and cited. The auto generated summaries were curated by the editor to meet Springer Nature publication standards. To cite this content, please refer to the original papers.

Machine generated keywords: line, assembly, assembly line, line balance, product, balance, mass, workstation, line balancing, balancing, customization, planning, plan, customize, mass customization.

A Skill- and Feature-Based Approach to Planning Process Monitoring in Assembly Planning [1]

This is a machine-generated summary of:

Gonnermann, Clemens; Hashemi-Petroodi, S. Ehsan; Thevenin, Simon; Dolgui, Alexandre; Daub, Rüdiger: A skill- and feature-based approach to planning process monitoring in assembly planning [1].

Published in: The International Journal of Advanced Manufacturing Technology (2022).

Link to original: https://doi.org/10.1007/s00170-022-09931-5.

If you want to cite the papers, please refer to the original.

For technical reasons we could not place the page where the original quote is coming from.

Abstract-Summary "The customized flexibility and several characteristics of RMSs provide many opportunities in terms of process and production planning."

"Those characteristics greatly increase the complexity of the design and planning of production systems."

"This paper presents a decision support system relying on a skill-based approach to design a reconfigurable assembly line considering the planning of assembly processes and monitoring."

"The main input data is a CAD model of a new product variant for the identification of the assembly and monitoring requirements."

"In the first developed module, assembly-by-disassembly and a skill-based approach are used to generate different assembly plans."

"In the second module, feature recognition and skill-based approaches generate process monitoring alternatives."

"The third module uses a linear program (LP) that aims to minimize the total cost of workstation activation and reconfiguration, as well as cycle time, and to maximize the process quality of the assembly tasks."

"In the fourth and final module, a simulation of the optimized assembly plan allows either the validation of the assembly plan and process monitoring plan or initiates a new iteration due to their infeasibility."

Introduction

"An RMS is defined as a manufacturing system where physical manufacturing components and process capabilities can be added, removed, or modified easily and efficiently to change the production capacity."

"To be able to handle such flexibility a great amount of manual effort and expert knowledge is needed to plan the production processes."

"Besides planning assembly processes including the assignment of production devices and stations, the quality of the produced goods must be considered."

"This paper addresses today's challenges in production planning by generating and analyzing various assembly and monitoring plans offered by an RMS."

"We identify the different alternatives for assembling a new product in an existing production system and simultaneously consider the possibility of introducing process monitoring."

"The developed automated planning methodology includes the optimization of the solution space (assembly plans) concerning production criteria (e.g., necessary reconfigurations, monitoring efficiency)."

Literature Review

"The literature on RMS in general and automated planning in combination with skill-based approaches and optimization approaches for process planning to efficiently use the flexibility of RMS are reviewed."

"They focus on skills in assembly so that an existing production system is able to offer these skills for the assembly and thereby satisfy product requirements [2] build upon this approach and defined skills that are relevant for monitoring processes in assembly."

"The generation of assembly plans via skill-based approaches allows to use the flexibility of RMS."

"In manufacturing planning, feature-based approaches are used to define product and process requirements [3, 4] describe an approach to identify critical manufacturing features on metallic aerospace components."

"There are few approaches to how to consider process monitoring in today's (semi-)automatic production or assembly planning."

Problem Description

"A set of automated and manual equipment pieces (i.e., tools) is considered in the resource library of the line (part of the digitized assembly system) which can be used to execute a task (i.e., process)."

"Valuable information for such decisions in a production graph is the feasibility of the task, necessary reconfigurations (including the additional time and costs), process parameters (e.g., assembly paths, cycle times, efficiency of process monitoring), and task types (e.g., screwing—primary; handling—secondary)."

"The objective of the optimization model is to determine the optimal solution with regard to individual criteria (e.g., cycle time and monitoring efficiency) for an assembly plan including monitoring tasks in order to enable high process quality."

System Overview

"These steps consist of importing data (e.g., CAD-model of the assembly, assembly plans), selection of monitoring requirement templates (i.e., solution neutral monitoring requirements for assembly processes), data extraction (i.e., rule-based feature recognition), and determination of monitoring requirements (i.e., filling the templates according to analyses)."

"The automated generation of process monitoring alternatives is executed by a comparison of product-specific monitoring features and assembly system-specific (i.e., resource-specific) monitoring skills [5, 6]."

"The result of the module (2) is a process graph and process monitoring graph containing valid assembly plans and process monitoring alternatives (i.e., allocated actorial and sensorial resources and stations)."

"Information about the individual resources and processes is provided by the assembly and monitoring plan identified by the optimizer."

Computational Experiments and Results

"The first case study shows a relatively simple product to be assembled, illustrating the flexibility available in creating an optimized assembly plan including a process monitoring alternative."

"Due to the fact that this methodology considers the planning of process monitoring alongside the assembly planning, the optimization and simulation-based validation are done after the allocation of resources/tools for monitoring."

"The comparison of sensorial skills and monitoring requirements, taking into account the ability to assemble the product at each station (from assembly plans), identifies different alternatives for process monitoring."

"Each assembly plan contains different alternatives for process monitoring as a result of module 2 and the allocation of resources with sensorial skills."

"Consideration of process monitoring allows the optimization module to use various criteria for selecting the best assembly plan, taking into account quality aspects when planning production."

"The results of the two case studies show how assembly plans, including alternatives for process monitoring, can be generated automatically."

Conclusion

"The proposed methodology represents an approach to integrating process monitoring planning into assembly planning in combination with optimization."

"Considering monitoring processes as a relevant aspect in process planning can result in higher quality in process execution and final products."

"To handle the complexity of various product variants, assembly sequences, assembly and monitoring resources, and assembly and monitoring plans, the proposed methodology has been implemented with proper feasibility and efficiency."

"Automated analysis of product CAD files to identify assembly and monitoring processes, and resource matching through a skill-based approach in an existing production system, enables more efficient process planning and use of RMS."

"The optimization module shows that additional criteria, by integrating process monitoring alternatives via a quality factor, can extend and improve the generation of assembly plans for RMS."

Energy-Efficient Integration of Assembly Line Balancing and Part Feeding with a Modified Genetic Algorithm [7]

This is a machine-generated summary of:

Chen, Junhao; Jia, Xiaoliang: Energy-efficient integration of assembly line balancing and part feeding with a modified genetic algorithm [7].

Published in: The International Journal of Advanced Manufacturing Technology (2022).

Link to original: https://doi.org/10.1007/s00170-022-09422-7.

If you want to cite the papers, please refer to the original.

For technical reasons we could not place the page where the original quote is coming from.

Abstract-Summary "The manufacturing industry has been seeking to design an energy-efficient assembly system."

"Reasonable assembly line balancing (ALB) can reduce the energy consumption of the assembly production subsystem."

"Appropriate part feeding (PF) can reduce the energy consumption of the assembly distribution subsystem."

"Existing research considers energy efficiency-related issues by treating ALB and PF separately, which misses the potential opportunity for energy savings for the entire assembly system."

"ALB and PF are closely interrelated, this study investigates the energy-efficient integration of assembly line balancing and part feeding (EIALBPF) problems by a modified genetic algorithm (MGA)."

"The results demonstrate that (1) the proposed method significantly reduces the total energy consumption of the entire assembly system compared with the traditional separate programming method and (2) the proposed MGA outperforms the nested genetic algorithm in terms of the approximation to the true frontier and computational efficiency."

Introduction

"No studies have been conducted to take ALB and PF as an integrated problem to save energy consumption of the entire assembly system."

"Different combinations of assembly task assignments in ALB and part supply relationships in PF concurrently affect the energy consumption and traditional performance indicators of entire assembly system."

"How can production energy consumption be reduced in the ALB-I problem?"

"Traditional energy efficiency-related ALB research focuses on production energy consumption in the ALB-II and ALB-E problems [8, 9]."

"When optimizing the production energy consumption in the ALB-I problem, it needs to consider not only how to allocate stations but also how to choose the appropriate stations."

"A new model should be established to save production energy consumption in the ALB-I problem."

Literature Review

"Various streams of literature inspire this study, including (1) ALB considering energy savings, (2) PF considering energy savings, (3) bi-level multi-objective optimization method, and (4) the IALBPF problem."

"A review of the literature indicates certain specific research gaps in the EIALBPF problem such as the inability to reduce the total energy consumption of the entire assembly system, the inability to save energy consumption in the ALB-I problem, ignoring the production energy consumption in the ADS of numerous supermarkets, and adopting the low-performance optimization method."

"The study differs from the previously mentioned research, as it incorporates energy consumption into the optimization objectives of the EIALBPF problem, considers the production energy consumption of the ALB-I problem, investigates the logistics energy consumption in the ADS of numerous supermarkets, and designs a mixed encoding and repair strategy to improve the performance of optimization method."

Mathematical Formulation

"The overall solution to the EIALBPF problem, i.e., the corresponding relations between all the assembly tasks, stations, and supermarkets, is called an EIALBPF solution."

"For APS production, the station consumes the energy to complete assembly task operations and maintain standby status, which are called operation energy consumption and standby energy consumption, respectively."

"Different stations have different levels of operation energy consumption to complete the same assembly task."

"The operation and standby energy consumption are both influenced by the corresponding relations between assembly tasks and stations."

"The part logistics energy consumption represents that a tow train consumes energy to transport required parts to corresponding stations."

"During the PF, the part logistics energy consumption of each service station of each supermarket depends on the moving distance and the weight of the required parts."

The Modified Genetic Algorithm for the EIALBPF Problem

"Step 1: After the crossover operation and mutation operation of the ALB segment and PF segment, the feasibility of the new individual and repair condition are checked."

"Step 2: If the new individual is infeasible and the repair condition is met, the ALB segment repair is executed until the ALB segment is feasible."

"Step 3: After the ALB segment repair, the PF segment repair is executed until the PF segment is feasible."

"During the repair procedure of the ALB segment, the number of tasks of each station is constant."

"PF segment repair is achieved by station movement."

"There are two rules for station movement of PF segment repair."

"Step 1: The PF segment and station subpart are separated from an infeasible chromosome of the EIALBPF problem."

"The second movement of PF segment repair in which station 2 is moved from supermarket 2 to supermarket 3 is achieved."

Computational Experiments and Results

"The separate programming method considering energy works as follows. (1) Without considering the PF, the method [8] obtains ALB solutions of minimization of the number of stations and production energy consumption. (2) For each ALB solution, the method [10] is adopted to obtain PF solutions of minimization of the number of supermarkets and logistics energy consumption. (3) ALB solutions and corresponding PF solutions are combined to obtain the solutions to the EIALBPF problem. (4) The Pareto-optimal solutions to the EIALBPF problem of the separate programming method are selected based on minimization of the number of stations, the number of supermarkets, and total energy consumption."

"The integrated programming method without considering energy works as follows. (1) According to the cost index, the number of stations, and supermarkets, the method in [11] obtains solutions to the EIALBPF problem. (2) The energy consumption of each solution to the EIALBPF problem is calculated. (3) The Pareto-optimal solutions to the EIALBPF problem of the integrated programming method without considering energy are selected based on minimization of the number of stations, the number of supermarkets, and total energy consumption."

Conclusion

"A new ALB-I problem with the objectives of minimizing the production energy consumption is defined."

"To the traditional energy-oriented ALB research that reduces energy consumption in ALB-II and ALB-E problems, the proposed model can reduce the production energy consumption in the ALB-I problem."

"Due to the assumptions and limitations of this study, future research can be conducted in the following aspects. (1) More types of constraints can be included, such as incompatible pair tasks or uncertain task times. (2) More types of assembly lines can be taken into consideration, such as U-shaped assembly lines or multi-product parallel assembly lines. (3) Other environmental protection issues can be explored in the IALBPF problem, such as reduction of carbon emissions and noise pollution. (4) The optimization objectives of this research are the number of stations, number of supermarkets, and total energy consumption."

A New Expert System for Active Vibration Control (AVC) for High-Speed Train Moving on a Flexible Structure and PID Optimization Using MOGA and NSGA-II Algorithms [12]

This is a machine-generated summary of:

Koç, Mehmet Akif: A new expert system for active vibration control (AVC) for high-speed train moving on a flexible structure and PID optimization using MOGA and NSGA-II algorithms [12].

Published in: Journal of the Brazilian Society of Mechanical Sciences and Engineering (2022).

Link to original: https://doi.org/10.1007/s40430-022-03441-x.

If you want to cite the papers, please refer to the original.

For technical reasons we could not place the page where the original quote is coming from.

Abstract-Summary "If the HST is moving on a structure such as a bridge, resonant frequencies due to train-bridge interaction (TBI) increase these vibrations."

"This study deals with the use of an active suspension system (ASS), artificial intelligence techniques (AIT) and PID optimization and control to reduce vibrations, which disturbs the comfort for passenger travel that occurs in HST traveling on a bridge."

"The moving train on the bridge is modeled as 10-DOF."

"A linear actuator (LA) is installed in the secondary suspension systems of the train to reduce the disturbing vertical and angular accelerations in the train body."

"To optimize the coefficients of the designed PID controller, a multi-objective genetic algorithm (MOGA) with six objective functions was used, taking into account passenger comfort (PC) and rail holding (RH)."

"A significant success was achieved in the development of the PC by reducing the RMS value calculated by the weighted frequency of the comfort disturbing acceleration on the passengers by 87%."

Introduction

"The results show that the newly designed absorber is more effective for HST in reducing train vibrations than conventional rubber absorber [13] proposed a decoupling optimization method to eliminate the damping loss due to changes in natural frequencies after determining UCE's natural frequencies according to the dynamic vibration absorber theory."

"Optimized suspension parameters achieved a 34% improvement in PC at high speeds [14] studied the effect of the lateral and longitudinal damping effect of the secondary suspension system of HST on dynamic behavior."

"The importance of active suspension application in vehicles with high-speed flexible bodies such as trains is not only increased the PC; it also allows the train body to be designed in a lighter fashion [15]."

"Adaptive control studies with HST for ASS are mainly aimed at improving PC and driving safety due to factors such as rail irregularities, wind impact, vehicle body flexibility."

Simulation Model

"These vibrations on the bridge affect the train in the opposite direction and adversely affect the driving safety of the HST and the passenger comfort on the train."

"[16] modeled the sophisticated train model as a mass spring model to examine the resonance effect of the train model consisting of multiple cars on the bridge beam."

"In the mathematical modeling for the TBI analysis the following assumptions will be performed: the moving HST has constant velocity and physical model of the train consist of 10-DOF with different parts such as train body, wheels and bogies."

"The motion equations for the train car body, front and rear wheels are expressed using the accelerations of the each DOF r_i, respectively, as follows: The response of the bridge beam, in the form of n second order differential equations, are formulated as below: Here D_1, D_2, D_3 and D_4 are variables that are defined according to the intervals given in study [17]."

PID Controller Designing for LAs of ASS

"PID coefficients were optimized by using the MOGA-NSGA-II combined algorithm to reduce the vertical and angular acceleration of the train body."

"In multi-objective optimization, the objective function is not to obtain a single scalar number representing the optimal solution, as in the SOO problem, but to obtain a vector with different possible optimal solutions [18]."

"After this step, the population is generated according to the ranking and crowding distance (CD) parameters following the selection procedure of the new population and then sorted according to the non-dominated approach."

"This algorithm creates individuals within the population (PID coefficients in this study), selects parents by mimicking natural choices to produce new children from the generated population, and the most appropriate individuals survive, while the worst ones are eliminated from the population."

"As for GA, the transposition operator increases the difference between solutions within the population, resulting in stronger generations."

Numerical Results and Discussion

"The maximum vertical acceleration of the train body was determined as 0.4283 m/s^2 as a result of the analysis made with PID coefficients of the optimum solution given by C_1 (f_1–f_2) for active controlled conditions."

"According to the PID coefficients obtained as a result of optimization performed according to C_1 f_1–f_2, total RMS value affecting driver's seat was determined as 0.0943 m/s^2 and decreased by 70%."

"Of optimization performed according to C_1 f_1–f_3, f_1–f_4, f_1–f_6 and C_2 F_{comf}–F_{hold} scenarios, the reduction rates of vertical displacement of train body were determined as 75%."

"In the figure, the increase in the speed of crossing the train bridge has led to an increase in the maximum values of the forces generated in both the front and rear actuators for all optimization scenarios."

Conclusions

"The application of ASS for the damping of TBI-induced vibrations between the 10-DOF half HST model and the Euler–Bernoulli flexible structure was successfully applied to the suspension system and optimized with PID and MOGA-NSGA-II algorithms and the PC and RH of the HST was developed."

"ASS reduced train body vertical displacement by 75% and secondary suspension displacement by 15%."

"The ASS train installed in the secondary suspension system is very effective in reducing body acceleration and displacement and improved the PC."

"If ASS is used in the primary suspension system as in the secondary suspension system and is included in the control algorithm, it will be even more effective in the development of RH."

"In the TBI problem caused by flexible structure interaction with HST, ASS is highly effective in damping vibrations and contributed significantly to the development of both PC and RH."

Immune Particle Swarm Optimization of SUV Rollover Crashworthiness and Weight [19]

This is a machine-generated summary of:

Jin, Zhilin; Zhang, Senhao; Weng, Jiansheng: Immune particle swarm optimization of SUV rollover crashworthiness and weight [19].

Published in: Structural and Multidisciplinary Optimization (2021).

Link to original: https://doi.org/10.1007/s00158-021-02906-2.

If you want to cite the papers, please refer to the original.

For technical reasons we could not place the page where the original quote is coming from.

Abstract-Summary "When designing a sport utility vehicle (SUV), designers strive to improve the vehicle's rollover crashworthiness while avoiding a significant increase in its weight."

"Building on the analysis results, the SUV's rollover crashworthiness and weight optimization model are proposed."

"The proposed algorithm then utilizes particle swarm and artificial immune to seek Pareto solutions that optimize SUV structure."

"Compared to previous studies, the results show that the proposed hybrid optimization algorithm improves the Pareto solution sets' diversity and distribution uniformity, enhances SUV rollover crashworthiness, and reduces SUV structure components' weight."

Introduction

"Due to its speed and good convergence properties, NSGA-II became a benchmark multi-objective optimization algorithm [20] utilized NSGA-II for multi-objective optimization of vehicle body's side parts."

"The body structures' thickness and material were optimized using a genetic algorithm, improving the vehicle body's energy absorption by 8.3% and reducing the weight by 33.32%."

"Wang et al. [21] employed a multi-objective collaborative optimization method to minimize the vehicle body weight while maintaining crashworthiness."

"Building on the SUV rollover crashworthiness analysis, a multi-objective and multi-disciplinary SUV rollover crashworthiness and weight optimization is performed using the material structure integrated design method."

"Radial basis function (RBF) neural network and a genetic algorithm are utilized to improve the overall accuracy of surrogate models describing the total weight, maximum contact force, and SUV structure's torsion frequency."

"A hybrid optimization algorithm that combines particle swarm with artificial immune is proposed to enhance SUV rollover crashworthiness and minimize the weight."

SUV Rollover Crashworthiness Analysis and Improvement

"Previous research utilized the FE method for roof crash simulation and determined the rigid plate speed could be set to 2235.2 mm/s. The maximum contact force between the rigid plate and vehicle roof could was as an important indicator of SUV rollover crashworthiness when the rigid plate moved forward within 127 mm [22]."

"Contribution analysis relies on statistics to determine the magnitude of each variable's influence on the performance, consequently enabling detection of components critical for the SUV rollover crashworthiness and weight."

"The maximum contact force between the rigid plate and vehicle roof (F_{max}) and the total SUV components' weight (W_s) are collected from each experiment."

"The maximum contact force between the rigid plate and vehicle roof increased by 16.7% (i.e., to 54.5 kN), while the total SUV components' weight increased by 1.05 kg."

SUV Rollover Crashworthiness and Weight Optimization

"The contribution analysis shows that, while increasing C_1–C_3 and C_6–C_9 thickness significantly improves crashworthiness, it also dramatically increases the total SUV weight."

"There is a trade-off between SUV rollover crashworthiness and weight."

"Since local improvements to A- and B-pillars increase the SUV rollover crashworthiness, the thickness of reinforcing plates added to A- and B-pillars may replace the C_3, C_7, and C_9 thickness."

"The multi-disciplinary and multi-objective optimization model for SUV rollover crashworthiness and weight is expressed as:"

SUV Rollover Crashworthiness and Weight Surrogate Model

"Since test data greatly affects the RBF neural network's accuracy, the OLHD sampling method was used to generate 90 data groups, of which 75 were used to train the network's weights and optimize the hyperparameters."

"The RBF neural network's output is obtained from the hidden layer as: where w_{ij} denotes the weight from the ith neuron to the jth output, and $\varphi_i(X)$ represents the radial basis function, i.e., the ith neuron's activation."

"Gaussian centers (S_i), variance (σ), and learning rate (η) greatly affect the RBF neural network's accuracy."

"A genetic algorithm can optimize variance (σ) and learning rate (η) to improve the surrogate model's accuracy."

"The RBF neural network is established, and the weights are trained 2000 times to obtain a surrogate model."

"Once established, the optimal parameters, σ^* and η^*, are used in the RBF neural network."

Hybrid Algorithm for SUV Rollover Crashworthiness and Weight Optimization

"The immune algorithm simulates the process of antibody recognition in an immune system to find the optimal solution."

"The hybrid optimization algorithm uses PSO to quickly search a group of non-dominated solutions on the Pareto front."

"In that case, the artificial immune algorithm obtains an optimal solution by limiting the non-dominated solutions to vibrate in a small range."

"The solutions obtained using the hybrid optimization algorithm are the closest to the Pareto front, indicating better SUV rollover crashworthiness and weight reduction."

"Since the hybrid optimization algorithm utilizes density variance in solution selection, its Pareto solution set is more uniform than that of PSO or immune algorithm."

"The number of solutions obtained using PSO is 29, while the hybrid optimization and immune algorithms created 60."

"The hybrid optimization algorithm effectively improves both PSO's and immune system's solutions' diversity, precision, and distribution uniformity."

Conclusions

"This paper studied the SUV rollover crashworthiness and weight reduction through the lenses of structure components improvement and optimization."

"The work analyzed the main structure components' contributions to SUV rollover crashworthiness and weight reduction."

"Based on the structure analysis, the influential parts are optimized, and crashworthiness is enhanced (the maximum contact force increased by 16.7%)."

"A multi-disciplinary and multi-objective model for SUV rollover crashworthiness and weight optimization is developed using the material structure integrated design method."

"This algorithm effectively improves the local search ability, decreases the total SUV components' weight by 34.5%, and increases the maximum contact force between the SUV roof and rigid plate by 18.7%."

Mobile Rack AS/RS Dimensions Optimization for Single Cycle Time Minimization [23]

This is a machine-generated summary of:

Kouloughli, Sihem; Korso Feciane, Malika; Sari, Zaki: Mobile rack AS/RS dimensions optimization for single cycle time minimization [23].

Published in: The International Journal of Advanced Manufacturing Technology (2022).

Link to original: https://doi.org/10.1007/s00170-022-09450-3.

If you want to cite the papers, please refer to the original.

For technical reasons we could not place the page where the original quote is coming from.

Abstract-Summary "The present work aims primarily to study the Automated Storage/Retrieval System (AS/RS) with sliding racks."

"It also has one single Storage/Retrieval machine."

"The racks can slide to free up space between two adjoining bins, hence allowing the Storage/Retrieval machine to access the desired rack."

"The analytical expressions for the single cycle time which is the basis of our optimization have three positive real variables, namely the travel time in the horizontal direction, the travel time in the vertical direction, and the travel time along the aisles, with a positive real parameter that represents the travel time of the sliding racks."

"To the causal constraints of horizontal and vertical transport times, we are also faced with a constraint that is related to the size of the system and the constancy of this size, which brings us back to an optimization problem involving real numbers of three-variable functions parameterized with constraints."

"Since the analytical optimization was difficult to achieve, it was deemed more interesting to perform a numerical simulation to determine the three optimal temporal dimensions, for any given value of the sliding time of the racks."

"The temporal dimensions obtained will make it possible to find the optimal length, optimal depth and optimal height of the system knowing the horizontal speed and the vertical speed of the storage/retrieval machine."

Introduction

"These systems are classified according to their physical configuration, namely: The number of Storage/Retrieval (S/R) machines used and their capacity, The arrangement of racks and aisles in the system, The positions of the Pickup/Delivery (P/D) stations, The depth of the racks (maximum number of products that can be stored in a single bin)."

"The present work is primarily concerned with optimizing the dimensions of the sliding rack AS/RS that would minimize the single cycle time."

"We tried to use the same analytical method that has been used for optimizing the dimensions of the multi-aisle AS/RS developed by Kouloughli [24]; however, this was not possible for a sliding rack AS/RS because the analytical model of the simple cycle time of this system depends on a parameter representing the sliding time of the racks."

"The results obtained can then be used to build a sliding rack AS/RS with optimal dimensions in order to minimize the cycle time."

Literature Review on AS/RS

"A large number of researchers have become increasingly interested in AS/RS, and more particularly in some of their important features such as the design, the mathematical modeling of cycle time of the S/R machine, the optimization of the AS/RS dimensions, the location of the dwell point of the S/R machine, the scheduling of storage and retrieval operations, the different types of storage methods used in these systems, etc. In this context, several research studies that have been carried out in the field of design, cycle time estimation, and optimization of AS/RS dimensions are worth mentioning."

"Among the research work carried out on the optimization of AS/RS dimensions, it is worth mentioning that of De Koster [25] who was particularly interested in designing an optimal storage rack for a compact 3D AS/RS in order to determine the optimal ratios between the three spatial dimensions of the system as this will ensure the efficient functioning of the system with regard to the cycle time, over the long term."

Description of a Mobile Rack AS/RS

"It is worth recalling that a sliding rack AS/RS, also called a mobile rack AS/RS, is a variation of the multi-aisle AS/RS."

"It is made up of a set of racks that are arranged in parallel with one single S/R machine."

"The racks forming the system slide sideways on rails so that the S/R machine can access an aisle between any two adjoining racks."

Single Cycle Time Model of a Mobile Rack AS/RS

"For ESC1, three cases arise: Regarding the first and second cases ($t_r < t_p$): The first and second cases are used to evaluate the average single cycle time of the sliding rack AS/RS, when the sliding time of the racks of the system is less than the travel time from the P/D station to the last aisle."

"The expression of ESC1 was developed by Guezzen et al. [26]: The second case is used when the travel time from the P/D station is less than the sum of the sliding time of the racks and the travel time in the horizontal direction from the first bin to the last bin of a rack."

"The ESC1 expression was developed by Kouloughli [27]: The third case is used to evaluate the average single cycle time of the sliding rack AS/RS, when the sliding time of the racks of the system is greater than the travel time from the P/D station to the last aisle."

Optimization of the Dimensions of the Sliding Racks AS/RS

"Based on the calculated minimum points, we try to look for mathematical expressions that allow us to directly obtain the values of the three optimal dimensions t_p, t_h, and t_v because the value of the sliding time t_r of the racks is already known."

"The calculated minimum points help us to look for mathematical expressions that allow us to obtain directly the values of the three optimal dimensions t_p, t_h and t_v, since the value of the sliding time t_r of the racks is known."

"For $t_r = 0.9$, the results are then obtained for each value of α between 45° and 85°, with a minimum single cycle time for $\alpha = 45°$.".

"Based on the minimum points that were calculated for the case ($t_r > t_p$), some mathematical expressions must be found in order to obtain the values of the optimal dimensions, since the sliding time t_r of the racks is known."

Summary of Results

"The results presented here, it can be said that the minimum single cycle time found is equal to 1.98 for the case where ($t_r < t_p$) and ($t_p < t_r + t_h$)."

"For the case ($t_r < t_p$) and ($t_p > t_r + t_h$), the minimum single cycle time is 2.32 per unit for a value of t_r around 0.856."

"For the case ($t_r > t_p$), the single cycle time is 2.26, which corresponds to values of t_r around 0.7 per unit."

Conclusion

"This research work aims primarily to optimize the dimensions of the AS/RS provided with sliding racks under the two causal constraints of transport times and size constancy of the system."

"Given the difficulty of the calculations due to the parameter t_r, which represents the sliding time of the racks, this optimization was not possible."

"It was decided to use a numerical simulation which allowed us to find the optimal ratios between the different dimensions of the system, for each value given to t_r."

"As a perspective, it would be highly desirable to optimize the dimensions of the AS/RS supplied with sliding racks for the minimum dual cycle time."

Disassembly Line Balancing Optimization Method for High Efficiency and Low Carbon Emission [28]

This is a machine-generated summary of:

Zhang, Lei; Zhao, Xikun; Ke, Qingdi; Dong, Wanfu; Zhong, Yanjiu: Disassembly Line Balancing Optimization Method for High Efficiency and Low Carbon Emission [28].

Published in: International Journal of Precision Engineering and Manufacturing-Green Technology (2019).

Link to original: https://doi.org/10.1007/s40684-019-00140-2.

If you want to cite the papers, please refer to the original.

For technical reasons we could not place the page where the original quote is coming from.

Abstract-Summary "When disassembling large quantities of products, the disassembly efficiency is crucial for enterprises."

"Disassembly line is the best choice for automated disassembly of disposal products."

"Disassembling products, especially complex structure of products, consume a lot of resources and energy and produce large amounts of carbon emissions."

"The mathematical model of carbon emission is built by quantifying the carbon emissions in the process of product disassembly."

"Taking into account the basic disassembly time, direction change time and tool change time, the mathematical model of high efficiency is presented for optimizing disassembly time."

"Based on the traditional multi-objective disassembly line balancing problem (DLBP), a novel multi-objective optimization model of the DLBP with shortest disassembly time and minimum carbon emissions is proposed for improving disassembly efficiency and reducing the carbon emissions in the process of disassembly."

Introduction

"McGovern and others solved the disassembly line balancing problem using combinatorial optimization techniques, and a genetic algorithm was presented for obtaining optimal or near-optimal solutions for the DLBP [29]."

"As of 2016, there were 635 enterprises with disassembly qualifications in China, and 2.8 million scrapped vehicles were recycled, a year-on-year increase of 7.7% [30].When disassembling a product, enterprises should consider not only disassembly sequence, but also how to select specific resources, such as robots, equipment, materials and staff to improve disassembly efficiency and reduce carbon emissions."

"It is obviously that energy consumption and carbon emissions, which generated by disassembling a large number of products, has great optimization significance."

"Based on the traditional multi-objective disassembly line balancing problem, a multi-objective model, which is affected by disassembly time and energy consumption, is proposed in this paper to optimize the disassembly sequence and obtain optimal or solutions."

Modeling of Disassembly Sequence

"All components are regarded as the same and there's no consideration for the disassembly relationship among components, the disassembly model cannot express the differences between components in constraint relationship definitely [31]."

"The hybrid graph model can elegantly solve the problems above and it can better describe product disassembly information and express the direct and indirect constraint relationship among components definitely."

"In this paper, the hybrid graph model is proposed for describing the complex disassembly relationships."

"Of product disassembly, when a function component i is disassembled, the relationship of precedence constraint and connection between the component and other components will be released."

Low Carbon Emission and High Efficiency Optimization Models of DLBP

"The disassembly line balancing problem consists of finding a feasible assignment of tasks to workstations in such a way that some objectives are met and the constraints of the disassembly process are satisfied [32]."

"The high efficiency function is mainly reflected in minimum total disassembly time in the disassembly process."

"The total disassembly time of DLBP is given as follows: 1."

"Direction change time A change of bearing means that the adjacent disassembly components require different disassembly directions."

"Carbon emissions of disassembly line are mainly related to the input of material consumption and energy consumption and output of GHG."

"P_i, t_i, M_j are the power of disassembly tool, welding cutting time and quality of material consumption."

"Disassembly cost refers to the cost produced in the process of disassembling specific components."

"There are three main kinds of cost in disassembly process."

Disassembly Sequence Optimizations Based on Genetic Algorithm

"The procedure of genetic crossover is given as follows: Step 1 In the interval [30, 33] take a random number k as the crossover point, set k = 4; Step 2 The genes of parent U_1 before the crossover point are copied to the offspring O_1 in the order of the parents' gene."

"Step 3 The other genes of offspring O_1 are taken from another parent U_2."

"Besides genetic crossovers, genic mutation is also applied to form new individuals by replacing one chromosome's gene with its allele."

"The procedure of genetic mutation is given as follows: Step 1 According to the mutation probability and the number of genes, calculate the number of genes that need be mutated."

"Step 2 In the interval $[1, n - 1]$ take a random integer m as the location of gene mutation."

Case Study

"The disassembly sequences of an automobile engine are analyzed to verify the effectiveness of the multi-objective optimization model of DLBP."

"A multi-objective optimal example that consists of thirty-two components is chosen to prove the effectiveness of the algorithm."

"With low carbon emission as an objective function, the optimal disassembly sequence has a more tool changes and direction changes, resulting in increasing disassembly time."

"With low carbon emission as well as high efficiency as objective function, multi-objective optimization can combine the advantages of the above two optimization solutions."

Conclusions and Future Work

"In order to disassemble components with less carbon emissions and less time, an optimization model is developed to quantify carbon emissions and disassembly time in the process of product disassembly."

"Based on the methods of genetic crossover and gene mutation, an improved genetic algorithm is presented for optimizing the disassembly sequences."

"While optimizing disassembly sequences, all objective functions of disassembly line such as high efficiency objective function, low-carbon objective function and basic objective function are integrated into one."

"The case study also demonstrated that this disassembly analysis obtained a disassembly sequence having a reduction in disassembly time and carbon emissions."

"Based on the optimal solutions of disassembly sequences, we will improve objective models."

Multi-Objective Mathematical Programming Framework for Integrated Oil Sands Mine Planning and Tailings Disposal Optimization [34]

This is a machine-generated summary of:

Maremi, Ahlam; Ben-Awuah, Eugene; Askari-Nasab, Hooman: Multi-objective Mathematical Programming Framework for Integrated Oil Sands Mine Planning and Tailings Disposal Optimization [34].

Published in: Mining, Metallurgy & Exploration (2021).

Link to original: https://doi.org/10.1007/s42461-021-00418-7.

If you want to cite the papers, please refer to the original.

For technical reasons we could not place the page where the original quote is coming from.

Abstract-Summary "This research introduces a goal programming framework that simultaneously optimizes the (1) production schedule with limited duration stockpiling and directional mining; (2) dyke construction schedule; (3) size, shape, and location of tailings-cells; and (4) mining and processing production targets."

"The results show that in creating in-pit tailings-cells designs, decreasing the mining-cells volume and increasing the number of mining-cells improve the net present value of the operation due to increased operational flexibility."

"The model generated a uniform schedule for ore and practical tailings-cells designs for backfilling activities."

"Integrating tailings-cells optimization into oil sands mine planning and waste management ensures the efficient use of in-pit mined-out areas required for sustainable operations."

Introduction

"Previous research efforts in mine planning and waste management proceeded by dividing the final pit area into pushbacks based on the footprint required by tailings dam engineers to set up in-pit tailings facility cells [35–39]."

"The focus of this study is to formulate and implement a mixed integer linear goal programming (MILGP) model that concurrently schedules for production, reclamation, and dyke material and optimizes the size, shape, and location of in-pit tailings-cells in addition to defining the optimal mining and processing annual targets in an integrated mine planning and waste management framework."

"The model integrates the concept of directional mining with determination of the optimal mining and processing annual targets and tailings-cells designs for waste management as part of the scheduling problem."

"The MILGP model also seeks to support effective reclamation planning in the mined-out areas using pseudo backfilling revenue generated through in-pit tailings management."

Summary of Literature Review

"Some of the primary limitations in literature with regards to integrated mine planning and waste management include one or more of the following: (1) handling waste disposal planning as a post-production scheduling optimization activity and not integrated directly in the strategic mine planning optimization process; (2) providing predefined mining and processing targets as inputs through trial and error; (3) not integrating limited duration stockpiling to manage oxidation of ore during production scheduling optimization; and (4) not incorporating all functional metallurgical variables including bitumen, fines, and organic-rich solids that impact processing and waste management."

"A new MILGP framework is proposed to concurrently schedule for production, reclamation, and dyke material and optimize the size, shape, and location of in-pit tailings-cells in addition to generating the optimal mining and processing annual targets in an integrated mine planning and waste management scheme."

Oil Sands Mining as a Case Study

"Each mining-cut contains one or more of ore, interburden and overburden dyke materials, reclamation material, or waste."

"These strategic production schedules and tailings-cells optimization impacts the profitability and sustainability of oil sands mining projects."

"Modeling mining with mining-panels, processing with mining-cuts, and tailings-cells creation with mining-cells ensure that the orebody extraction progresses with practical mining width and the proposed mathematical model formulation size is computationally tractable."

"While increasing the number of mining-panels, mining-cuts, and mining-cells in the formulation improve selectivity and hence NPV during optimization, that may sometimes result in a problem size that cannot be deployed on regular computers."

"A good balance is required as to what is the appropriate number of mining-panels, mining-cuts, and mining-cells that facilitates practical mining, generates a good NPV, and can be easily deployed on regular computers."

MILGP Theoretical Model Formulation for Integrated Production and Waste Disposal Planning

"The proposed oil sands production scheduling model integrates waste management through dyke construction in the mined-out areas for tailings containment."

"The production schedule is controlled by a variety of economic, technical, physical, and regulatory constraints, which enforce mining extraction sequence, dyke construction capacities, blending, and backfilling requirements."

"The economic mining-cut value is determined based on the costs incurred in mining from a designated location and revenue from a processing destination, as well as dyke construction cost at a specified destination."

"The objective function also aims to minimize the reclamation and dyke materials mining cost and deviations from the production goals."

"These constraints include: (a) Variables logic control: the logic of the mining, processing, stockpiling, reclamation and dyke materials, backfilling, and deviational variables with regard to their limits and definitions are within acceptable ranges."

"(b) Non-negativity control: all decision variables for mining, processing, stockpiling, reclamation and dyke materials, backfilling, and deviational variables are non-negative."

Application of the MILGP Framework for Integrated Strategic Mine Planning

"Tailings-cells optimization constraints are used to control the backfilling in the in-pit mined areas."

"The performance of the proposed MILGP framework was analyzed based on the NPV, mining and processing production targets, smoothness and practicality of the generated schedules, availability of tailings containment areas at the required time, percentage of remaining backfilling volume, pseudo backfilling revenue, and computational time required for convergence."

"The implemented scenarios focus on a practically integrated oil sands production planning and tailings-cells optimization strategy that generates value and is sustainable."

"This includes directional mining and the availability of tailings-cells for in-pit dyke construction and subsequently tailings deposition."

"Dyke material is needed when all mining-cells required to create a specific tailings-cell are mined out and dyke construction is ready to take off."

"Increasing the number of mining-cells in the east-west direction gives the optimizer more flexibility when generating the integrated production schedule."

Conclusions and Recommendations

"The integrated oil sands mine planning problem involves the incorporation of waste management into the production planning process in an optimization framework that maximizes the project value and generates a sustainable waste management plan."

"The model also optimizes tailings-cells designs for waste disposal planning in line with regulatory requirements."

"Four waste management scenarios based on variations in the number of mining-cells are investigated."

"The results show that decreasing the mining-cell size and increasing the number of mining-cells used to create in-pit tailings-cells increase the NPV of the operation."

"Decreasing the number of mining-cells in the vertical direction (north-south) increases the CPU time as demonstrated in scenarios 2 and 4."

"This integrated mine planning framework can be implemented for various directions of mining and different sizes and shapes of tailings-cells."

Mathematical Model for Robotic Two-Sided Assembly Line Balancing Problem with Zoning Constraints [40]

This is a machine-generated summary of:

Yadav, Ashish; Agrawal, Sunil: Mathematical model for robotic two-sided assembly line balancing problem with zoning constraints [40].

Published in: International Journal of System Assurance Engineering and Management (2021).

Link to original: https://doi.org/10.1007/s13198-021-01284-8.

If you want to cite the papers, please refer to the original.

For technical reasons we could not place the page where the original quote is coming from.

Abstract-Summary "In, the recent scenario, the utilization of robots in different areas such as packaging, loading/unloading, transportation, and especially assembly lines enhance the productivity of production systems."

"Robotic two-sided assembly line balancing problem (RTALBP) generally occurs in plants producing high volume, large-sized products, where there is a process of installing more than one robot on every single station of the assembly line for manufacturing the product."

"Robotic two-sided assembly line balancing problem is well known NP-hard problem that's why an exact solution approach is proposed to solve the problem."

"real-life assembly line balancing based production case study data is collected, and a proposed mathematical model is applied to get a feasible solution."

Introduction

"Robotic assembly line balancing problems basically divide into simple and general robotic assembly line balancing problems."

"In a robotic two-sided assembly line product can be manufactured or assembled in a parallel direction both at the left and right sides of the lines."

"The task in robotic two sided assembly line will have direction restriction to perform because the use of both sides of the line."

"The major difference between a simple robot and a two-sided robotic assembly line is that the sequence of the tasks within a workstation is not important in a simple robot assembly line but in a robotic two-sided assembly line plays a crucial role for an efficient assignment of tasks."

"The robotic two-sided assembly line has many advantages over a simple robot assembly line those are mentioned below: These advantages motivate authors to work in this direction [41]."

Literature Review

"Nilakantan et al. [8, 42] presented a mathematical programming model with a Particle swarm optimization (PSO) algorithm with the objective of minimizing cycle time and minimize energy consumption for solving robotic assembly line balancing problems."

"Zhou and Wu [43] presented a mathematical programming model with Adaptive Simulated Annealing and genetic algorithm (ASAGA) algorithm with the objective of minimizing cycle time for solving robotic assembly line balancing problems."

"Nilakantan et al. [44, 45] In this paper author presented robotic assembly line balancing (RALB) problems with the mathematical formulation of two objectives including minimizing the cycle time and minimizing the cost are considered under precedence, capacity constraints and developed differential evolution (DE) optimization method to solve it optimally."

"presented mathematical programming model with simulated annealing optimization (SA) algorithm with an objective of minimizing cycle time and minimize energy consumption for solving robotic two-sided assembly line balancing (RTALB) problem."

Problem Structure and Mathematical Model

"The real world assembly line balancing problems such as resource optimization, idle time minimization etc. motivates the author to think about the improvement in this direction."

"Based on the recent research work in the area of robotic two-sided assembly line balancing, the authors are motivated to improve the existing assembly line of the 'ABC' plant by converting the same into the robotic two-sided assembly line."

"The line efficiency, robot utilization, and total productivity of the plant increase when the plant is designed with a robotic two-sided assembly line balancing approach."

"The mathematical model and its accuracy are verified using the test problems and are applied to solve the assembly line balancing problem of the existing plant."

"The robotic two sided assembly line problem is solved by mixed integer mathematical programming model."

Data Collection and Solution Approach

"This section indicates the benchmark problem data with the solutions that indicate feasibility of the mathematical model."

"Authors in the further process considered production case study data and explain the methodology, solution approach steps to solve production case study problem."

"Benchmark problem tested with mathematical models to check the accuracy of mathematical models."

"In further steps, when authors get the optimal and feasible results of the benchmark problem, authors applied this concept to solve the production case study problem with step one to three that is explained below."

"Lingo 16 solver is an operation research software tool that use mathematical model concept to solve optimization problems."

"In a further step, when authors get the optimal results of the benchmark problem, authors applied this concept to solve the production case study problem that is explained below."

Computational Results and Discussion

"Results indicate that Lingo 16 solver uses branch and bound methodology in solver execution and authors check its capability to get optimal solutions and computational time taken for solution convergence."

"The output reveals that this solver is not capable to provide optimal solutions for large size problem and the solver time increase exponentially with the problem dimensions."

"Robot B efficiency is 71.43% where number of robots reduces from 7 to 2. In solution, process result indicates that all robots are trained to perform any task in any side and any workstation play a major role to increase efficiency."

"Results indicate that Lingo 16 solver uses branch and bound methodology which indicates that computational time to find optimal solutions using this solver is very much slower."

"The computational time grows along with problem size in branch and bound optimization approach that's why this solver is not capable to provide optimal solutions for large-size problems."

Conclusion and Future Research

"The proposed mathematical programming formulation is able to solve small-size benchmark problems, in reasonable time with optimal solutions therefore, this paper solves the production case study problem."

"The proposed mathematical programming formulation is able to solve the numerical problem, with good solutions therefore authors applied this approach to solve the production case study problems."

"Computational results indicate that the reduced number of workstations increases the efficiency of the plant, now the robotic line efficiency of the case study problem is 70.59%."

"The computational results indicate that a proposed mathematical model provides a good solution for the minimized number of robots and minimized number of workstations, this will helps the plant indirectly to reduce the energy consumption and cost of purchase more robots."

"This mathematical model can be applied with new meta-heuristic methods such as whale optimization and grey wolf optimization to get the optimal solution for the robotic two-sided assembly line balancing problem."

Hybrid Genetic Algorithm for a Type-II Robust Mixed-Model Assembly Line Balancing Problem with Interval Task Times [46]

This is a machine-generated summary of:

Zhang, Jia-Hua; Li, Ai-Ping; Liu, Xue-Mei: Hybrid genetic algorithm for a type-II robust mixed-model assembly line balancing problem with interval task times [46].

Published in: Advances in Manufacturing (2019).

Link to original: https://doi.org/10.1007/s40436-019-00256-3.

If you want to cite the papers, please refer to the original.

For technical reasons we could not place the page where the original quote is coming from.

Abstract-Summary "A robust optimization model for this problem is formulated to hedge against uncertainty."

"The counterpart of the robust optimization model is developed by duality."

"A hybrid genetic algorithm (HGA) is proposed to solve this problem."

"An adaptive local search procedure and a discrete Levy flight are hybridized with the genetic algorithm (GA) to enhance the performance of the algorithm."

"The effect of uncertainty parameters on production efficiency is also investigated."

Introduction

"Both stochastic mixed-model assembly line balancing [47, 48] and fuzzy mixed-model assembly line balancing [49] have been proposed to deal with uncertain task times."

"In stochastic mixed-model assembly line balancing, it is assumed that task times are subject to probability distribution, generally normal distribution."

"In fuzzy mixed-model assembly line balancing, task times are assumed to be fuzzy numbers with given membership."

"Many researchers have adopted the robust optimization approach to study the assembly line balancing problem with uncertain task times: Al-e-hashem et al. [50] considered the type-I robust mixed-model assembly line balancing problem (RMMALBP-I)."

"Only Al-e-hashem et al. [50] have used the robust optimization method to study the mixed-model assembly line balancing problem with uncertain task times, in addition to formulating a mathematical model of RMMALBP-I. However, there is no related work on RMMALBP-II, which is directly related to the production rate of the mixed-model assembly line."

Problem Description and Model Formulation

"A mixed-model assembly line with m workstations is considered."

"p similar models are assembled simultaneously in an intermixed sequence."

"The precedence relationship of each model is predefined and all relationships can be combined into only one precedence graph with n tasks."

"The aim is to minimize cycle time C. The RMMALBP-II has the following assumptions: (1) The task time of each model is uncertain and is represented by a given interval dataset. (2) Each common task for different models must be assigned to the same workstation for economy. (3) The precedence graphs for different models are predefined, and a combined precedence graph can be obtained. (4) The line is a paced line with a fixed cycle time. (5) The line is serial with no feeder lines or parallel workstations. (6) There are no assignment restrictions of tasks except precedence constraints. (7) There are no buffers between workstations."

Proposed Hybrid Genetic Algorithm for RMALBP-II

"Empty A, and go to Step 5. (2) If it is the heuristic method, generate a random combination of input parameters a, b, and c, and choose the task that maximizes the priority rule from A. The task is assigned for the jth gene in the forward direction search or for the (n − j + 1)th in the backward direction search."

"With each chromosome being a feasible task sequence, the procedure decides which tasks can be assigned to predefined workstations respecting the cycle time constraint."

"Step 1 k chromosomes in the population are selected at random."

"Step 1 The mutation point is generated randomly, and one chromosome from the population is divided into head and tail."

"Step 4 Obtain the feasible task set S using the precedence matrix, which makes the Levy flight obtain a feasible solution."

Numerical Experiments

"In the tables, the "OCT" column is the optimal solution found by the exact algorithm; "CPU/s" column refers to the average processing time in seconds spent by the algorithms; "Improvement/%" column presents the comparative results of the performance of the HGA with GA."

"For the small size instances, both the HGA and GA converge to the same optimal solutions found by the integer solver."

"They spend more CPU time on small size instances, the GA and HGA spend less CPU time than LINGO on medium and large-size instances."

"Compared with the processing time of the GA, 22.22%, 69.35%, and 52.66% increases in average CPU times of the HGA are observed, respectively, for small, medium, and large-size instances."

"With the HGA and GA converging to the same optimal solutions for the small size instances and only one instance having the OCT solution among the large size instances, the statistical test is implemented for the medium size instances."

Conclusions

"Mixed-model assembly lines are widely used in industries at present."

"With its practical benefits, the consideration of uncertain task times in the mixed-model assembly balancing problem is important."

"The robust model of this problem and its counterpart are formulated."

"For future research, the mixed-model assembly line balancing and sequencing with interval uncertainty could be considered simultaneously."

An Integrated Approach to Optimize the Configuration of Mass-Customized Products and Reconfigurable Manufacturing Systems [51]

This is a machine-generated summary of:

Sabioni, Rachel Campos; Daaboul, Joanna; Le Duigou, Julien: An integrated approach to optimize the configuration of mass-customized products and reconfigurable manufacturing systems [51].

Published in: The International Journal of Advanced Manufacturing Technology (2021).

Link to original: https://doi.org/10.1007/s00170-021-06984-w.

If you want to cite the papers, please refer to the original.

For technical reasons we could not place the page where the original quote is coming from.

Abstract-Summary "The current global unpredictable market is characterized by increasing demand for highly customized products."

"MPD's configuration phase represents a key step for mass customization because it allows customers to be integrated into the value-creation process."

"Reconfigurable manufacturing systems (RMS) seem to be the most appropriate manufacturing system to manufacture mass-customized products due to their ability to be quickly reconfigured, adjusting their production capacity and functionality to fit new market demands."

"Aiming to integrate individual customer needs with product and manufacturing decisions, this paper proposes a new 0–1 nonlinear integer programming model to concurrently optimize the configuration of modular products and RMS (including machine and layout configuration levels), driven by individual customer requirements."

"An example of customizable office chairs is used to illustrate the proposition, and different scenarios of customer requirements and RMS configurations were presented."

"This work confirms the relevance of integrating modular product and RMS configuration decisions for minimizing costs of producing highly mass-customized products."

Introduction

"According to Koren [52], the production of highly customized or personalized products requires a new business model with a closer interaction between the product development process and the manufacturing system, keeping the main focus on the customer."

"Regarding the modular products manufacturing in RMS, this becomes still more critical since the RMS's reconfigurability added to the wide product modules choice and individual customer requirements can considerably increase the system complexity regarding product differentiation [53, 54]."

"Investigating the concurrent optimization of product's and RMS's configurations for mass-customizing products to meet customer uniqueness seems to be of great importance."

"This paper aims to fill these gaps by integrating the optimization of mass-customized product configuration—driven by individual customer requirements—with its process planning (i.e., operations sequence and machine-configuration selection) and layout configuration in an RMS."

Background and Literature Review

"Although the previous works addressed layout configuration issues, they mainly placed machines into different stages, according to the production demand, without considering costs associated with machines' displacement from one stage to another neither material handling costs."

"Although some papers did not address layout issues, they worried about material handling costs when generating their process plan according to the machine-configuration selection and operations sequencing for planning the process [55, 56] or production [57]."

"Although there is a considerable quantity of papers addressing the process plan generation and layout configuration in RMS, the number of works directly addressing the RMS layout configuration issues (e.g., costs of displacing machines, handling material) is still limited."

"Most of the papers addressing process/production plan and layout issues treated both problems sequentially (into separated steps), or they did not address machine configuration or layout configuration issues in a concrete way (i.e., detailing costs of displacing machines or handling material)."

Problem Statement

"The research question addressed in this paper is: How to integrate the product configuration, driven by individual customer requirements, with its process planning and layout configuration in a reconfigurable manufacturing system to minimize the overall manufacturing cost?"

"It means that production cost cannot be dependent only on the module variant since it varies according to the type of machine configuration assigned to an operation."

"The decision for process planning consists of identifying which pair of machine configuration will perform each required operation to produce the module instances and in which sequence these operations will be performed."

"These decisions are to be made while considering the current configuration of machines and layout, in order to minimize all manufacturing costs, including costs of (1) raw material of selected module instances, (2) manufacturing operations, and reconfiguration of (3) machine and (4) layout as well as (5) material handling."

Proposition

"The candidate module instances, i.e., those capable of satisfying at least one PFC, will filter the solution space of RMS configuration, which includes the process planning (operations sequence and machine-configuration assignment) and machine layout configuration."

"RMS configuration information for process planning and layout design: the machine configurations' cost/time required to accomplish each operation p is supposed to be available."

"Up to our knowledge, there is no work addressing a GA-based approach to solve an optimization problem that concurrently makes the decisions of (1) product configuration, (2) operations identification, (3) process planning, and (4) layout configuration, driven by individual customer requirements, as presented in this paper."

"All individuals randomly generated must present the operations in part 2 correspondent to the operations required by module instances selected in part 1, even if these module instances do not correspond to a valid product variant regarding customer satisfaction needs."

Illustrative Example

"In this approach, the product configuration is optimized according to the customer requirements, meaning that the product module instances need to be selected to meet the product functions required by the customer."

"In example 1, the product variant corresponds to an office chair with a high back and lumbar support (M1I1), standard height (M2I1), and adjustable arms (M4I2), made in vinyl (M315) equipped with casters adapted for hard floor and carpet (M5I1)."

"Results for case 1 evidence that even when the initial layout configuration is changed the final layout configuration keeps the same, meaning that modifying machines' positions, in this case, does not contribute to reducing material handling costs."

"This confirms this method's stability in terms of calculation time since no matter the example used or the initial configuration of machines and layout, it keeps a computation time close to 10 min, which is relatively low considering the problem size."

Conclusion and Future Work Directions

"This paper proposed an approach to optimize the integration of the product configuration with the process planning and layout configuration in an RMS, driven by individual customer requirements."

"After GA being set up with the optimal parameters, it ran 30 independent times for two different examples of customer required functions, and then, one example of customer required functions was used to compare two different cases in which the initial configurations of the (1) layout and (2) machines were changed."

"The results evidenced the GA-based approach's ability to find good solutions in a reasonable computation time for finding an optimal configuration for product and RMS when different customer requirements are considered."

"Future research aims to further investigate this trade-off by comparing this integrated approach to the separated ones (i.e., single optimization of product configuration, process planning, and layout configuration) considering different scenarios from the product and process levels."

A Multi-Objective Two-Sided Disassembly Line Balancing Optimization Based on Artificial Bee Colony Algorithm: A Case Study of an Automotive Engine [58]

This is a machine-generated summary of:

Zhang, Lei; Wu, Yuanfeng; Zhao, Xikun; Pan, Shiwen; Li, Ziqi; Bao, Hong; Tian, Yongtin: A Multi-objective Two-sided Disassembly Line Balancing Optimization Based on Artificial Bee Colony Algorithm: A Case Study of an Automotive Engine [58].

Published in: International Journal of Precision Engineering and Manufacturing-Green Technology (2021).

Link to original: https://doi.org/10.1007/s40684-021-00394-9.

If you want to cite the papers, please refer to the original.

For technical reasons we could not place the page where the original quote is coming from.

Abstract-Summary "To optimize the disassembly system and the use of resources, the disassembly line balancing problem (DLBP) has attracted much attention."

"Compared with the conventional one-sided straight disassembly line, the two-sided disassembly line can use both the left and right sides of a conveyor belt for disassembly operation, thereby improving production efficiency, especially for large-sized and complicated products."

"Due to the constraints and precedence among parts, it is a challenge to plan the disassembly scheme for a two-sided disassembly line."

"A model is established to solve a two-sided disassembly line balancing problem (TDLBP)."

"A hybrid graph is utilized to express constraints and precedence relationships, and a novel encoding and decoding method is developed for the disassembly scheme planning of a two-sided line for handling the challenge caused by constraints and precedence among parts."

Introduction

"To improve the disassembly system and optimize the use of resources, the researchers proposed the disassembly line balancing problem (DLBP), that is, to reasonably assign disassembly tasks to each workstation, evenly distribute tasks, and achieve a balancing state while satisfying the constraint and precedence relationship among parts [59]."

"To reduce construction costs and improve disassembly efficiency, the number of workstations and the idle time are the preferred optimization objectives."

"The conventional straight disassembly line follows a simple linear process mode where workers can only disassembly one part at a time, and only one side is utilized, and the majority of current DLBP researches are based on it [60]."

"A multi-objective two-sided disassembly line model is established to optimize the number of mated-workstations, idle time, smoothness, and auxiliary processes, and the ABC algorithm is adapted and applied to the current model."

Precedence Model and Disassembly Scheme Generation

"The disassembly task assignment method and following multi-objective models are based on these assumptions for problem simplification [61, 62]: (a) The disassembly time of each part is known in advance; (b) Completely disassemble the product and disassemble a single product on the disassembly line; (c) The supply of end-of-life product is infinite; (d) Each task can only be assigned to one station; (e) The total disassembly time of tasks assigned to one side of mated-workstation shall not exceed cycle time CT; (f) All the parts in a product are available and recyclable, without deletions, additions, modifications, physical or functional defects."

"The decoding refers to plan a disassembly scheme based on the type of the disassembly line and disassembly sequence, i.e. assigning each disassembly task to a certain workstation and determining its start and end time."

Multi-objective Functions and Optimization Method

"The employed bees search for new food sources (new solutions) and share information with onlooker bees."

"According to the available disassembly scheme, the balancing state of a disassembly line can be evaluated by the objective functions, including the number of mated-workstations (f_1), idle time (f_2), auxiliary operation indicator (f_3), smoothness index (f_4)."

"Perform the selected operator on solution X and get new solution X′; Step 4."

"If a solution is dominated by the solution before its neighbor search, put this solution into inferior set Q. In the neighborhood search, the new solution X′ may be dominated by the original solution X, and collect all the X′, namely inferior set

Q. In this phrase, the two-point crossover of GA is adopted to explore solution space and perform a two-point crossover on the inferior set."

Case Study

"To test the optimization performance of IMOABC, three multi-objective algorithms are introduced for comparison, i.e., the famous NSGA-II [63], HABC [64] (an ABC recently applied to DLBP), and the multi-objective flower pollination algorithm (MOFPA) [65] (a recent TDLBP optimization method)."

"In terms of mated-workstation, compared with MOFPA and IMOABC, all methods can get the 6 mated-workstations, while HABC and NSGA-II have fewer solutions of six mated-workstations."

"It means that the optimization results of the IMOABC's Pareto solution on the idle time objective is significantly better than the other three methods, which improves work efficiency significantly."

"Whether it is the quality of the Pareto solution set measured by the hypervolume indicator or the overall optimization effect of each objective, IMOABC performs better than the other three algorithms."

"The IMOABC has a better performance on two-sided disassembly line scheme optimization compared with these three methods."

Conclusion and Future Work

"The majority of published DLBP works are mainly based on a straight one-sided disassembly line, only one part can be removed at a time, and this linear production mode leads to a long work time."

"This work studied DLBP based on a two-sided disassembly line for disassembly efficiency improvement and construction costs reduction."

"Aiming to reduce the number of mated-workstations, idle time, auxiliary indicator, and smooth index, a multi-objective two-sided disassembly line balancing model is established, and an improved multi-objective artificial bee colony algorithm which based on Pareto non-dominated relationship, crossover and mutation operators of GA is proposed for disassembly scheme optimization."

"Besides, the result indicates that the developed methodology can guide the disassembly of end-of-life products on a two-sided disassembly line."

"It is necessary to introduce a mixed-model for a two-sided disassembly line, which can handle various products by one line."

Mixed Model Two Sided Assembly Line Balancing Problem: An Exact Solution Approach [66]

This is a machine-generated summary of:

Yadav, Ashish; Verma, Pawan; Agrawal, Sunil: Mixed model two sided assembly line balancing problem: an exact solution approach [66].

Published in: International Journal of System Assurance Engineering and Management (2020).

Link to original: https://doi.org/10.1007/s13198-020-00956-1.

If you want to cite the papers, please refer to the original.

For technical reasons we could not place the page where the original quote is coming from.

Abstract-Summary "In MTALBP there is a procedure of introducing two single stations in each position left and right of the assembly line for the combined product model."

"The proposed mathematical model is solved with benchmark test problems mentioned in research papers and applied to solve the case study problem of a turbocharger assembly line plant."

"The experimental results of the case study problem show that line efficiency is obtained 86.50% for model A and 80.75% for model B and the number of single and mated stations of the assembly line is close to the theoretical minimum number of stations."

"Results indicate that applying boundary conditions reduce the computational time to solve the case study problem as well as minimizes the number of stations, reduces idle time and reduces the length of the assembly line for the MTALBP."

Introduction

"In two-sided assembly line balancing (TALB) product can be manufactured or assembled in the parallel direction both at the left and right sides of the lines."

"In the TALB problem interference phenomenon occurs, where the left side task can interfere with the right side task through precedence restrictions that might create idle time on that condition station and worker requirements wait till the past assignment completion process at the reverse side of the line [67, 68]."

"This paper generally presents the following input that can help the researcher to solve assembly line problems: (1) The main aim of this work is to propose a mathematical model for workload maximization on mated stations for the feasible task allotment in the MTALB problem. (2) The new mathematical model is verified on T(9), T(12) and T(16) benchmark problem after that solve the turbocharger assembly line case study problem to obtain the optimal solutions."

Literature Review

"Simaria and Vilarinho [69] presented a mathematical model with an ant colony based optimization algorithm for tackling a MTALB problem with an objective of limits the quantity of stations of the line."

"The author considered first objective minimizes the number of MS and the second objective limits the quantity of stations with the consideration of cycle time."

"Chutima and Chimklai [70] presented a negative knowledge-based particle swarm optimization to solve the multi-objective problem of MTALBP with the objective of limits the quantity of MS when the cycle time is given."

"Delice et al. [71] presented an improved PSO algorithm to solve the MTALBP by considering minimize the number of MS as the first objective and limits the quantity of stations as the second objective when the cycle time is given."

"The author considered the objective minimizes the number of MS with the consideration of cycle time to get the optimal result."

Problem Statement

"They told various problems like production rate, quality assembly line balancing, and non-arrival of material in time."

"In unstructured ALB task are not systematically distributed so that cycle time is high and directly line efficiency reduced."

"In the case of mixed model based two-sided line more attention is needed in precedence constraints since tasks are assigned on both sides of the line as well as different models."

"The main objective of the proposed model is to assign the set of tasks of model A and model B to the mated stations of two-sided assembly line so that the number of single stations, idle time are reduce and indirectly increase space and reduce computational time on the MTALB problem."

"Since assembly line balancing is an np hard problem that's why increase a single variable or constraint puts a lot of increment in time."

Problem Data and Method

"This section indicates the benchmark problem data with the solutions that indicate feasibility of the mathematical model."

"Lingo 16 solver is an operations research software tool that use mathematical model concept to solve optimization problems."

"Theoretical least number of stations can be calculated by the equation mentioned below: In equation fifteen (Z) + indicates the lowest integer greater than or equals to Z. Benchmark problem is solved by taken the maximum value of task time for model A and model B that maximum value of task time considered as a mixed model task time."

"After considering the mixed model task time, cycle time is calculated three for problem T(9) and T(12)."

"Problem T(16) is solved based on cycle time ten which the authors get by the calculation of mixed model task time."

"Result also indicates that the idle time of mated-stations for both models A, B is optimized."

Computational Result and Analysis

"A total of 5 numbers of mated stations is utilized to assigned 22 tasks."

"Results indicate that for the case study problem, the total numbers of mated stations are less as compared to the theoretical least number of stations."

"According to the equation, sixteen theoretical least numbers of stations are ten that is higher as compared to the total number of mated stations that are five for the turbocharger case study problem."

"Left side model B shows maximum idle time as compared to the right side model A. In this industrial case study problem idle time are reduced because of two reasons first one minimization of station where maximization of workload on each assembly line station."

"One assignment of multiple task on a single station where reduced number of stations assigns their tasks to other stations that had sufficient time to perform another task."

Conclusions and Future Research

"A real-life case study with the MTALB concept is presented in this paper, which considers two different assembly line models with a straight layout."

"Numerical example problem such as T(9), T(12) and T(16) is solved utilizing the proposed way to validate the line efficiency of the proposed model."

"The proposed mathematical model can explain the numerical problem, in a reasonable time; therefore the authors applied this methodology to resolve the case study problem."

"86.50% for model A and 80.75% for model B. Results indicates that applying boundary conditions reduce the computational time to solve the case study problem."

"According to the industrial scenario different real world case study based assembly line balancing problems can be resolved by the help of this approach."

An Optimization Approach for Setup Planning and Operation Sequencing with Tolerance Constraints [72]

This is a machine-generated summary of:

Wu, Wenbo; Huang, Zhengdong; Wu, Kangxiang; Chen, Yongfu: An optimization approach for setup planning and operation sequencing with tolerance constraints [72].

Published in: The International Journal of Advanced Manufacturing Technology (2020).

Link to original: https://doi.org/10.1007/s00170-019-04791-y.

If you want to cite the papers, please refer to the original.

For technical reasons we could not place the page where the original quote is coming from.

Abstract-Summary "Setup planning and operation sequencing are two interrelated problems in computer-aided process planning."

"They are addressed by most traditional methods separately, in which setup planning focuses on precision assurance, while operation sequencing is devoted to cost reduction."

"A constrained optimization approach is proposed to address the setup planning and operation sequencing problems in a combined way."

"With an objective to minimize the manufacturing cost, the optimization model includes various constraints, especially the precision constraints on setup error, tool error, and stacking error."

Introduction

"CAPP mainly includes three tasks: (1) translating a part's geometric model into machining features; (2) deciding available machining resources and possible machining operations for each feature and enumerating its candidate setup plans; and (3) selecting an appropriate setup plan and operation sequence that has the lowest processing cost under the constraints from both design and manufacturing practice."

"Plenty of approaches have been developed for solving the problems of setup planning and operation sequencing."

"An ideal optimal process plan should be the joint planning result of both setup planning and operation sequencing, since joint planning is beneficial to tackling the constraints and achieving the objective in the planning model."

"From the perspective of constraints, setup plan and operation sequence are mutually constrained by each other."

"To solve the joint optimization problem for setup planning and operation sequencing problem, this paper utilizes a heuristic method called the modified particle swarm optimization (PSO)."

Related Works

"Huang et al. [73, 74] proposed a classic graph-theoretical approach for rotational parts based on tolerance analysis, in which design specifications are presented as a graph and the setup planning problem is converted into a graph searching problem."

"Rule-based reasoning is another kind of setup planning method, in which researchers should organize existing machining knowledge and experience beforehand in certain way."

"Huang [75] developed such a method for automatic setup planning of lathe machining."

"As the precedence and tolerance relations are not considered, this method can only handle single TAD machining problems."

"Besides the methods mentioned above, fuzzy approach [76–78] and artificial neural network [79–81] are also developed for setup planning and operation sequencing problem."

"Setup planning and operation sequencing are involved in multiple machining factors."

"This article adopts a particle swarm optimization (PSO) algorithm to solve the tolerance-constrained setup planning and operation sequencing problem after it is formulated as an optimization model with multiple constraints."

Model Formation

"Operation is the basic element in the setup planning and operation sequencing model, in which TAD, machine tool (MT), cutting tool (CT), datum face (DF), and clamping face (CF) are five variables that need to be designated."

"Given the total number of cutting tools n_c, CL can be represented as: In addition, operations can be separated into different setups, according to TAD, MT, DF, and CF in PS."

"For the i-th operation, the constraint condition can be given as: Here, λ_1 is set as the maximum tool error of the process system that can be accepted in the process procedure."

"Given the i-th operation with a selected setup datum in setup S_k, the constraint condition for the stacking error can be given as: Here, $T_{k,i}$ is the accumulative error from the given datum of S_k to the i-th operation."

Model Solution

"The values saved in d include the nominal dimension, upper deviation, and lower deviation: Step 2: Identify the datum feature p and the machining feature q, and set the matrix entries Dim(p,q) and Dim(q,p) as 1."

"In the algorithm, matrix dim keeps the signed dimensions; the dimensions forward from the datum to the feature are positive while those backward are negative."

"The accumulative error from datum 1 to feature 3 is calculated as follows: The PSO algorithm uses a population of random candidate solutions called particles in the searching space to find the objective optimum location."

"They are calculated according to the particle matrix PS and its setup matrix s. To evaluate whether the generated process solution meets the precision requirements, setup error, stacking error, and tool error are calculated in this step."

Case Study

"The total cost changes from 2527 to 2427. (2) Comparing plan B with plan C, the operation distribution in each setup is changed owing to the value variation of λ_0."

"Since the λ_0 in plan C is smaller than both T_{f3} and T_{f4}, these two pairs of operations are put in the same setup, respectively. (3) Comparing plan A with plan B, the machining resources selection is changed owing to the variation of λ_1."

"Since λ_1 equals to 0.008, the machine and tool combination of f_4, f_9, f_{12}, and f_{13} in plan A are changed in plan B. The tool error of each operation in plan B satisfies the constraint after the adjustment. (4) Comparing plan B with plan C, the datum selection for each operation is changed owing to the variation of λ_2."

Conclusion and Future Work

"A multi-constraint optimization model has been developed to generate the cost-optimum solution with guaranteed precision for the setup planning and operation sequencing problem."

"One can obtain the most efficient process plan at an achievable precision level by quantitatively setting the scalar values of the error constraints. (3) A PSO method is adopted to solve this multi-constrained optimization model, which not only provides the global optimal solution but also improves the computational efficiency."

"In this point of view, the future work would concentrate on improving the error predictions of machining processes, extending the quality constraints to be more comprehensive for a product, and adapting the constrained optimization model proposed in this paper to other types of process planning problems."

A Decision Support Methodology for Integrated Machining Process and Operation Plans for Sustainability and Productivity Assessment [82]

This is a machine-generated summary of:

Hatim, Qais Y.; Saldana, Christopher; Shao, Guodong; Kim, Duck Bong; Morris, K. C.; Witherell, Paul; Rachuri, Sudarsan; Kumara, Soundar: A decision support methodology for integrated machining process and operation plans for sustainability and productivity assessment [82].

Published in: The International Journal of Advanced Manufacturing Technology (2020).

Link to original: https://doi.org/10.1007/s00170-019-04268-y.

If you want to cite the papers, please refer to the original.

For technical reasons we could not place the page where the original quote is coming from.

Abstract-Summary "This paper presents a systematic methodology to enable environmental sustainability and productivity performance assessment for integrated process and operation plans at the machine cell level of a manufacturing system."

"Integration of process and operation plans allows globalized assessment of sustainability and productivity, while development of a multi-criteria decision-making method leads to optimization of process planning activities based on the impact of conflicting sustainability and productivity metrics."

"A case study is detailed to demonstrate the sustainability-focused methodology, wherein integrated simulation and optimization techniques are used to support analysis of candidate scenarios and selection of preferred alternatives from a finite set of alternate process and operation plans."

"The outcomes of this work include determination of optimized feature sequence plans which optimize various key performance indicators depending on stakeholder interest based on time, sustainability and production cost."

Introduction

"To assist in planning manufacturing processes with consideration beyond traditional productivity factors, methods and tools are required for assessing impact on environment, cost, and productivity."

"The primary challenge is to simultaneously assess and optimize the combined impact of various process and operation plans with sustainability and productivity performance considerations."

"The combined influence of the selection of process and process settings affects tool and lubricant use, energy consumption, emissions, cost, and quality of the product, as well as overall productivity measures such as throughput."

"A simulation model of a production shop allows investigation of the combined impacts of process selection and process settings."

"Further, the interactions between selection of processes, machines, and process parameters can help make complex decisions to balance sustainability and productivity [83, 84]."

"We propose an integrated methodology to aid the assessment of sustainability performance and decision-making for process planning of machined parts."

Related Work

"Several modeling techniques are attractive for integrating optimization of process and operation planning with regard to productivity and sustainability."

"Zhang and Liling [85] proposed a method for process planning optimization with energy efficiency consideration and used the network part program technique to generate multiple alternative process plans."

"Wang et al. [86] proposed a systematic approach for milling process planning and scheduling optimization by using multiple objectives (e.g., energy efficiency, productivity) and constraints (e.g., surface quality)."

"The integrated and systematic methodology proposed in this paper can provide the optimal operation sequence, as well as the optimal process settings, by integrating non-linear process planning, DES modeling, and MCDM techniques."

"A generalized data model for the integrated process planning approach is proposed."

Methodology for Integrating Process and Production Plans

"By defining the problem goal and scope, relevant information for assessing the impact of different scenarios with respect to sustainability and productivity is provided to stakeholders (e.g., process planners)."

"Dispersed data are then gathered and tabulated according to the goal of the assessment and system boundary conditions should be integrated to provide required information for designing planning strategies and evaluating their performance."

"Simulating integrated process and production planning scenarios to determine the impacts on different performance indicators could be time-consuming because of extensive comparisons needed across alternative planning options."

"An attractive solution is to combine simulation and optimization methods through mathematical formulation of process and production planning activities, control variables, data, and constraints for supporting various KPI assessments."

"Before determining the process plan that has the optimal impact on environmental sustainability and productivity, a range of process parameters is assigned according to the operation-method matrix entities, and the relevant data are retrieved from the data model."

Case Study

"The following are necessary criteria for evaluating the energy consumption of various types of processes: (1) the set of suitable manufacturing processes, (2) setup requirements, (3) values of process-related material information (e.g., formability, machinability, weldability, hardenability), (4) process effects on product durability and desirability, and (5) effects of design requirements (e.g., tolerances, surface roughness) on process selection."

"To explicitly identify the unique resources for each part feature, analysis models (i.e., combined simulation and optimization models) were developed to investigate the sustainability and productivity impacts for alternative processes and operation plans scenarios."

"The returned information populates the selection lists in the section of process/machine/tool. (2) A combination of material, process, machine, tool, and their related data is input into the simulation and optimization models. (3) Based on the objective of the stakeholder, the simulation and optimization model determines the optimal process settings with respect to both sustainability and productivity considerations."

Conclusion and Future Work

"We developed a systematic methodology for enabling sustainability and productivity performance assessment for integrated process and operation plans at the machine cell level of manufacturing systems."

"Various scenarios involving sequencing the processes and optimizing machine settings were analyzed to determine optimal performance for selected KPIs."

"Sensitivity analysis was performed to determine the importance of process parameter settings on these metrics for different process and production alternatives."

"Although a machining-focused case study was considered in this paper, the methodology can be used for a wide range of production applications for assessing sustainability and productivity impacts when both process and operation plans are integrated into a G2G life cycle."

Hybrid Frameworks for Flexible Job Shop Scheduling [87]

This is a machine-generated summary of:

Bharti, Puja; Jain, Sushma: Hybrid frameworks for flexible job shop scheduling [87].

Published in: The International Journal of Advanced Manufacturing Technology (2020).

Link to original: https://doi.org/10.1007/s00170-020-05398-4.

If you want to cite the papers, please refer to the original.

For technical reasons we could not place the page where the original quote is coming from.

Abstract-Summary "Improvisation of schedule by simultaneous optimization of the performance measures is imperative for the manufacturing firms to stay ahead of competition and is one of the key responsibilities of shop floor managers."

"This research approaches the problem by developing two hybrid frameworks."

"The first framework (Hybrid-I) is a co-evolutionary combination of the social thinking capability of particle swarm optimization and local search capability of gravitational search algorithm."

"The second approach (Hybrid-II), hybridize the effectiveness of genetic algorithm in finding global best region with PSO's cluster interactions to improve the search for an optimal solution."

"With the view to make a reasonable comparison between both approaches and with the state-of-the-art methods, tests have been conducted on 28 benchmark instances taken from three different data sets."

"The substantial performance of the proposed approach on benchmark instances as well as real-life industrial data supports a strong candidature for optimization of FJSSP."

Introduction

"Lately, the research community has shown deeper interest in hybridization, i.e., combining two or more approaches to solve such optimization problems."

"The novelty of this paper consists of the following four points: (1) to deal with the limitations of traditional algorithms for scheduling optimization, this paper presents hybrid algorithms capable of rapidly identifying the global optimum solution. (2) Efficiency of PSO is enhanced by adopting variable random function strategy along with the mutation strategy; improved version of PSO is named as ePSO. (3) As GA and GSA both suffer from the inherent disadvantages of trapping in local minima and the slow convergence rate resulting in low solution quality, the proposed techniques overcome this problem by integrating them with ePSO. (4) The managerial implication of the proposed work is verified using a real-life industrial case."

Literature Review

"Nagata and Ono proposed a local search-based method that works in partial solution space to solve JSSP [88]."

"Some of the recent and most popular approaches in this category are genetic algorithm (GA), particle swarm optimization (PSO), artificial bee colony (ABC), gravitational search algorithm (GSA), ant colony optimization (ACO), etc. Rossi and Dini solved job shop with routing flexibility, sequence-dependent setup, and transportation time using ACO-based approach with makespan objective."

"Initial solutions were generated using localization methods and the multi-objective FJSSP was solved by combining hybrid fuzzy logical algorithm with EA."

"Hosseinabadi and others introduced a multi-parametric objective function with dynamic and flexible parameters to develop a new method called TIME_GELS that used the gravitational emulation local search algorithm (GELS), to solve multi-objective flexible dynamic job shop scheduling problem [89]."

"Zhang and others and Moslehi and Mahnam [90, 91] applied particle swarm optimization (PSO) algorithms to solve the multi-objective FJSSP."

Problem Formulation

"Each job consists of operations, denoted by O_{ij}, that are to be performed in a predefined sequence, where ni is total number of operations for i job $\{O_{ij}|i \in n, j \in ni\}$."

"Each operation of job i is executed by any machine defined in set M. The following assumptions are considered: setup times of machines and move times between operations are negligible; all jobs and machines are available at time 0; and each operation must be completed without interruption once it starts, i.e., preemption is not allowed."

"Machine selection variable For the selection of the machine M_k, for the execution of the operation O_{ij},for j sequence of the job i, switching variable, Z_{ijk} is used which carries the value either 0 or 1 for the selection of machines."

Proposed Work

"Particle swarm optimization approach, given by Kennedy and Eberhart in 1975 [92, 93], mimics the social behavior of swarm of bees/flock of birds/school of fish, etc. In PSO, each solution is represented as a particle having two properties: position and velocity."

"The masses attract each other by gravitational force that exist between them and is mathematically shown in equations below: Total force acting on ith agent in d-dimensional problem space is given by: Acceleration of an agent is calculated using law of motion as follows: Velocity and position of agents is calculated as follows: Like any other evolutionary approach, performance of GSA can be improved by controlling exploration and exploitation capability."

"PSO is similar to GA in random initialization of the population and global search through succeeding generations, it does not use crossover and mutation operations."

"As the PSO algorithm does not require sorting of fitness values of solutions in any process, this offers a significant computational advantage over GA, particularly in the case of large population size."

Experiment Results

"The results show that for small instance, all the approaches performed at par as far as MS, MW, and TW are concerned; however, both the hybrid approaches performed better on idle time value (12 as compared to 13 for standard approaches)."

"On comparing the schedules generated by the proposed approaches, we observe that for ka 4 × 5 problem, both approaches performed at par but for ka 10 × 10 and ka 10 × 15 problems Hybrid-II offers better allocation of operations with superior objective values."

"In case of 10 × 10 problem, Hybrid-I performs better than temporal decomposition approach on all objectives, while Hybrid-II outperformed "temporal decomposition," "classic GA," and "approach by localization" on all objectives and reduced TW [50] under the same value of MS and MW in comparison with "PSO + SA" and "PSO + TS."".

Industrial Case Study

"The company has experienced a steady increase in their orders, but due to lack of proper scheduling, the company faces long production hours, resulting in extra costs in terms of extended labor working hours, increased energy consumption, high in-process inventory, etc. At present, the company schedules the manufacturing operations manually based on intuition and experience."

"According to it, order 1 consists of three jobs, job J1 has 100 units and requires four operations O_{11}, O_{12}, O_{13}, O_{14} which can be carried out on machines of groups G1/G3, G2, G6, and G1/G2/G6 respectively."

"As every job has a specific number of units to be processed, the total processing time of an operation O_{ij} on machine k corresponds to the time taken to process all the units of that job."

Conclusion

"This paper presents two PSO-based hybrid algorithms to solve the scheduling problem corresponding to a FJSSP with three objectives, namely, makespan, maximal machine workload, and total workload."

"The computational results obtained indicate that Hybrid-I performs well for small-scale instance whereas Hybrid-II outperforms standard approaches as well as existing approaches in majority of the cases."

"The industrial application of Hybrid-II demonstrates 69.3%, 52.8%, and 31.5% improvement in MS, MW, and TW with regard to the former manual scheduling method used in the plant."

"Possible extensions of this study include the expansion of the proposed design to solve problems by incorporating stochastic measures such as due dates, machine breakdown, and other uncertainties faced by various manufacturing industries as well as to implement it in scheduling areas other than manufacturing, like transportation and logistics."

An Industry 4.0 Approach to Assembly Line Resequencing [94]

This is a machine-generated summary of:

Rossit, Daniel Alejandro; Tohmé, Fernando; Frutos, Mariano: An Industry 4.0 approach to assembly line resequencing [94].

Published in: The International Journal of Advanced Manufacturing Technology (2019).

Link to original: https://doi.org/10.1007/s00170-019-03804-0.

If you want to cite the papers, please refer to the original.

For technical reasons we could not place the page where the original quote is coming from.

Abstract-Summary "These assembly systems will be heavily influenced by the advent of Industry 4.0 technologies, enabling the proposal of business models that allow the late customization of the products, i.e., the customer can modify attributes of its product once the production of it is started."

"This business model requires that the manufacturing tools are able to make decisions online and negotiate with the customer the changes that can be carried out, according to the workload flowing through the production system."

"We show that Industry 4.0 systems can autonomously manage the production management process, and then, we present a framework based on tolerance planning strategies (tolerance scheduling problem), to determine which changes can be carried out."

"The ability of resequencing the production process is also implemented in the case that the operations associated with late customization allow it (i.e., when intermediate buffers are available)."

Introduction

"It is natural to think that the entire flow of tasks in these mass customized production processes, involving the planning of operations and their implementation, can be handled in a decentralized and autonomous way by the CPS themselves."

"Bortolini et al. [95] state that the products produced in ALs may not only be personalized but also that late customization will be possible."

"Late customization can be seen as a postponement production strategy [96] as well as a leagile framework [97]."

"We intend to show how late customization may be implemented in a production plan."

"Under a business model that allows late customization, the events that modified the initial scenario (with which production was initially scheduled) are of another nature: these events are modifications of operations and specifications."

"This proposal works by combining problems of tolerance scheduling problem [98] and of non-permutation sequences [99], and allows the company's management system to analyze the late customizations of customers regarding the state of its production system."

Assembly Line Problems and the Industry 4.0 Approach

"We present succinctly some Industry 4.0 notions and review the main contributions of Industry 4.0 in AL systems found in the literature to set our contribution in context."

"With the enhanced connectivity of production systems, this information can be transmitted to processing centers with higher capacity, endowing decision support systems (DSS) with data traditionally restricted to control levels and thus becoming useful for CPS [100–102]."

"We will now review the main contributions in the literature on the application of Industry 4.0 features to AL systems."

"Bortolini et al. [95] propose a framework to develop AL systems in Industry 4.0 environments and the future production paradigm of "personalized production," allowing the customer to participate both in the design stage of the product and in the incorporation of modifications to the final product."

The Late Customization Model

"These manufacturing models intend to face the uncertainty of the market by increasing the response capability of the organization in order to satisfy the customers with similar costs to mass production industries [103]."

"Leagile systems get the best of lean and agile production processes, for both the generic components and the customized final products [104]."

"Bortolini et al. [95] indicate that this opens the possibility to offer the client the possibility of late customizing their product since the availability of real-time information on the state of the production process allows late modifications (i.e., after placing the order)."

"The product platform, by handling materials or components in late customization scenarios, improves the efficiency of the process."

"This indicates that the new technologies allow producing low volumes or even single units without increasing the costs over mass production."

Production Planning and Control in Industry 4.0 Environments

"This means that these technologies have to handle production orders and the flow of material, as well as program production operations, execute the orders, control the execution, etc. These functions are usually hierarchically structured, with higher level functions, restricting lower level functions, according to standards as for instance ANSI/ISA 95 or ISA 95."

"This will be due to the ability of CPPS to carry out a wide spectrum of activities, ranging from the physical operations of production (level 0) to planning, evaluating, and managing the entire production process (level 3), by controlling the actions and systems on levels 1 and 2 (i.e., the measurement and sensing instruments as well as the control systems)."

"We have shown that CPPS can autonomously handle several levels of ISA-95 from the physical process to the master production plan."

"The scheduling process gets embedded in those levels, indicating that CPPS could solve autonomously the scheduling problem."

Resequencing Assembly Lines in Industry 4.0 Environments

"For the first case, of consecutive workstations, the customer should have the possibility of modifying the product within the tolerances defined by the tolerance scheduling problem."

"These late customizations may imply, for example, varying the operating times on a workstation, according to the tolerances defined by the system when solving the tolerance scheduling problem at the beginning of production."

"The second case of late customizations contemplated in our design is when the product arrives in an intermediate storage buffer."

"If the customer, after having agreed to place the order, wants to make a modification to its product can choose the option of late customization."

"Even if the current tolerances cannot absorb the customer's late requirement, it is passed to a second evaluation instance: resequencing the products in the intermediate buffer (in our example, reordering the production sequence from the buffer $B_{3/4}$)."

Future Work

"To advance in the design and implementation of a framework to offer late customization services to the customer, it is necessary to develop further two aspects of this framework: on the one hand, the assignment of tolerance to any given production order and, on the other, generating efficient calculation tools for resequencing the production process."

"These mechanisms will depend on the company, which may opt for an auction, where all customers bid to obtain the margin of tolerance to make their late modifications."

"A vital aspect in all these analyses will be the value of the factor δ for the calculation of the tolerance, since a high value will allow more slack to the company for the incorporation of late modifications of the customers, but at the expense of the efficiency of the sequence for that δ (the larger the δ, the larger the difference between the optimal sequence and the selected one) [98]."

Conclusion

"A framework was presented that allows supporting late customization strategies in AL systems."

"This framework presents as an innovation the possibility of implementing it on an autonomous system, since it can determine whether a late modification required by a client can be processed by the AL given the current production conditions, without requiring a scheduler to manage the system."

"It also allows incorporating the possibility of resequencing the workflow when possible, increasing the possibility of incorporating late modifications."

References

1. Gonnermann C, Hashemi-Petroodi S, Thevenin S, Dolgui A, Rüdiger D (2022) A skill- and feature-based approach to planning process monitoring in assembly planning. Int J Adv Manuf Technol. https://doi.org/10.1007/s00170-022-09931-5
2. Gonnermann C, Weth J, Reinhart G (2020) Skill modeling in cyber-physical production systems for process monitoring. Proc CIRP 93:1376–1381. https://doi.org/10.1016/j.procir.2020.03.095
3. Garcia F, Lanz M, Järvenpää E, Tuokko R (2011) Process planning based on feature recognition method. In: 2011 IEEE international symposium on assembly and manufacturing (ISAM). IEEE, Piscataway, pp 1–5. https://doi.org/10.1109/ISAM.2011.5942296
4. Jones T, Reidsema C, Smith A (2006) Automated feature recognition system for supporting conceptual engineering design. KES J 10:477–492. https://doi.org/10.3233/KES-2006-10606
5. Gonnermann C, Reinhart G (2019) Automatized setup of process monitoring in cyber-physical systems. Proc CIRP 81:636–640. https://doi.org/10.1016/j.procir.2019.03.168

6. Gonnermann C, Zels B, Reinhart G (2021) Automatized generation of alternatives for process monitoring in cyber-physical assembly systems. Proc CIRP 104:732–737. https://doi.org/10.1016/j.procir.2021.11.123
7. Chen J, Xiaoliang J (2022) Energy-efficient integration of assembly line balancing and part feeding with a modified genetic algorithm. Int J Adv Manuf Technol. https://doi.org/10.1007/s00170-022-09422-7
8. Nilakantan JM, Huang GQ, Ponnambalam SG (2015) An investigation on minimizing cycle time and total energy consumption in robotic assembly line systems. J Clean Prod 90:311–325
9. Nilakantan JM, Ponnambalam SG, Jawahar N (2016) Design of energy efficient RAL system using evolutionary algorithms. Eng Comput 33(2):580–602
10. Zhou BH, Shen CY (2018) Multi-objective optimization of material delivery for mixed model assembly lines with energy consideration. J Clean Prod 192:293–305
11. Nourmohammadi A, Eskandari H, Fathi M, Bourani MR (2018) An integrated model for cost-oriented assembly line balancing and parts feeding with supermarkets. Proc CIRP 72:381–385
12. Koç MA (2022) A new expert system for active vibration control (AVC) for high-speed train moving on a flexible structure and PID optimization using MOGA and NSGA-II algorithms. J Braz Soc Mech Sci Eng. https://doi.org/10.1007/s40430-022-03441-x
13. Xia Z, Gong D, Zhou J et al (2018) Decoupling optimization design of under-chassis equipment suspension system in high-speed trains. Shock Vib. https://doi.org/10.1155/2018/6292595
14. Jin T, Liu Z, Ren Z, Li X (2018) Study of combination damping characteristics effect of high-speed Train Damper. Huanan Ligong Daxue Xuebao 46:116–124. https://doi.org/10.3969/j.issn.1000-565X.2018.09.016
15. Foo E, Goodall RM (2000) Active suspension control of flexible-bodied railway vehicles using electro-hydraulic and electro-magnetic actuators. Control Eng Pract 8:507–518. https://doi.org/10.1016/S0967-0661(99)00188-4
16. Yang YB, Yau JD (2017) Resonance of high-speed trains moving over a series of simple or continuous beams with non-ballasted tracks. Eng Struct. https://doi.org/10.1016/j.engstruct.2017.04.022
17. Koç MA (2021) Finite element and numerical vibration analysis of a Timoshenko and Euler-Bernoulli beams traversed by a moving high-speed train. J Braz Soc Mech Sci Eng 43:165. https://doi.org/10.1007/s40430-021-02835-7
18. Koulocheris D, Papaioannou G, Christodoulou D (2017) An approach for multi-objective optimization of vehicle suspension system. IOP Conf Ser Mater Sci Eng 252:012037. https://doi.org/10.1088/1757-899X/252/1/012037
19. Jin Z, Zhang S, Weng J (2021) Immune particle swarm optimization of SUV rollover crashworthiness and weight. Struct Multidiscip Optim. https://doi.org/10.1007/s00158-021-02906-2
20. Hou S, Liu T, Dong D (2014) Factor screening and multivariable crashworthiness optimization for vehicle side impact by factorial design. Struct Multidiscip Optim 49:147–167
21. Wang C, Wang D, Zhang S (2016) Design and application of weight reduction multi-objective collaborative optimization for a parametric body-in-white structure. Proc Inst Mech Eng Part D 230(2):273–288
22. Walczak J, Guillermin O, Bathe KJ (1999) Advances in crush analysis. Comput Struct 72:31–47
23. Kouloughli S, Korso Feciane M, Zaki S (2022) Mobile rack AS/RS dimensions optimization for single cycle time minimization. Int J Adv Manuf Technol. https://doi.org/10.1007/s00170-022-09450-3
24. Kouloughli S (2015) Multi-aisle AS/RS dimensions optimization for cycle time minimization. Int J Adv Manuf Technol 79:675–692
25. De Koster MBM, Le-Anh T, Yu Y (2007) Optimal storage rack design for a 3 dimensional compact AS/RS. Int J Prod Res 46(6):1495–1514
26. Guezzen AH, Sari Z, Castagna P, Cardin O (2013) Travel time modeling and simulation of a mobile racks automated storage/retrieval system. Int J Eng Technol 5(3):420–423

27. Kouloughli S (2013) Optimisation de systèmes automatisés de stockage/déstockage multi-allées et à racks glissants. Manufacturing Engineering University of Tlemcen (Doctoral thesis)
28. Zhang L, Zhao X, Ke Q, Dong W, Yanjiu Z (2019) Disassembly line balancing optimization method for high efficiency and low carbon emission. Int J Precis Eng Manuf-Green Technol. https://doi.org/10.1007/s40684-019-00140-2
29. McGovern SM, Gupta SM, Kamarthi SV (2003) Solving disassembly sequence planning problems using combinatorial optimization. In: Proceedings of the 2003 Northeast Decision Sciences Institute conference. Northeast Decision Sciences Institute, Hershey, pp 178–180
30. Ministry of Commerce of the People's Republic of China (2017) Report of China renewable resource recycling industry development. Resour Recycl 5:16–25
31. Wang YL, Li FY, Chen J et al (2009) Hybrid graph disassembly model and sequence planning for product maintenance. In: Technology and innovation conference. ITIC, Hyderabad, pp 515–519
32. Gungor A, Gupta SM (2001) A solution approach to the disassembly line problem in the presence of task failures. Int J Prod Res 39(7):1427–1467
33. Kalayci CB, Gupta SM (2014) A tabu search algorithm for balancing a sequence-dependent disassembly line. Prod Plan Control 25(2):149–160
34. Maremi A, Ben-Awuah E, Askari-Nasab H (2021) Multi-objective mathematical programming framework for integrated oil sands mine planning and tailings disposal optimization. Mining Metallurgy Explor. https://doi.org/10.1007/s42461-021-00418-7
35. Ben-Awuah E, Askari-Nasab H (2013) Incorporating waste management into oil sands long term production planning. Min Technol 122(1):33–45. https://doi.org/10.1179/1743286312Y.0000000031
36. Ben-Awuah E, Askari-Nasab H, Awuah-Offei K (2012) Production scheduling and waste disposal planning for oil sands mining using goal programming. J Environ Inf 20(1):20–33
37. Ben-Awuah E, Askari-Nasab H, Maremi A, Seyed Hosseini N (2018) Implementation of a goal programming framework for production and dyke material planning. Int J Min Reclam Environ 32(8):536–563. https://doi.org/10.1080/17480930.2017.1331082
38. Syncrude Canada Ltd (2011) Annual tailings plan submission syncrude Aurora North. Energy Resources Conservation Board, Fort McMurray
39. Syncrude Canada Ltd (2012) Annual tailings plan submission syncrude mildred lake mine. Energy Resources Conservation Board, Fort McMurray
40. Yadav A, Agrawal S (2021) Mathematical model for robotic two-sided assembly line balancing problem with zoning constraints. Int J Syst Assur Eng Manag. https://doi.org/10.1007/s13198-021-01284-8
41. Li Z, Kucukkoc I, Nilakantan JM (2017) Comprehensive review and evaluation of heuristics and meta-heuristics for two-sided assembly line balancing problem. Comput Oper Res 84:146–161
42. Nilakantan JM, Ponnambalam SG, Jawahar N, Kanagaraj G (2015) Bio-inspired search algorithms to solve robotic assembly line balancing problems. Neural Comput & Applic 26(6):1379–1393
43. Zhou B, Wu Q (2018) A novel optimal method of robotic weld assembly line balancing problems with changeover times: a case study. Assem Autom 38(4):376–386
44. Nilakantan JM, Nielsen I, Ponnambalam SG, Venkataramanaiah S (2017) Differential evolution algorithm for solving RALB problem using cost-and time-based models. Int J Adv Manuf Technol 89(1–4):311–332
45. Nilakantan JM, Li ZX, Tang QH, Nielsen P (2017) Multi-objective co-operative co-evolutionary algorithm for minimizing carbon footprint and maximizing line efficiency in robotic assembly line systems. J Clean Prod 156:124–136
46. Zhang J-H, Li A-P, Liu X-M (2019) Hybrid genetic algorithm for a type-II robust mixed-model assembly line balancing problem with interval task times. Adv Manuf. https://doi.org/10.1007/s40436-019-00256-3

47. Zhang WQ, Xu WT, Liu G et al (2017) An effective hybrid evolutionary algorithm for stochastic multiobjective assembly line balancing problem. J Intell Manuf 28(3):783–790
48. Dong JT, Zhang LX, Xiao TY (2018) A hybrid PSO/SA algorithm for bi-criteria stochastic line balancing with flexible task times and zoning constraints. J Intell Manuf 29(4):737–751
49. Van Hop N (2006) A heuristic solution for fuzzy mixed-model line balancing problem. Eur J Oper Res 168(3):798–810
50. Al-e-hashem SMJM, Aryanezhad MB, Malekly H et al (2009) Mixed model assembly line balancing problem under uncertainty. In: Paper presented at the 2009 international conference on computers and industrial engineering. IEEE, pp 233–238
51. Sabioni RC, Daaboul J, Le Duigou J (2021) An integrated approach to optimize the configuration of mass-customized products and reconfigurable manufacturing systems. Int J Adv Manuf Technol. https://doi.org/10.1007/s00170-021-06984-w
52. Koren Y (2010) The global manufacturing revolution: product-process-business integration and reconfigurable systems. Wiley, Hoboken
53. Tseng MM, Jiao J, Merchant ME (1996) Design for mass customization. CIRP Ann 45:153–156
54. Koren Y, Gu X, Guo W (2018) Choosing the system configuration for high-volume manufacturing. Int J Prod Res 56:476–490
55. Touzout FA, Benyoucef L (2019) Multi-objective sustainable process plan generation in a reconfigurable manufacturing environment: exact and adapted evolutionary approaches. Int J Prod Res 57:2531–2547
56. Touzout FA, Benyoucef L (2019) Multi-objective multi-unit process plan generation in a reconfigurable manufacturing environment: a comparative study of three hybrid metaheuristics. Int J Prod Res 57:7520–7535
57. Bortolini M, Galizia FG, Mora C, Pilati F (2019) Reconfigurability in cellular manufacturing systems: a design model and multi-scenario analysis. Int J Adv Manuf Technol 104:4387–4397
58. Zhang L, Wu Y, Zhao X, Pan S, Li Z, Bao H, Tian Y (2021) A multi-objective two-sided disassembly line balancing optimization based on artificial bee colony algorithm: a case study of an automotive engine. Int J Precis Eng Manuf-Green Technol. https://doi.org/10.1007/s40684-021-00394-9
59. Gupta S, Gungor A (1999) Disassembly line balancing. In: Proceedings of the 1999 annual meeting of the Northeast Decision Sciences Institute. Northeast Decision Sciences Institute, Hershey, pp 193–195
60. Laili Y, Li Y, Fang Y, Pham DT, Zhang L (2020) Model review and algorithm comparison on multi-objective disassembly line balancing. J Manuf Syst 56:484–500
61. McGovern SM, Gupta SM (2007) A balancing method and genetic algorithm for disassembly line balancing. Eur J Oper Res 179(3):692–708
62. Zhang Z, Wang K, Zhu L, Wang Y (2017) A Pareto improved artificial fish swarm algorithm for solving a multi-objective fuzzy disassembly line balancing problem. Expert Syst Appl 86:165–176
63. Deb K, Pratap A, Agarwal S, Meyarivan T (2002) A fast and elitist multiobjective genetic algorithm: NSGA-II. IEEE Trans Evol Comput 6(2):182–197. https://doi.org/10.1109/4235.996017
64. Wang S, Guo X, Liu J (2019) An efficient hybrid artificial bee colony algorithm for disassembly line balancing problem with sequence-dependent part removal times. Eng Optim 51(11):1920–1937
65. Wang K, Li X, Gao L (2019) A multi-objective discrete flower pollination algorithm for stochastic two-sided partial disassembly line balancing problem. Comput Ind Eng 130:634–649
66. Yadav A, Verma P, Agrawal S (2020) Mixed model two sided assembly line balancing problem: an exact solution approach. Int J Syst Assur Eng Manag. https://doi.org/10.1007/s13198-020-00956-1

67. Loganathan MK, Gandhi OP (2016) Maintenance cost minimization of manufacturing systems using PSO under reliability constraint. Int J Syst Assur Eng Manag 7(1):47–61
68. Ighravwe DE, Oke SA, Adebiyi KA (2017) Preventive maintenance task balancing with spare parts optimisation via big-bang big-crunch algorithm. Int J Syst Assur Eng Manag 8(2):811–822
69. Simaria AS, Vilarinho PM (2009) 2-ANTBAL: an ant colony optimization algorithm for balancing two-sided assembly lines. Comput Ind Eng 56(2):489–506
70. Chutima P, Chimklai P (2012) Multi-objective two-sided mixed-model assembly line balancing using particle swarm optimisation with negative knowledge. Comput Ind Eng 62(1):39–55
71. Delice Y, Aydogan EK, Ozcan U, İlkay MS (2017) A modified particle swarm optimization algorithm to mixed-model two-sided assembly line balancing. J Intell Manuf 28(1):23–36
72. Wu W, Huang Z, Wu K, Chen Y (2020) An optimization approach for setup planning and operation sequencing with tolerance constraints. Int J Adv Manuf Technol. https://doi.org/10.1007/s00170-019-04791-y
73. Huang SH, Zhang HC (1996) Tolerance analysis in setup planning for rotational parts. J Manuf Syst 15:340–350
74. Huang SH, Zhang HC, Oldham WJB (1997) Tolerance analysis for setup planning: a graph theoretical approach. Int J Prod Res 35:1107–1124
75. Huang SH (1998) Automated setup planning for lathe machining. J Manuf Syst 17:196–208
76. Zhang HC, Huang SH (1994) A fuzzy approach to process plan selection. Int J Prod Res 32:1265–1279
77. Sun KK, Mou SS, Qiu JB (2019) Adaptive fuzzy control for non-triangular structural stochastic switched nonlinear systems with full state constraints. IEEE Trans Fuzzy Syst 27:1587–1601
78. Qiu JB, Sun KK, Wang T, Hu HJ (2019) Observer-based fuzzy adaptive event-triggered control for pure-feedback nonlinear systems with prescribed performance. IEEE Trans Fuzzy Syst. https://doi.org/10.1109/TFUZZ.2019.2895560
79. Chang CA, Angkasith V (2001) Using Hopfield neural networks for operational sequencing for prismatic on NC machines. Eng Appl Artif Intell 14:357–368
80. Chen J, Zhang YF, Nee AYC (1998) Setup planning using Hopfield net and simulated annealing. Int J Prod Res 36:981–1000
81. Mei JN, Zhang HC, Oldham WJB (1995) A neural network approach for datum selection in computer-aided process planning. Comput Ind 27:53–64
82. Hatim QY, Saldana C, Shao G, Kim DB, Morris KC, Witherell P, Rachuri S, Kumara S (2020) A decision support methodology for integrated machining process and operation plans for sustainability and productivity assessment. Int J Adv Manuf Technol. https://doi.org/10.1007/s00170-019-04268-y
83. Rajemi M, Mativenga P, Jaffery S (2009) Energy and carbon footprint analysis for machining titanium Ti-6AI-4V Alloy. J Machine Eng 9(1):103–112
84. Gutowski T, Dahmus J, Thiriez A (2006) Electrical energy requirements for manufacturing processes. In 13th CIRP International Conference on Life Cycle Engineering, Leuven, Belgium
85. Zhang Y, Liling G (2015) Method for process planning optimization with energy efficiency consideration. Int J Adv Manuf Technol 77(9–12):2197–2207
86. Wang S, Lu X, Li X, Li W (2015) A systematic approach of process planning and scheduling optimization for sustainable machining. J Clean Prod 87:914–929
87. Bharti P, Jain S (2020) Hybrid frameworks for flexible job shop scheduling. Int J Adv Manuf Technol. https://doi.org/10.1007/s00170-020-05398-4
88. Nagata Y, Ono I (2018) A guided local search with iterative ejections of bottleneck operations for the job shop scheduling problem. Comput Oper Res 90:60–71
89. Hosseinabadi AAR, Siar H, Shamshirband S, Shojafar M, Nasir MHNM (2015) Using the gravitational emulation local search algorithm to solve the multi-objective flexible dynamic job shop scheduling problem in small and medium enterprises. Ann Oper Res 229(1):451–474

90. Zhang G, Shao X, Li P, Gao L (2009) An effective hybrid particle swarm optimization algorithm for multi-objective flexible job shop scheduling problem. Comput Ind Eng 56(4):1309–1318
91. Moslehi G, Mahnam MA (2011) Pareto approach to multi-objective flexible job-shop scheduling problem using particle swarm optimization and local search. Int J Prod Econ 129(1):14–22
92. Kennedy J, Eberhart RC (1995) Particle swarm optimization. IEEE IJCNN 4:1942–1948
93. Eberhart R, Kennedy J (1995) A new optimizer using particle swarm theory. In: Proceedings of the Sixth International Symposium on Micro Machine and Human Science. SCIRP, Irvine, pp 39–43
94. Rossit DA, Tohmé F, Frutos M (2019) An industry 4.0 approach to assembly line resequencing. Int J Adv Manuf Technol. https://doi.org/10.1007/s00170-019-03804-0
95. Bortolini M, Ferrari E, Gamberi M, Pilati F, Faccio M (2017) Assembly system design in the industry 4.0 era: a general framework. IFAC-PapersOnLine 50(1):5700–5705
96. Bucklin LP (1965) Postponement, speculation and the structure of distribution channels. J Mark Res 2:26–31
97. Naylor JB, Naim MM, Berry D (1999) Leagility: integrating the lean and agile manufacturing paradigms in the total supply chain. Int J Prod Econ 62(1–2):107–118
98. Rossit DA, Tohme F, Frutos M (2018) Industry 4.0: smart scheduling. Int J Prod Res 57(12):3802–3813. https://doi.org/10.1080/00207543.2018.1504248
99. Rossit DA, Tohmé F, Frutos M (2018a) The non-permutation flow-shop scheduling problem: a literature review. Omega 77:143–153
100. Rossit D, Tohmé F (2018) Scheduling research contributions to smart manufacturing. MFGLET 15(B):111–114
101. Yao X, Zhou J, Lin Y, Li Y, Yu H, Liu Y (2019) Smart manufacturing based on cyber-physical systems and beyond. J Intell Manuf 30:2805–2817. https://doi.org/10.1007/s10845-017-1384-5
102. Wang L, Törngren M, Onori M (2015) Current status and advancement of cyber-physical systems in manufacturing. J Manuf Syst 37(Part 2):517–527
103. Yusuf YY, Gunasekaran A, Adeleye EO, Sivayoganathan K (2004) Agile supply chain capabilities: determinants of competitive objectives. Eur J Oper Res 159(2):379–392
104. Mason-Jones R, Naylor B, Towill DR (2000) Lean, agile or leagile? Matching your supply chain to the marketplace. Int J Prod Res 38(17):4061–4070

Chapter 6
Optimization Methods in Fault Detection and Diagnosis

Introduction by the Editor

For any system to run seamlessly and without failure is very important in several aspects including overall cost, market value and competition as well as overall performance of the entire process. This is very essential for any manufacturing system that the faults can be predicted/detected and diagnosed, which may help for scheduling preventive maintenance as well as finding the right solution in right time. The chapter discusses the artificial intelligence and machine learning methods such as genetic algorithm, Bayesian classifier, decision tree, random forest, k-nearest neighbour, support vector machine as well as artificial neural networks, Long Short-Term Memory that have been used to detect, classify, and locate faults a variety of faults for the real-world cases of conditioning monitoring of the roller bearing systems, short-circuit fault detection, traction motor fault detection, etc. The chapter also reviews the self-supervised leaning method such as data driven Kernel Principal Component Analysis for the fault prognosis usable for the small scale as well as large scale manufacturing system. In addition, the chapter also describes generative adversarial network uses labelled and unlabelled images for classifying die cast products and helps in detecting the fault in the system.

A critical survey of applying machine learning methods in fault detection and classification of power transmission lines is elaborated in very details. Along with the generic machine learning methods, it discusses presents a comprehensive review of various machine learning methodologies including naive Bayesian classifier, decision tree, random forest, k-nearest neighbour, and support vector machine as well as artificial neural networks such as feedforward neural network, convolutional neural network, and adaptive neuro-fuzzy inference system that have been used to detect, classify, and locate faults in transmission lines. The review associated with promoting accurate condition monitoring of bearing such as intelligent instance annotation and automatic feature extraction and selection is also discussed in details.

A. J. Kulkarni (ed.), *Optimization Methods in Manufacturing Processes*,
https://doi.org/10.1007/978-981-96-5257-0_6

The elaborated comparison of the results revealed the superior performance of the gaussian mixture models and convolutional neural networks compared to existing advanced classifiers. It is important to know that the fault detection and diagnosis is a quite challenging field and constitutes a crucial factor in industrial systems. It is in details reviewed in accordance with fuzzy cognitive networks with functional weights and multiple classifier systems for a case study of squirrel cage induction generator. The comparison results with several machine learning approaches show evidence that our implementations classify more accurately the induction generator operating conditions.

An important approach reviewed is a fault diagnosis method for electromechanical actuators based on machine learning. It involves the neural network learning algorithm and network parameters and improves the classification effect of the neural network. It is important to mention that the work reviews the motor fault detection model and application based on machine learning and wavelet analysis. The simulation results reveals that the diagnostic accuracy is more efficient than the traditional support vector machine and back propagation type of the neural network methods. In addition, the analysis concludes proposing a new idea and method for the fault diagnosis of permanent magnet fault-tolerant motors. A self-supervised learning approaches for fault detection is discussed and reviewed in very details. The applicability of the approach demonstrated based on the effectiveness of the proposed algorithm. It involves the evaluation of the algorithm on a benchmark dataset for fault detection, and the results show that the algorithm outperforms other fault detection methods. The case study involves evaluation and experimentation on a benchmark dataset for fault detection highlighting that the deep learning exhibited its advantages on several tasks, and further underscored the need to explore the research of self-supervised deep learning method on fault detection.

The chapter also reviews deep learning models such as autoencoder, convolutional neural networks, and long short-term memory can automatically extract suitable features for detecting faults from raw sensor trace data. It highlights that the semiconductor manufacturing field struggle with the accuracy paradox, which occurs when a training dataset has substantially more normal data than fault data, making the classifier predict most new data as normal. In such situation, one-class learning can be a solution for training fault detection models using only normal data and classifying a new wafer instance as a fault if the instance is located outside the decision boundary in the feature space. The work reviewed on the self-organizing fuzzy classifier suggests a data-driven framework based on real-time fault detection of dynamically evolving large-scale data streaming systems. The main contributions of this work are development of a general framework for real-time fault detection to be applied to dynamically evolving large-scale data streaming systems that allow identification of possible abnormal conditions before a fault occurs. It may help in preventing more serious consequences for the system. In addition, it concludes that the work implements a method that can adapt quickly to changes in streaming data patterns and does not require large databases to be used in the model update even for large-scale data systems.

Machine Generated Summaries

Disclaimer: The summaries in this chapter were generated from Springer Nature publications using extractive AI auto-summarization: An extraction-based summarizer aims to identify the most important sentences of a text using an algorithm and uses those original sentences to create the auto-summary (unlike generative AI). As the constituted sentences are machine selected, they may not fully reflect the body of the work, so we strongly advise that the original content is read and cited. The auto generated summaries were curated by the editor to meet Springer Nature publication standards. To cite this content, please refer to the original papers.

Machine generated keywords: fault, detection, fault detection, diagnosis, signal, feature, classifier, fault diagnosis, motor, bear, classification, feature extraction, monitor, extraction, sensor.

Application of Machine Learning Methods in Fault Detection and Classification of Power Transmission Lines: A Survey [1]

This is a machine-generated summary of:

Shakiba, Fatemeh Mohammadi; Azizi, S. Mohsen; Zhou, Mengchu; Abusorrah, Abdullah: Application of machine learning methods in fault detection and classification of power transmission lines: a survey [1].

Published in: Artificial Intelligence Review (2022).

Link to original: https://doi.org/10.1007/s10462-022-10296-0

If you want to cite the papers, please refer to the original.

For technical reasons we could not place the page where the original quote is coming from.

Abstract-Summary "This paper presents a survey on recent machine learning-based techniques for fault detection, classification, and location estimation in transmission lines."

"Costly consequences of probable faults motivate the need for immediate actions to detect them using intelligent methods, especially emerging machine learning approaches that are powerful in solving diagnosis problems."

"This paper presents a comprehensive review of various machine learning methodologies including naive Bayesian classifier, decision tree, random forest, k-nearest neighbor, and support vector machine as well as artificial neural networks such as feedforward neural network, convolutional neural network, and adaptive neuro-fuzzy inference system that have been used to detect, classify, and locate faults in transmission lines."

Introduction

"An approach is chosen for fault identification (detection and classification of TL faults) and/or fault location estimation."

"Traveling wave (TW)-based approaches are used in industry to address the problem of detection, classification, and location estimation of the faults [2–8]."

"Although such techniques have advantages such as their independence from the network configuration, remaining unaffected by load variations, high grounding resistance, and series capacitor, they suffer from being expensive, requiring high sampling frequency for capturing the high frequency fault transients, the limited capability to distinguish between waves reflected from a fault and the remote end of a TL, and the fundamental limitation of not detecting the faults at zero crossing of the voltage waveform [9–11]."

"The machine learning (ML) techniques discussed in this study make use of the amplitudes of the fundamental frequency component of the voltage and current signals to avoid dealing with the transient waves generated from faults."

Generic Machine Learning Methods

"In Aker et al. [12] voltage and current signals are given to a DWT for feature extraction, and NBC is used to classify TL faults."

"k-NN is a classifier that usually performs detection and classification of faults in TLs, and only a few papers have presented location estimation of faults [13, 14]."

"The study [15] proposes a scheme for fault classification based on fuzzy k-NN and continuous wavelet features."

"An approach for TL fault detection and classification is presented by Dasgupta et al. [16] using cross-correlation and k-NN."

"A semi-supervised ML methodology is proposed in Abdelgayed et al. [17] that co-trains DTC and k-NN to generate the classification model for labeled and unlabeled data to automate the fault classification process."

Artificial Neural Networks (ANNs)

"Besides their similar structures, every type of ANN has its own characteristics such as computational methods for updating the value of weights, the connection of neurons among layers, storing significant values, removing the unwanted data, etc.

Among various types of ANNs, some of them are used in TL fault detection, classification and localization such as feedforward neural network (FNN), convolutional neural network (CNN), adaptive neuro-fuzzy inference system (ANFIS), extreme learning machine (ELM), and probabilistic neural network (PNN)."

"In one study [18], a self attention CNN framework and a time series image-based feature extraction model is presented for fault detection and classification of TLs using a DWT for denoising the faulty voltage and current signals."

"The work [19] presents an ML-based CNN for TL fault detection and classification that takes advantage of DWT for feature extraction."

"The study [20] proposes an application of ANFIS for fault classification and location estimation of TLs based on measured data from one terminal."

Hybrid Methods

"The study [21] proposes a hybrid methodology for detection, classification and location estimation of short-circuit faults in power TLs."

"A two-stage finite impulse response filter integrated with SVMs is used for detection and classification of the faults, and the SVRs are implemented to estimate the location of them."

"Another study [22] presents an identification and categorization methodology for TL fault types and location estimation of the faults."

"Balakrishnan and Gopinath [23] present a hybrid methodology for detection and classification of TL faults based on RF algorithm."

"In this methodology, the fault location of extra high voltage TLs is estimated by using an radial basis function-based SVM applied to the reconstructed input data points, and a gradient based NN (GNN) using scaled conjugate gradient."

Comparisons and Discussions

"The third column shows the complexity of the training step for each method based on its requirements such as the number of training samples, number of features, number of trees (for RF), and number of layers for deep learning approaches."

"To solve TL fault classification and location estimation problems, DWT is needed to do data preprocessing for the Bayesian classifiers."

"This approach cannot work effectively with noisy datasets, which undermines the reported accuracy for TL fault classification using SVM."

"The only concerns about FNNs in the TL fault identification process are its time-consuming training step, and the need for data preprocessing methods such as DFT and DWT that makes the whole process complex."

"Considering the advantages and disadvantages of all the above-mentioned classifiers, it can be concluded that the deep learning-based methods are superior solutions to TL fault detection, classification, and location estimation problems."

Future Research Directions

"Not only the datasets for TL fault detection and location estimation are time series and unbalanced, but also there exist numerous different types of power system distribution networks."

"This methodology enables the processing of data in real-time which is a required feature in solutions for TL fault detection problems [24]."

"Edge ML assists the complex implementation of recently emerged power TL technologies to become efficient by using cloud computing [25]."

"Some recent studies such as [26] have used edge computing methods to overcome the TL fault detection issues such as the low capability of finding fault features for line-to-ground or line-to-line faults."

"Implementation of computationally expensive power TL technologies can be done efficiently by using edge computing techniques [25]."

Conclusion

"A comprehensive review of ML methods for TL fault detection, classification, and location estimation has been presented."

"The existing methodologies are divided into three main categories called generic ML, ANNs, and hybrid methods."

"In this survey, generic ML approaches are categorized into five subsections namely Naive Bayes Classifier, Decision Tree, Random Forest, Support Vector Machine, and k-Nearest Neighbour."

"Artificial Neural Networks are divided into four main categories based on their various structures, namely, Feed forward Neural Network, Convolutional Neural Network, Adaptive Neuro Fuzzy Inference System, and other small groups such as Extreme Learning Machine and Probabilistic Neural Network, and the third category is assigned to hybrid methods."

Roller Bearing Failure Analysis Using Gaussian Mixture Models and Convolutional Neural Networks [27]

This is a machine-generated summary of:

Rathore, Maan Singh; Harsha, S. P.: Roller Bearing Failure Analysis using Gaussian Mixture Models and Convolutional Neural Networks [27].

Published in: Journal of Failure Analysis and Prevention (2022).

Link to original: https://doi.org/10.1007/s11668-022-01469-8

If you want to cite the papers, please refer to the original.

For technical reasons we could not place the page where the original quote is coming from.

Abstract-Summary "To promote accurate condition monitoring of bearing following two conditions must be achieved (1) intelligent instance annotation and (2) automatic feature extraction and selection."

"To validate the proposed method, experimentation is conducted on a high-speed rotor-supported bearing test rig to acquire run-to-failure roller bearing lifetime data."

"Bearing lifetime responses are segregated into different operating conditions, namely normal, slight degradation (SLD), severe degradation (SVD), and failure using t-SNE and GMM techniques."

"The comparisons are conducted with the existing advanced classifiers based on assessment metrics ROC (receiver operating characteristic) curve, AUC values, precision, recall, and F1-score."

"The comparison results revealed the superior performance of the proposed method compared to existing advanced classifiers."

"The proposed methodology precisely classifies the bearing lifetime data into different operating conditions using vibration acceleration responses."

Introduction

"To implement the machine learning algorithm on vibration data, relevant features are required to be extracted that best represent the degradation with time [28]."

"Li and others [29] estimated RUL by implementing multi-scale DCNN (MS-DCNN) that efficiently extracts features of different scales by utilizing three different sizes of convolution operations on preprocessed data, and then these features are later used for training the model."

"The proposed method effectively utilizes the data reduction and unsupervised learning models for health monitoring of bearing."

"A one-dimensional CNN (1DCNN) model is proposed that is capable of directly processing the raw vibration signal with automatic feature extraction and learning capability."

"In [30, 31], one-dimensional CNN is successfully implemented for the feature extraction and classification of the electrocardiogram (EEG) data and motor faults directly from the raw sensor data."

Theoretical Overview

"ReLU is implemented as an activation function to map the nonlinear transformation features [32]."

"Due to the vanishing gradient problem during backpropagation with ReLU a problem arises known as 'dying ReLU', which restricts its implementation."

"Maaten and Hinton first introduced t-SNE in 2008 [33], which is suitable for dimensionality reduction of the high-dimensional data to visualize it in the low-dimensional space by assigning each point a location in a two-dimensional space or map."

"The t-SNE is a developed form of SNE technique that suffers from over-crowded points in low-dimensional space that are far apart in high-dimensional space and the difficulty in optimizing the complex cost function [34]."

"t-SNE attempts to minimize the single Kullback–Leibler divergence between joint probability distributions P, and Q in the high-dimensional and low-dimensional space, respectively [35], the relation for this is as shown below GMM is described as the parametric probability density function of a particular random variable represented as the weighted sum of multiple Gaussian components [36]."

Experimental Facilities

"To validate the effectiveness of the proposed method, experiments are performed on the bearing test rig and data related to test-to-failure for a roller bearing are acquired."

"The vibration acceleration responses are recorded until the bearing fails using a data acquisition system."

"The noise and vibration levels are reached to a considerably high value where the bearing is said to reach its failure point."

"After a prolonged operating period, the bearing RMS values begin to fluctuate as the experiment progresses, and their amplitude steadily increases."

"The bearing expected life may be calculated considering the real-life operating conditions such as the effects of misalignment, temperature, contamination, exterior vibration, and looseness."

"The statistical time-domain features are extracted from the vibration responses, and the variations of these features provide enough useful information regarding the operating condition of the bearing."

Experimental Validation and Results

"This section describes the validation of the proposed methodology for classifying the different operating conditions of bearing by performing the accelerated test to acquire bearing lifetime data."

"Another most utilized classifier performance assessment metric ROC (receiver operating characteristic) curve, and the AUC (area under the curve) values are presented to provide a fair comparison of classification ability among different classes."

"Enabling total segregation of the lifetime data into different operating conditions of bearing."

"The partition of bearing lifetime data into different operating conditions is achieved to a greater extent by implementing PCC."

"The implementation of GMM to the bearing data provides sufficient segregation of data into different operating conditions."

"To assess the performance of different classifiers under the influence of noise, variation of magnitude is set from −10 dB to 10 dB. The classification accuracy values are obtained for different magnitudes of noise, and the classification accuracy of different classifiers is compared."

Conclusions

"A one-dimensional CNN model is used to classify bearing lifetime data into different operating states using the data dimension reduction technique, t-SNE, and unsupervised learning method, GMM."

"The results revealed that the discriminative ability of the proposed methodology using one-dimensional CNN is quite superior in classifying bearing lifetime data into different operating conditions."

"The following conclusions may be drawn from the analysis of the results: (1) The classification accuracy obtained for the one-dimensional CNN model is considerably higher than the accuracy values obtained for existing advanced classifiers."

"The one-dimensional CNN model and the data segregation based on GMM and t-SNE show better performance than state-of-the-art methods for bearing condition monitoring."

An Accurate Multiple Cognitive Classifier System for Incipient Short-Circuit Fault Detection in Induction Generators [37]

This is a machine-generated summary of:

Karatzinis, Georgios D.; Boutalis, Yiannis S.; Karnavas, Yannis L.: An accurate multiple cognitive classifier system for incipient short-circuit fault detection in induction generators [37].

Published in: Electrical Engineering (2021).

Link to original: https://doi.org/10.1007/s00202-021-01445-9

If you want to cite the papers, please refer to the original.

For technical reasons we could not place the page where the original quote is coming from.

Abstract-Summary "Fault detection and diagnosis is an identified challenging field and constitutes a crucial factor in industrial systems."

"We introduce, for the first time, a multiple cognitive classifier system (MCCS), which incorporates fuzzy cognitive networks with functional weights (FCN-FW) in distinct multi-classifier structures."

"The proposed distinct MCCS structures assume either a parallel topology of combined FCNs, or serial topology in which two classifiers form a simple multi-layer cascade classification system or finally a hierarchical topology."

"The comparison provides clear evidence that the proposed FCN-based MCCS produces superior performance in terms of the overall accuracy (96.9%) while also discriminates perfectly the induction generator's normal condition (100%) over six faulty states."

"The use of functional weights alleviate the overall scheme from the need to store and process the acquired knowledge in large fuzzy rules databases, resulting in low execution times."

"Due to its performance and the low execution time, the proposed scheme is appropriate for implementation in real-life systems."

Introduction

"The present study aims at a reliable fault detection of inter-turn short circuits in stator windings of squirrel cage induction generator (SCIG) using an intelligent diagnostic system."

"Distinct fault detection methods in the area of induction machines have been reported in the literature [38–42]."

"We introduce FCNs equipped with functional weights (FCN-FW) for the effective participation of these networks in three multi-classifier system (MCS) topologies."

"The combined models aim at an enhanced detection and diagnosis of short circuits in squirrel cage induction generator to provide the aforementioned potential benefits in wind turbine systems."

"These networks use different polynomial weight structures, and their connection in the MCS varies between parallel, serial and hierarchical topology in three distinct approaches."

"The methodology is applied and tested for the accurate detection of incipient faults in wind turbines' induction generators."

Related Work

"A plethora of studies associated with ITSC faults in induction machines have been proposed in the literature including distinct fault detection techniques in terms of signal preprocessing, multiple feature extraction/selection and classification approaches."

"In [43], a signal-based algorithm was presented for the stator inter-turn fault detection and the faulty phase identification of an induction motor."

"A model-based ITSC fault detection framework was proposed in [44] for squirrel cage induction motors using on-line and off-line parameters estimation to perform faults detection, localization and real-time monitoring of minor ITSC faults in stator winding."

"Different methods of feature extraction and selection of the most informative ones have been incorporated in a data-driven on-line ITSC fault detection framework of induction motors under varying operating conditions using an ensemble approach [45]."

"In [46], an approach based on Random Forest (RF) and Park's Vector has been presented to detect stator winding short-circuit faults in squirrel-cage induction motors."

Problem Statement and Data Set Characteristics

"Faults in the turns of the stator winding can lead to catastrophic failure, as the increased current flow through the shorted turns produces excessive heating which in turn causes critical insulation breakdown [47]."

"The present study focuses on fault detection and diagnosis in electric generators and especially in the squirrel cage induction generator (SCIG) following the rationale of detecting and identifying ITSC faults in their incipient stage."

"In this context, the conducted here work focuses on the detection of incipient short circuits in SCIG."

"Recent works utilize this dataset in a multi-sensor architecture for the identification of incipient short circuits in wind turbine generators [48] and IoT-based framework for wind farms [49]."

"A suitable short-circuit test board was employed to emulate the short-circuit faults in one of the SCIG stator windings."

Fuzzy Cognitive Networks with Functional Weights

"After the training phase, the acquired knowledge is stored in a fuzzy if-then rules database which associates, in a fuzzy manner, the various weights with the corresponding equilibrium node values."

"The knowledge storage mechanism through a fuzzy rule database is completely dropped out and replaced by weights that do not assume simple numerical values, but they are functions which are continuously updated during the training phase of the FCN framework."

"Since FCNs with functional weights do not require the existence and processing of a large fuzzy rules database, as conventional FCNs do, the proposed schemes are appropriate to be used in real-life applications."

"Weights affect the activation degree of nodes and their absolute value demonstrates how strongly a concept node affects its interconnected one."

"In conventional FCNs, the cumulative knowledge of the trained classifier would be stored in fussy if-then rules associating the desired equilibrium concept node values with the corresponding trained weights."

Multiple Classifier Systems

"Crucial criteria for constructing MCS, incorporate: (1) proper generation and selection of base classifiers and adjustment of their parameters; (2) appropriate design of the architecture or topology in order to interconnect and integrate individual models; (3) suitable selection of fusion model or combination function, where all models contribute to the final decision."

"The main six strategies for generating a diverse set of classifiers are based [50] on different: (a) initializations—This stands for training models like neural networks that their result is initialization dependent; (b) parameters—Diverse parameter configurations for each classifier in the pool; (c) architectures—This is related with the size of a network model or the number of layers it is constituted of; (d) classifiers—In this case, each model is different from the remaining involved models; (e) training sets—This is referred to the exploitation of bootstrapping methods with or without replacement."

Description of the Proposed Methodology, Implementation and Results

"By using imbalanced data between classes, we expect a lower performance for our single FCN-based classifiers compared with the case of balanced trained classifier."

"Each classifier is provided with different training sets which have been generated during the preprocessing stage using Fourier transform, HOS and SCM methods."

"It has been observed that classifiers which have been fed with Fourier features perform the best accuracy in recognizing normal condition."

"For each of the three feature sets, we choose to use two FCN classifiers, which use the best and the second best polynomial order of the functional weights, respectively."

"A FCN with Fourier features and polynomial order equals to 3 is selected for the first layer and a FCN classifier which is fed with HOS features and polynomial order equals to 4 is preferred for the second layer implementation."

"Normal conditions were classified with 100% accuracy, and the overall performance of the cascade FCN-based system is 96.68%."

Results Comparison and Discussion

"The evaluated performance of the proposed multi-classifier system using FCN-based classifiers in hierarchical topology is 96.9% in terms of accuracy."

"Taking into account the aforementioned analysis in terms of (1) the overall performance (96.68%-serial topology, 96.90%-hierarchical topology); (2) clear discrimination of healthy conditions over faulty states (100%); (3) the inherent causal representation of FCNs which satisfy naturally the diagnosis causal requirements between faults and their symptoms-locations; (4) the compact network's knowledge storage in the form of functional weights, it can be concluded that the proposed MCS FCN-based framework is quite appropriate for the ITSC fault detection and diagnosis in a practical real-time implementation in wind turbine induction generators."

"In respect to execution time, indicative experimental results are as follows: The average required time for feature extraction and classification of one sample by the proposed case of two distinct layers (serial topology) is 50.13 ms (first layer classifier: 22.89 ms, second layer classifier:27.24 ms)."

Conclusions and Future Work

"The proposed system uses different multi-classifier structures, which employ fuzzy cognitive networks with functional weights as core individual classifiers."

"The efficiency and reliability of our approaches are highlighted by their overall accuracy (96.90%), as well as from the excellent discrimination properties that have been presented between normal operating conditions and induction generator's faulty states (100%)."

"The comparison results with several machine learning approaches show evidence that our implementations classify more accurately the induction generator operating conditions."

"Due to its high performance and the minimal storage requirements implied by the use of functional weights, the proposed methodology is appropriate for implementation in real-life induction machines."

Exploring Real-Time Fault Detection of High-Speed Train Traction Motor Based on Machine Learning and Wavelet Analysis [51]

This is a machine-generated summary of:

Li, Yanshu: Exploring real-time fault detection of high-speed train traction motor based on machine learning and wavelet analysis [51].

Published in: Neural Computing and Applications (2021).

Link to original: https://doi.org/10.1007/s00521-021-06284-0

If you want to cite the papers, please refer to the original.

For technical reasons we could not place the page where the original quote is coming from.

Abstract-Summary "The implementation of real-time fault detection technology for key components of high-speed train traction electromechanical system is of great significance for improving train motor reliability and reducing guarantee costs."

"Machine learning is a new and powerful means of researching fault detection technology."

"Based on machine learning, this paper conducts real-time fault detection technology research on the electromechanical actuators of the key components of electromechanical systems."

"Based on wavelet packet decomposition and reconstruction, the fault simulation data of electromechanical actuators and hydraulic pumps are analyzed, the wavelet packet energy distribution is calculated, and time-domain statistics are combined to extract energy feature vectors that can reflect component fault characteristics."

"This paper proposes a fault diagnosis method for electromechanical actuators based on machine learning, designs, and improves the neural network learning algorithm and network parameters, and improves the classification effect of the neural network; proposes a fault diagnosis method based on the GA-SVM algorithm, using real values."

Introduction

"Due to the complex fault mechanism of the motor, the fault characteristic signal is not obvious and the fault signal is nonlinear and non-stationary, and the traditional feature extraction method has some defects, such as the complex calculation of the feature vector, excessive reliance on prior knowledge, etc. [52], and fault feature extraction technology based on signal processing methods such as Hilbert Huang transform and wavelet transform."

"Aimming at the many shortcomings of traditional diagnostic technology, our work applies methods in the field of machine learning, and studies Long Short-Term Memory (LSTM) and Stacked Sparse Autoencoders (Stacked Sparse Auto) in machine learning, and applied to motor fault diagnosis."

"In the fault diagnosis of the motor, the running motor cannot be in a static state, so whether it is the vibration signal of the rolling bearing or the current signal of the stator winding will have the characteristics of non-stationary and sudden change, and is affected by the collection environment."

Related Work

"The collection of motor fault signals is the first step in diagnosis."

"In order to realize the fault online diagnosis, Zaman et al. [53] proposed a motor rotor load fault diagnosis based on the stator current quadratic method."

"By amplifying the characteristic signal in the stator current, the amplitude of the fault is also amplified, and the position of the frequency changes, which is conducive to the establishment of motor fault diagnosis model."

"Motor vibration monitoring is an important method for motor fault diagnosis."

"Deng et al. [54] combined the wavelet packet progressive extraction method with the wavelet packet frequency band energy extraction method, which was used to diagnose the fault of the rotor balance signal of the motor and accurately locate the fault location; some researchers put the wavelet packet."

"It is used to extract the stator current of the motor rotor fault and weaken the influence of noise and load changes on the diagnosis result."

Motor Fault Detection Model Based on Machine Learning and Wavelet Analysis

"The specific steps are described as follows: (1) Establish a sample set: the collected electromagnetic torque data of the six-phase permanent magnet fault-tolerant motor under normal and various fault states, use Redundant second-generation wavelet analysis extracts characteristic signals, which are used as training and testing samples for establishing LS-SVM failure mode identifiers. (2) LS-SVM structure selection: first select the appropriate form from the commonly used kernel functions of support vector machines to construct the LS-SVM classifier, and then select the appropriate multi-class coding based on the comprehensive consideration of the fault category, the complexity of the algorithm and the training time algorithm. (3) Parameter optimization: according to the selected kernel function, use training samples and test samples to optimize the structural parameters in the LS-SVM classifier to achieve the best classification and recognition effect. (4) Test verification: We use "unknown" fault category samples, and input them into the established LS-SVM classifier, and compare the identified results with the true categories of the samples, and test the classification capabilities of the established classifiers, and verify all Establish the effectiveness and generalization ability of LS-SVM fault identifier."

Application and Analysis of Motor Fault Detection Model Based on Machine Learning and Wavelet Analysis

"The recognition accuracy of support vector machine is high and stable, and it is less affected by the number of external training samples; in the case of the same or different training samples, IS-SVM decision-making time is the shortest, and the decision-making speed is significantly better than traditional SVM and BP neural network methods."

"It can be found that after the test of 51,005 groups of samples, both the decision tree and KNN algorithm have a high accuracy rate and can better identify the fault type, but the KNN algorithm takes less than half of the decision tree, and the accuracy rate for the SVM vector machine."

"It uses the decision tree, KNN, logistic regression and SVM vector machine linear classifier algorithms in the Scikit-Learn library to classify faults, 80% of the sample book is used as the learning set and 20% as the test set."

Conclusion

"According to the wavelet analysis algorithm and the KNN machine learning algorithm theory, the fault situation awareness system of the six-phase high-speed train traction motor is constructed, and the different working conditions of the motor are simulated offline and the system is guided to learn, and then the online characteristic signal is used to identify the actual fault."

"A high-speed train traction motor fault diagnosis method combining least square support vector machine and redundant second-generation wavelet is proposed."

"The simulation results show that the diagnostic accuracy is more efficient than the traditional support vector machine and BP neural network methods, and it provides a new idea and method for the fault diagnosis of permanent magnet fault-tolerant motors."

"The proposed fault situation awareness method is effective and stable, and can provide a guarantee for the fault-tolerant operation of high-speed train traction motors on different occasions."

Intelligent Active Fault-Tolerant System for Multi-Source Integrated Navigation System Based on Deep Neural Network [55]

This is a machine-generated summary of:

Guo, Chengjun; Li, Feng; Tian, Zhong; Guo, Wei; Tan, Shusen: Intelligent active fault-tolerant system for multi-source integrated navigation system based on deep neural network [55].

Published in: Neural Computing and Applications (2019).

Link to original: https://doi.org/10.1007/s00521-018-03975-z

If you want to cite the papers, please refer to the original.

For technical reasons we could not place the page where the original quote is coming from.

Abstract-Summary "An active fault-tolerant integrated navigation system is established by adding neural network to the fault-tolerant integrated navigation system based on one-class support vector machine fault detection algorithm."

"When there is no fault, the neural network trains each sub-filter; when there is a fault, the neural network which has been in the training state will predict the fault time data and use the neural network prediction data to replace the fault data into the main filter for fusion."

"It can be seen from the simulation analysis that the system can detect the fault of the navigation sub-filtering system well, and when the fault occurs, the prediction data of the neural network is used for information fusion."

Introduction

"Besides Kalman filtering and other traditional methods for state estimation, neural network is also proposed for state estimation."

"In paper [56], the state estimation of integrated navigation system using Elman neural network is proposed."

"The paper [57] proposes the state estimation of integrated navigation using BP neural network."

"According to the navigation characteristics of inertia, terrain and geomagnetism in the environment that multi-source integrated navigation system needs to face, this paper proposes a new navigation state estimation by combining Laplace feature map (LE) with deep neural network model."

"The simulation results show that the proposed method can detect the abrupt and slow change faults of the navigation sub-filtering system with short delay, strong real-time performance and analysis."

"This paper proposes an active fault-tolerant integrated navigation system, in which the neural network is added to the fault-tolerant integrated navigation system based on one-class SVM fault detection algorithm."

Problem Description and the Proposed Method

"To improve the accuracy and real-time performance of the fault-tolerant integrated navigation system, aiming at the timing signal of navigation system, the training samples need to be preprocessed: phase space reconstruction and support vector preselection to improve modeling efficiency."

"Phase space reconstruction One-class SVM has a faster speed for system modeling, which can be directly used for system data samples with several dimensional features, but for time series navigation signals, the need for phase space reconstruction [58] and then establish fault forecast model using one-class SVM model [59]."

"Support vector preselection One-class SVM has the sparsity, that is, the final form is only determined by the support vector, so the complexity of the algorithm can be reduced by reducing the training sample set, which can effectively improve the efficiency of fault detection and improve the real-time performance of the fault-tolerant integrated navigation system."

Simulation and Analysis

"In order to verify that the one-class SVM-based fault detection algorithm is superior to the neural network-based fault detection algorithm in training sample numbers, the neural network and one-class SVM are trained with different numbers of identical training sample sets, and then, the same test sample and error amplitude are used to compare the fault detection rates of the two models."

"The simulation results show that the one-class SVM fault detection algorithm can detect the abrupt and slow change faults of the navigation sub-filter system very well, with short delay and strong real-time performance."

"Compared with the fault detection algorithm based on neural network, it is proved that the detection rate is higher and the performance is better in small sample."

"Simulation and analysis of single subsystem without fault Each sub-filter estimates one or several states of the overall velocity, position, and attitude."

Conclusion

"This paper proposes an active fault-tolerant integrated navigation system, that is, adds neural network to help establish active fault-tolerant integrated navigation in fault-tolerant integrated navigation based on one-class SVM fault detection algorithm."

"During the running of the active fault-tolerant integrated navigation system, each sub-filter enters the main filter for information fusion, and the one-class SVM fault detection algorithm detects the fault of the system and as the signal for the neural network starts to predict."

"It can be known that, compared with the general fault-tolerant integrated navigation, the active fault-tolerant integrated navigation system can make each sub-filter work uninterruptedly and the overall state can be optimally estimated regardless of whether the sub-navigation system fails."

Data-Driven Prognostic Method Based on Self-Supervised Learning Approaches for Fault Detection [60]

This is a machine-generated summary of:

Wang, Tian; Qiao, Meina; Zhang, Mengyi; Yang, Yi; Snoussi, Hichem: Data-driven prognostic method based on self-supervised learning approaches for fault detection [60].

Published in: Journal of Intelligent Manufacturing (2018).

Link to original: https://doi.org/10.1007/s10845-018-1431-x

If you want to cite the papers, please refer to the original.

For technical reasons we could not place the page where the original quote is coming from.

Abstract-Summary "As a part of prognostics and health management (PHM), fault detection has been used in many fields to improve the reliability of the system and reduce the manufacturing costs."

"Due to the complexity of the system and the richness of the sensors, fault detection still faces some challenges."

"The fault is detected by the reconstruction error in the feature space."

"To demonstrate the effectiveness of the proposed algorithm, we evaluate our algorithm on a benchmark dataset for fault detection, and the results show that our algorithm outperforms other fault detection methods."

Introduction

"If the situation of the production can be interpreted in time, the proper fault detection method can find the abnormal state to ensure the valid production process [61]."

"The machine learning methods have been adopted to deal with the fault detection problem."

"Faced with seriously class-imbalance samples, this paper studies the feature extraction method based on KPCA (Kernel Principal Component Analysis), a self-supervised learning method, to achieve the rapid detection of the faulty or abnormal situation."

"We show how the proposed KPCA strategy can be used for fault detection for the raw sensor data in the production process."

"The contributions of this paper lie in: We handle the fault detection problem by a self-supervised learning method, which is a novel method in fault detection and one of the novel scientific contributions in our paper."

Related Work

"In our proposed fault detection method, only the normal samples are learned in the training process."

"These machine learning methods have been used widely for fault detection problem."

"In our work, the self-supervised method is proposed for fault detection, which is different from the state-of-the-art algorithms which learned both the normal and abnormal samples."

"Besides these machine learning approaches for classification or clustering, the feature representation methods were also crucial in the fault detection problem."

"In our proposed fault detection method, only the normal samples are learned with one whole architecture."

"Due to the imbalance between the normal and faulty samples, the general methods based on supervised learning of the data in two-classes cannot work well."

"In our work, only the statistic characteristics of the raw data are learned in the proposed fault detection algorithm, and outperform the state-of-the-art methods."

Self-Supervised Learning Based on KPCA

"Dissimilar to those methods, the Kernel Principal Component Analysis (KPCA) is applied as a self-supervised learning method for fault detection without other assistant algorithms."

"Faced with seriously class-imbalanced samples in the monitoring data of industrial manufacturing, this paper studies the feature extraction method in a self-supervised learning manner."

"The efficient and effective detection of the abnormal or fault situation is accomplished by comparing reconstruction error of the sample data with the threshold."

"The threshold of the fault detection algorithm is preset as the maximum reconstruction error of the training normal samples."

"The reconstruction error L is:In our paper, we use KPCA for fault detection in a self-supervised learning manner."

"The KPCA for fault detection algorithm is trained by feeding only one-class samples, i.e. the normal samples."

"In the training process, the normal data are learned by the KPCA, and the eigenvectors are calculated."

Fault Detection Based on Self-Supervised Learning

"Different from other fault detection methods, our proposed algorithm is achieved only by KPCA, rather than the fusion of different machine learning methods."

"The training data are the set of normal samples, where the status of the machine or the course of the manufacturing is normal."

"Step 2 As the manufacturing course is usually recorded in a period, each subset consists of samples with several batches."

"As the proposed method based on the KPCA for fault detection method is robust, the index of the mean of one batch is good enough in our work to distinguish the fault from the normal samples."

Case Study

"In order to evaluate our proposed self-supervised algorithm, we do experiments on a benchmark dataset for fault detection."

"The proposed self-supervised algorithm does not detect the fault with high accuracy."

"For the fault detection of the industrial processing, the sampled data are collected in a consecutive time."

"The fault can be detected by our self-supervised algorithm using the mean of each sample."

"For fault detection objective, taking Experiment 29th as an example, the samples consist of 33 normal batches and 9 faulty batches."

"In the testing step, each incoming batch of data is handled in the same way as the training samples by KPCA."

"The reconstruction error of each testing sample in the feature subspace is measured for fault detection."

"The results show that the mean index of the batch with the KPCA for fault detection method can distinguish the fault status from the normal ones."

Conclusion

"We propose a data-driven method for fault detection."

"Our proposed method is a self-supervised algorithm based on KPCA, which is only trained by normal samples."

"In our proposed algorithm, we first map the training samples which contain normal samples to a subspace with high dimension, and then extract principal components in the subspace."

"The fault is detected as the sample with higher reconstruction error than training samples."

"Deep learning has shown its advantages on several tasks, and we will also explore the research of self-supervised deep learning method on fault detection."

A Variational Autoencoder for a Semiconductor Fault Detection Model Robust to Process Drift due to Incomplete Maintenance [62]

This is a machine-generated summary of:

Kim, Youngju; Lee, Hoyeop; Kim, Chang Ouk: A variational autoencoder for a semiconductor fault detection model robust to process drift due to incomplete maintenance [62].

Published in: Journal of Intelligent Manufacturing (2021).

Link to original: https://doi.org/10.1007/s10845-021-01810-2

If you want to cite the papers, please refer to the original.

For technical reasons we could not place the page where the original quote is coming from.

Abstract-Summary "In the semiconductor manufacturing field, few studies on fault detection (FD) models have considered process drift due to incomplete maintenance."

"Process drift refers to the shift in sensor measurements over time due to tool aging, and it leads to defective production when it is severe."

"We propose an FD model robust to process drift by modeling process drift with a variational autoencoder (VAE)."

"Because process drift is characterized by time-varying information, the proposed model encodes some time-varying information through separate hidden layers."

"By adopting a strategy that combines information separately encoded in a feature vector, the proposed model successfully models process drift."

Introduction

"FD models monitor trace data observed by sensors that measure process variables in milliseconds over a processing period."

"FD models detect process faults using trace data at early processing steps to save manufacturing costs and maintain high wafer quality."

"Process drift is caused by two incomplete maintenance mechanisms: chamber cleaning and tool repair [63]."

"Incomplete chamber cleaning generates lot-to-lot drift, which is the gradual shift in process variables."

"We consider a variational autoencoder (VAE) model, a one-class learning model that uses only normal process data for model training."

"VAE-IDF is designed for modeling the process drift."

"It trains the trace dataset of several repair cycles and models patterns of the process drift that not only naturally occur but also change according to the repair."

"VAE-IDF's great strength lies in the ability to generate unseen levels of drifted trace data based on process drift modeling."

Related Works

"Deep learning models (e.g., AEs, convolutional neural networks (CNNs), and long short-term memory (LSTM)) can automatically extract suitable features for detecting faults from raw sensor trace data."

"Lee and Kim [64] introduced a CNN model that analyzes sensor traces to diagnose the root cause of wafer faults in trace data."

"Data scientists in the semiconductor manufacturing field struggle with the accuracy paradox, which occurs when a training dataset has substantially more normal data than fault data, making the classifier predict most new data as normal."

"In this situation, one-class learning can be a solution for training FD models using only normal data and classifying a new wafer instance as a fault if the instance is located outside the decision boundary in the feature space."

"The studies handle process drift in an adaptive manner that regularly updates the fault criterion with new data."

VAE

"The VAE is a stochastic generative model that can produce new data from the estimated probability distribution of given datasets [65]."

"Unlike an AE, which outputs compacted input information as the feature vector, a VAE's encoder outputs the parameters of the distribution generating the feature vector."

"If we input a randomly sampled value from the estimated distribution into the decoder, we can obtain reconstructed data."

"VAEs not only reconstruct inputs but also act as generators."

"The filter moves along only the time axis, and a one-dimensional feature map is generated."

"The architecture enables the filter to extract time-invariant features considering the whole sensor correlation."

Proposed Method

"It is not reasonable to assume that the drift coefficient functions follow specific forms, such as linear functions, because the drift pattern varies according to the manufacturing step, tool, engineer repair skill, process procedure, etc. Therefore, we propose a novel structure to automatically learn the drift coefficient functions."

"As an extension of this perspective, the proposed model, VAE-IDF, combines hidden layer units with different tasks to learn the process drift information from drifted trace data."

"This structure allows the VAE-IDF to learn the drift coefficient functions and generate drifted trace data."

"This structure enables the VAE-IDF to extract the robust representative feature vector not only by interpolation in the range of the training data but also by extrapolation beyond the range based on the learned drift coefficient functions."

"We focus on the difference (i.e., error) between the actual trace data and the predicted normal trace data by the VAE-IDF, and the proposed algorithm can detect wafer faults based on the error."

Experiment

"It was assumed that drift by the cleaning operation occurred in each lot and continuously for one tool repair cycle."

"The drift degree means a maximum reduced amount of sensor measurements in one repair cycle and was defined as the ratio of the difference between the maximum and minimum values in the drifted process step."

"In the underrun phase, model training and testing were set to the same drift degree, whereas in the overrun phase, the drift degree was set to double the value defined in the training data."

"The scenario (1%, 100 lots, linear, improper repair, underrun) means that when 100 lots were produced, the drift occurred linearly up to 1% from the initial state, the drift speed was accelerated after each tool repair, and the drift degree in the test data did not exceed 1% of the drift degree of the training data."

Results and Analysis

"The nonsignificant factors meant that the proposed model showed robust prediction performance even when the intensity level changed."

"The difference in the intensity level of the running phase factor had a strong effect on the prediction performance because the change in the prediction performance was severe when a drift that was not used to train the proposed model appeared in the test data (overrun phase)."

"In terms of the false alarm rate, considering that the rate of deep SVDD was abnormally low, the proposed model showed the reasonably best performance."

"In the case of the VAE-IDF, although a new repair cycle appeared in the test data and the VAE-IDF had to predict untrained information, the model achieved a prediction performance as high as that of the VAE."

"In the overrun phase, the traces between the training and test data were not the same, and the VAE could not reconstruct them, thus significantly decreasing the prediction performance."

Conclusion

"We propose an FD model using semiconductor manufacturing process data in which process drifts appear due to incomplete chamber cleaning and tool repair."

"We successfully modeled the process drift using the newly proposed VAE-IDF structure that received the lot and repair cycle IDs as well as the sensor data through the units of the hidden layers and combined the outputs in the feature vectors."

"The proposed model achieved a higher prediction performance than the comparison models and could reduce false alarms due to process drift in a manufacturing environment in which various process drifts occurred."

"The first is to apply various process drift factors in the proposed model."

"Adding other factors that influence the degradation of the tool as input information enables more elaborate drift modeling."

A Data-Driven Time-Series Fault Prediction Framework for Dynamically Evolving Large-Scale Data Streaming Systems [66]

This is a machine-generated summary of:

Hell, Michell; Pestana de Aguiar, Eduardo; Soares, Nielson; Goliatt, Leonardo: A Data-Driven Time-Series Fault Prediction Framework for Dynamically Evolving Large-Scale Data Streaming Systems [66].

Published in: International Journal of Fuzzy Systems (2022).

Link to original: https://doi.org/10.1007/s40815-022-01292-1

If you want to cite the papers, please refer to the original.

For technical reasons we could not place the page where the original quote is coming from.

Abstract-Summary "This paper presents a framework to tackle the problem of real-time fault classification for systems that can change their operating conditions dynamically over time."

"This framework is based on a nonparametric fuzzy-based classifier built to deal with large-scale, streaming, and dynamically evolving problems called Self-Organizing Fuzzy (SOF) classifier."

"The SOF classifier processes the selected features to determine the existence of a fault pattern in the time-series data stream."

"The approach presented in this paper was tested in the real-world problem of real-time fault classification of railway switch machines."

Introduction

"This work suggests a data-driven framework based on SOF classifier to the problem of real-time incipient fault detection of dynamically evolving large-scale data streaming systems."

"The main contributions of this work are summarized as follows: We propose a new general framework for real-time fault detection to be applied to dynamically evolve large-scale data streaming systems that allow identification of possible abnormal conditions before a fault occurs, thus preventing more serious consequences for the system; We identified the most suitable feature extraction and selection algorithms to be applied to time series obtained from evolving large-scale data streaming systems; The SOF-based approach suggested in this work implements a method that can adapt quickly to changes in streaming data patterns and does not require large databases to be used in the model update even for large-scale data systems."

Fault Detection in Time Series for Large-Scale Data Streaming Systems

"These time intervals can vary from a few milliseconds to weeks depending on the element being monitored generating, in this way, a constant stream of time series data that need to be constantly analyzed to determine the current operating condition of the element under observation [67]."

"Depending on the number of variables under observation and/or the time interval adopted between the observations, these time series data can comprise huge amounts of data, making fault detection in these processes time-consuming and expensive computationally."

"The choice of both classifiers and strategies for extraction and selection of the feature set depends heavily on the type of process being observed."

"This work focused on choosing strategies that could handle large amounts of data presented in the form of a data stream and update the fault detection system when environmental changes occur in the process under analysis."

The SOF-based Fault Detection Proposed Framework

"The next section presents a brief review of SOF classifiers' theoretical basis, such as the AnYa type fuzzy rule-based systems and the Empirical Data Analytics (EDA) Computational Framework [68]."

"This type of calculation is responsible for endowing the proposed approach with streaming data processing, the SOF classifier's key feature."

"The offline training of the SOF classifier is then performed according to the following steps 1 [69]."

"Algorithm 1 presents a pseudo-code for the offline training procedure of the SOF classifier."

"Online Self-evolving Training During the online training, the meta-parameters and structure of the SOF classifier are continuously updated with the streaming data on a sample-by-sample basis."

"The online training of SOF classifier is then performed according to the following steps 1 [70]."

"The AnYa type fuzzy rule is then updated according to the updated meta-parameters, and the SOF is ready for processing the next data sample or conducting classification."

Experimental Results

"For all implemented methods, the training dataset (offline dataset for the SOF classifier) was formed by the operations of switch machines #1, #2, and #3, totaling 210 samples."

"After the training/offline process, all methods were validated with a dataset composed of the switch machines #4, #5 and #6 in order to identify abnormal operating conditions of these switch machines as soon as possible."

"This fact indicates that scheduled preventive maintenance was performed after operation 54, correcting a switch machine problem and preventing the fault."

"Switch machine, we can see that between operations 16 and 32, the SOF classifier identifies the operating condition in an alternated way during successive operations surpassing all its counterparts."

"The results show that the SOF classifier was able to detect incipient faults with several operations in advance of the occurrence of faults, demonstrating that the SOF classifier represents a promising approach to the proposed problem."

Conclusion

"In large-scale problems with streaming and dynamically evolving characteristics, such as the proposed problem, the SOF classifier proved to be an attractive alternative to deal with real-time railway switch machine problems."

"Despite the encouraging results, further research is needed to assess the SOF performance on switch machines from other manufacturers."

"It is also worth highlighting that the switch machines are located in a region with similar weather conditions."

"The following steps of this research involve comprehensive testing and analysis of the proposed SOF classifier to accurately release a robust and reliable tool to prevent switch machine failures accurately."

Deep Generative Model with Time Series-Image Encoding for Manufacturing Fault Detection in Die Casting Process [71]

This is a machine-generated summary of:

Song, Jiyoung; Lee, Young Chul; Lee, Jeongsu: Deep generative model with time series-image encoding for manufacturing fault detection in die casting process [71].

Published in: Journal of Intelligent Manufacturing (2022).

Link to original: https://doi.org/10.1007/s10845-022-01981-6

If you want to cite the papers, please refer to the original.

For technical reasons we could not place the page where the original quote is coming from.

Abstract-Summary "The increasing demand for advanced fault detection in manufacturing processes has encouraged the application of industrial intelligence based on deep learning."

"Implementing deep learning technology at actual manufacturing sites remains challenging because the data acquired during the manufacturing process are not only unlabeled but also imbalanced time series data."

"We constructed semi-supervised manufacturing fault detection methods to deal with the imbalanced time series data obtained from manufacturing applications, based on recently proposed deep generative models: variational autoencoder-reconstruction along projection pathway (VAE-RaPP) and Fence generative adversarial network (Fence GAN)."

"The deep generative models learned the features of good product from these labeled samples during training."

"We converted the time series data acquired during the manufacturing process into images to improve the feature extraction capability of deep neural networks based on three encoding methods: Gramian angular difference field (GADF), Markov transition field (MTF), and recurrence plot (RP)."

"This research provides a novel approach that combines the encoding of time series into images and deep generative models for manufacturing fault detection."

Introduction

"Lee et al. [72] proposed a hybrid approach that combines time series prediction based on deep neural networks and conventional one-class classification algorithms for semi-supervised fault detection in die casting."

"In spite of the fact that the semi-supervised and unsupervised fault detection modules have succeeded in diagnosing abnormal operations in machines, the recent state-of-the-art anomaly detection algorithms based on generative models have not yet been tested in manufacturing applications."

"We focused on developing a fault detection module based on a deep generative model using time series data acquired during the actual manufacturing process."

"We proposed a semi-supervised learning approach for manufacturing fault detection that combines encoding methods that transform time series data into images to change the input types of the deep generative model."

"This study is the first attempt to utilize these encoding methods to improve the feature extraction performance of the deep generative model in the fault detection module."

Characteristics of Casting Defects

"During the die casting process, various casting defects such as cold shots, blow holes, shrinkages, and cracks are inevitable [73, 74]."

"We classified the major types of casting defects and defined the prime parameters that cause the casting defects as follows: geometric complexity-driven, pressure-driven, and temperature-driven defects."

"Inadequate pressure applied to the molten flow results in a lack of fluidity in the molten metal and can form cavities (called cold shots or blow holes) in the product, while inadequate temperature supplied to the product can result in cracks or tears in the product, as well as shrinkage defects."

"Our group developed an advanced data acquisition system that incorporates three pressure sensors and seven thermocouples to obtain physical information during the die casting process."

"Using this data acquisition system, we collected both pressure and temperature data during the die casting process."

Proposed Approach

"The GADF, MTF, and RP were utilized to convert the time series data into images."

"Both raw data and converted images were used as inputs for the deep generative models of the VAE-RAPP and Fence GAN."

"From the time series-to-imaging process, the input dimension of the encoder is set to 104 × 104 × 10."

"The number of epochs for the raw data and the time series-to-imaging were set to 500 and 300, respectively."

"The VAE-RaPP accurately reconstructed the raw data based on the time series as well as the converted image."

"The image was generated by a sequential 2-D convolutional-transpose layer with various numbers of filters, batch normalization, and ReLU activation."

"We compared the performance of the Fence GAN with the same time series data and converted the image with the VAE-RaPP."

"We validated the performance of the Fence GAN with the images converted from the time series data."

Data Acquisition

"The data obtained from the actual die casting process [75] were used to evaluate the proposed fault detection method, where the pressure supplied to the equipment and temperature over the metal mold were measured at each processing step."

"All the data signals were recorded with a sampling rate of 2 Hz, which resulted in approximately 60 numeric values for each of 10 data channels corresponding to the process time of about 30 s. The measured data were preprocessed in the following sequence before being trained on the deep neural networks."

"Because the exact process time varied by each production step, a zero padding was applied at the end of each time series data to equalize the dimensions of the signals, which resulted in the input dimensions of 104×10 corresponding to the time series length and number of channels, respectively."

Results

"After the training, the performance of the classifier was assessed by AUROC metrics, AP score, PR curve, and accuracy."

"The AUROC score of the raw data with the VAE-RaPP was 95.88%, which improved to 96.68% for the MTF and 96.28% for the RP."

"The AP score of the VAE-RaPP improved from 0.9588 in the raw data to 0.9699 in the MTF and 0.9718 in the RP."

"Of the AUROC score, AP score, and PR curve, we confirmed that the time series–image encoding improved the fault detection performance using the deep generative model."

"We also validated the effect of the sampling frequency and the number of points on the MTF with the VAE-RaPP, which showed the highest score on the AUROC score."

"We confirmed that the sampling rate of the measurement data affected the performance of fault detection using the deep generative model."

Conclusions and Future Works

"We introduced a novel approach that integrated time series-to-image encoding into the state-of-the-art deep generative models of the VAE-RaPP and Fence GAN to detect defective products in manufacturing processes."

"The capability of applying deep generative models to fault detection was validated using industrial manufacturing data."

"We adopted three encoding methods, namely the GADF, MTF, and RP, for time series to imaging; we also attempted to optimize the combination of encoding methods and deep generative models for the best performance."

"Although there was a subtle difference in the performance of different encoding methods, the VAE-RaPP is more suitable for detecting defective product, whereas the Fence GAN has the benefit of ensuring the product quality."

"Our results confirmed that the convolutional deep generative model can be used as a powerful toolkit for fault detection in industrial processes."

Deep Learning-Based Fault Diagnostic Network of High-Speed Train Secondary Suspension Systems for Immunity to Track Irregularities and Wheel Wear [76]

This is a machine-generated summary of:

Ye, Yunguang; Huang, Ping; Zhang, Yongxiang: Deep learning-based fault diagnostic network of high-speed train secondary suspension systems for immunity to track irregularities and wheel wear [76].

Published in: Railway Engineering Science (2021).

Link to original: https://doi.org/10.1007/s40534-021-00252-z

If you want to cite the papers, please refer to the original.

For technical reasons we could not place the page where the original quote is coming from.

Abstract-Summary "Fault detection and isolation of high-speed train suspension systems is of critical importance to guarantee train running safety."

"The existing methods concerning fault detection or isolation of train suspension systems are briefly reviewed and divided into two categories, i.e., model-based and data-driven approaches."

"A 1D convolution network-based fault diagnostic method for high-speed train suspension systems is designed."

"To show the performance of this method, a multibody dynamics simulation model of a high-speed train is built to generate the lateral acceleration of a bogie frame corresponding to different track irregularities, wheel profiles, and secondary suspension faults."

"The 1DCNN method is further validated using tracking data of a CRH3 train running on a high-speed railway line."

Introduction

"The approaches for fault detection or isolation (FD/I) of railway vehicle suspension systems can be classified into two main categories, i.e., the model-based approach and data-driven approach [77]."

"Two model-based approaches (robust observer, and the KF combined with the generalized likelihood ratio test (GLRT) and two data-driven approaches (dynamical principal components analysis (DPCA), and the dynamical canonical variate analysis (CVA)) were, respectively, introduced in [78] to detect the faults in the primary and secondary suspensions of an urban rail vehicle."

"The main contribution of this work is summarized as follows: (1) A vehicle–track model considering track flexibilities, track irregularities, and wheel wear is built. (2) Two strategies for improving the robustness of the diagnostic network of train suspension systems are proposed, i.e., a Gaussian white noise strategy (GWN-strategy) and an edge sample training strategy (EST-strategy). (3) A DL-based fault diagnostic network for RVSFD is established."

Vehicle–Track Coupled Model

"The vehicle model built in our work consists of three sub-structures, one for the carbody and two for the bogies, where each bogie consists of one bogie frame, two wheelsets, and four axleboxes."

"To investigate the impact of track irregularities on the robustness of the fault diagnosis method, track irregularities measured on three different high-speed railway lines are introduced, namely Wuhang–Guangzhou railway line (WG-line), Beijing–Tianjin railway line (JJ-line), and Zhengzhou–Xian railway line (ZX-line)."

"When a train is running, the wheel profile will change continuously due to wear, which will always affect wheel–rail contact, including contact force, contact patch size, etc., and further affects the dynamic characteristics of the bogie, including the bogie frame acceleration."

"To analyze the impact of wheel wear on the robustness of the diagnostic method, the wheel profile evolution of a CRH3 high-speed train running on the WG-line is introduced in our work for analysis."

Diagnostic Network Design

"The proposed model includes 7 main blocks, the first 5 blocks are designed for feature extraction, among which each block consists of a 1DCNN layer, an advanced activation function (AAF) layer, and a max-pooling layer."

"After the 1DCNN and activation function layers of each block in the first 5 blocks, a local max-pooling layer is added to extract the key features of the 1DCNN layer output and reduce the model parameters."

"In order to make the fault diagnostic network immune to these relatively high-frequency and low amplitude impact components, a strategy of adding Gaussian white noise to the original signal is proposed to overcome these impact components caused by the track irregularities, i.e., the GWN-strategy."

"To overcome the above problem, an EST-strategy is proposed in this paper, i.e., during the phase of training dataset establishment, the dataset corresponding to the new wheel profile (S1002CN) and the dataset corresponding to the most worn wheel profile (S1002W190K) are used as the training dataset for the fault diagnosis method."

Simulation

"The strategy of adding Gaussian white noise (GWN-strategy) to the original acceleration signals can improve the immunity of the diagnostic method to track irregularities since this strategy reduces the sensitivity of diagnostic methods to changes in track spectrum. (2) The wheel profile will change as the train running mileage increases."

"The EST-strategy can improve the immunity of the diagnostic method to wheel wear mainly due to two reasons: (I) The training dataset of the diagnostic method covers a wider range of samples, which can identify the testing dataset to a certain extent more accurately. (II) Training the diagnostic method with the datasets corresponding to the new wheel profile and the most worn wheel profile makes the method immune to the changes in the acceleration amplitude caused by wheel wear to a certain extent."

Experimental Verification

"To verify the performance of the 1DCNN method in real-life conditions, the field tracking data of a CRH3 train running on a high-speed railway line are applied."

"When processing the tracking data, it was found that the acceleration signal of front bogie at the third car was abnormally vibrating, and its amplitude, for most of the time, was usually greater than that of the acceleration signal of other front bogies."

"The experiment results obtained using the field tracking data further verify that the proposed 1DCNN-based method can accurately identify the faults of the vehicle suspension systems at different speeds."

"Due to the limitation of experimental resources, this paper only verified the method of 1DCNN to identify the secondary lateral damper failure using the field tracking data."

Conclusion

"The developed fault diagnosis method for train suspension systems, therefore, must be guaranteed to be immune to changes resulting from the track irregularities and wheel wear before being put into use."

"Aiming at solving this issue, a GWN-EST-1DCNN-based method for high-speed train suspension systems is proposed."

"In the third phase (training and recognition), a 1DCNN-based fault diagnostic network of high-speed train suspension systems is built."

"The field tracking data of a CRH3 train running on a high-speed railway line are used to further verify the effectiveness of the 1DCNN method."

"The degradation of suspension systems, including dampers, will be studied in the following-up work. (3) In the field experimental part, due to the limitation of experimental resources, this paper only verified the method of 1DCNN to identify the secondary lateral damper failure using the field tracking data."

Real-Time Broken Rotor Bar Fault Detection and Classification by Shallow 1D Convolutional Neural Networks [79]

This is a machine-generated summary of:

Ince, Turker: Real-time broken rotor bar fault detection and classification by shallow 1D convolutional neural networks [79].

Published in: Electrical Engineering (2019).

Link to original: https://doi.org/10.1007/s00202-019-00808-7

If you want to cite the papers, please refer to the original.

For technical reasons we could not place the page where the original quote is coming from.

Abstract-Summary "There has been a significant increase in application of various data-driven deep learning models for motor fault detection and diagnosis problems."

"Due to high computational complexity and large training dataset requirements of deep learning models, in this study, shallow and adaptive 1D convolutional neural networks (CNNs) are applied to real-time detection and classification of broken rotor bars in induction motors."

"The other advantages of the proposed approach are (1) its compact architecture configuration performing only 1D convolutions with a set of filters and subsampling, making it suitable for implementing with real-time circuit monitoring, (2) its requirement for a limited size of training dataset for efficient training of the classifier and (3) its cost-effective implementation."

Introduction

"For complex fault detection problems, the hybrid systems utilizing the features extracted from statistical methods like principal component analysis (PCA) and partial least squares (PLS) or signal processing methods such as fast Fourier transform (FFT) and discrete wavelet transform (DWT) can also be employed."

"Most of prior works focused on utilizing various feature extraction and classification techniques for motor fault detection and diagnosis problem in general [80–93], and specifically for diagnosis of broken rotor bars (BRBs) in induction machines (IMs) [94–107]."

"In [105], advanced wavelet-based signal processing method is applied to extract effective features for broken rotor bar detection which are then input to a multilayer perceptron classifier for fault classification."

"Although most of these prior works reported successful fault detection performance, researchers spent significant time to manually design the best features and classifiers to discriminate the applied motor signals."

"After the network is properly trained using motor current signal dataset, this system can optimally perform feature extraction and classification tasks for the given fault diagnosis problem."

Adaptive 1D CNN Classifier

"There are two types of layers in the adaptive 1D CNNs: (1) CNN layers where both 1D convolutions and subsampling occur and (2) MLP layers that are equivalent to the hidden and output layers of a standard MLP."

"In [31], both feature extraction and classification processes are merged in a single adaptive 1D CNN learner and during training each can be optimized by using the BP algorithm to maximize the classification performance."

"Compared to a regular MLP classifier, the differences are that (1) instead of multiplication with a scalar, 1D convolution of 1D signals with kernel filters are performed, (2) the connectivity of the previous layer neuron outputs to the current layer input map are limited by the size of the kernel, and (3) the same weights are used for all connections (weight sharing) as opposed to massive number of 1-to-1 connections in a regular MLP implementation."

Broken Rotor Bar Diagnosis from Motor Current Signals

"Broken rotor bar is the most common rotor-related failure that affects induction motors and coupled mechanisms because, under this problem, the motor operates apparently in normal conditions [95]."

"Under ideal conditions, the instantaneous stator current can be written as, Electrical rotor asymmetries of squirrel-cage induction motors such as broken rotor bars result in extra frequency components in the frequency spectrum of stator current."

"An automated and intelligent fault detection and diagnosis system can be employed based on automatic learning of characteristic fault effects (extra frequency components in spectral amplitude response) in the measured current signals depending on the type of each fault in a specific induction motor."

"The adaptive 1D CNN classifier is applied for diagnosis of broken rotor bars in induction motors from preprocessed sensor current data."

"Preprocessing includes necessary notch filtering (at line frequency of 60 Hz), decimation and standard normalization of captured motor current signals."

Experimental Evaluation

"Performance of the adaptive 1D CNN classifier-based real-time fault diagnosis technique is evaluated using the real broken bar(s) dataset from a squirrel-cage induction motor."

"Shallow training of 1D CNN networks is performed (as opposed to usual deep training of CNNs or DNNs), for which case the two termination criteria are set as follows: The maximum iterations for back-propagation (BP) algorithm is set to 100, and the minimum train classification error level of 1% is set to avoid overfitting in training data."

"These results verify that the proposed shallow 1D CNN classifier can be effectively used for the diagnosis of broken bars in induction motors from the input stator current data."

"Based on our implementation with Intel® OpenMP API, the (offline) training time of the 1D CNN classifier-based broken bar fault diagnosis system is measured as 0.5 min, which will be performed only once per motor."

Conclusions

"The main advantages of the 1D CNN-based approach are (1) its compact architecture configuration (rather than the complex deep architectures) which performs only 1D convolutions with a set of filters (kernels) and subsampling, (2) its cost-effective and practical real-time hardware implementation, (3) that it does not require any pre-determined transformation (such as FFT or DWT), hand-crafted feature extraction and feature selection and (4) limited size of training dataset and limited number of BP iterations are sufficient for efficient training of the classifier."

"Classification results demonstrated that 1D CNNs can learn highly discriminative features directly from the raw input sensor data and a competitive performance can be achieved with a shallow CNN configuration (rather than the complex CNNs commonly used for deep learning) that performs only 1D convolutions with a set of filters (kernels) and subsampling."

Networked Fault Detection of Field Equipment from Monitoring System Based on Fusing of Motion Sensing and Appearance Information [108]

This is a machine-generated summary of:

Wu, Chunxue; Guo, Shengnan; Wu, Yan; Ai, Jun; Xiong, Neal N.: Networked Fault Detection of Field Equipment from Monitoring System Based on Fusing of Motion Sensing and Appearance Information [108].

Published in: Multimedia Tools and Applications (2020).

Link to original: https://doi.org/10.1007/s11042-020-08885-8

If you want to cite the papers, please refer to the original.

For technical reasons we could not place the page where the original quote is coming from.

Abstract-Summary "A sensitive online fault detection for industrial equipment is of utmost significance to improve industrial intelligence based on video sensor network."

"We propose the Appearance and Motion SVM (AMSVM), an online fault detection method fusing multimodal information based on One-Class SVM (OCSVM), to monitor the working conditions of unsupervised equipment in the field."

"It utilizes multimodal features to detect faults in terms of the appearance and motion patterns of equipment."

"Decision level information (i.e., appearance and motion patterns) are combined to produce a more precise characteristic for fault detection."

Introduction

"To overcome these shortcomings, the intelligent computing, which refers to real-time data tracking and statistical analysis of industrial field devices through Internet technology, was considered and kept performing outstandingly in fault detection."

"Anomalous event detection aiming at automatic identification of anomalous events in surveillance videos, plays an irreplaceable role in both maintaining public safety and securing industrial production."

"We address these issues in anomaly detection within industrial video to meet the industrial production's needs."

"The proposed AMSVM is evaluated using industrial surveillance video datasets, which reveals its superior performance over other unitary feature methods."

"The novelty of this paper includes: To the best of our knowledge, this paper denotes the first attempt to settle the anomalous event detection task in networked monitoring system designed for working equipment in industrial production filed."

"As to motion fault detection, HOFO and OCSVM as motion feature description and behavior model, respectively, collectively contribute to abnormal event detection."

Related Works

"Given the type of the samples required for the learning process (normal or abnormal), classification methods can be categorized into supervised, semi-supervised and unsupervised methods [109, 110]."

"In [111], An abnormal behavior method using an unsupervised learning framework based on Dominant Set was proposed."

"Makris and others proposed a Bayesian approach for detecting abnormal event based on semantic scene model [112]."

"A rule based method,sparse coding was proposed method to detect abnormal behaviors [113–115]."

"Nannan and others [116] proposed an approach to detect abnormality by adopting Gaussian Process based model."

"Then, the nonlinear one-class support vector machine classification algorithm, following a learning period characterizing the normal behavior of training frames, detects abnormal events in the current frame."

"In [110], Zhang and others present a novel anomaly detection framework which integrates motion and appearance cues to detect abnormal events in video."

AMSVM for Fault Detection

"In the first phase, histogram of optical flow orientation (HOFO) and spatiotemporal gradients are adopted as the feature descriptors reflecting malfunctions in motion and appearance, respectively."

"In order to get good feature descriptors and reduce time complexity,actually, LBPH (Local Binary Patterns Histograms) algorithm is adopted to extracted the appearance features in this paper,and the thresholds of block and cell are set to 16 × 13 and 3 × 3 according to experiments,respectively."

"Compared with the general support vector machine (SVM), the One-class SVM (OCSVM) is adopted in this paper in the following reasons: (1) it only uses samples from one-class, instead of multiple classes for training, which saves the training resources; (2) though the mechanical equipment abnormalities are diverse, the one-class SVM frameworks suit the specificity of the abnormal event detection where one can just obtain samples in the normal scenes [117]."

Experiment Results and Analysis

"We evaluate the performance of the proposed AMSVM framework for detection of mechanical fault on video surveillance datasets, to demonstrate the superiority of our method over other alternatives."

"The experiment is conducted on the dataset EWVD (Equipment Work Video Dataset) from industrial production intelligent monitoring system, which contains diverse sets of field mechanical equipment work scene videos."

"In testing sets, the videos are divided into multiple subsets that contain examples of anomalies, suggesting that the pulley cannot work well with other mechanical parts, e.g. stopping working subset (SW), and belt wearing or unidentified objects subset (WU), and the pulley's speed is slower or faster than normal event (SF)."

"To implement motion fault detection proposed in this study, we first compute optical flow for ROI, which is labeled in training phase, in video frames, where directions of optical flow vectors are quantized into 8 directions from 0° to 360° in the direction histogram."

Conclusions and Future Works

"To better monitor unattended equipment in the field, the present study proposes an online fault detection method in intelligent monitoring system based-IoT technologies considering both motion and appearance faults."

"HOFO and LBP are utilized as feature descriptors for motion patterns and appearance cues, respectively."

"The proposed method is validated using monitoring video sets containing various normal or abnormal scenes of field equipment."

"As for network security, more reliable technology, e.g. fault tolerant technology, are expected to prevent systems from cyber-attacks."

"If the fault model parameters are more complicated, incorporating more appropriate IoT techniques, e.g. denoising technology, to maintain or increase system robustness can be taken into account."

A Generic Intelligent Bearing Fault Diagnosis System Using Compact Adaptive 1D CNN Classifier [118]

This is a machine-generated summary of:

Eren, Levent; Ince, Turker; Kiranyaz, Serkan: A Generic Intelligent Bearing Fault Diagnosis System Using Compact Adaptive 1D CNN Classifier [118].

Published in: Journal of Signal Processing Systems (2018).

Link to original: https://doi.org/10.1007/s11265-018-1378-3

If you want to cite the papers, please refer to the original.

For technical reasons we could not place the page where the original quote is coming from.

Abstract-Summary "Performance of a generic real-time induction bearing fault diagnosis system employing compact adaptive 1D Convolutional Neural Network (CNN) classifier is extensively studied."

"The main advantages of the 1D CNN based approach are (1) its compact architecture configuration (rather than the complex deep architectures) which performs only 1D convolutions making it suitable for real-time fault detection and monitoring, (2) its cost effective and practical real-time hardware implementation, (3) its ability to work without any pre-determined transformation (such as FFT or DWT), hand-crafted feature extraction and feature selection, and (4) its capability to provide efficient training of the classifier with limited size of training data set and limited number of BP iterations."

"Effectiveness and feasibility of the 1D CNN based fault diagnosis method is validated by applying it to two commonly used benchmark real vibration data sets and comparing the results with the other competing intelligent fault diagnosis methods."

Introduction

"Performance of a generic real-time bearing fault diagnosis approach from raw time-series sensor data which utilizes the adaptive 1D Convolutional Neural Network (CNN) classifier for feature extraction and pattern recognition is studied."

"After the network is properly trained using motor vibration signal dataset, this system can optimally perform feature extraction and classification tasks for the bearing fault diagnosis problem."

"The main advantages of the 1D CNN based approach are (1) its compact architecture configuration (rather than the complex deep architectures) which performs only 1D convolutions making it suitable for real-time fault detection and monitoring, (2) its cost effective and practical real-time hardware implementation, (3) its ability to work without any pre-determined transformation (such as FFT or DWT), hand-crafted feature extraction and feature selection, and (4) its capability to provide efficient training of the classifier with limited size of training data set and limited number of BP iterations."

Adaptive 1D CNN Classifier

"There are two types of layers in the adaptive 1D CNNs: (1) CNN-layers where both 1D convolutions and sub-sampling occur, and (2) MLP-layers that are equivalent to the hidden and output layers of a standard MLP."

"The CNN-layers basically process the raw data and learn to extract such features that can be used by the classification performed by the MLP-layers."

"Compared to a regular MLP classifier, the differences are that (1) instead of multiplication with a scalar, 1D convolution of 1D signals with kernel filters are performed, (2) the connectivity of previous layer neuron outputs to the current layer input map are limited by the size of the kernel, (3) the same weights are used for all connections (weight sharing) as opposed to massive number of 1-to-1 connections in a regular MLP implementation."

Induction Motor Faults Detection

"Bearing faults detection and classification from raw vibration data is considered."

"Inner race defect frequency f_{ID}, is given by Ball defect frequency f_{BD}, is determined from In this study, adaptive, compact 1D CNN classifiers are applied for fault classification of two benchmark bearing datasets: Intelligent Maintenance System (IMS) [119] and Case Western Reserve University (CWRU) bearing datasets [120]."

"The raw input data (vibration signals) are preprocessed before being input to the adaptive 1D CNN classifier for training and testing phases."

"The sampling rate of raw input signals can be reduced by down-sampling (decimation preceded by low-pass antialiasing filtering) in order to use more compact CNN configurations that improve training performance and classification speeds."

"Normalization of the train and test datasets are done to present zero mean and unity standard deviation input with linearly scaled into [−1, 1] interval to the 1D CNN classifier."

Experimental Evaluation

"Performance of the adaptive 1D CNN classifier based real-time bearing fault detection technique is extensively evaluated using the IMS and CWRU bearing datasets."

"IMS bearing dataset is used to validate the 1D CNN classifier and to test its performance."

"Since a compact 1D CNN with shallow training is applied (as opposed to deep training of a more complex classifier), only one of every four available training samples is used to train the classifier (total of 1224 samples for all four classes) and testing data consists of 5440 samples for all four classes."

"The compact 1D CNN classifier can effectively extract discriminative features directly from the preprocessed raw vibration data and achieve a competitive bearing fault classification performance."

"For each class, 605 samples with 200 data points (corresponding to a half revolution) are used for more efficient training (and testing) of the classifier."

Discussion

"The conventional intelligent fault detection systems include separately designed feature extraction/selection and classifier blocks."

"From the results, fault classification performance of the conventional methods depends highly on the specifically designed parameters of the feature extraction (such as choice of a wavelet kernel) and classifier algorithms, and they need fine tuning for different application."

"Extracting high dimensional features along with necessary post-processing or feature selection methods can significantly increase cost and computational complexity of the whole system."

"The advantages of adaptive 1D CNN classifier over DNN-based model for fault detection are (1) its compact architecture configuration (rather than the complex deep architectures) which performs only 1D convolutions with a set of filters (kernels) and subsampling and making it suitable for real-time fault detection and monitoring, (2) cost effective and practical real-time hardware implementation, and (3) limited size of training data set and limited number of BP iterations are sufficient for efficient training of the classifier."

Conclusions

"The main advantages of the 1D CNN based approach are (1) its compact architecture configuration (rather than the complex deep architectures) which performs only 1D convolutions making it suitable for real-time fault detection and monitoring, (2) its cost effective and practical real-time hardware implementation, (3) its ability to work without any pre-determined transformation (such as FFT or DWT), hand-crafted feature extraction and feature selection, and (4) its capability to provide efficient training of the classifier with limited size of training data set and limited number of BP iterations."

"The 1D CNN classifier based fault diagnosis system is tested for bearing fault diagnosis from two commonly used real benchmark vibration datasets The experimental results validate the effectiveness and feasibility of the 1D CNN model."

Monitoring and Detecting Faults in Wastewater Treatment Plants Using Deep Learning [121]

This is a machine-generated summary of:

Mamandipoor, Behrooz; Majd, Mahshid; Sheikhalishahi, Seyedmostafa; Modena, Claudio; Osmani, Venet: Monitoring and detecting faults in wastewater treatment plants using deep learning [121].

Published in: Environmental Monitoring and Assessment (2020).

Link to original: https://doi.org/10.1007/s10661-020-8064-1

If you want to cite the papers, please refer to the original.

For technical reasons we could not place the page where the original quote is coming from.

Abstract-Summary "Several different statistical and learning methods are proposed in the literature which can automatically detect faults."

"While these methods have shown promising results, the nonlinear dynamics and complex interactions of the variables in wastewater data necessitate more powerful methods with higher learning capacities."

"This study investigates a method based on deep neural networks (specifically, long short-term memory) compared with statistical and traditional machine-learning methods."

"The network is specifically designed to capture temporal behaviour of sensor data."

"The proposed method is evaluated on a real-life dataset containing over 5.1 million sensor data points."

Introduction

"As WWTPs generate a large amount of data, a promising solution lies in the automatic detection of such faults in the system, using machine-learning methods and algorithms to automatically process the data."

"Neural networks such as multi-layer perception, self-organizing maps, radial bases functions and functional-link neural networks are the most successful learning methods in fault detection of WWTP data [122]."

"This work is the first to evaluate a fully automatic fault detection method using a LSTM network, which learns the relevant features in WWTP sensor data without manual intervention."

"A stacked LSTM network is used to detect collective faults in wastewater sensor data at runtime."

"This contrasts with the proposed method whereby the LSTM network automatically learns relevant features, consequently reducing domain experts' time and providing superior fault performance detection."

Methods

"LSTM-based methods have shown breakthrough results in dealing with temporal data, such as audio, video, and general time series data."

"Sensor data typically have several challenges that must be addressed before using them in a learning system."

"Sensor data are a continuous time series where the data at each time step are related to the previous values in time."

"This characteristic of the time series data leads the solution into a recursive approach where a window of data is processed to understand each time step."

"This work addresses this problem using LSTM units which receive a relatively large window of data and automatically learn the effective windows of the problem at hand using training data."

"The data flow in the LSTM layers through time, and the output is a set of carefully extracted features which is given to a softmax classification layer."

Results and Discussion

"The dataset includes also these load shifts that allowed us to investigate whether the proposed method can capture atypical variations."

"Data from 12 different sensors (both chemical and operational sensors), including ammonia, have been collected from 20 January to 20 December 2017 at 1-min intervals."

"The data are segmented to a maximum window size to create the sequences for the LSTM neural network."

"The larger the window size, the higher the dimensions of the network and its parameters would become."

"The size of the window is considered as a hyperparameter for the model and a grid search is applied to find the optimal value which was found to be 60 min."

"The samples with at least 10 min of faults are labelled as faults and the rest of the data are labelled as normal."

Experiments and Evaluation

"ARIMA is a statistical univariate model that learns the normal sequence of a time series to predict its next value in time."

"Since ARIMA is univariate, the data from an ammonia sensor which includes both faults and normal values is set as its input."

"The best parameters are derived as ARIMA (4,1,4) and the model is trained on normal data to set the coefficient for predicting future values."

"To predict the next value in the sequence of data, the data from 4 previous steps are integrated once and multiplied by the learned coefficients in addition to 4 error terms with their learned coefficients which are all summed."

"The ARIMA model is tested on the test data which contain both normal data and faults and the overall root-mean-square error (RMSE) between the predictions and the real data is 0.07."

"It is not possible to detect the collective fault behaviour with the ARIMA model in the test data."

Discussion

"The measures on each class are presented separately, highlighting results pertaining to fault detection."

"This has resulted in increased automation of the operation of these plants and, consequently, an increase in the number of measurement sensors."

"These sensors are being increasingly used, not only for the environmental monitoring but they are also becoming an important tool in the management of the plants, and the detection of sensor faults is essential in ensuring correct operation of the plant."

"While the current systems are very efficient, there is a clear need to develop methods that can reliably detect sensor faults and provide ample time to the plant operators, such that environmental damage is limited when faults occur."

"A system such as the work presented in this paper is the first step towards implementing a fully automated fault detection system that can address the issues arising from automatic management of WWTPs."

Conclusions

"One important aspect is the detection and management of faults in a timely manner."

"The results presented in this paper have shown that there is a vast potential in using deep neural networks in managing WWTP faults, and this work is only the first step in this direction."

"The proposed method not only outperformed traditional methods but the performance achieved a fault detection (recall) of over 92% which will enable a new class of WWTP monitoring and management that requires very little human supervision."

"It is expected that the work will further encourage the use of deep neural networks, not only in WWTP management but also in the general field of environmental protection."

Enhanced Bearing Fault Detection Using Multichannel, Multilevel 1D CNN Classifier [123]

This is a machine-generated summary of:

Ozcan, Ibrahim Halil; Devecioglu, Ozer Can; Ince, Turker; Eren, Levent; Askar, Murat: Enhanced bearing fault detection using multichannel, multilevel 1D CNN classifier [123].

Published in: Electrical Engineering (2021).

Link to original: https://doi.org/10.1007/s00202-021-01309-2

If you want to cite the papers, please refer to the original.

For technical reasons we could not place the page where the original quote is coming from.

Abstract-Summary "Mechanical bearing faults have the highest statistical occurrence percentage among all of the motor fault types."

"Accurate and advance detection of the bearing faults is critical to avoid unpredicted breakdowns of electric motors."

"Most of the fault detection works in the literature attempted to detect binary {healthy, faulty} motor fault case based on a single input."

"Effectiveness and feasibility of the proposed method are validated by applying it to the benchmark IMS bearing vibration dataset for inner race and rolling element faults and comparing the results with the conventional single-axis data-based fault detection."

Introduction

"Motor fault detection and diagnosis systems can be classified into three groups: (1) Model-based techniques are used to diagnose errors by monitoring the consistency between measured outputs and the predicted outputs of developed model for the system [124]. (2) Signal-based methods are widely used to obtain information about the health of system by processing measured signals via time-domain [125, 126], frequency-domain [127–129] and time–frequency analysis technique [130–133]. (3) The knowledge-based systems comprise both qualitative methods and quantitative methods."

"The knowledge-based methods require high-quality and labeled data for their training to detect motor faults with high performance."

"Most of the studies employed only single sensor data at a time at input and detected only single level of fault from healthy samples, and their results are only applicable for relatively small train/test data sets."

"We first employ a multi-channel 1D Convolutional Neural Network (CNN) classifier to detect multiple levels of bearing faults in real time from multi-channel time-domain vibration data."

Induction Motor Bearing Fault Detection

"Common sources of induction motor related problems include stator winding faults, misalignment, static and/or dynamic air-gap irregularities, rotor bar breakage, bearing failures, shaft deflection, and eccentricity problems [134]."

"Motor faults can be classified as stator faults, rotor electric faults, rotor mechanical faults, and power-related faults."

"According to the study of Electric Power Research Institute (EPRI), among all the most common motor fault types, mechanical bearing faults take the highest rate with 41%."

"Statistical occurrence possibility of the electrical stator faults is 36%, electrical rotor faults is 9% and the other motor fault types is 14%."

"Induction machines widely use ball or rolling element bearings and the most common motor failures are due to these components."

"Study of both EPRI and IEEE states that among all the induction machine failures, bearing faults occurrence is over 40%."

Multichannel, Multilevel 1D-CNN Classifier

"In place of using 2D matrices, 1D arrays are used at each neuron that are used as inputs, output and kernels (weights) at either forward propagation (FP) which is applied to find output of each neuron at each layer and back propagation (BP) which is used for computing change in error at output layer during the training process."

"The proposed system is simultaneously trained with the two-channel vibration data applied as an input to an adaptive 1D-CNN classifier to detect bearing faults at different severity levels."

"A typical CNN network consists of convolutional layers followed by subsampling layers, and for classification its 1D feature-map output is fed into a dense multi-layer perceptron."

"In our recent works [135–141], the adaptive 1D CNNs have been successfully applied for feature extraction as well as classification of the raw biomedical (ECG) signals, motor current and vibration signals, and remote sensing (synthetic aperture radar images) data."

Experimental Evaluation

"As we used the IMS bearing faults dataset containing x- and y-axis vibration signals, the two-channel 1D-CNN classifier is designed and applied for detecting two different bearing faults (inner race and rolling element faults)."

"The proposed two-channel 1D-CNN classifier model is designed to learn different bearing faults from the abovementioned x-axis and y-axis vibration signals."

"These hierarchical convolutional layers process input two-channel accelerometer signals (in parallel) with their optimized (learned) filters to extract the best possible features for the bearing fault classification."

"Two identical multi-channel 1-D CNN classifier structures consist of six convolutional and two scalar MLP layers, to be used in all experiments for real-time condition monitoring of the two specific motor faults."

"Under the same training and testing conditions, we compare performance of a single-channel (using x- or y-axis vibration data) CNN classifier to a two-channel one (both channels are applied simultaneously)."

Conclusions

"We implemented and tested a new bearing fault diagnosis system based on the idea of utilizing multiple channels of sensor data simultaneously using a multi-channel 1D CNN architecture."

"The proposed classifier can process multiple axis vibration data simultaneously to achieve enhanced fault detection performance."

"The collected vibration signals from multiple independent axis of the motor are being input to the proposed system to be processed simultaneously for inner race and rolling element fault classification."

"The experimental results show that 1D CNNs can learn highly distinctive features from multiple channels of raw data without any manual design for parameters or hand-designed features and the proposed classifier achieves significant classification performance with reasonable computational time."

"The proposed multi-channel CNN with the standard SGD optimizer processing two- axis data simultaneously the fault detection accuracy of 100.00% is achieved."

A Review on Fault Detection and Diagnosis Techniques: Basics and Beyond [142]

This is a machine-generated summary of:

Abid, Anam; Khan, Muhammad Tahir; Iqbal, Javaid: A review on fault detection and diagnosis techniques: basics and beyond [142].

Published in: Artificial Intelligence Review (2020).

Link to original: https://doi.org/10.1007/s10462-020-09934-2

If you want to cite the papers, please refer to the original.

For technical reasons we could not place the page where the original quote is coming from.

Abstract-Summary "Malfunction monitoring capabilities are instilled in the system for detection of the incipient faults and anticipation of their impact on the future behavior of the system using fault diagnosis techniques."

"This paper presents developments within Fault Detection and Diagnosis (FDD) methods and reviews of research work in this area."

"The review presents both traditional model-based and relatively new signal processing-based FDD approaches, with a special consideration paid to artificial intelligence-based FDD methods."

"Typical steps involved in the design and development of automatic FDD system, including system knowledge representation, data-acquisition and signal processing, fault classification, and maintenance related decision actions, are systematically presented to outline the present status of FDD."

Introduction

"A wide variety of FDD techniques are found, ranging from model-based approaches (including observer based [143, 144], structural graphs [145, 146] to the data-driven approaches (including classifiers [147], pattern recognition [148], and neural networks [149])."

"Most of the initial work on FDD was carried on model-based approaches utilizing the adaptive observers and system identification models of the processes [150, 151]."

"Model-based FDD requires accurate mathematical model of the process and is the most suitable candidate for small systems with fewer inputs and explicit mathematical model; however, its performance is drastically affected in the presence of unmodeled disturbances and uncertainties [152]."

"Signal-based approaches can be divided into statistical [153] and non-statistical approaches [154]."

"Others have focused on audio and vibration-based FDD [155]."

"This review presents both traditional model-based and relatively new signal processing-based FDD approaches, with a special consideration paid to artificial intelligence-based FDD methods."

Prior Knowledge of System

"Physical redundancy refers to the inclusion of additional components in the design of system/process e.g. identical or diverse actuators and/or sensors that are capable of working in parallel and may provide backup in case of malfunction [156]."

"After system representation and system redundancy, the subsequent significant factor in selection of appropriate class of FDD systems is the type of faults."

"Hardware faults [157–160] can further be classified according to type of component failure i.e. sensor faults, actuator faults, plant/process faults, structural faults (controller faults)."

"Another classification can be based on nature and dynamics of faults such as transient, permanent, intermittent and incipient faults [161, 162]."

"Incipient (drift like) faults exhibit slow and gradual changes in state of faulty component variables [163]."

Data Acquisition and Signal Processing

"In Cheng et al. [159], Mahgoun et al. [164], Jin et al. [165] gearbox fault diagnosis methods are developed using vibration signal analysis."

"With the advancement of deep learning methods [166, 167], have presented fault diagnosis methods with automatic feature selection from a list of well-established features."

"Several current signal feature extraction methods could be found in literature such as time-domain features [168, 169], frequency-domain features [170] and time-frequency feature [171, 172], which are employed in the development of fault diagnosis schemes."

"Zhang et al. [173] presents an image processing approach to extract features from wavelet time-frequency spectra image obtained from vibration signal of motor."

"Vibration signal denoising methods using wavelet transform have been presented for detection of tooth faults in gearboxes [165] and alternator faults [174]."

"In Huang et al. [175], a blind source separation technique using Fast Independent Component Analysis is implemented on vibration signals of ball screw fault modes."

Fault Classification

"In modern industrial environment, large amount of online and historic industrial data is available and thus data-driven methods are employed for monitoring, fault detection and diagnosis purposes."

"Supervised learning methods require data samples along with their corresponding labels and the goal is to approximate a mapping for the prediction of output value or data label."

"In supervised learning methods, FDD scheme is trained to detect and consequently classify system status into known fault classes using designed decision criteria."

"Although supervised learning methods achieve high diagnosis accuracy, their application is limited due to significant gap in the explicit data label requirements of supervised learning methods and the real industrial data because the historical data from real plants commonly lack consistent data labeling procedures."

"In real-world problem a better choice may be semi-supervised learning methods taking leverage on both the label-related information using supervised learning as well as the complex data representations with unsupervised outliers' detection methods."

Decision Support and Maintenance

"The primary objective of fault diagnosis system is health-status monitoring which is accompanied with related operational maintenance responsibilities such as fault prediction, fault avoidance/prevention, and can be extended towards the development of a fault tolerant system."

"In Wang et al. [176], the degradation process is modeled using Weiner filter."

"In Li et al. [177], a convolution neural network (CNN) approach is employed for prognostics and RUL estimation on NSAA's turbofan engine degradation problem."

"Hardware FT was achieved in actuator faults [178] and fault switch process in PM synchronous motors [179]."

"A unified architecture for active fault detection and partial active fault tolerant control is presented for incipient parameter faults [162]."

"A stator turn fault detection and fault-tolerant PM drive method is presented in Wang et al. [170] employing second harmonic information as fault signature indicator."

"Other fault-tolerant frameworks for open and short circuit switch faults in converters include [180, 181]."

Conclusion

"As an improvement towards FDD, hybrid FDD methods have great scope in future real-world applications."

"Multiple model banks [182] and multi-stage filtering schemes [150] which combine parameter and state estimation techniques or hybrid methods combining model-based and signal processing approaches for more robust FDD design."

"With the growing scale and complexity of systems the design of unified framework and mutual interaction of diverse classifiers at different abstraction levels would be primary challenges whilst the development of automatic hybrid FDD schemes."

"FDD development for process systems [183, 184] will be more challenging due to its large scale and complex correlated relationships between sub-systems."

"Future FDD methods should be capable of dealing with inaccurate and uncertain information more effectively which is quite essential for their application in the industrial site."

References

1. Shakiba FM, Azizi SM, Zhou M, Abusorrah A (2022) Application of machine learning methods in fault detection and classification of power transmission lines: a survey. Artif Intell Rev. https://doi.org/10.1007/s10462-022-10296-0
2. Pei X et al (2019) A novel ultra-high-speed traveling-wave protection principle for vsc-based dc grids. IEEE Access 7:119765–119773
3. Cervantes M et al (2019) A traveling wave based fault location method using unsynchronized current measurements. In: IEEE power & energy society general meeting (PESGM). IEEE, Piscataway, p 1
4. Ding J et al (2018) Distributed traveling-wave-based fault-location algorithm embedded in multiterminal transmission lines. IEEE Trans Power Deliv 33(6):3045–3054
5. Ma Y et al (2018) Fault analysis and traveling-wave-based protection scheme for double-circuit lcc-hvdc transmission lines with shared towers. IEEE Trans Power Deliv 33(3):1479–1488
6. Felipe VL et al (2019) Practical methodology for two-terminal traveling wave-based fault location eliminating the need for line parameters and time synchronization. IEEE Trans Power Deliv 34(6):2123–2134
7. Zhang C et al (2019) Single-ended traveling wave fault location method in dc transmission line based on wave front information. IEEE Trans Power Deliv 34(5):2028–2038
8. Lopes FV, Leite EJ (2018) Traveling wave-based solutions for transmission line two-terminal data time synchronization. IEEE Trans Power Deliv 33(6):3240–3241
9. Koley E, Kumar R, Ghosh S (2016) Low cost microcontroller based fault detector, classifier, zone identifier and locator for transmission lines using wavelet transform and artificial neural network: a hardware co-simulation approach. Int J Electr Power Energy Syst 81:346–360
10. Gururajapathy SS (2017) Fault location and detection techniques in power distribution systems with distributed generation: a review. Renew Sust Energ Rev 74:949–958
11. De Andrade L, de Leão TP (2012) Impedance-based fault location analysis for transmission lines. IEEE, Piscataway, pp 1–6

12. Aker E, Othman ML, Aris I, Wahab NIA, Hizam H, Emmanuel O (2020) Transmission line fault identification and classification with integrated facts device using multiresolution analysis and Naïve Bayes classifier. Int J Power Electron Drive Syst 11(2):907
13. Swetapadma A et al (2018) A novel single-ended fault location scheme for parallel transmission lines using k-nearest neighbor algorithm. Comput Electr Eng 69:41–53
14. Farshad M, Sadeh J (2012) Accurate single-phase fault-location method for transmission lines based on k-nearest neighbor algorithm using one-end voltage. IEEE Trans Power Deliv 27(4):2360–2367. https://doi.org/10.1109/TPWRD.2012.2211898
15. Mahmud MN, Ibrahim MN, Osman MK, Hussain Z (2016) Fault classification in transmission line using wavelet features and fuzzy-knn. Appl Mech Mater 850:112–117
16. Dasgupta A, Debnath S, Das A (2015) Transmission line fault detection and classification using cross-correlation and k-nearest neighbor. Int J Knowl Based Intell Eng Syst 19(3):183–189
17. Abdelgayed TS, Morsi WG, Sidhu TS (2017) Fault detection and classification based on co-training of semisupervised machine learning. IEEE Trans Ind Electron 65(2):1595–1605
18. Fahim SR et al (2020) Self attention convolutional neural network with time series imaging based feature extraction for transmission line fault detection and classification. Electr Power Syst Res 187:106437
19. Fuada S, Shiddieqy HA, Adiono T (2020) A high-accuracy of transmission line faults (tlfs) classification based on convolutional neural network. Int J Electr Electron Eng Telecommun 66(4):655–664
20. Huan VP, Le Kim Hung NHV (2015) Fault classification and location on 220 kv transmission line hoa khanh–hue using anfis net. J Autom Control Eng 3:2
21. Fathabadi H (2016) Novel filter based ann approach for short-circuit faults detection, classification and location in power transmission lines. Int J Electr Power Energy Syst 74:374–383
22. Moradzadeh A, Teimourzadeh H, Mohammadi-Ivatloo B, Pourhossein K (2022) Hybrid CNN-LSTM approaches for identification of type and locations of transmission line faults. Int J Electr Power Energy Syst 135:107563
23. Balakrishnan P, Gopinath S (2020) A new intelligent scheme for power system faults detection and classification: a hybrid technique. Int J Numer Model 33(5):2728
24. Guo Y et al (2020) An improved alexnet for power edge transmission line anomaly detection. IEEE Access 8:97830–97838
25. Qian Y, Shi L, Li J, Zhou X, Shu F, Wang J (2020) An edge-computing paradigm for internet of things over power line communication networks. IEEE Netw 34(2):262–269
26. Wang M, Chai W, Xu C, Dong L, Li Y, Wang P, Qin X (2020) An edge computing method using a novel mode component for power transmission line fault diagnosis in distribution network. J Ambient Intell Humaniz Comput 2020:1–14
27. Rathore MS, Harsha SPR (2022) Bearing failure analysis using gaussian mixture models and convolutional neural networks. J Fail Anal Prev. https://doi.org/10.1007/s11668-022-01469-8
28. Kaplan K, Kaya Y, Kuncan M, Minaz MR, Ertunç HM (2020) An improved feature extraction method using texture analysis with LBP for bearing fault diagnosis. Appl Soft Comput 87:106019
29. Li H, Zhao W, Zhang Y, Zio E (2020) Remaining useful life prediction using multi-scale deep convolutional neural network. Appl Soft Comput 89:106113
30. Ince T, Kiranyaz S, Eren L, Askar M, Gabbouj M (2016) Real-time motor fault detection by 1-D convolutional neural networks. IEEE Trans Ind Electron 63(11):7067–7075
31. Kiranyaz S, Ince T, Gabbouj M (2015) Real-time patient-specific ECG classification by 1-D convolutional neural networks. IEEE Trans Biomed Eng 63(3):664–675
32. Maas AL, Hannun AY, Ng AY (2013) Rectifier nonlinearities improve neural network acoustic models. Proc ICML 30(1):3
33. Maaten LVD, Hinton G (2008) Visualizing data using t-SNE. J Mach Learn Res 9:2579–2605

34. Alabsi M, Liao Y, Nabulsi AA (2021) Bearing fault diagnosis using deep learning techniques coupled with handcrafted feature extraction: a comparative study. J Vib Control 27(3–4):404–414
35. Hsu CC, Huang WH (2016) Integrated dimensionality reduction technique for mixed-type data involving categorical values. Appl Soft Comput 43:199–209
36. Ju Z, Liu H (2012) Fuzzy gaussian mixture models. Pattern Recogn 45(3):1146–1158
37. Karatzinis GD, Boutalis YS, Karnavas YL (2021) An accurate multiple cognitive classifier system for incipient short-circuit fault detection in induction generators. Electr Eng. https://doi.org/10.1007/s00202-021-01445-9
38. Martínez-Morales JD, Palacios-Hernández ER, Campos-Delgado D (2018) Multiple-fault diagnosis in induction motors through support vector machine classification at variable operating conditions. Electr Eng 100(1):59–73
39. Toma S, Capocchi L, Capolino GA (2012) Wound-rotor induction generator inter-turn short-circuits diagnosis using a new digital neural network. IEEE Trans Ind Electron 60(9):4043–4052
40. Verma A, Sarangi S, Kolekar MH (2014) Stator winding fault prediction of induction motors using multiscale entropy and grey fuzzy optimization methods. Comput Electr Eng 40(7):2246–2258
41. Rothenhagen K, Fuchs FW (2009) Doubly fed induction generator model-based sensor fault detection and control loop reconfiguration. IEEE Trans Ind Electron 56(10):4229–4238
42. Khelfi H, Hamdani S (2020) Induction motor rotor fault diagnosis using three-phase current intersection signal. Electr Eng 102(2):539–548
43. Devi NR, Sarma DV, Rao PV (2015) Detection of stator incipient faults and identification of faulty phase in three-phase induction motor-simulation and experimental verification. IET Electr Power Appl 9(8):540–548
44. Abdallah H, Benatman K (2017) Stator winding inter-turn short-circuit detection in induction motors by parameter identification. IET Electr Power Appl 11(2):272–288
45. Xu Z, Hu C, Yang F, Kuo SH, Goh CK, Gupta A, Nadarajan S (2017) Data-driven inter-turn short circuit fault detection in induction machines. IEEE Access 5:25055–25068
46. dos Santos T, Ferreira FJ, Pires JM, Damásio C (2017) Stator winding short-circuit fault diagnosis in induction motors using random forest. In: 2017 IEEE international electric machines and drives conference (IEMDC). IEEE, Piscataway, pp 1–8
47. Bellini A, Filippetti F, Tassoni C, Capolino GA (2008) Advances in diagnostic techniques for induction machines. IEEE Trans Ind Electron 55(12):4109–4126
48. Xu Y, Nascimento NMM, de Sousa PHF, Nogueira FG, Torrico BC, Han T, Jia C, Rebouças Filho PP (2021) Multi-sensor edge computing architecture for identification of failures short-circuits in wind turbine generators. Appl Soft Comput 101:107053
49. de Sousa PHF, Navar de Medeiros M, Almeida JS, Rebouças Filho PP, de Albuquerque VH et al (2019) Intelligent incipient fault detection in wind turbines based on industrial iot environment. J Artif Intell Syst 1(1):1–19
50. Duin RP (2002) The combining classifier: to train or not to train? In: Object recognition supported by user interaction for service robots, vol 2. IEEE, Piscataway, pp 765–770
51. Li Y (2021) Exploring real-time fault detection of high-speed train traction motor based on machine learning and wavelet analysis. Neural Comput & Applic. https://doi.org/10.1007/s00521-021-06284-0
52. Toma RN, Prosvirin AE, Kim JM (2020) Bearing fault diagnosis of induction motors using a genetic algorithm and machine learning classifiers. Sensors 20(7):1884
53. Zaman SM, Liang X (2021) An effective induction motor fault diagnosis approach using graph-based semi-supervised learning. IEEE Access 9:7471–7482
54. Deng W, Zhang S, Zhao H et al (2018) A novel fault diagnosis method based on integrating empirical wavelet transform and fuzzy entropy for motor bearing. IEEE Access 6:35042–35056

55. Guo C, Li F, Tian Z, Guo W, Tan S (2019) Intelligent active fault-tolerant system for multi-source integrated navigation system based on deep neural network. Neural Comput & Applic. https://doi.org/10.1007/s00521-018-03975-z
56. Li-Jia XU, Yang-Zhou C, Ping-Yuan C (2004) State estimation of integrated navigation system based on neural network. J Chin Inert Technol 12(2):40–46
57. Wang M, Fu Y (2008) State estimation of ALV integrated navigation system based on BP neural network. In: Eighth international conference on intelligent systems design & applications. IEEE, Piscataway
58. Li YC, Lu XL, Wang HX, Bao Z (2007) Research on positioning and measuring speed in the high speed sar system based on high precision map matching. J Syst Eng Electron. https://doi.org/10.3233/ATDE230514
59. Yan X, Ouyang Y, Sun F et al (2013) Kalman filter applied in underwater integrated navigation system. Geodesy Geodyn 4(1):46–50
60. Wang T, Qiao M, Zhang M, Yang Y, Snoussi H (2018) Data-driven prognostic method based on self-supervised learning approaches for fault detection. J Intell Manuf. https://doi.org/10.1007/s10845-018-1431-x
61. Garca V, Snchez JS, Rodrguez-Picn LA, Mndez-Gonzlez LC, de Jess Ochoa-Domnguez H (2018) Using regression models for predicting the product quality in a tubing extrusion process. J Intell Manuf. https://doi.org/10.1007/s10845-018-1418-7
62. Kim Y, Lee H, Kim CO (2021) A variational autoencoder for a semiconductor fault detection model robust to process drift due to incomplete maintenance. J Intell Manuf. https://doi.org/10.1007/s10845-021-01810-2
63. Moyne J, Iskandar J (2017) Big data analytics for smart manufacturing: case studies in semiconductor manufacturing. PRO. https://doi.org/10.3390/pr5030039
64. Lee KB, Kim CO (2020) Recurrent feature-incorporated convolutional neural network for virtual metrology of the chemical mechanical planarization process. J Intell Manuf 31(1):73–86. https://doi.org/10.1007/s10845-018-1437-4
65. Kingma DP, Max W (2013) Auto-encoding variational bayes. arXiv:1312.6114 (preprint)
66. Hell M, Pestana Aguiar E, Soares N, Goliatt L (2022) A data-driven time-series fault prediction framework for dynamically evolving large-scale data streaming systems. Int J Fuzzy Syst. https://doi.org/10.1007/s40815-022-01292-1
67. Sánchez-Fernández A, Baldán F, Sainz-Palmero G, Benítez J, Fuente M (2018) Fault detection based on time series modeling and multivariate statistical process control. Chemom Intell Lab Syst 182:57–69
68. Angelov PP, Gu X, Kangin D (2017) Empirical data analytics. Int J Intell Syst 32:1261–1284
69. Gu X, Angelov PP (2018) Self-organising fuzzy logic classifier. Inf Sci 447:36–51
70. Gu X, Angelov P, Zhao Z (2021) Self-organizing fuzzy inference ensemble system for big streaming data classification. Knowl-Based Syst 218:106870. https://doi.org/10.1016/j.knosys.2021.106870
71. Song J, Lee YC, Lee J (2022) Deep generative model with time series-image encoding for manufacturing fault detection in die casting process. J Intell Manuf. https://doi.org/10.1007/s10845-022-01981-6
72. Lee J, Lee YC, Kim JT (2020) Fault detection based on one-class deep learning for manufacturing applications limited to an imbalanced database. J Manuf Syst 57:357–366. https://doi.org/10.1016/j.jmsy.2020.10.013
73. Fiorese E, Bonollo F, Timelli G, Arnberg L, Gariboldi E (2015) New classification of defects and imperfections for aluminum alloy castings. Int J Met 9(1):55–66. https://doi.org/10.1007/BF03355602
74. Ingle V, Sorte M (2017) Defects, root causes in casting process and their remedies: review. Int J Eng Res Appl 7(3):47–54. https://doi.org/10.9790/9622-0703034754
75. Lee J, Lee YC, Kim JT (2021) Migration from the traditional to the smart factory in the die-casting industry: novel process data acquisition and fault detection based on artificial neural network. J Mater Process Technol. https://doi.org/10.1016/j.jmatprotec.2020.116972

76. Ye Y, Huang P, Zhang Y (2021) Deep learning-based fault diagnostic network of high-speed train secondary suspension systems for immunity to track irregularities and wheel wear. Railway Eng Sci. https://doi.org/10.1007/s40534-021-00252-z
77. Ye Y, Zhang Y, Wang Q, Wang Z, Teng Z, Zhang H (2020) Fault diagnosis of high-speed train suspension systems using multiscale permutation entropy and linear local tangent space alignment. Mech Syst Signal Process 138:106565
78. Wei XK, Jia LM, Liu H (2013) A comparative study on fault detection methods of rail vehicle suspension systems based on acceleration measurements. Veh Syst Dyn 51(5):700–720
79. Ince (2019) Turker real-time broken rotor bar fault detection and classification by shallow 1D convolutional neural networks. Electr Eng. https://doi.org/10.1007/s00202-019-00808-7
80. Di Stefano R, Meo S, Scarano M (1994) Induction motor fault diagnostic via artificial neural network. In: Proceedings of the IEEE international symposium on industrial electronics (ISIE' 94). IEEE, Piscataway, pp 220–225
81. Li B, Chow M-Y, Tipsuwan Y, Hung JC (2000) Neural network based motor rolling bearing fault diagnosis. IEEE Trans Ind Electron 47:1060–1069
82. Bin GF, Gao JJ, Li XJ, Dhillon BS (2012) Early fault diagnosis of rotating machinery based on wavelet packets—empirical mode decomposition feature extraction and neural network. Mech Syst Signal Process 27:696–711
83. Tung VT, Yang B-S, Oh M-S, Tan ACC (2009) Fault diagnosis of induction motor based on decision trees and adaptive neurofuzzy inference. Expert Syst Appl 36:1840–1849
84. Wu S-D, Wu P-H, Wu C-W, Ding J-J, Wang C-C (2012) Bearing fault diagnosis based on multiscale permutation entropy and support vector machine. Entropy 14:1343–1356
85. Kang M, Kim J, Kim J-M, Tan ACC, Kim EY, Choi B-K (2014) Reliable fault diagnosis for low-speed bearings using individually trained support vector machines with kernel discriminative feature analysis. IEEE Trans Power Electron 30:2786–2797
86. Kowalski CT, Kowalska TO (2003) Neural network application for induction motor faults diagnosis. Math Comput Simul 63:435–448
87. Ballal MS, Khan ZJ, Suryawanshi HM, Sonolikar RL (2007) Adaptive neural fuzzy inference system for the detection of inter-turn insulation and bearing wear faults in induction motor. IEEE Trans Ind Electron 54:250–258
88. Zidani F, Diallo D, Benbouzid MEH, Nait-Said R (2008) A fuzzy based approach for the diagnosis of fault modes in a voltage-fed PWM inverter induction motor drive. IEEE Trans Ind Electron 55(2):586–593
89. Aydin I, Karakose M, Akin E (2014) An approach for automated fault diagnosis based on a fuzzy decision tree and boundary analysis of a reconstructed phase space. ISA Trans 53(2):220–229
90. Van M, Kang H-J (2016) Bearing defect classification based on individual wavelet local fisher discriminant analysis with particle swarm optimization. IEEE Trans Industr Inform 12:124–135
91. Liu R, Yang B, Zhang X, Wang S, Chen X (2016) Time-frequency atoms-driven support vector machine method for bearings incipient fault diagnosis. Mech Syst Signal Process 75:345–370
92. Filippetti F, Franceschini G, Tassoni C, Vas P (2000) Recent developments of induction motor drives fault diagnosis using AI techniques. IEEE Trans Ind Electron 47(5):994–1004
93. Thomson WT, Fenger M (2001) Current signature analysis to detect induction motor faults. IEEE Ind Appl Mag 7(4):26–34
94. Aydin I, Karakose M, Akin E (2011) A new method for early fault detection and diagnosis of broken rotor bars. Energy Convers Manag 52:1790–1799
95. Martin-Diaz I, Morinigo-Sotelo D, Duque-Perez O, Arredondo-Delgado PA, Camarena-Martinez D, Romero-Troncoso RJ (2017) Analysis of various inverters feeding induction motors with incipient rotor fault using high-resolution spectral analysis. Electr Power Syst Res 152:18–26

96. Ordaz-Moreno A, Romero-Troncoso RD, Rivera-Guillen JR, Vite-Frias JA, Garcia-Perez A (2008) Automatic online diagnosis algorithm for broken-bar detection on induction motors based on discrete wavelet transform for FPGA Implementation. IEEE Trans Ind Electron 55:2193–2202
97. Eltabach M, Charara A, Zein I (2004) A comparison of external and internal methods of signal analysis for broken rotor bars detection in induction motors. IEEE Trans Ind Electron 51(1):107–121
98. Bachir S, Tnani S, Trigeassou J, Champenois G (2006) Diagnosis by parameter estimation of stator and rotor faults occurring in induction machines. IEEE Trans Ind Electron 53(3):963–973
99. Tan WW, Huo H (2005) A generic neurofuzzy model-based approach for detecting faults in induction motors. IEEE Trans Ind Electron 52(5):1420–1427
100. Bellini A, Filippetti F, Franceshini G, Tassoni C (2001) Quantitative evaluation motor broken bars by means of electrical signature analysis. IEEE Trans Ind Appl 37:1248–1254
101. Ayhan B, Chow M, Song M (2005) Multiple signature processing-based fault detection schemes for broken rotor bar in induction motors. IEEE Trans Energy Convers 20:336–343
102. Sadeghian A, Ye Z, Wu B (2009) Online detection of broken rotor bars in induction motors by wavelet packet decomposition and artificial neural networks. IEEE Trans Instrum Meas 58(7):2253–2262
103. Keskes H, Braham A, Lachiri Z (2013) Broken rotor bar diagnosis in induction machines through stationary wavelet packet transform and multiclass wavelet SVM. Electr Power Syst Res 97:151–157
104. Briz F, Degnert MW, Garcia P, Bragado D (2008) Broken rotor bar detection in line-fed induction machines using complex wavelet analysis of startup transients. IEEE Trans Ind Appl 44(3):760–768
105. Zolfaghari S, Noor SBM, Mehrjou MR, Marhaban MH, Mariun N (2017) Broken rotor bar fault detection and classification using wavelet packet signature analysis based on fourier transform and multi-layer perceptron neural network. Appl Sci 8(25):1–21. https://doi.org/10.3390/app8010025
106. Kurek J, Osowski S (2010) Support vector machine for fault diagnosis of the broken rotor bars of squirrel-cage induction motor. Neural Comput & Applic 19:557. https://doi.org/10.1007/s00521-009-0316-5
107. Singh H, Seera M, Abdullah MZ (2013) Detection and diagnosis of broken rotor bars and eccentricity faults in induction motors using the fuzzy min-max neural network. In: 2013 international joint conference on neural networks (IJCNN). IEEE, Piscataway, pp 1–5
108. Wu C, Guo S, Wu Y, Ai J, Xiong NN (2020) Networked fault detection of field equipment from monitoring system based on fusing of motion sensing and appearance information. Multimed Tools Appl. https://doi.org/10.1007/s11042-020-08885-8
109. Mabrouk AB, Zagrouba E (2018) Abnormal behavior recognition for intelligent video surveillance systems: a review. Expert Syst Appl 91:480–491
110. Zhang Y, Lu H, Zhang L, Ruan X (2016) Combining motion and appearance cues for anomaly detection. Pattern Recogn 51(C):443–452
111. Alvar M, Torsello A, Sanchez-Miralles A, Armingol JM (2014) Abnormal behavior detection using dominant sets. Mach Vis Appl 25(5):1351–1368
112. Jørgensen EK (2014) Fall detection for the elderly using microsoft kinect. Department of Engineering Cybernetics
113. Cong Y, Yuan J, Liu J (2011) Sparse reconstruction cost for abnormal event detection. In: IEEE conference on computer vision and pattern recognition. IEEE, Piscataway
114. Lu C, Shi J, Jia J (2014) Abnormal event detection at 150 FPS in MATLAB. In: IEEE international conference on computer vision. IEEE, Piscataway
115. Zhao B, Li FF, Xing EP (2011) Online detection of unusual events in videos via dynamic sparse coding. In: IEEE conference on computer vision and pattern recognition. IEEE, Piscataway

116. Li N et al (2015) Anomaly detection in video surveillance via Gaussian process. Int J Pattern Recognit Artif Intell 29(6):150426191333005
117. Wang T, Snoussi H (2012) Histograms of optical flow orientation for visual abnormal events detection. In: IEEE ninth international conference on advanced video and signal-based surveillance. IEEE, Piscataway
118. Eren L, Ince T, Kiranyaz S (2018) A generic intelligent bearing fault diagnosis system using compact adaptive 1D CNN classifier. J Signal Process Syst. https://doi.org/10.1007/s11265-018-1378-3
119. Lee J, Qiu H, Yu G, Lin J (2017) Rexnord technical services, IMS, University of Cincinnati. Bearing data set, NASA Ames prognostics data repository. NASA Ames Research Center. https://ti.arc.nasa.gov/tech/dash/pcoe/prognostic-data-repository/#bearing. Accessed 15 March 2017
120. Case Western Reserve University Bearing Data Center Website. http://csegroups.case.edu/bearingdatacenter/home. Accessed 15 March 2017
121. Mamandipoor B, Majd M, Sheikhalishahi S, Modena C, Osmani V (2020) Monitoring and detecting faults in wastewater treatment plants using deep learning. Environ Monit Assess. https://doi.org/10.1007/s10661-020-8064-1
122. Maier HR, Dandy GC (2000) Neural networks for the prediction and forecasting of water resources variables: a review of modelling issues and applications. Environ Model Softw 15(1):101–124
123. Ozcan IH, Devecioglu OC, Ince T, Eren L, Askar M (2021) Enhanced bearing fault detection using multichannel, multilevel 1D CNN classifier. Electr Eng. https://doi.org/10.1007/s00202-021-01309-2
124. Gao Z, Cecati C, Ding SX (2015) A survey of fault diagnosis and fault-tolerant techniques—part I: fault diagnosis with model-based and signal-based approaches. IEEE Trans Ind Electron 62:3757–3767
125. Filippetti F, Bellini A, Capolino GA (2013) Condition monitoring and diagnosis of rotor faults in induction machines: state of art and future perspectives. In: Proceedings of the IEEE WEMDCD. IEEE, Piscataway, pp 196–209
126. Zhou W, Habetler T, Harley R (2008) Bearing fault detection via stator current noise cancellation and statistical control. IEEE Trans Ind Electron 55:4260–4269
127. Kral C, Habetler TG, Harley RG (2004) Detection of mechanical imbalances of induction machines without spectral analysis of time domain signals. IEEE Trans Ind Appl 40:1101–1106
128. Schoen RR, Habetler TG, Kamran F, Bartheld RG (1995) Motor bearing damage detection using stator current monitoring. IEEE Trans Ind Appl 31:1274–1279
129. Kliman GB, Premerlani WJ, Yazici B, Koegl RA, Mazereeuw J (1997) Sensorless online motor diagnostics. IEEE Comput Appl Power 10:39–43
130. Pons-Llinares J, Antonino-Daviu JA, Riera-Guasp M, Lee SB, Kang TJ, Yang C (2015) Advanced induction motor rotor fault diagnosis via continuous and discrete time–frequency tools. IEEE Trans Ind Electron 62:1791–1802
131. Li DZ, Wang W, Ismail F (2015) An enhanced bispectrum technique with auxiliary frequency injection for induction motor health condition monitoring. IEEE Trans Instrum Meas 67:2279–2287
132. Eren L, Devaney MJ (2004) Bearing damage detection via wavelet packet decomposition of the stator current. IEEE Trans Instrum Meas 53:431–436
133. Yan R, Gao RX, Chen X (2014) Wavelets for fault diagnosis of rotary machines: a review with applications. Signal Process 96:1–15
134. Grubic S, Aller JM, Lu B, Habetler TG (2008) A survey on testing and monitoring methods for stator insulation systems of low-voltage induction machines focusing on turn insulation problems. IEEE Trans Ind Electron 55:4127–4136
135. Kiranyaz S, Ince T, Gabbouj M (2015) Real-time patient-specific ECG classification by 1D convolutional neural networks. IEEE Trans Biomed Eng 63:664–674

136. Ince T, Kiranyaz S, Eren L, Askar M, Gabbouj M (2016) Real-time motor fault detection by 1D convolutional neural networks. IEEE Trans Ind Electron 63:7067–7075
137. Eren L, Ince T, Kiranyaz S (2019) A generic intelligent bearing fault diagnosis system using compact adaptive 1D CNN classifier. J Signal Process Syst 91(2):179–189
138. Ince T (2019) Real-time broken rotor bar fault detection and classification by shallow 1D convolutional neural networks. Electr Eng 101(2):599–608
139. Eren L (2017) Bearing fault detection by one-dimensional convolutional neural networks. Math Probl Eng 2017:1–9
140. Ahishali M, Kiranyaz S, Ince T, Gabbouj M (2019) Dual and single polarized SAR image classification using compact convolutional neural networks. Remote Sens 11(11):1340. https://doi.org/10.3390/rs11111340
141. Eren L, Cekic Y, Devaney M (2009) Broken rotor bar detection via wavelet packet decomposition of motor current. Int Rev Electr Eng 4:844–850
142. Abid A, Khan MT, Iqbal J (2020) A review on fault detection and diagnosis techniques: basics and beyond. Artif Intell Rev. https://doi.org/10.1007/s10462-020-09934-2
143. Capisani LM, Ferrara A, Alejandra Ferreira DL, Fridman ML (2011) Manipulators fault diagnosis via higher order sliding mode observers. IEEE Trans Ind Electron 59(10):3979–3986
144. Salmasi FR, Najafabadi TA, Maralani PJ (2010) An adaptive flux observer with online estimation of DC-link voltage and rotor resistance for VSI-based induction motors. IEEE Trans Power Electron 25(5):1310–1319
145. Low CB, Wang D, Member S, Arogeti S, Luo M (2010) Quantitative hybrid bond graph-based fault detection and isolation. IEEE Trans Autom Sci Eng 7(3):558–569
146. Benmoussa S, Bouamama BO, Merzouki R (2014) Bond graph approach for plant fault detection and isolation: application to iIntelligent autonomous vehicle. IEEE Trans Autom Sci Eng 11(2):585–593
147. Zhou S, Qian S, Chang W, Xiao Y, Cheng Y (2018) A novel bearing multi-fault diagnosis approach based on weighted permutation entropy and an improved SVM ensemble classifier. Sensors 18(6):1–23
148. Soualhi A, Clerc G, Razik H (2013) Detection and diagnosis of faults in induction motor using an improved artificial ant clustering technique. IEEE Trans Ind Electron 60(9):4053–4062
149. Malhi A, Gao RX (2004) PCA-based feature selection scheme for machine defect classification. IEEE Trans Instrum Meas 53(6):1517–1525
150. Ben Hmida F, Khémiri K, Ragot J, Gossa M (2012) Three-stage Kalman filter for state and fault estimation of linear stochastic systems with unknown inputs. J Franklin Inst 349(7):2369–2388
151. Huang S, Tan KK, Lee TH (2012) Fault diagnosis and fault-tolerant control in linear drives using the Kalman filter. IEEE Trans Ind Electron 59(11):4285–4292
152. Zhang Y, Fan Y, Du W (2016) Nonlinear process monitoring using regression and reconstruction method. IEEE Trans Autom Sci Eng 13(3):1343–1354
153. Abbasi AR, Mahmoudi MR, Avazzadeh Z (2018) Diagnosis and clustering of power transformer winding fault types by cross-correlation and clustering analysis of FRA results. IET Gener Transm Distrib 12(19):4301–4309
154. Abaei G, Selamat A (2014) A survey on software fault detection based on different prediction approaches. Vietnam J Comput Sci 1(2):79–95
155. Henr P, Alonso B, Ferrer MA, Travieso CM (2014) Review of automatic fault diagnosis systems using audio and vibration signals. IEEE Trans Syst Man Cybern Syst 44(5):642–652
156. Abid A, Khan MT (2017) Multi-sensor, multi-level data fusion and behavioral analysis based fault detection and isolation in mobile robots. In: IEEE 8th annual information technology. Electronics and mobile communication conference (IEMCON). IEEE, Piscataway, pp 40–45
157. Tabbache B, El M, Benbouzid H, Kheloui A, Bourgeot J-M (2013) Virtual-sensor-based maximum-likelihood voting approach for fault-tolerant control of electric vehicle powertrains. IEEE Trans Veh Technol 62(3):1075–1083

158. Laurentys CA, Ronacher G, Palhares RM, Caminhas WM (2010) Design of an artificial immune system for fault detection: a negative selection approach. Expert Syst Appl 37(7):5507–5513
159. Cheng G, Cheng YL, Shen LH, Qiu JB, Zhang S (2013) Gear fault identification based on Hilbert–Huang transform and SOM neural network. Measurement 46(3):1137–1146
160. Jiang G, Xie P, He H, Yan J (2018) Wind turbine fault detection using a denoising autoencoder with temporal information. IEEE/ASME Trans Mechatron 23(1):89–100
161. Zhao Y, Lam J, Gao H (2009) Fault detection for fuzzy systems with intermittent measurements. IEEE Trans Fuzzy Syst 17(2):398–410
162. Wang J, Zhang J, Qu B, Wu H, Zhou J (2017) Unified architecture of active fault detection and partial active fault-tolerant control for incipient faults. IEEE Trans Syst Man Cybern Syst 47(7):1688–1700
163. Abid A, Khan MT, Silva CWD (2015) Fault detection in mobile robots using sensor fusion. In: 10th international conference on computer science and education (ICCSE 2015). Cambridge University, Cambridge, pp 8–13
164. Mahgoun H, Bekka RE, Felkaoui A (2013) Gearbox fault detection using a new denoising method based on ensemble empirical mode decomposition and Fft. In: 4th international conference on integrity, reliability and failure (IRF2013). Universidade do Porto, Porto, pp 1–11
165. Jin S, Kim JS, Lee SK (2015) Sensitive method for detecting tooth faults in gearboxes based on wavelet denoising and empirical mode decomposition. J Mech Sci Technol 29(8):3165–3173
166. Xia M, Li T, Xu L, Liu L, De Silva CW (2018) Fault diagnosis for rotating machinery using multiple sensors and convolutional neural networks. IEEE/ASME Trans Mechatron 23(1):101–110
167. Abid A, Khan MT, Khan MS (2020b) Multidomain features-based GA optimized fault detection. IEEE Trans Syst Man Cybern Syst 50(1):348–359
168. Abid A, Khan MT, Ullah A, Alam M, Sohail M (2017) Real time health monitoring of industrial machine using multiclass support vector machine. In: 2nd International conference on control and robotics engineering, vol 2. Springer, Cham, pp 77–81
169. Abid A, Khan MT, de Silva CW (2018) Layered and real-valued negative selection algorithm for fault detection. IEEE Syst J 12(3):2960–2969
170. Wang B, Wang J, Griffo A, Sen B (2018) Stator turn fault detection by second harmonic in instantaneous power for a triple-redundant fault-tolerant PM drive. IEEE Trans Ind Electron 65(9):7279–7289
171. Strangas EG, Aviyente S, Zaidi SSH (2008) Time-frequency analysis for efficient fault diagnosis and failure prognosis for interior permanent-magnet AC motors. IEEE Trans Ind Electron 55(12):4191–4199
172. Blödt M, Chabert M, Regnier J, Faucher J (2006) Mechanical load fault detection in induction motors by stator current time-frequency analysis. IEEE Trans Ind Appl 42(6):1454–1463
173. Zhang J, Wang P, Gao RX, Yan R (2018) An image processing approach to machine fault diagnosis based on visual words representation. Proc Manuf 19:42–49
174. Abad MRAA, Moosavian A, Khazaee M (2016) Wavelet transform and least square support vector machine for mechanical fault detection of an alternator using vibration signal. J Low Freq Noise Vib Active Control 35(1):52–63
175. Huang H, Ouyang H, Gao H (2015) Blind source separation and dynamic fuzzy neural network for fault diagnosis in machines. J Phys 628:012070
176. Wang Z-Q, Hu C-H, Fan H-D (2018) Real-remaining useful life prediction for a nonlinear degrading system in service: application to bearing data. IEEE/ASME Trans Mechatron 23(1):211–222
177. Li X, Ding Q, Sun JQ (2018) Remaining useful life estimation in prognostics using deep convolution neural networks. Reliab Eng Syst Saf 172:1–11
178. Zhang G, Zhang H, Huang X, Wang J, Yu H, Graaf R (2016) Active fault-tolerant control for electric vehicles with independently driven rear in-wheel motors against certain actuator faults. IEEE Trans Control Syst Technol 24(5):1557–1572

179. Guo H, Xu J, Chen YH (2015) Robust control of fault-tolerant permanent-magnet synchronous motor for aerospace application with guaranteed fault switch process. IEEE Trans Ind Electron 62(12):7309–7321
180. Salehifar M, Arashloo RS, Moreno-equilaz JM, Sala V, Romeral L (2014) Fault detection and fault tolerant operation of a five phase PM motor drive using adaptive model identification approach. IEEE J Emerg Select Top Power Electron 2(2):212–223
181. Yang S, Member S, Tang Y (2018) Seamless fault-tolerant operation of a modular multilevel converter with switch open-circuit fault diagnosis in a distributed control architecture. IEEE Trans Ind Electron 33(8):7058–7070
182. Abid A, Khan MT, Lang H, Silva CWD (2019) Adaptive system identification and severity index-based fault diagnosis in motors. IEEE/ASME Trans Mechatron 24(4):1628–1639
183. Cheng F, He QP, Zhao J (2019) A novel process monitoring approach based on variational recurrent autoencoder. Comput Chem Eng 129:1–14
184. Wu H, Zhao J (2020) Fault detection and diagnosis based on transfer learning for multimode chemical processes. Comput Chem Eng 135:1–27

Chapter 7
Optimization Methods in Material Waste Management

Introduction by the Editor

The efficiency of the manufacturing systems helps in reducing the greenhouse gases, material waste, human hours waste and eventually helps in maximum utilization of the available resources. So, the efforts need to be towards devising processes for minimum wastage as well as efficient and effective systems in the management of the waste. The chapter reviews the optimal policies for production system including quality inspection that can be classified as a green production system. It is supported by several numerical illustrations and sensitivity analysis. The chapter also reviews a multi-product multi-supplier multi-stage multi-period mixed-integer linear programming model considering technology effects and raw materials, semi-finished products, and end products waste in a supply chain minimizing cost elements, such as raw materials inspection waste, manufacturing waste, ordering, purchasing manufacturing technologies, transportation, production variable, production inventory holding, raw material inventory holding, etc. Scatter search algorithm, genetic algorithms and linear programming approach for minimizing multiple recycling, manufacturing/remanufacturing, and efficient distribution centers have also been discussed with regard to the industry sustainability.

The chapter discusses the review of the policy for a green production system for the necessary quality inspection and warranty cost under environmental regulations. The work demonstrates how environmental policies can be incorporated with inventory production and inspection policies for imperfect quality items. The key decisions are to determine the optimal production and inspection policies by jointly optimising the total cost function. The NSGA-II and Monte Carlo simulation approaches have been discussed for integrated production and distribution planning problem of a multi-level green closed-loop SC system, which includes multiple recycling, manufacturing/remanufacturing, and distribution centres. A scatter search algorithm is discussed for the improvement method with the aim of being able to

A. J. Kulkarni (ed.), *Optimization Methods in Manufacturing Processes*,
https://doi.org/10.1007/978-981-96-5257-0_7

determine the best results for solution to our multi objective model. It is concluded that the classification model can be applied on the large inventory items that holding in the warehouse for example pharmaceutical items, spare parts, electronical components, and electromechanical items, etc. Two contributions have also been discussed, such as establishing models for energy consumption embodied in parts during the manufacturing phase and achieving the trade-off between energy saving and high production efficiency in the workshop. This helped in exploring the impact of uncertainty distribution of duration and machine failure rate on energy consumption. It is important to mention that the innovation of energy-saving optimization in the workshop is that the feasible solutions are ranked after obtaining the Pareto set to quickly obtain the most suitable solution.

As the local authorities for environmental protection and urban planners are interested in performing detailed investigations using traffic and air pollution simulations for testing various urban scenarios and raising citizen awareness where necessary. The chapter reviews an integrated simulation model which predicts and visualizes various environmental changes inside the neighbourhood such as: air pollution, traffic flow or meteorological information. It includes integrated traffic and air pollution simulation model, regression models. The limitations of this work elaborated are such that the restricted data sets concerning the air pollutant concentrations and traffic counts which have been received from the traffic management centre and having access to longer periods of data would help to build more traffic demand scenarios. In addition, integrating the emissions from the train station would also improve the current model. The future directions found in these lines are associated with extending the hybrid model to the whole eco-neighbourhood and building more detailed air pollution estimation models which would take into consideration all the types of vehicles on the roads.

Machine Generated Summaries

Disclaimer: The summaries in this chapter were generated from Springer Nature publications using extractive AI auto-summarization: An extraction-based summarizer aims to identify the most important sentences of a text using an algorithm and uses those original sentences to create the auto-summary (unlike generative AI). As the constituted sentences are machine selected, they may not fully reflect the body of the work, so we strongly advise that the original content is read and cited. The auto generated summaries were curated by the editor to meet Springer Nature publication standards. To cite this content, please refer to the original papers.

Machine generated keywords: emission, inventory, item, air, supply chain, raw material, investment, energy, raw, fuel, defective, weight, inspection, gas, workshop.

Solving Material Handling Equipment Selection Problems in an Industry with the Help of Entropy Integrated COPRAS and ARAS MCDM Techniques [1]

This is a machine-generated summary of:

Goswami, Shankha Shubhra; Behera, Dhiren Kumar: Solving Material Handling Equipment Selection Problems in an Industry with the Help of Entropy Integrated COPRAS and ARAS MCDM techniques [1].

Published in: Process Integration and Optimization for Sustainability (2021).

Link to original: https://doi.org/10.1007/s41660-021-00192-5.

If you want to cite the papers, please refer to the original.

For technical reasons we could not place the page where the original quote is coming from.

Abstract-Summary "Material handling equipment (MHE) selection is a rapidly increasing multi-criteria decision-making (MCDM) issue with a plethora of factors affecting the selection process."

"The selection of the most suitable MHE for a specific engineering application is a costly as well as time-consuming process in which many candidate alternatives available on the market are considered as preliminary choices."

"All the three MCDM problems are evaluated and alternative rankings are proposed using these two methods, while the objective weighted system, entropy is used to assess the relative weights of the criteria."

"Each example provides a listing of all the possible options from optimum to worst MHE that could potentially improve the previously proposed rankings and sensitivity analysis is computed to demonstrate that these two MCDM methods are robust enough and capable of being used effectively in solving MHE selection problems."

"Conveyor 3, AGV 5, and ASEA-IRB 60/2 robot are found to be the best choice, while conveyor 1, AGV 3, and Unimation PUMA 500/600 robot are the worst alternative for the three MHE selection issues respectively."

Introduction

"Robots are also computer-programmed automated devices that are used for doing several tasks like pick and place operations, loading, and unloading; hence, efficient and well-organized handling systems are required to increase the system flexibility and production, material flow efficiency, facility utilization improvement, and minimization of lead time as well as handling and labor cost [2]."

"Such three MHE selection MCDM problems are taken into account in this present research work and evaluated using COPRAS and newly developed ARAS methodologies."

"The main purpose of this article is to choose the best alternatives for three distinct MHE selection issues, namely conveyor selection, AGV selection, and robot selection, and to suggest a preference ranking order of these three MHE alternatives."

Literature Survey

"Athawale and Chakraborty [3] (2 studied the ranking performance of 10 well-known MCDM methods while solving an industrial robot selection problem."

"Karande and Chakraborty [2] applied weighted utility additive (WUTA) method for solving an MHE selection problem and compared the alternative ranking with PROMETHEE, VIKOR, and ELECTRE."

"There are very few research works ever documented related to the MHE selection problems that implemented ARAS and COPRAS together as ranking tools and entropy as a weighting tool within a single article, which is the primary focus of this article and the author's field of investigation."

"It is clear from the above literatures that many researchers have done a lot of investigations on MHE selection problems in previous years using various MCDM approaches mainly, complex and time-consuming old traditional tools like, TOPSIS, PROMETHEE, ELECTRE, etc., but still there is a need to use an easy, clear, and systematic mathematical approach to direct the decision-maker in making an effective MHE selection decision for a particular engineering application."

Theoretical Analysis

"Zavadskas and Turskis [4] stated that "According to the ARAS method, a utility function value determining the complex relative efficiency of a feasible alternative is directly proportional to the relative effect of values and weights of the main criteria considered in a project"".

"p Step 2: Generate an optimal alternative ideally (A0) by considering the best values from all criteria."

""A_{ij}" are the weighted values of the jth criteria and ith alternative."

"COPRAS method considers the effect of beneficial and non-beneficial criteria individually to determine the relative significances of the alternatives followed by quantitative utility based on which the best option is proposed."

"It must be noted that, in case of COPRAS method, the minimizing criteria are not converted into maximizing criteria by taking the reciprocal values like entropy and ARAS."

"n. "A_{+ij}" and "A_{-ij}" are the weighted values of the maximizing and minimizing criteria respectively."

Research Methodology

"Entropy is applied to evaluate the criteria weights then ARAS and COPRAS are applied to choose the best alternative."

"Rao [5, 6] considered the same problem and applied Graph Theory and Matrix Approach (GTMA), SAW, WPM, TOPSIS, and modified TOPSIS method for the selection process."

"Rao [5, 6] assigned equal weights to the attributes and to remove this biasedness; Karande and Chakraborty [2] calculated the attributes weights using AHP method and again extended the selection process by WUTA method."

"Since the entropy approach is unaffected by the DM's thoughts and opinions, the weights are recalculated, and newly established ARAS and COPRAS due to their specific merits over other MCDM tools are used to solve the current conveyor selection issue."

"This AGV selection MCDM problem was previously solved by Maniya and Bhatt [7] using M-GRA considering eight alternatives and six criteria."

"Several researchers adopted this same robot selection problem and solved using WSM, WPM, WASPAS, MOORA, MULTIMOORA, VIKOR, ELECTRE, etc. utilizing the same AHP weights derived by Rao [5, 6]."

Result and Discussion

"The alternative having the highest utility degree and the quantitative degree can be termed as the best one and the ranking can be proposed according to the decreasing order."

"The outcomes from these three MCDM problems and their alternative rankings are presented in the following sub-section."

"Despite the fact that COPRAS, ARAS, SAW, and WASPAS provide the exact same AGV rankings, there are some variations in the middle-order alternative rankings offered by TOPSIS, WPM, MOORA, and MULTIMOORA, but the best and worst choice is same for all the eight MCDM techniques."

"Mathew and Sahu [8] were able to conclude that AGV 5 is the best option by all the five adopted methods, but they were unable to recommend the worst AGV alternative, because different techniques yielded dissimilar worst choices."

"The current decision on the AGV alternative ranking is more genuine and reliable than the previous proposed results."

Sensitivity Analysis

"Sensitivity analysis is carried out to study the effect of differences in criteria weights on the final alternative ranking [9]."

“A single dimensional weight sensitivity study is performed in this paper to analyze the effect of various criteria weights on the alternative final ranking obtained using eight adopted MCDM methods [9].”

“The criterion with the highest weight means that it has the greatest effect on the alternative ranking, so that it can be viewed as the most important criterion [9].”

“It is important to determine the possible range within which the weight of the chosen criterion can be varied [9].”

“The weight of the most important criterion can be reduced to 0 and increased to w_j* [9].”

Conclusion

“An entropy-embedded ARAS and COPRAS MCDM system is built to investigate three MHE selection issues and to substantially improve the previous proposed rankings by eliminating the ambiguity, uncertainty, and vagueness associated with the existing literatures.”

“ARAS and COPRAS proves to be very stable and consistent decision-making tools, both providing the exact same ranking order for all the three MHE selection instances.”

“The sensitivity analysis also shows that the weight variation has a very small or negligible impact on the final ranking order and selection of three MHE alternatives, proving ARAS and COPRAS as one of the most robust MCDM techniques among the tools adopted.”

“The following three MHE selection problems are performed on a limited number of essential criteria, but there are other parameters as well that can also be considered in addition to these, which can also cause the results to deviate [10].”

Optimal Policy for a Green Production System with Quality Inspection and Warranty Cost Under Environmental Regulations [11]

This is a machine-generated summary of:

Rini; Kishore, Aakanksha: Optimal Policy for a Green Production System with Quality Inspection and Warranty Cost Under Environmental Regulations [11].

Published in: Process Integration and Optimization for Sustainability (2021).

Link to original: https://doi.org/10.1007/s41660-021-00182-7.

If you want to cite the papers, please refer to the original.

For technical reasons we could not place the page where the original quote is coming from.

Abstract-Summary "Apart from dealing with environmental policies, the manufacturers need to look after their brand image by constantly putting efforts towards quality of the product being sold in the market."

"Production of imperfect items during the production process, the manufacturer implements product inspection policy on the produced lot."

"Type I error, mistakenly marking a non-defective unit as defective, is considered to be manufacturer's risk and leads to direct monetary loss to his firm."

"Type II error, mistakenly marking a defective item as non-defective, is considered to be customer's risk which hampers the brand image of the firm due to customer frustration."

"It is assumed that the production process gradually shifts from in-control state to out of control state, where the latter produces more defective items."

"Numerical analysis is conducted to validate the hypothesis by jointly optimising non-inspected fraction and the fraction of cycle time for production by minimising the expected total cost function."

Introduction and Literature Review

"Saga et al. [12] studied an imperfect production system with inspection errors, reduction in set up cost, and also incentive and penalty-based carbon emissions."

"The present study contributes in the literature to the following ways: (1) Consideration of two-tier supply chain where manufacturer encounters imperfect items during production process. (2) Partial product inspection policy is considered with inspection errors, namely, type I and type II at manufacturer's end. (3) The model considers environmental considerations with utilization of carbon cap-and-trade policy."

"This policy puts a restriction/cap on the percentage of carbon emissions coming out of various industrial processes. (4) The existing literature ignores various cases of partial inspection policy where the time duration of the production process shifting from in-control state to out-of-control state varies along with the consideration of environment and hence the policies derived to control the harmful effect of carbon emissions."

Assumptions

"The manufacturer opts for a partial inspection policy, i.e. leaving initial u proportion to be uninspected, to be released to the market with associated warranty cost C_w for defectives."

"Due to type I error, the manufacturer suffers direct monetary loss by discarding few perfect items as defectives by mistake."

"Due to type II error, the manufacturer suffers more goodwill loss rather than financial as the customers are returning the defective products which had wrongly been sent to the mainstream. (7) Green production has been considered in terms of a manufacturing process which focuses on emitting less carbon emissions in overall process."

"It incorporates carbon cap-and-trade policy as a marketing mechanism between the supplier and manufacturer."

Problem Description and Mathematical Model Formulation

"Such items amount to $(1 - \theta_1)p(t - ut)m_1$ units and are thus entitled to salvage cost C_s. Moreover, there is also a portion of defective items without any misclassification error, i.e. $\theta_1 p(t - ut)(1 - m_2)$ units."

"A total of $(1 - \theta_1)p(t - ut)m_1$ units plus $\theta_1 p(t - ut)(1 - m_2)$ units are considered as defectives and will be considered for salvage cost C_s."

"All the units in sub-intervals [ut, X] (which is in in-control state) and [X, t] (which is in out-of-control state) are inspected and the defectives found in the duration [ut, t] are salvaged at cost C_s."

"As partial inspection policy is employed in the study, where u is the non-inspected fraction in a batch, so, product inspection starts from (ptu)th item and continues until the end of production length t. Thus, all the inspected pX units are produced in in-control state and p(ut – X) units are produced in out-of-control state."

Numerical Illustration

"The optimal policy calls for an optimum production-run length of l. $T = 0.108580 \times 60 \approx 7$ h along with the optimum non-inspected fraction as 0.018770."

"The optimum production-run length in the absence of carbon trade-and-cap is l. $T = 0.276614 \times 60 \approx 17$ h with optimal non-inspected fraction and per unit cost as 0.007525 and \$1.52258 respectively."

"The implementation of carbon cap-and-trade leads to two major aspects: (1) Lesser carbon emissions in order to combat the costs incurred due to regulations and probable profit by selling the unused allowed limit, which is beneficial for environment and also fulfils corporate social responsibility. (2) Increased sustainability with reference to production process through effective, optimal utilization of resources along with sustainable demand by focusing on quality and post-sales service."

Sensitivity Analyses

"When the percentage of defective items in in-control state increases, the manufacturer needs to raise the production levels to match the demand with perfect items."

"When the the percentage of defective items in out-of-control state increases, the manufacturer may encounter problem of very few perfect items at hand."

"When the proportion of type II error increases, this suggests that more items from defective lot are getting accepted to be sold as non-defective items in the mainstream."

"When the value of scale parameter increases, the non-inspected proportion in a production lot increases, whereas the total cost per unit decreases."

"If the production process is error-prone, an increase in the value of α decreases the defectives produced, thereby leading to an increase in uninspected lot."

"With an increase in the value of shape parameter, the non-inspected proportion on a production lot decreases while the per unit cost increases."

Summary and Conclusion

"Businesses play a significant role in emitting harmful carbon emissions through various production and warehousing activities."

"Cap-and-trade policy is an approach that harnesses market forces to reduce emissions cost-effectively."

"The present paper demonstrates how environmental policies can be incorporated with inventory production and inspection policies for imperfect quality items."

"The key decisions are to determine the optimal production and inspection policies by jointly optimising the total cost function."

"The present model assists the managers to handle carbon emission constraints to make better inventory production decisions in their own favour."

Quality-Based Design of Supply Chains Considering Supplier and Technology Effects [13]

This is a machine-generated summary of:

Hejazi, Taha-Hossein; Asadi Zeidabadi, Sepideh; Abbasi, Mozhgan: Quality-based design of supply chains considering supplier and technology effects [13].

Published in: Soft Computing (2022).

Link to original: https://doi.org/10.1007/s00500-022-06967-3.

If you want to cite the papers, please refer to the original.

For technical reasons we could not place the page where the original quote is coming from.

Abstract-Summary "Producing high-quality products and reducing supply chain costs are strategic goals that help any organization to achieve the appropriate portion of the global competitive market."

"If the waste is grown in a supply chain, the quality level and efficiency of the firms will be declined and extra costs are provided."

"This paper presents a multi-stage multi-product multi-supplier multi-period supply chain model considering the wastes of raw materials, end products, and semi-finished products."

"In the proposed model, the manufacturer can select the production technology considering cost and efficiency."

"The proposed model is a mixed-integer linear programming (MILP) model considering minimization of cost elements, including raw materials inspection waste, manufacturing waste, ordering, purchasing manufacturing technologies, transportation, production inventory holding, and raw material inventory holding."

"A numerical example is developed to validate the model, and a sensitivity analysis is provided to investigate fluctuation effects of parameters, including the waste rate of products, waste rate of purchased raw materials, raw materials purchasing price, transportation cost, and demand on the model."

Introduction

"One of the essential principles for sustainable business competition is to decrease manufacturing costs and increase the quality of products in a supply chain."

"Quality of products and total costs have a noticeable impact on the pricing."

"Manufacturing costs are divided into three categories: supplying raw materials, production, and distribution."

"If the order quantity of raw materials exceeds the manufacturer's requirement, it causes additional costs for the manufacturer, such as raw materials inventory holding and raw materials purchasing."

"When the raw materials have poor quality, it causes extra inspection costs and reworking costs."

"The costs of this process include the inspection cost and the raw materials waste cost."

"In the storage and inventory control process, the raw materials inventory holding cost plays an important role."

"After the production process, the end products are inspected to confirm their qualities, which accompanies inspection costs [14]."

Literature Review

"Mirzapour Al-e-hashem et al. [15] presented a robust multi-objective mixed-integer nonlinear programming model for aggregate production planning problem considering minimization of costs, including production, hiring, firing, training, raw material and production inventory holding, transportation, and shortage costs."

"Khakdaman et al. [16] presented a 'Make-to-Stock-Make-to-Order' model under robust scenario-based optimization considering minimization of costs, including the cost of raw material, production, backorder, inventory holding, workforce, subcontracting, and transportation."

"Moslemi and Pasandideh [17] presented a facility location model for a dynamic blood supply chain under uncertainty considering quality, cost, and time."

"Mansouri and Sahraeian [18] presented a multi-objective tri-echelon closed-loop supply chain model considering disruptions in the distribution centers, cost minimization, warranty policies, reworking strategy, and returning the used items."

"The proposed model is a mixed-integer linear programming model considering total costs of the supply chain, technology effects, raw materials and productions inventory planning, inspection policy, and capacity planning."

Proposed Model

"The process of this research can be explained as follows: first, suppliers deliver the raw materials to the manufacturer."

"In the production process, multiple stages and multiple technologies are considered."

"The assumptions of the model are as follows: The responsibility of raw materials transportation costs is by suppliers."

"The material flow is considered continuous, and we ignore time and cost waste between manufacturing levels."

"Each of the manufacturing technologies can produce all products."

"Multiple transportation modes, multiple manufacturing technologies, multiple manufacturing levels, multiple suppliers, multiple raw materials, multiple products, and multiple periods are considered."

Computational Results

"In larger instances, the exact method cannot find solutions in a reasonable time while the GA and the GASA could."

"It can be observed that the GA and the GASA find near-optimal solutions for all problems, and even in some cases, the proposed algorithms could find better solutions than the exact method."

"Considering the comparison of the GA and the exact method, there are 15 instances (38%) in which the GA performs better in both the solution quality and time than the exact method."

"Considering the comparison of the GASA and the exact method, there are 26 instances (65%) in which the GASA performs better in both the solution quality and time than the exact method."

"Considering the comparison of the GASA and the GA, there are 31 instances (78%) in which the GASA performs better in both the solution quality and time than the GA."

Conclusion

"In a supply chain, reduction of the generated waste during manufacturing stages is as vital as raw materials and end products waste reduction."

"In this research, we presented a multi-product multi-supplier multi-stage multi-period mixed-integer linear programming (MILP) model considering technology effects and raw materials, semi-finished products, and end products waste in a supply chain."

"The proposed model objective function deals with minimization cost elements, including raw materials inspection waste, manufacturing waste, ordering, purchasing manufacturing technologies, transportation, production variable, production inventory holding, and raw material inventory holding."

"In terms of solution methods, first, a sensitivity analysis was used to investigate how the fluctuations of the parameters can affect the model results."

"This method was investigated on the products waste rate, purchased raw materials waste rate, raw materials purchasing price, transportation cost, and demand."

"For the future researches, it is suggested to consider the waste of productions during the transportation and the waste time between manufacturing stages."

Simulation–Optimization Approach for the Multi-Objective Production and Distribution Planning Problem in the Supply Chain: Using NSGA-II and Monte Carlo Simulation [19]

This is a machine-generated summary of:

Kabiri, Niloofar Nadim; Emami, Saeed; Safaei, Abdul Sattar: Simulation–optimization approach for the multi-objective production and distribution planning problem in the supply chain: using NSGA-II and Monte Carlo simulation [19].

Published in: Soft Computing (2022).

Link to original: https://doi.org/10.1007/s00500-022-07152-2.

If you want to cite the papers, please refer to the original.

For technical reasons we could not place the page where the original quote is coming from.

Abstract-Summary "We consider the integrated production and distribution planning problem of a multi-level green closed-loop SC system, which includes multiple recycling, manufacturing/remanufacturing, and distribution centers."

"A hierarchical iterative approach utilizing the LP-metric method and the non-dominated sorting genetic algorithm (NSGA-II) is introduced to solve the proposed model."

"Monte Carlo (MC) simulation is applied to tackle uncertainty in demand, and the NSGA-II algorithm is integrated with MC simulation (MCNSGA-II)."

"Through numerical experiments on randomly generated instances, we observe that the average of the relative gaps NSGA-II and LP-metric method are 7%, 7%, and 1% for the three levels, respectively."

"It is observed that the value of convergence (C) and spacing (S) metrics of the MCNSGA-II algorithm are 0 and 1.82×10^7, which are better than NSGA-II."

Introduction

"In this study, we investigate the integrated PD planning problem for the GCLSC, which is multi-product, multi-period, and multi-level."

"In the GCLSC, the multiple recycling centers disassembled the returned products to components."

"Each distribution center can select one or more remanufacturing/manufacturing centers for satisfying demand, and provide remanufactured or new products to retailers for sales."

"There are many questions about the considered GCLSC, e.g., amount of returned products that are disassembled and tested in recycling centers, amount of components that are disposed of in recycling centers, inventory quantity of returned products at recycling centers, amount of remanufactured products that are assembled in factories, inventory quantity of remanufactured products in factories, amount of returned products that are collected by distributors from downstream markets."

"To answer the questions and other ones, we present a multi-level bi-objective programming model that includes multiple recycling, manufacturing/remanufacturing factories, and distribution centers."

Literature Review

"Geoffrion and Graves [20] studied a single-period PD problem and suggested a solution approach based on the Benders decomposition algorithm."

"Devapriya et al. [21] considered a PD planning problem of a perishable product and provided a MIP model to minimize cost."

"Zheng et al. [22] presented a penalty function-based method to solve a risk-averse PD planning problem."

"Ma et al. [23] proposed a PD planning model applying bi-level programming for SCM and developed a GA to solve the model."

"Rezaeian et al. [24] suggested an MINLP model for the integrated PD and inventory planning for perishable products with a fixed lifetime all over a two-echelon SC by integrating production centers and distributors."

"Wei et al. [25] studied the integrated PD planning problem and presented a model with a two-stage production structure."

"They applied a Monte Carlo simulation-based multi-objective programming model to solve the integrated optimization problem."

Mathematical Modeling

"A three-level bi-objective programming model is presented for a GCLSC, which includes multiple recycling centers, multiple manufacturing/remanufacturing factories, and multiple distributors."

"In the considered GCLSC, returned products are disassembled to components in the multiple recycling centers."

"The objective function [3] minimizes the amount of greenhouse gas emissions of each vehicle per kilometer for a unit of eligible components between the recycling centers and remanufacturing/manufacturing factories."

"The objective function [17] minimizes the amount of greenhouse gas emissions of each vehicle per kilometer for a unit of new products and remanufactured products between remanufacturing/manufacturing factories and distribution centers, as well as the amount of greenhouse gas emissions during processing and reprocessing."

"The objective function [26] minimizes the amount of greenhouse gas emissions of each vehicle per kilometer for a unit of returned product between distribution and recycling centers."

Solution Approach

"The proposed three-level bi-objective programming formulation is solved based on a hierarchical iterative approach using the LP-metric method for small-size instances and the NSGA-II algorithm for large-size instances, respectively."

"The NSGA-II algorithm [27] is applied for solving the considered bi-objective models at each level."

"In this simulation approach, a single value is selected for each input random variable based on the best guess by the modeler."

"MC simulation randomly samples values from every input variable distribution and applies the sample to compute the model's output."

"MC simulation is applied to appraise the objective functions of each chromosome, considering the probability distribution of the demand that is obtained from historical data."

"We apply the following metrics to assess the obtained non-dominated solutions of the NSGA-II and MCNSGA-II algorithms."

Computational Results

"Numerical experiments are conducted to investigate the performance of the three-level bi-objective MIP model, the NSGA-II algorithm, and the MCNSGA-II algorithm based on random data."

"The Taguchi approach is then employed to tune the parameters of the NSGA-II algorithm, and the best level of each parameter is obtained by the signal-to-noise diagram in the Minitab software."

"The results obtained by the LP-metric method and the NSGA-II algorithm are compared, and the relative distance of each sample is determined."

"That the LP-metric method and the NSGA-II algorithm are implemented in GAMS and MATLAB, respectively, and are tested on a computer with a 2.3 GHz CPU and 8.0 GB of RAM."

"The average run-time of the LP-metric method and the NSGA-II algorithm is 300 s and 180 s, respectively."

"The medium- and large-size problems are solved to compare the NSGA-II and MCNSGA-II algorithms."

Conclusions

"A hierarchical iterative approach using the LP-metric method and the NSGA-II algorithm is suggested to solve the proposed model."

"Studying the literature on multi-level production–distribution planning shows that fewer study goals, such as customer satisfaction and reduction of delay time in the model, were considered."

"In this and other studies, multi-level planning has been solved at two or three levels."

"Adding the fourth level as the supplier level can be an appropriate area for future research."

"Solving this model with other multi-objective metaheuristic algorithms such as multi-objective particle swarm optimization (MOPSO) algorithm, Pareto simulated annealing (PSA) algorithm, multi-objective gray wolf optimization (MOGWO) algorithm, multi-objective gravitational search algorithm (MOGSA) would be another future scope."

A Scatter Search Algorithm for Multi-Criteria Inventory Classification Considering Multi-Objective Optimization [28]

This is a machine-generated summary of:

Saracoglu, Ilkay: A Scatter Search Algorithm for Multi-Criteria Inventory Classification considering Multi-Objective Optimization [28].

Published in: Soft Computing (2022).

Link to original: https://doi.org/10.1007/s00500-022-07227-0.

If you want to cite the papers, please refer to the original.

For technical reasons we could not place the page where the original quote is coming from.

Abstract-Summary "Classification of the items influences the results of inventory management."

"This is usually classified with considering an annual dollar usage criterion but maybe other criteria such as lead time, criticality, perishability, inventory cost, and demand type can be affected on that classification."

"Inventory items that have more than one criterion are discussed under multi-criteria inventory classification (MCIC) methods in the literature."

"The MCIC problem is considered with two objectives as follows: (1) minimization of total inventory relevant cost and (2) minimization of the dissimilarity index."

"The proposed Mixed Integer Nonlinear Programming (MINLP) model of the MCIC problem is formulated using Scatter Search Algorithm (SSA)."

"The proposed model generated a Pareto frontier solution with the maximum satisfaction level and minimum distance from ideal point."

Introduction

"For a company beginning inventory management studies, the first procedure is to classify items with ABC analysis."

"The traditional ABC classification had weighted criteria applied with the Analytic Hierarchic Process (AHP) by Flores and Clay Whybark [29] and items were listed according to the score values obtained as they began the first multi-criteria classification studies."

"Inventory items were classified with annual dollar usage (ADU), average unit cost (ACU), criticality (CF) and lead time (LT), and they determined the weights of the criteria values by using AHP method."

"Park et al. [30] developed a new model called cross-evaluation-based weighted linear optimization (CE-WLO) to find the optimal efficiency score for inventory items."

"Some studies decide the criteria weights when giving classification weights, while some studies focus on methods to minimize total inventory costs."

Model Description

"The Total Relevant Cost (TRC), which is one of the objective functions of our algorithm used in studies by Mohammaditabar et al. [31], Ramanathan [32], and Kaabi et al. [33] to minimize inventory costs."

"The second objective function is defined to minimize dissimilarity index value between items in the same class."

"We divided into three model's type: minimize the TRC, minimize the dissimilarity index, and maximize satisfaction level function, respectively."

"Two-objective functions were considered, including the total relevant cost value and the dissimilarity of the items to each other."

"The TRC function was chosen as the main objective function and the dissimilarity index function was considered as a constraint function."

"According to Mohammaditabar et al. [31], this model is very difficult to solve optimally since the objective functions are not linear and the value of T_k is dependent on the value of x_{ik}."

Scatter Search Algorithm

"Improvement Method (IM) is applied to develop new individuals from the initial solutions."

"Initial Trial Solution: (2,3,3,1,2,3) TRC = 308.378 $ If we swap the first element and the second element the new solution will be (3,2,3,1,2,3) and the result is 309.562$. If we swap fourth item class and fifth item class, another new solution will be (2,3,3,2,1,3) and TRC is calculated of 306.430$. So, the fourth item class and fifth item class will be the tabu list and not changed in the next iteration because of the better TRC value."

"TS algorithm continues to get better solutions and generated new solution vectors."

"The reference set consisting of high quality solutions and the sum of various solutions is used to derive new solutions by applying Refset."

"Attempts are made to obtain new solutions with the subset solutions created with this method."

Illustrative Example

"With the aim of being able to determine the best results for solution to our multi objective model, the optimized minimum TRC value was $1098.165, and dissimilarity index was 195.026 in Model 1."

"By applying the SSA, the satisfaction level values of the TRC and the dissimilarity index were found to be 0.984 and 0.996, respectively."

"Of Douissa and Jabeur [34], the TRC value is $1,122,779 as a result of the classification obtained by the ELIII-CVNS method, while the cost obtained in this study is calculated as $1,102,539 with two objectives."

"It is seen that the dissimilarity index within classes was obtained by the SSA method applied in this study."

"The same item is assigned to Class A in the second model and the TRC values and dissimilarity index changes as 60.66 and 2.95, respectively."

Experimental Results with Liu et al.'s Dataset

"TRC is equal to $58,669.31 with the proposed method in this study when only a single objective function to minimize the TRC is considered."

"When only a single objective function is considered, the TRC will be decreased by 4.8% according to the ELIII-CVNS method and dissimilarity index will be decreased by 44.1%."

"Item 5 was assigned to Class B in Douissa and Jabeur's study and Class C in Liu et al.'s study."

"In the models proposed in this study, the fifth item is calculated as Class A if the objective is to minimize the TRC or maximize the satisfaction level, Class B if the objective is to minimize dissimilarity index."

"The inventory cost of fifth item will be $1142.51 if it is taken as Class B, whereas, if it is taken as Class A, the cost will be $979.03."

Conclusion

"MCIC problem is formulated with the objectives of minimizing total inventory cost and dissimilarity index."

"In the third model, multi-objective optimization problem is constructed to maximize satisfaction level of the minimization of TRC and dissimilarity index."

"A SSA has been proposed and illustrated in this paper for classifying inventory items with multiple criteria and multiple objectives."

"Three different methods are used to solve the multi-objective optimization problem."

"After selecting the multi-objective method, different criteria weights are examined to find the better result."

"The algorithm can be used to outperform existing methods for different models of classification of items."

"The proposed classification model can be applied on the large inventory items that holding in the warehouse for example pharmaceutical items, spare parts, electronical components, and electromechanical items, etc. Studies are performed with fuzzy along with different inventory policies such as (s, S) and (Q, r)."

Embodied Energy of Parts in Sheet Metal Forming: Modeling and Application for Energy Saving in the Workshop [35]

This is a machine-generated summary of:

Xiong, Wei; Huang, Haihong; Li, Lei; Gan, Lei; Zhu, Libin; Gao, Mengdi; Liu, Zhifeng: Embodied energy of parts in sheet metal forming: modeling and application for energy saving in the workshop [35].

Published in: The International Journal of Advanced Manufacturing Technology (2021).

Link to original: https://doi.org/10.1007/s00170-021-08209-6.

If you want to cite the papers, please refer to the original.

For technical reasons we could not place the page where the original quote is coming from.

Abstract-Summary "Parts or products can be considered as the carrier of energy consumption during manufacturing since they are the final output of workshops."

"Previous work mainly discussed methods and technologies to apply the embodied energy of parts for selecting materials and processes in product design; the role that embodied energy of parts can play in production optimization was seldom investigated."

"Different from machining workshops, which has been widely discussed, there is less research on energy saving in the sheet metal forming workshop."

"Modeling and application of embodied energy for energy saving in the sheet metal forming workshop is presented."

"Energy-oriented scheduling in the workshop is implemented, and a two-stage approach based on ranks of EEPM that can quickly select more efficient production solutions is proposed."

"This study provides a practical approach to seek a trade-off between energy saving and high production efficiency for energy-intensive discrete workshops."

[Section 1]

"The low energy efficiency of forming equipment and inappropriate scheduling of the equipment in a production line/chain or a workshop make sheet metal forming energy-intensive [36]."

"The energy consumption reduction of sheet metal forming has become one of the urgent issues."

"Energy saving of forming equipment and processes has made some progress."

"Schenke and Weber [37] studied the energy efficiency of displacement control drive systems in hydraulic forming presses."

"The dynamic punch force was analyzed during the process to optimize the energy consumption of the drawing process."

"Existing studies for energy saving in workshops focus on processes and equipment while ignoring the role that parts or products which are the final output of workshops can play."

"The research presented in this paper explores the role that embodied energy of forming parts can play in achieving energy saving in the sheet metal forming workshop."

Literature Review

"Kara et al. [38] presented a model to assess the impact of global manufacturing (especially supply chains) on embodied energy."

"The embodied energy of products that were performed by multi-processes was modeled and evaluated to support the design for energy minimization and operational improvements."

"As for optimization methods, production scheduling is direct and useful to achieve energy-efficient manufacturing."

"The role that embodied energy (especially EEPM) can play in production optimization was seldom investigated. (2) In terms of saving energy in the workshop, most of the studies aimed at machining workshops, some studies aimed at workshops that produce solidified parts and bulk-forming parts."

"The two contributions of this study are included as follows: (1) establishing models for energy consumption embodied in parts during the manufacturing phase and (2) achieving the trade-off between energy saving and high production efficiency in the workshop by implementing energy-oriented scheduling based on a two-stage solution selection approach."

[Section 3]

"When process i of part x is performed by press m, the energy consumption is obtained by summing the integrated values of the corresponding power over time."

"The considered production information includes: (1) The type of parts and the planned and completed processing quantity of each part; (2) The process required for each part, the available equipment, power, and duration of processing; (3) Power of transportation and TBS equipment in the workshop."

"The status of presses is identified by reading the identification bits in the machine control device; power data of transportation and TBS equipment is acquired based on their handbooks; the corresponding start and turned-off durations are calculated in combination with the system clock."

"The fixed RFID tag reader reads the production number in the RFID tag and displays the relevant information on the operator's information interaction device by searching the database, when the material is transported to a press to be processed."

[Section 4]

"The press is the basic processing unit in a sheet metal forming workshop, which can perform a variety of processes as needed by consuming energy and materials."

"The specific objective of saving energy in the sheet metal forming workshop is reducing total energy consumption (TE) and makespan (T) in the workshop under the premise of ensuring the quality of parts."

"The value-added efficiency ratio can reflect the processing energy efficiency of the workshop; the specific energy consumption can reflect the energy efficiency of forming parts. (1) Value-added efficiency ratio The value-added efficiency ratio refers to the ratio of value-added energy consumption to total energy consumption in a production cycle."

"The process of smearing lubricating oil is neglected since it has a minor impact on the current embodied energy model of forming parts and the scheduling model of the forming workshop."

Case Study

"Different duration and machine failure rates lead to different calculation results of energy consumption."

"The impact of uncertainty distribution of duration and machine failure rate on energy consumption is explored."

"The reason is that the higher the machine failure rate, the higher the uncertainty of production, which leads to a larger dispersion of energy consumption."

"Configuring suitable and reliable presses in production can effectively reduce energy consumption and improve the stability of production."

"In order to acquire lower energy consumption and makespan and reduce the proportion of E_{NVA}, energy-oriented scheduling in the stamping workshop was implemented."

"Since the tonnage of presses for different stamping parts to complete their processing is different, the match between installed and demanded power in presses is performed before scheduling to avoid energy waste [39]."

"The proposed methods are verified to be effective for reducing energy consumption in the workshop and achieving lower embodied energy of parts."

Limitations

"Some major limitations that could be addressed in future researches are listed below: The innovation of energy-saving optimization in the workshop is that the feasible solutions are ranked after obtaining the Pareto set to quickly obtain the most suitable solution."

"Better optimization results are expected, combining these advanced algorithms with the ranking strategy is based on the embodied energy of parts."

"In order to make the models and methods more general, hot forming processes (research on hot stamping has been a trend recently) should be taken into consideration in the future."

Conclusions

"Modeling and application of embodied energy of parts for energy saving in the sheet metal forming workshop have been studied."

"Energy distribution, energy-intensive processes, and bottlenecks have been identified based on discrete event simulation."

"A two-stage solution selection approach based on ranks of EEPM has been developed for better energy-oriented scheduling."

"The trade-off between energy saving and high production efficiency has been found."

Predicting Air Quality by Integrating a Mesoscopic Traffic Simulation Model and Simplified Air Pollutant Estimation Models [40]

This is a machine-generated summary of:

Mihăiţă, Adriana Simona; Ortiz, Mirian Benavides; Camargo, Mauricio; Cai, Chen: Predicting Air Quality by Integrating a Mesoscopic Traffic Simulation Model and Simplified Air Pollutant Estimation Models [40].

Published in: International Journal of Intelligent Transportation Systems Research (2018).

Link to original: https://doi.org/10.1007/s13177-018-0160-z.

If you want to cite the papers, please refer to the original.

For technical reasons we could not place the page where the original quote is coming from.

Abstract-Summary "More than ever, local authorities for environmental protection and urban planners are interested in performing detailed investigations using traffic and air pollution simulations for testing various urban scenarios and raising citizen awareness where necessary."

"The main objective of this work is to build an integrated simulation model which would predict and visualize various environmental changes inside the neighbourhood such as: air pollution, traffic flow or meteorological information."

"We conduct a data profiling analysis on the received data sets together with a discussion on the daily and hourly traffic patterns, average nitrogen dioxide concentrations and the regional background concentrations recorded in the eco-neighbourhood for the study period."

"We build the 3D mesoscopic traffic simulation model using real data sets from the local traffic management centre."

"By using reliable data sets from the local air-quality management centre, we build a regression model to predict the evolution of nitrogen dioxide concentrations, as a function of the simulated traffic flow and meteorological data."

Introduction

"This health scourge causes a huge loss of revenue for the global economy: 225 billion dollars loss of revenue per year according to the 2016 report of the World Bank, which wanted to raise high awareness on the financial burden of poor air quality around the globe [41]."

"Although they require a high amount of data and can be computationally intensive, they can be efficiently combined with various tools for air pollution monitoring; a recent comparison study of various integrated traffic simulation and air pollution tools is provided in [42]."

"While the construction of the traffic simulation or air pollution models can follow standard modelling techniques, integrating both modules together requires an increased complexity and data analysis."

"Although various research studies consider other factors, such as speed and acceleration to model and predict the air pollution concentrations, these data sets are not always available in many urban areas with limited traffic detectors or monitoring capabilities, this being the case for Nancy Grand Cœur as well."

Integrated Traffic and Air Pollution Simulation Model

"By using the provided real data sets we build the 3D mesoscopic traffic simulation model, which allows us to obtain statistical measures of the traffic model, such as mean number of cars/average stay-time (average travel-time on each road section), and also graphically visualise the evolution of meteorological conditions and NO_2 concentrations."

"Apart from constructing a clear insight of the traffic conditions in this area during congested periods, our efforts have been also oriented towards the integration of pollution data and meteorological information in the simulation model, which can be a complementary source of information in the process of traffic and air quality dynamic monitoring."

"Although it can offer precise estimations of various pollutant concentrations, it needs large amount of input data, such as the topography and height of the buildings, the annual traffic emissions computed with Circul'Air [43], meteorological data including the direction and wind speed, etc. The main limitation we encountered for integrating MISKAM and FlexSim outputs was mainly related to the differences in the static and dynamic behaviour of the tools."

Regression Model of the NO_2

"We start by training a least squares multiple regression model (LSMRM) on the first 20 days of our data set from January 2015, and we further test the results on the temporally holdout 10 days."

"In order for the prediction model to work accurately, one needs to use continuous data sets for training the model."

"By analysing the current results, we observe that one of the independent predictors (traffic volume) is sensitive to time changes during the whole data set we are testing."

"Validating the model on longer periods needs a more detailed analysis and verification, especially if other predictors would be available for the study."

Limitations and Further Studies on SO_2 and PM_{10}

"The main limitations of this work are: (a) restricted data sets concerning the air pollutant concentrations and traffic counts which have been received from the traffic management centre; having access to longer periods of data would help to build more traffic demand scenarios and analyse for example how traffic control measures would improve the air quality in the neighbourhood, (b) integrating the emissions from the train station would also improve the current model."

"In [44] the authors enhance on the fact that sulphur dioxide is almost exclusively issued by the manufacturing-waste-construction industry to an extent of almost 98%, which makes any prediction or environmental analysis very challenging as the data sets will be highly influenced by the industrial activity in the region where the air pollution monitoring takes place and the season when the data has been collected."

Conclusions and Future Perspectives

"We proposed an integrated air pollution and traffic simulation model for building a simplified NO_2 estimation model, inside an eco-neighbourhood from north-eastern part of France."

"The second step was to construct the 3D mesoscopic traffic simulation model of the study area using real data sets provided by the Grand Nancy council consisting in profiled traffic counts (every 15 min), traffic light signal controls and plans, pedestrian information and manual counts verifications."

"The third step consisted in using the simulated traffic volumes to build the regression model and estimate the NO_2 concentrations over the last 10 days of the model."

"We are currently working on extending the hybrid model to the whole NGC eco-neighbourhood and build more detailed air pollution estimation models which would take into consideration all the types of vehicles on the roads."

Effects of (Co-)Combustion Techniques and Operating Conditions on the Performance and NO Emission Reduction in a Biomass-Fueled Twin-Cyclone Fluidized-Bed Combustor [45]

This is a machine-generated summary of:

Ninduangdee, Pichet; Arromdee, Porametr; Se, Chhaina; Kuprianov, Vladimir I.: Effects of (Co-)Combustion Techniques and Operating Conditions on the Performance and NO Emission Reduction in a Biomass-Fueled Twin-Cyclone Fluidized-Bed Combustor [45].

Published in: Waste and Biomass Valorization (2020).

Link to original: https://doi.org/10.1007/s12649-020-01124-z.

If you want to cite the papers, please refer to the original.

For technical reasons we could not place the page where the original quote is coming from.

Abstract-Summary "In the (co-)combustion study, rice husk (a base fuel) was first co-fired with high-moisture sugarcane bagasse (sharing 15% of the total heat input) using an air staging technique, whereas in the second group of experiments, the base fuel was combusted alone with flue gas recirculation."

"Detailed investigations of (co-)combustion and emission characteristics of the two proposed techniques was performed with a 100 kW heat input to the combustor for variable operating parameters: 30–60% excess air, 0–0.3 secondary-to-total air ratio in the tests for air staging, and 5–20% fraction of the recycling flue gas when using the second technique."

"In all test runs, both (co-)combustion techniques ensured NO reducing conditions in the lower combustor's chamber, leading eventually to the NO emission reduction, compared to burning the base fuel alone."

"At optimal operating parameters, the combustor showed high (~99%) combustion efficiency and reduction of NO emission: by 30% for husk-bagasse co-firing with air staging, and by 38% when burning rice husk with flue gas recirculation, as compared to conventional combustion of the base fuel."

Statement of Novelty

"A novel twin-cyclone fluidized-bed combustor with a compact design, ensuring high-efficiency combustion of rice husk with elevated fuel-ash content, has been proposed in this work."

"A highly intensive combustion process is achieved in the lower chamber of the combustor, due to the swirling fluidization regime of the gas–solid bed induced by an annular spiral air distributor."

"To operate this twin-cyclone combustor at the reduced NO and acceptable CO and C_xH_y emissions, rice husk can be either (1) co-fired with a high-moisture, low-calorific biomass (e.g., sugarcane bagasse) using air staging or (2) combusted alone with flue gas recirculation, as shown in this work."

Introduction

"Among the most appropriate technologies for energy conversion from biomass, fluidized- bed combustion systems (combustors/furnaces) showed a number of advantages over stoker-firing and pulverized fuel-firing systems, such as flexibility for fuel type, excellent mixing, relatively low combustion temperature, thermal homogeneity, lower pollutant emissions, and higher combustion efficiency [46–50]."

"Despite that the vortexing and cyclonic fluidized-bed techniques ensure high (~99%) combustion efficiency when firing various biomasses, the combustion of biomass with elevated fuel-N in "short" and compact combustors operated with high combustion intensity is reported to result in rather high NO_x emissions [51, 52]."

"Some research studies reported the impacts of air staging on formation and emission of NO in fluidized-bed systems fired with a single biomass fuel [53–55]."

"This work was performed on a novel twin-cyclone fluidized-bed combustor with a swirling fluidized bed (referred to as 'twin-cyclone FBC'), to explore the effectiveness of two (co-)combustion techniques in reducing NO emission from this biomass-fueled combustor."

Materials and Methods

"Total air (TA) required for burning the fuel/blend with bottom air injection (or primary air, in the tests with air staging), was supplied to the combustor via a pipe of 0.1 m inner diameter and injected into the lower chamber through an annular spiral air distributor (also termed 'swirler')."

"In the tests with air staging, secondary air (SA) was tangentially injected into the conical section of the upper combustion chamber, via two air pipes of 0.06 m inner diameter, at Z = 1.9 m above the air distributor."

"During a test run of each (co-)combustion option, temperature, O_2, CO, C_xH_y (as CH_4), and NO were measured along the combustor centerline, using the above-mentioned gas analyzer, to study the effects of the (co-)combustion methods and operating parameters (EA, SA/TA, and FGR fraction) on combustion and emission characteristics of the twin-cyclone FBC, in particular on formation and oxidation/reduction of major gaseous pollutants in the combustor."

Results and Discussion

"As compared to conventional combustion of RH, with increasing SA/TA in the co-firing tests with air staging, the temperature increased at all points inside the lower chamber (despite that the moisture content in the RH–SB blend was higher), mainly due to the lowered proportion of primary air injected through the swirler."

"With increasing EA at fixed SA/TA in the test group for air staging (or fixed r in the experiments for FGR), the two emissions decreased, mainly due to the increased rates of CO and C_xH_y oxidation in different combustor regions."

"In the co-firing tests (at $EF_2 = 0.15$) with air staging, the two combustion-related heat losses increased with higher SA/TA, resulting in the lowered combustion efficiency of the combustor, as compared to conventional combustion of pure RH at similar EA."

Conclusions

"A novel twin-cyclone (two-chamber) fluidized-bed combustor with a swirling bed has been successfully tested using two NO reducing techniques: (1) co-firing rice husk (a base fuel) and high-moisture sugarcane bagasse (sharing 15% heat input) with an air staging technique and (2) burning the base fuel alone using flue gas recirculation, for the ranges of operating parameters (excess air, secondary-to-total air ratio, and flue gas recirculation fraction)."

"With the optimal operating parameters, about 30% NO emission reduction is possible through co-firing pre-mixed rice husk and sugarcane bagasse using air staging, whereas it can be as high as 38% during combustion of pure rice husk with flue gas recirculation."

"The use of the NO reducing techniques lead to some increase in the CO and C_xH_y emissions (within the corresponding emission limits), deteriorating insignificantly the combustion efficiency which nevertheless remains at a high level (~99%) in optimal combustor operation."

References

1. Goswami SS, Behera DK (2021) Solving material handling equipment selection problems in an industry with the help of entropy integrated COPRAS and ARAS MCDM techniques. Process Integr Optimiz Sustain. https://doi.org/10.1007/s41660-021-00192-5
2. Karande P, Chakraborty S (2013) Material handling equipment selection using weighted utility additive theory. J Ind Eng 2013:268708. Retrieved January 20, 2021, from http://downloads.hindawi.com/archive/2013/268708.pdf
3. Athawale V, Chakraborty S (2011) A comparative study on the ranking performance of some multi-criteria decision-making methods for industrial robot selection. Int J Ind Eng Comput 2(4):831–850. Retrieved January 20, 2021, from http://www.growingscience.com/ijiec/Vol2/IJIEC_2011_22.pdf
4. Zavadskas EK, Turskis Z (2010) A new additive ratio assessment (ARAS) method in multicriteria decision-making. Technol Econ Dev Econ 16(2):159–172. Retrieved January 20, 2021, from https://journals.vgtu.lt/index.php/TEDE/article/view/5850/5093
5. Rao RV (2007) Selection of material handling equipment. In: Decision making in the manufacturing environment: using graph theory and fuzzy multiple attribute decision making methods, vol 1. Springer, London, pp 187–195. Retrieved January 20, 2021, from https://www.springer.com/gp/book/9781846288180
6. Rao RV (2007) Robot selection for a given industrial application. In: Decision making in the manufacturing environment: using graph theory and fuzzy multiple attribute decision making methods, vol 1. Springer, London, pp 176–179. Retrieved January 20, 2021, from https://www.springer.com/gp/book/9781846288180
7. Maniya KD, Bhatt MG (2011) A multi-attribute selection of automated guided vehicle using the AHP/M-GRA technique. Int J Prod Res 49(20):6107–6124. https://doi.org/10.1080/00207543.2010.518988
8. Mathew M, Sahu S (2018) Comparison of new multi-criteria decision making methods for material handling equipment selection. Manag Sci Lett 8(3):139–150. Retrieved January 20, 2021, from http://www.growingscience.com/msl/Vol8/msl_2018_4.pdf
9. Karande P, Zavadskas E, Chakraborty S (2016) A study on the ranking performance of some MCDM methods for industrial robot selection problems. Int J Ind Eng Comput 7(3):399–422. Retrieved January 20, 2021, from http://www.growingscience.com/ijiec/Vol7/IJIEC_2016_1.pdf
10. Goswami SS, Behera DK (2021) Evaluation of the best smartphone model in the market by integrating fuzzy-AHP and PROMETHEE decision-making approach. Decision. https://doi.org/10.1007/s40622-020-00260-8
11. Rini KA (2021) Optimal policy for a green production system with quality inspection and warranty cost under environmental regulations. Process Integr Optimiz Sustain. https://doi.org/10.1007/s41660-021-00182-7
12. Saga RS, Jauhari WA, Laksono PW, Dwicahyani AR (2019) Investigating carbon emissions in a production-inventory model under imperfect production, inspection errors and service-level constraint. Int J Log Syst Manag 34(1):29–55

13. Hejazi T-H, Asadi Zeidabadi S, Mozhgan A (2022) Quality-based design of supply chains considering supplier and technology effects. Soft Comput. https://doi.org/10.1007/s00500-022-06967-3
14. Sarkar B, Saren S (2016) Product inspection policy for an imperfect production system with inspection errors and warranty cost. Eur J Oper Res 248(1):263–271
15. Mirzapour Al-E-Hashem S, Malekly H, Aryanezhad M (2011) A multi-objective robust optimization model for multi-product multi-site aggregate production planning in a supply chain under uncertainty. Int J Prod Econ 134(1):28–42
16. Khakdaman M, Wong KY, Zohoori B, Tiwari MK, Merkert R (2015) Tactical production planning in a hybrid make-to-stock–make-to-order environment under supply, process and demand uncertainties: a robust optimisation model. Int J Prod Res 53(5):1358–1386
17. Moslemi S, Pasandideh SHR (2021) A location-allocation model for quality-based blood supply chain under IER uncertainty. RAIRO 55:967
18. Mansouri Y, Sahraeian R (2021) Facility disruptions in a closed-loop supply chain featuring warranty policy and quality-based segmentation of returns. Scientia Iranica, Tehran
19. Kabiri NN, Emami S, Sattar SA (2022) Simulation–optimization approach for the multi-objective production and distribution planning problem in the supply chain: using NSGA-II and Monte Carlo simulation. Soft Comput. https://doi.org/10.1007/s00500-022-07152-2
20. Geoffrion AM, Graves GW (1974) Multicommodity distribution system design by benders decomposition. Manag Sci 20(5):822–844
21. Devapriya P, Ferrell W, Geismar N (2017) Integrated production and distribution scheduling with a perishable product. Eur J Oper Res 259(3):906–916
22. Zheng Y, Zhang G, Han J, Lu J (2016) Pessimistic bilevel optimization model for risk-averse production-distribution planning. Inf Sci 372:677–689
23. Ma Y, Yan F, Kang K, Wei X (2016) A novel integrated production-distribution planning model with conflict and coordination in a supply chain network. Knowl-Based Syst 105:119–133
24. Rezaeian J, Haghayegh S, Mahdavi I (2016) Designing an integrated production/distribution and inventory planning model of fixed-life perishable products. J Optimiz Ind Eng 9(19):47–60
25. Wei W, Guimarães L, Amorim P, Almada-Lobo B (2017) Tactical production and distribution planning with dependency issues on the production process. Omega 67:99–114
26. Mihăiţă A-S, Dupont L, Chery O, Camaego M, Cai C (2018) Evaluating air quality by combining stationary, smart mobile pollution monitoring and data-driven modelling
27. Deb K, Pratap A, Agarwal S, Meyarivan T (2002) A fast and elitist multiobjective genetic algorithm: NSGA-II. IEEE Trans Evol Comput 6(2):182–197
28. Saracoglu I (2022) A scatter search algorithm for multi-criteria inventory classification considering multi-objective optimization. Soft Comput. https://doi.org/10.1007/s00500-022-07227-0
29. Flores BE, Clay Whybark D (1986) Multiple criteria ABC analysis. Int J Oper Prod Manag. https://doi.org/10.1108/eb054765
30. Park J, Bae H, Bae J (2014) Cross-evaluation-based weighted linear optimization for multi-criteria ABC inventory classification. Comput Ind Eng 76:40–48. https://doi.org/10.1016/j.cie.2014.07.020
31. Mohammaditabar D, Ghodsypour HS, O'Brien C (2012) Inventory control system design by integrating inventory classification and policy selection. Int J Prod Econ 140:655–659
32. Ramanathan R (2006) ABC inventory classification with multiple-criteria using weighted linear optimization. Comput Oper Res 33:695–700. https://doi.org/10.1016/j.cor.2004.07.014
33. Kaabi H, Jabeur K, Ladhari T (2018) A genetic algorithm-based classification approach for multicriteria ABC analysis. Int J Inf Technol Decis Mak 17:1805–1837. https://doi.org/10.1142/S0219622018500475
34. Douissa MR, Jabeur K (2019) A non-compensatory classification approach for multi-criteria ABC analysis. Soft Comput. https://doi.org/10.1007/s00500-019-04462-w

35. Xiong W, Huang H, Li L, Gan L, Zhu L, Gao M, Zhifeng L (2021) Embodied energy of parts in sheet metal forming: modeling and application for energy saving in the workshop. Int J Adv Manuf Technol. https://doi.org/10.1007/s00170-021-08209-6
36. Gao MD, He K, Li L, Wang QY, Liu CH (2019) A review on energy consumption, energy efficiency and energy saving of metal forming processes from different hierarchies. PRO 7:357. https://doi.org/10.3390/pr7060357
37. Schenke C, Weber J (2019) Energy efficiency of displacement control drive systems in hydraulic forming presses. J Manuf Sci Eng 141:041013. https://doi.org/10.1115/1.4042604
38. Kara S, Manmek S, Herrmann C (2010) Global manufacturing and the embodied energy of products. CIRP Ann Manuf Technol 5:29–32. https://doi.org/10.1016/j.cirp.2010.03.004
39. Li L, Huang HH, Liu ZF, Li X, Triebe MJ, Zhao F (2016) An energy-saving method to solve the mismatch between installed and demanded power in hydraulic press. J Clean Prod 139:636–645. https://doi.org/10.1016/j.jclepro.2016.08.063
40. Mihăiţă AS, Ortiz MB, Camargo M, Chen C (2018) Predicting air quality by integrating a mesoscopic traffic simulation model and simplified air pollutant estimation models. Int J Intell Transp Syst Res. https://doi.org/10.1007/s13177-018-0160-z
41. W. B. Group and IHME (2016) The cost of air pollution, strengthening the economic case for action. The World Bank and Institute for Health Metrics and Evaluation, University of Washington, Seattle
42. Fontes CM (2015) How to combine different microsimulation tools to assess the environmental impacts of road traffic? Lessons Direct 34:293–306
43. Galineau J (2012) Bilan des émissions atmosphériques du transport routier en Lorraine. Air Lorraine, Nancy
44. Lorraine A (2016) Caractérisation de la qualité de l'air ambiant à Nancy en 2015 en contexte de proximité trafic. Air Lorraine, Nancy
45. Ninduangdee P, Arromdee P, Se C, Kuprianov VI (2020) Effects of (Co-)combustion techniques and operating conditions on the performance and NO emission reduction in a biomass-fueled twin-cyclone fluidized-bed combustor. Waste Biomass Valoriz. https://doi.org/10.1007/s12649-020-01124-z
46. Werther J, Saenger M, Hartge E-U, Ogada T, Siagi Z (2000) Combustion of agricultural residues. Prog Energy Combust Sci 26:1–27
47. Mazza GD, Soria JM, Gauthier D, Reyes Urrutia A, Zambon M, Flamant G (2016) Environmental friendly fluidized bed combustion of solid fuels: a review about local scale modeling of char heterogeneous combustion. Waste Biomass Valoriz 7(2):237–266
48. Chaivatamaset P, Tia S, Methaviriyasilp W, Pumisampran W (2018) Bed particle agglomeration and defluidization in the rubber wood and coir-fired fluidized bed. Waste Biomass Valoriz 10:3457
49. Liu C, Huang Y, Dong L, Duan L, Xu L, Wang Y (2018) Combustion characteristics and pollutants in the flue gas during shoe manufacturing waste combustion in a 2.5 MWth pilot-scale circulating fluidized bed. Waste Biomass Valoriz 11:1603–1614. https://doi.org/10.1007/s12649-018-0476-6
50. El-Sayed SA, Mostafa ME (2019) Combustion and emission characteristics of Egyptian sugarcane bagasse and cotton stalks powders in a bubbling fluidized bed combustor. Waste Biomass Valoriz 10(7):2015–2035
51. Madhiyanon T, Sathitruangsak P, Soponronnarit S (2010) Combustion characteristics of rice-husk in a short-combustion-chamber fluidized-bed combustor (SFBC). Appl Therm Eng 30:347–353
52. Madhiyanon T, Lapirattanakun A, Sathitruangsak P, Soponronnarit S (2006) A novel cyclonic fluidized-bed combustor (ψ-FBC): combustion and thermal efficiency, temperature distributions, combustion intensity, and emission of pollutants. Combust Flame 146:232–245

53. Duan F, Chyang C-S, Wang Y-J, Tso J (2014) Effect of secondary gas injection on the peanut shell combustion and its pollutant emissions in a vortexing fluidized bed combustor. Bioresour Technol 154:201–208
54. Okasha F (2007) Staged combustion of rice straw in a fluidized bed. Exp Thermal Fluid Sci 32:52–59
55. Atimtay AT, Varol M (2009) Investigation of co-combustion of coal and olive cake in a bubbling fluidized bed with secondary air injection. Fuel 88:1000–1008

Chapter 8
Optimization Methods in Staff Scheduling and Allocation

Introduction by the Editor

As of today, human resource management is an inherent part of any manufacturing systems. This includes scheduling, assignment, etc. of the staff. The efficient and optimum solution to these problems is quite important as it may save a considerable amount of cost and efforts. The chapter reviews several scheduling cases as well as algorithms addressing the associated problems. The non-preemptive scheduling workflow planning and scheduling process in automotive manufacturing as well as genetic algorithms and tabu search methods and mixed integer linear programming method for matching and assignment approach considering flexibility of assignment of workforce have been described. The complex applications of multi-resource scheduling problem using Dijkstra and multi-ant colony optimization algorithms, technician routing and scheduling problem using genetic algorithm and particle swarm optimization, memetic global and local search, and greedy algorithm for hybrid flow shop with multiprocessor task scheduling problem have also been addressed. In addition, single objective as well as multi-objective multi-period job assignment and scheduling problem solution using integer programming as well as non-dominated sorting genetic algorithm.

The contribution associated with the nursing rescheduling problem with multiple rescheduling methods under uncertainty. Two kinds of departments are involved in this study, such as general the nurses in the departments can be rescheduled between each other, and special department in which nurses require special skills and can only be rescheduled between each other in special departments. The work builds upon and complements previous work by employing stochastic programming and allocation robust optimization methods to optimize the problem of rescheduling nurses between different departments in different hospitals to proactively consider such uncertainty, and assessing the different methods to deal with uncertainty differences. The flexible scheduling of diagnostic tests in automotive manufacturing is

A. J. Kulkarni (ed.), *Optimization Methods in Manufacturing Processes*,
https://doi.org/10.1007/978-981-96-5257-0_8

one of the major studies reviewed here. It highlights that the manual scheduling is complicated due to constraints on time, predecessor-successor relationships, mutual exclusion criteria, resources and status conditions on the car engineering and assembly line. The work presents a procedure for automated and non-preemptive scheduling in the testing and commissioning of cars, which is built on a Boolean satisfiability problem on parallel and identical machines with temporal and resource constraints. It is important to mention that, it is the starting point for an automated workflow planning and scheduling process in automotive manufacturing. The work discusses model formulation and solution for flexible scheduling as well as a novel automated scheduling procedure and industrial evaluation.

One of the important objectives discussed is find to an optimal scheduling solution that minimizes the maximum completion time for the whole set jobs. It elaborates the application of a combination between these proposed heuristics and metaheuristics such as tabu search and genetic algorithm. It concludes that the evolutionary algorithm can be applied to determine job scheduling solution by the composition of a metaheuristic and a proposed heuristic. A model design of mobile resource scheduling in large scale activities is reviewed in detail. This underscores that the large-scale exhibition is a complex crowd system, providing quality service for many tourists is an important issue that managers must pay attention to. It utilizes ant colony optimization algorithm and Dijkstra algorithm to solve the large-scale exhibition resource scheduling problems. It is important to mention here that an example is given to solve the problem of multi-resource scheduling, and the feasibility and effectiveness of the algorithm has been verified. One of the important works reviewed involved an improved genetic algorithm to provide feasible solutions to the technician routing and scheduling problem. It discusses the importance of the mathematical formulation of the technician routing and scheduling problem.

An energy-efficient scheduling in intelligent production systems is critically reviewed and elaborated. The work aimed to provide a comprehensive literature review of production scheduling for intelligent manufacturing systems with the energy-related constraints and objectives. The main goals are summarized, such as analyze, discuss, and synthesize the existing achievements, current research status, and ongoing studies, and to give useful insight into future research, especially intelligent strategies for solving the energy-efficient scheduling problems. Furthermore, hybrid flow shop with multiprocessor task scheduling, memetic global and local search algorithms have been discussed. The solution approach to solving technician routing and scheduling problem using improved particle swarm optimization is elaborated. The study involves additional decisions that require to be considered, e.g., the assignment of technicians into teams, the assignment of teams to tasks, the construction of routes, and the selection of the day on which a service is provided. The work aims to route technicians to execute tasks at various client locations to maximize the total profit associated with served customers, minus the operations costs.

Machine Generated Summaries

Disclaimer: The summaries in this chapter were generated from Springer Nature publications using extractive AI auto-summarization: An extraction-based summarizer aims to identify the most important sentences of a text using an algorithm and uses those original sentences to create the auto-summary (unlike generative AI). As the constituted sentences are machine selected, they may not fully reflect the body of the work, so we strongly advise that the original content is read and cited. The auto generated summaries were curated by the editor to meet Springer Nature publication standards. To cite this content, please refer to the original papers.

Machine generated keywords: scheduling, scheduling problem, job, resource, shop, service, personnel, assignment, flow shop, day, constraint, project, route, task, genetic algorithm.

Nursing Rescheduling Problem with Multiple Rescheduling Methods Under Uncertainty [1]

This is a machine-generated summary of:

Long, Zhiren; Wen, Xianxiu; Lan, Mei; Yang, Yongjian: Nursing rescheduling problem with multiple rescheduling methods under uncertainty [1].

Published in: Complex & Intelligent Systems (2021).

Link to original: https://doi.org/10.1007/s40747-021-00554-z

If you want to cite the papers, please refer to the original.

For technical reasons we could not place the page where the original quote is coming from.

Abstract-Summary "This decision-making needs to be made in a stochastic setting to meet uncertain demand with insufficient historical data or inaccurate forecasting methods."

"A stochastic programming model and a distributionally robust model are developed for the nurse rescheduling problem with multiple rescheduling methods under uncertain demands."

"The results show that the stochastic programming model and the distributionally robust model can reduce the cost by 78.71% and 38.92%, respectively."

"We also evaluate the benefit of the distributionally robust model against the stochastic model and perform sensitivity analysis on important model parameters to derive some meaningful managerial insights."

Introduction

"The reasons for employing distributionally robust optimization are threefold: (1) it allows uncertain parameters to follow an arbitrary distribution resided in the ambiguity set, instead of a given distribution, or fuzzy number that is needed for stochastic optimization and fuzzy optimization which are two popular methods for handing uncertainty; (2) the solution obtained from the distributionally robust model is usually robust and flexible when encountering uncertainty in future, although the negative effect of the worst-case scenarios with low possibility and high penalty is substantial; and (3) the decision-makers can control the conservatism level by varying the parameters in the ambiguity set."

"The main contributions of this study are as follows: For the first time, we present a novel nurse rescheduling problem with multiple rescheduling methods under uncertain demands of the nurse."

"For the uncertain demand of the nurse, we consider the stochastic programming model and the distributionally robust model."

Literature Review

"Maenhout and Vanhoucke [2] propose a more compact integrated staffing and scheduling model for a long-term nurse management problem over multiple departments."

"Bard and Purnomo [3] consider the problem of short-term nurse rescheduling for daily fluctuation in patient demand, where a given mid-term schedule is revised to cover the shortage."

"He et al. [4] formulate an integrated nurse staffing and scheduling model under patient demand uncertainty into a two-stage stochastic programming model with an emphasis on understaffing risk control."

"No existing models in the literature cover all aspects of our problem (especially multiple hospitals and departments, and uncertainty of demand using a distributed robust approach)."

"Our work builds upon and complements previous work by employing stochastic programming and allocation robust optimization methods to optimize the problem of rescheduling nurses between different departments in different hospitals to proactively consider such uncertainty, and assessing the different methods to deal with uncertainty differences."

Problem Description and Model Formulations

"The decision-makers, for each time period, need to determine the amounts of nurses rescheduling to the general department and the special department."

"We first introduce the nurse rescheduling model with stochastic nurse demand."

"For the uncertain demands of the nurse, we use the sample mean approximation (SAA) method to construct stochastic programming models."

"To formally set up the stochastic model (SM) for the problem we consider, we introduce the following variables."

"Section, we propose a distributionally robust model to mitigate the effects of these drawbacks by dealing with problem setups in which partial information on the demands of nurses is available in advance."

"This results in an unpredictable number of demands of nurses for each department in each hospital."

"To formally set up the distributionally robust model (DRM) for the problem we consider, we introduce the following variables."

Case Study

"We collected data on the demand for nurses in each department in each hospital for each day of the year and will present the relevant parameters in detail in the following."

"Although the time to solve DRM is slower than RM, the speed difference is not significant, and DRM is less than RM on ATC, which shows the advantage of DRM to help decision-makers control hospital costs without affecting the solution time."

"If decision-makers can add hospitals with more vacant nurses to the joint hospitals, then the cost of ATForRM can be reduced significantly, which in turn can further reduce the total costs."

"These vacant nurses will go to other hospitals by the second rescheduling method, which leads to an increase in the cost of the second rescheduling."

Conclusion

"Nurse rescheduling problems show non-stationary and uncertain demand."

"This study extends the literature on nurse rescheduling problem in three ways: (1) using multiple rescheduling methods to meet the demand for nurses in each hospital department as much as possible; (2) constructing stochastic programming models based on historical data; and (3) using distributionally robust optimization to address the difficulties of high uncertainty in demand and the limitations of the lack of sufficient historical data."

"We develop a novel stochastic model and a distributionally robust model of nurse rescheduling problem under uncertain demand of nurses."

"We construct the ambiguity sets of uncertain emergence demands and travel times using only the first-order moment knowledge."

"For the uncertainty of demand, this paper uses the estimation based on the historical data, and the corresponding ambiguity set may be inaccurate when the variance of the data is large."

Flexible Scheduling of Diagnostic Tests in Automotive Manufacturing [5]

This is a machine-generated summary of:

König, Simone; Reihn, Maximilian; Abujamra, Felipe Gelinski; Novy, Alexander; Vogel-Heuser, Birgit: Flexible scheduling of diagnostic tests in automotive manufacturing [5].

Published in: Flexible Services and Manufacturing Journal (2022).

Link to original: https://doi.org/10.1007/s10696-021-09438-3.

If you want to cite the papers, please refer to the original.

For technical reasons we could not place the page where the original quote is coming from.

Abstract-Summary "Enabling these customisation options forces modern automotive manufacturers to update their standardised scheduling concepts for testing and commissioning cars."

"This concept is essential to provide individual testing workflows where the time and resources are optimised for every car."

"Manual scheduling is complicated due to constraints on time, predecessor-successor relationships, mutual exclusion criteria, resources and status conditions on the car engineering and assembly line."

"This paper presents a procedure for automated and non-preemptive scheduling in the testing and commissioning of cars, which is built on a Boolean satisfiability problem on parallel and identical machines with temporal and resource constraints."

"This paper is the starting point for an automated workflow planning and scheduling process in automotive manufacturing."

Introduction and Motivation

"Automotive software allows customers to use more assisted, software-linked car functions during their car journey."

"Automotive manufacturing should therefore be able to produce, test, and commission software-linked, customised cars."

"As every chosen configuration code may need its own testing procedure to complete the car, such as the commissioning of a seat massage, it is essential to provide time and resource-optimised individual testing workflows for any given car."

"The main contribution of this paper is a method for the automated and flexible scheduling of diagnostic tests on cars subject to several constraints."

"This is the first use case to describe flexible scheduling on parallel identical machines with complex constraints for testing and commissioning in car manufacturing."

Problem Definition

"We explain the technical requirements arising from expert discussions that need to be fulfilled for scheduling in testing and commissioning."

"In test location 1, the maximum number of tests is conducted on the fully equipped car with seven scheduled tests."

"A fully equipped car has more electrical and electronic components than others and thus, for cars with less equipment codes, fewer tests need to be scheduled."

"These explicit requirements arise from expert knowledge in automotive diagnostic test scheduling."

"The finite set of Z different status conditions, which must be considered in the test schedule, is defined asFor example, the car engine must be turned on before testing the air conditioning."

"That the more accurate the time predictions are, the more efficient an optimised test schedule will be in the production line."

Related Work

"Shi et al. [6], for example, developed test plans with tight schedules that combine multiple diagnostic tests on cars to fully utilise all of the available time in prototype car test scheduling."

"Spieckermann et al. [7] dealt with the scheduling problem of operating paint shop systems with sequence-dependent set-up costs."

"As the algorithm engineering of modern SAT solvers has improved over recent years, the real world running time in the present use case of static scheduling is of negligible importance [8]."

"Weckenborg et al. [9] discussed the automotive capacity scheduling of prototype vehicle production at the Volkswagen Pre-Production Center."

"Buergin et al. [10] introduced an optimisation model for the local order scheduling of mixed-model aircraft assembly lines covering both assignment to lines as well as sequencing with respect to cost."

"Dörmer et al. [11] covered the problem of master production scheduling for high-variant, mixed-model automotive assembly lines."

Model Formulation and Solution for Flexible Scheduling

"An intuitive explanation will follow for these conditions based on the Pseudo code illustrated in Algorithm 1 of Appendix A. At the start of the testing procedure, all status conditions are 'off' by definition."

"This has the disadvantage of having an exponential number of constraints depending on the number of tests that use a resource."

"An output format is needed that still forces the test schedule to adhere to all requirements."

"An adjacency list is introduced which contains a list of predecessors for every test."

"In order for test i to be started, every test in the predecessor list of test i must have finished."

A Novel Automated Scheduling Procedure

"The procedure for scheduling diagnostic tests on a selected car starts with Step 1."

"In Step 2, the location of the car on the assembly line is determined with the help of a production system at the supervisory level (ISA 10)."

"In Step 3, a production system at the planning level offers the constraints C for the configuration code-dependent tests of the car on the assembly line."

"In Step 5, the resulting schedule of N tests for the car on the assembly line is transferred for execution with a controller at the supervisory level."

"If the repetition is feasible, the tests with the result equal to 'not OK' are repeated for the car on the given assembly line by going back to Step 2."

"In Step 7, the N test processes have been carried out on the car."

Industrial Evaluation

"As suggested in the benchmark test of Da Col and Teppan [12], there is no real-world run time problem, even if one would expect the number of constraints to rise gradually with the development of electrified cars and more complex assembly lines."

"Real-world case study, we therefore look at the procedure from two angles: first, we select a specific test location at the automotive assembly line."

"The test of the exit lights is performed in tests 9 and 10, and can be carried out automatically depending on a specific configuration code."

"The test of the exit lights in tests 9 and 10 can be carried out automatically depending on a specific configuration code."

"The exit lights are executed manually in tests 3 and 6 if the specific configuration code is not present for a car."

Conclusion and Future Work

"We have described the development, implementation and successful evaluation of a mathematical optimisation model as basis for the flexible scheduling of automotive diagnostic tests on an assembly production line."

"We embedded the model into a procedure to schedule the tests for each car with configuration codes at the test location."

"By integrating the model into the procedure, we showed that complex scheduling with multi-constraints can be flexible and automated."

"This is the first use case that describes flexible scheduling on parallel identical machines with complex constraints for testing and commissioning in car manufacturing."

Resolution for Bounded-Splitting Jobs Scheduling Problem on a Single Machine in Available Time-Windows [13]

This is a machine-generated summary of:

Son, Trang Hong; Van Lang, Tran; Huynh-Tuong, Nguyen; Soukhal, Ameur: Resolution for bounded-splitting jobs scheduling problem on a single machine in available time-windows [13].

Published in: Journal of Ambient Intelligence and Humanized Computing (2020).

Link to original: https://doi.org/10.1007/s12652-020-02162-0.

If you want to cite the papers, please refer to the original.

For technical reasons we could not place the page where the original quote is coming from.

Abstract-Summary "The objective aims to find an optimal scheduling solution that minimizes the maximum completion time for the whole set jobs."

"Besides that we propose some effective heuristics concerned about assignment strategy such as assignment heuristic, heuristic based on shortest/longest processing time rules, heuristic based on max flow resolution, matching and assignment approach."

"In order to improve solution quality, we also propose to apply a combination between these proposed heuristics and metaheuristics such as tabu search and genetic algorithm."

"In the hope of achieving the better solutions in acceptable time, we introduce another approach called exact for subset-jobs matheuristic which is combined between mathematical programming and heuristic derived from single-attribute priority rule."

"Experimental results show the performance of proposed approximating approaches in comparing between them and the exact method based on the MILP model which is implemented by CPLEX solver."

"Among proposed effective algorithms, the matheuristic showed the superiority in terms of solution quality and execution time."

Introduction

"There has been a lot of research to consider the job shop scheduling problem [14–16]."

"In most research results, however, the jobs with splitting constraint processed independently on different kinds of parallel machines. Tahar et al. [17] proposed a resolution approach based on linear programming techniques for identical parallel machine scheduling problem involving job splitting and sequence-dependent setup."

"With the makespan criterion, the authors used a lower bound to evaluate the performance of their new approach on a large number of randomly generated instances. Wang et al. [18] proposed a hybrid differential evolution algorithm for parallel machine scheduling with splitting jobs."

"Wang et al. [19] considered parallel machine scheduling problems with job splitting and learning for minimization total completion time."

"One of the last related results considered the scheduling problem with job-splitting considering learning and vital-few law [19, 20]."

Problem Definition

"The bounded-splitting jobs scheduling problem on a single machine in available time-windows has the following components are the tasks need to be scheduled (called jobs); each task has execution time (called processing-time); blank time window frames can organize jobs into it (called available time-window) and full time or non-scheduled window frames (called unavailable time-window)."

"This considered problem is arranging jobs depending on the available time-windows and basing on different criteria so as to maximize their productivity."

"This scheduling problem was proved to be strongly NP-Hard in the following steps [21]: define the decision problem (called SPLITSCHE) corresponding to this optimization problem (called BSJSP)."

"Prove that SPLITSCHE is in strongly NP-complete class by reducing from the 3-PARTITION problem."

"The time-windows in this problem corresponding to the bin in the Bin-packing problem."

Mathematical Model

"Properties of a particular optimal solution were proposed in Nguyen et al. [22]."

"Determining a solution of the scheduling problem is based on the above optimal structures by answering two following questions: 1."

"Each created solution is considered as a feasible schedule and of course there are many solutions from this problem."

"This MILP model is represented in terms of decision variables, intermediate variables, constraints, and objective function."

Proposed Approaches

"In the matching step (Algorithm 5): the objective aims to find some jobs whose total processing time is equal to the size of the window."

"Due to this characteristic, another proposed approach to solve this problem is trying to change the order of input jobs for finding the better possible solutions."

"There are many metaheuristics that can be applied to solve this combination optimization problem by changing the the order of input jobs such as stochastic hill climbing, random restart hill climbing, simulated annealing, tabu search, etc. in neighborhood search algorithms and genetic algorithm, memetic algorithm, etc. in evolutionary algorithms."

"Metaheuristic such as genetic algorithm (GA) and tabu search (TABU) can be implemented by considering job order as a chromosome in GA or solution encoding in TABU."

"Evolutionary algorithm can be applied to determine job scheduling solution by the composition of a metaheuristic and a proposed heuristic."

Experimental Results

"While metaheuristic specifically GA on MAAS algorithm, the quality of the solution is more improved than heuristic (percentage gap decreased from 4.48 to 3.37%) but the processing time is much more (average is more than 1 min)."

"The matheuristic with E4SSJ algorithm is still the best in the heuristics group because the gap remains stable at about 1% and the time to find the solution is less than 10 s. In summary, for scheduling applications that require an optimal solution with a small number of input jobs, the exact method for solving the MILP model by the solver should be chosen."

"To the large number of input jobs, the matheuristic method with E4SSJ algorithm is the best choice because the quality of the solution found to be very good and fast processing time."

Conclusion

"In order to determine the solution of the problem, we proposed a general MILP model which be solved by solvers to find the optimal solution in 0.37% average gap from lower-bound (for dataset DS1), however this takes a lot of time."

"To reduce the computation time, some effective heuristics concerned about assignment strategy such as Assignment algorithm, based on Shortest/Longest processing time rules, based on Max Flow resolution, MAtching and ASsignment approach were also proposed to find the better solutions in ranges from 8 to 5% average gap from lower-bound."

"They be able to find more better solutions in about 2% average gap from lower-bound."

"We also present a simple matheuristic to achieve more and more better solutions in about 1% average gap from lower-bound."

An Enhanced Load Balancing Approach for Dynamic Resource Allocation in Cloud Environments [23]

This is a machine-generated summary of:

Praveenchandar, J.; Tamilarasi, A.: An Enhanced Load Balancing Approach for Dynamic Resource Allocation in Cloud Environments [23].

Published in: Wireless Personal Communications (2021).

Link to original: https://doi.org/10.1007/s11277-021-09110-x.

If you want to cite the papers, please refer to the original.

For technical reasons we could not place the page where the original quote is coming from.

Abstract-Summary "Platform as a service, Infrastructure as a service, and Software as a Service are services provided by the Cloud."

"Each end user's Quality of service must be ensured by the cloud service provider."

"Work, a new load balancing approach is proposed for the dynamic resource allocation process to improve stability and to increase profit."

"PBMM algorithm is devised for an effective load balancing process through which, resource scheduling is performed."

"To optimize the waiting time, resource tables and task tables are employed."

"The simulation results show that the proposed load balancing technique ensures the maximum profit and it enhances load balancing stability by increasing the number of special users."

Introduction

"In a Distributed Computing Environment (DCE), cloud computing is the on-demand availability of computing resources such as memory, CPU, network, etc. Resource and data are shared across the network."

"It creates the layer between physical components and virtualized resources."

"The Load balancing is necessitated for heterogeneous workloads operating in the network."

"In data centers, physical servers are not fully utilized."

"To improve resource utilization, an efficient load distribution technique needs to be adapted to a network."

"Generating more revenue through load balancing by ensuring Quality of service to each end-user is a most challenging task."

"Another solution is to implement an effective load balancing approach to cloud networks through which the dynamic workloads are distributed across the networks."

"It also reduces the unevenness in running loads which improves resource utilization."

Related Work

"Mobile cloud resource allocation is optimized using a cross-layer load-balancing technique [24]."

"A new load balancing approach is proposed in [25]."

"In [26], a novel approach for task scheduling and load balancing process is proposed."

"In [27], a new load balancing approach using an improved RR algorithm is discussed."

"Based on the capabilities of each VM(virtual machines) load balancing and resource scheduling is done."

"In [28], for efficient load balancing a new resource allocation is proposed which is working in a dynamic fog environment."

"In [29], An improved load-balancing algorithm is proposed to manage the virtualized resources named as IES (Improved Efficient Scheme) algorithm."

"In [30], a new load balancing approach is discussed based on the users."

"In [31], to enhance the performance of a Wireless Local Area Network, a new load balancing approach is discussed."

Proposed Methodology

"Work, we propose an improved load-balancing approach to distribute the massive workloads during the dynamic resource allocation process in a cloud computing environment."

"All incoming user requests are denoted as All incoming requests are represented as tasks T, this is denoted as Size of the arrived tasks (S) can be represented as follows Inter arrival time of the tasks could be represented as follows Execution time of all arrived tasks are represented as Available resources A in the system is represented as follows The completion time C of each arrived tasks, could be represented as Then finally newly arrived tasks Y during the execution is represented as Each request that arrived from the user side is represented as a task in this dynamic scheduling environment."

Profit Based Load Balancing Approach

"Balancing the workload is one of the major tasks to ensure the QoS. There are various load balancing algorithms exists in a cloud system, like load balancing for multiservice, policy-based load balancing, WLAN Load balancing, etc. still, some instability exists during balancing the massive workloads."

"A novel load balancing-based resource allocation approach is proposed to ensure stability during load balancing and to maximize the profit."

"In a normal process, two factors are taken into consideration for Virtual machine allocation. (1) Size of a workload (2) Completion time of a task in VM A task with the longest execution time is assigned to the fastest VM and the remaining tasks are allocated based on the same scenario."

"Newly arrived tasks and the running tasks are distributed among the available resource based on the proposed PBMM load balancing algorithm."

Experimentation and Results

"In terms of average waiting time, load balancing stability, profit, successful Task completion rate, and average response time for premium clients, the comparative results show that the suggested PBMM Load balancing delivers the best outcomes."

"Most load balancers achieve limited task migration because they employ a static method to select the most suitable VM for a certain task that is waiting in a heavily loaded VM."

"The total profit earned by the PBMM load balancing method is estimated based on the number of tasks completed successfully with high bid values and the task's inter arrival time."

"For the proposed load balancing algorithm, the average waiting time of tasks with larger bid amounts (M (0…p/2) is computed."

"The Task's arrival time (T_{arr}) and the resource allocated time (T_{all}) is used to calculate average waiting time (T_w)."

Conclusion

"The proposed PBMM Load Balancing method increases the load balancing stability."

"The load is distributed evenly depending on the available VM length and the bidding amount of each user."

"Users who are willing to pay extra for a specific resource are treated as special users, and they are given preference during the load balancing process to allocate the resource."

"When compared to previous load balancing methods, the suggested method yields better outcomes in terms of higher profit and load balancing stability."

The Model Design of Mobile Resource Scheduling in Large Scale Activities [32]

This is a machine-generated summary of:

Zhou, Cai-Qiu; Yang, Yu-Wang; Gong, Bei; Li, Meng-Kun; Li, Jin; Wen, Ren-Qiang: The Model Design of Mobile Resource Scheduling in Large Scale Activities [32].

Published in: Mobile Networks and Applications (2018).

Link to original: https://doi.org/10.1007/s11036-018-1023-1.

If you want to cite the papers, please refer to the original.

For technical reasons we could not place the page where the original quote is coming from.

Abstract-Summary "Large scale exhibition is a complex crowd system, providing quality service for many tourists is an important issue that managers must pay attention to."

"In order to carry out the management of tourist service resources in the large scale exhibition, the concept model of mobile resource scheduling is put forward, and the detailed introduction of each dimension is presented."

"Multi-resource scheduling model based on Dijkstra and multi-ant colony optimization algorithms are established respectively."

"The research has practical significance for the development of scientific and effective population service operation plan and service management plan for large scale exhibition activities."

Introduction

"Effective smart tourists service supply urgently needed in-depth study on the resource service scheduling problem in a mobile environment."

"Dai Gengxin and Da Qingli established the mathematical model of the single disaster point multi-resource emergency problem under the earliest target of emergency start time and solved it [33]."

"The resource continuous consumption emergency systems scheduling problem in a save point to the emergency location uncertainty is proposed based on the multi resource continuous consumption emergency system response time the earliest models and algorithms by Gao Shuping and Liu Sanyang [34]."

"Multi-resource demand supply point for resource service scheduling problem is less, especially the demand of service resources is inevitably involved in manpower, material resources and other various resources."

"In order to deal with the crowd service problems in the large exhibitions, scheduling service personnel and equipment for guidance and control of the crowds, two kinds of multi-resource scheduling model and algorithm were proposed [35, 36]."

Resource Scheduling in Mobile Environment

"In a mobile environment, resources (Mobile Resource) are a mobile state of personnel or resources, which can be dynamically scheduled according to the needs of different places."

"Mobile resource scheduling is what kind of resources, when and where it goes through what kind of the best scheduling task."

"Mobile staff is object mobile resource scheduling task assignment, through the effective allocation of resources to achieve them [37]. (2) Mobile task."

"Mobile task differs from the static task, move the task has two important features: time and space dependent and multi-task processing."

"The situation of mobile work refers to the mobile staff to carry out the environment of the mobile task."

"Time structure not only includes time, but also relates to workplace, because they are closely linked with work activities. (4) Mobile technology."

Ant Colony Optimization Algorithm

"That the effect of the ant colony h pheromone on the other ant colony has a positive effect, the other ants for the optimal solution group of ants in the city to choose the next city of the alternative path will be relatively concentrated in the future search which contributes to the results of the ant colony tends to be consistent, causing premature, stagnation of the possibility of the greater the possibility; on the other hand, the poly of the ant colony h is small in the city i, the distribution of the pheromone emitted by the ant colony h from the city i to the route of the other city is relatively dispersed, and the influence of the ant colony h on the other ant colony is positive or negative what has no obvious effect, but it is not conducive to the optimal information of the ant colony h, which leads to the slow convergence of the algorithm as a whole."

Dijkstra Algorithm

"We will combine the genetic algorithm and Dijkstra algorithm to solve the large-scale exhibition resource scheduling problem."

"When the population size is too small, the genetic algorithm optimization process variables are too large what lead to the search may stop in the immature stage, that is to say that the phenomenon of immature convergence is easy to optimize the optimization; when the group size is too large, the diversity of groups can avoid the phenomenon of immature convergence and reduce the probability of local solution, but lead to an increase in the amount of calculation at the same time."

"The corresponding constraints and conditions are usually designed, next the solution of the objective function is solved according to these constraints, and then optimize the combination of practical conditions to get the best for large-scale resource scheduling optimal program."

"Select the objective function of large-scale exhibition resource scheduling, that is, the fitness function involved in the genetic algorithm by evaluating each sequence, selecting a certain number of sequences according to the selection operator for genetic operation, and eliminating the remaining sequences."

Example Analysis

"Summary of the heuristic algorithm is about to get the objective function value Z = 11,480, the operation is over. (1) Route reliability In the multi-ant colony optimization algorithm, the reliability of the system is proposed, which is considered to be a resource scheduling path from any number of road arcs."

"By comparing the time ratios of the two algorithms in the algorithm, it is found that the Dijkstra algorithm is better than the multi-ant colony optimization algorithm when the number of iterations is small."

"With the number of iterations of 300 times, the calculation time of multi-ant colony optimization algorithm is better than that of Dijkstra algorithm."

"When the number of iterations reaches 500, the calculation time of multi-ant colony optimization algorithm is more than Dijkstra algorithm, the relative error is 2.4%, and the computation time is 25%."

Conclusion

"According to the previous shortage of resource scheduling, we put forward the existing problems through the analysis of mass service in large-scale exhibition activities."

"In order to carry out the management of the tourist service resources in the large-scale exhibition activities, a conceptual model of mobile resource scheduling is proposed and each dimension is introduced in detail."

"An example is given to solve the problem of multi-resource scheduling, and the feasibility and effectiveness of the algorithm has been verified."

A Simple Solution to Technician Routing and Scheduling Problem Using Improved Genetic Algorithm [38]

This is a machine-generated summary of:

Pekel, Engin: A simple solution to technician routing and scheduling problem using improved genetic algorithm [38].

Published in: Soft Computing (2022).

Link to original: https://doi.org/10.1007/s00500-022-07072-1.

If you want to cite the papers, please refer to the original.

For technical reasons we could not place the page where the original quote is coming from.

Abstract-Summary "This paper proposes an improved genetic algorithm (IGA) to provide feasible solutions to the technician routing and scheduling problem (TRSP)."

"Three neighborhood structures improve genetic algorithm (GA) by searching for better solutions."

"Further, the performance of the proposed algorithm is experimentally compared with GA and improved particle swarm optimization (IPSO) algorithm by providing the TRSP solutions on the generated benchmark instances."

"The numerical results indicate that IGA offers fast and better solutions considering GA and IPSO algorithms."

Introduction

"Technician routing and scheduling problem (TRSP) includes decisions such as the assignment of technicians into teams, the assignment of feasible groups to suitable tasks, the routings, the selection of the visiting day, and technicians that must meet a proficiency level [39, 40]."

"Chen et al. [41] have presented a technician routing model that explicitly forms individualized, experience-based learning by minimizing the fulfillment time of the final task and performing the record-to-record travel (RTR) method."

"Papers [42, 43] worked on the Service Technician Routing and Scheduling Problem, where tasks have different priorities and time windows and technicians have different skills and working hours."

"The algorithm-based genetic has been preferred in this paper because it contains solution spaces that could not be searched before, thanks to the crossover tools it has in technician routing and scheduling-type problems."

Mathematical Model of TRSP

"Scheduled team members must meet the skill requirements of each task."

"There is no extra cost if the team created is more proficient than the task sequence needs."

"At least one scheduled team is allocated to a day."

"The waiting customer applies a penalty if a team begins a task after the latest start time."

Methodology

"These algorithms on a simple chromosome-like data structure encode a potential solution to a particular problem and apply recombination operators to these structures in a way that preserves essential details."

"The initial values of chromosomes are assigned to 0 or 1, and this assignment procedure should generate a feasible solution using the greedy randomized adaptive search procedure (GRASP)."

"While drop and swap are used only to obtain better solutions, the body change operator has been used to achieve better solutions and perform crossover operations."

"By selecting two different particles on a chromosome, the body change operator works."

"After IGA obtains the initial population according to the GRASP approach, the crossover and mutation mechanisms work by providing the determined probability states and providing the opportunity to search for different solution spaces."

"Improving the different solution spaces obtained is carried out in the order shown by the three operators in the pseudo-code."

Results

"Performing the IGA method in 29 test sets offers the best results in 27 test sets, although the IPSO method gives the best solution in only 9 test sets."

"The IGA hybrid algorithm provides a 0.26% average gap applying to 29 test sets, while the IPSO offers a 1.11% average gap."

"It is advantageous to perform the IGA algorithm in this problem type considering the found average gap, the best number of solutions, and solution time."

"The solutions obtained by applying the IPSO algorithm to C107, C108, C-109, R-110, RC-101, RC-102, RC-106, and RC-107 data sets were incorrect to violation technician skill constraint."

"Tukey, LSD (least significant difference) and Scheffe tests were applied to statistically evaluate whether the results obtained by running 5 times for all three algorithms create a significant difference."

"To the better numerical results of the IGA algorithm, it is seen that the results obtained statistically make significant differences."

Conclusion

"The experimental findings indicate that the performance of the proposed algorithm outperforms GA and IPSO algorithm by reaching the feasible solutions of the TRSP on Pekel's test sets."

"The IGA algorithm consists of the GA method and three neighborhood structures."

"Performing the IGA algorithm to 29 test sets offers the best results in 27 test sets, although the IPSO algorithm provides the best solution in only 9 test sets."

"Considering the average of all data sets, the IGA algorithm offers a 0.26% average gap, while the IPSO algorithm offers a 1.11% average gap."

"The IGA and IPSO algorithm provides the feasible solutions of all sets on average 38.06 and 126.80 min, respectively."

"The numerical findings illustrate that IGA offers fast, less deviated, and better solutions considering the IPSO algorithm."

A Review of Energy-Efficient Scheduling in Intelligent Production Systems [44]

This is a machine-generated summary of:

Gao, Kaizhou; Huang, Yun; Sadollah, Ali; Wang, Ling: A review of energy-efficient scheduling in intelligent production systems [44].

Published in: Complex & Intelligent Systems (2019).

Link to original: https://doi.org/10.1007/s40747-019-00122-6.

If you want to cite the papers, please refer to the original.

For technical reasons we could not place the page where the original quote is coming from.

Abstract-Summary "Energy-efficient scheduling of production systems is an effective way to improve energy efficiency and to reduce energy cost."

"During the past 10 years, a large amount of literature has been published about energy-efficient scheduling, in which more than 50% employed swarm intelligence and evolutionary algorithms to solve the complex scheduling problems."

"This paper aims to provide a comprehensive literature review of production scheduling for intelligent manufacturing systems with the energy-related constraints and objectives."

"The main goals are to summarize, analyze, discuss, and synthesize the existing achievements, current research status, and ongoing studies, and to give useful insight into future research, especially intelligent strategies for solving the energy-efficient scheduling problems."

"The energy efficiency-related publications are classified and analyzed according to five criteria."

Introduction

"It is essential to reduce the manufacturing industry's energy consumption and demand."

"Many researchers use scheduling approaches to improve energy efficiency in manufacturing industries and energy-efficient scheduling has been proved to be an effective way of reducing the energy consumption with none or little capital investment [45, 46]."

"Energy efficiency or energy consumption is considered as constraints or scheduling objectives, like makespan, machine workload, and due-date objectives."

"Swarm intelligence and evolutionary algorithms are becoming more and more popular for solving large-scale scheduling and optimization problems with time constraints, including energy-efficiency scheduling in complex production."

"The design of intelligent scheduling strategies should consider energy efficiency and reducing energy consumption which is an important scheduling objective in current production scheduling area."

"Detail discussion and analyzing up to date and ongoing research in the field of production/shop scheduling with energy efficiency/consumption and constraints/objectives are given in "Research trends" section."

Scope

"The topic of this review is production/shop scheduling with energy efficiency, consumption, or cost as constraints or objectives, given in the following: (1) the category of the shop floor, (2) model of the scheduling problems, (3) objectives (e.g., energy, completion time, machine workload, due date, and so forth), (4) research approaches (strategies) or algorithms, especially swarm intelligence and evolutionary algorithms, and (5) the aspects of energy consumption."

"In some publications, energy efficiency or consumption is considered as constraints."

"We include some newest publications about the reviewed topic, which are not yet included in the Web of Science database at this moment."

"To review the literature related to the "production/shop scheduling with energy efficiency or consumption as constraints or objectives" topic, 90 publications are analyzed, discussed, and synthesized till present."

Classification, Analysis, and Synthesis of Existing Achievements

"To solve energy-efficiency scheduling problems, many articles employed or improved swarm intelligence and evolutionary algorithms, e.g., genetic algorithm (GA), simulated annealing (SA), particle swarm optimization (PSO), shuffled frog-leaping algorithm (SFLA), grey wolf optimizer (GWO), and so forth."

"To solve energy-efficiency scheduling problem, swarm intelligence and evolutionary algorithms do not need standard mathematical models; indeed, they need mathematical formulations to compute objective functions."

"It means that swarm intelligence and evolutionary algorithms are mainstream approaches for energy-efficiency scheduling, which has affected the problem modeling."

"Zhang et al. [47] developed an intelligent strategy to reduce energy consumption and improved energy efficiency in a production sequence fixed solution, by controlling the additional tardiness allowed for each job in the solution."

"More than 50% of publications have assumed energy efficiency as scheduling objective; however, the aspects of energy consumption or energy demand are not described clearly in these articles."

Research Trends

"Till April 2019, more than ten production scheduling articles for energy-related objectives are published."

"Energy-related objectives become a new trend of production scheduling and will play vital and important role in production scheduling."

"Compared to traditional scheduling objectives, energy-related objectives are novel, however, important to economic indicators."

"Many publications have considered energy-related objectives with traditional objectives simultaneously."

"It means that energy-related scheduling objectives should be considered together with traditional objectives for obtaining better decision-making by different performance indicators."

"More than 13 algorithms are employed or improved for solving production scheduling problems with energy-related objectives in the past 5 years."

"Swarm intelligence and evolutionary algorithms are effective and widely used for solving energy-efficiency scheduling problem."

"Energy-related objectives are aware of production scheduling from 2012, especially in the past 5 years."

"Energy-related objectives are optimized with traditional scheduling objectives, simultaneously."

Future Directions

"Based on the previous analysis of research trends of production scheduling with energy-related objectives, we consider and indicate some future research directions for this topic."

"Based on the synthesis and analysis in above two sections, swarm intelligence and evolutionary algorithms are efficient and effective to solve energy-efficiency scheduling problems, especially for the large-scale problems."

"How to design and develop more high-quality algorithms, especially non-dominated multi-objective algorithms, for solving energy-efficiency scheduling problem which is an important direction."

"Energy-efficiency scheduling is an important objective in intelligent production system."

"How to design energy efficiency-based multi-objective scheduling strategy is a novel and interesting direction."

"The existing modeling strategies, solving approaches and algorithms (including swarm intelligence and evolutionary algorithms), and benchmark instances for general production scheduling can be directly applied to energy-efficiency scheduling or be used after appropriate adjustments."

Conclusions

"The growing number of published articles, especially in the past 5 years, makes energy-efficiency scheduling a hot research topic."

"This study presented the review of five indicators, including shop floor, models, approaches and algorithms, objectives, and aspects of energy consumption in the literature for solving energy-efficiency scheduling problems."

"The current research trends, especially the intelligent strategies, are analyzed and summarized."

"For the continued study of this topic, five instructive directions, including modeling, objectives, intelligent strategies, intelligent scheduling framework, and practice, are given which provide an insight for future studies."

A New Memetic Global and Local Search Algorithm for Solving Hybrid Flow Shop with Multiprocessor Task Scheduling Problem [48]

This is a machine-generated summary of:

Engin, Batuhan Eren; Engin, Orhan: A new memetic global and local search algorithm for solving hybrid flow shop with multiprocessor task scheduling problem [48].

Published in: SN Applied Sciences (2020).

Link to original: https://doi.org/10.1007/s42452-020-03895-5.

If you want to cite the papers, please refer to the original.

For technical reasons we could not place the page where the original quote is coming from.

Abstract-Summary "A new memetic algorithm which combined the global and local search methods is proposed to solve the HFSMT scheduling problems."

"The developed new memetic global and local search (MGLS) algorithm consists of four operators."

"A preliminary test is performed to set the best values of these developed new MGLS algorithm operators for solving HFSMT scheduling problem."

"The proposed new MGLS algorithm is applied the 240 HFSMT scheduling instances from the literature."

"The performance of the generated new MGLS algorithm is compared with the genetic algorithm (GA), parallel greedy algorithm (PGA) and efficient genetic algorithm (EGA) which are used in the previous studies to solve the same set of HFSMT scheduling benchmark instances from the literature."

"The results show that the proposed new MGLS algorithm provides better makespan than the other algorithms for HFSMT scheduling instances."

"The developed new MGLS algorithm could be applicable to practical production environment."

Introduction

"Oğuz and Ercan [49] developed a constructive heuristic algorithm to solve the multiprocessor tasks in a two stage flow shop environment scheduling problems."

"Oğuz et al. [50] proposed a constructive heuristic algorithm to solve the multiprocessor task scheduling in a two stage hybrid flow shop scheduling problems."

"Şerifoğlu and Ulusoy [51] generated a genetic algorithm for solving the multiprocessor task scheduling in multistage hybrid flow shop scheduling problems."

"Engin et al. [52] generated a data mining to find patterns in genetic algorithm for solving hybrid flow shops scheduling with multiprocessor task problems."

"Tseng and Liao [53] developed a particle swarm optimization algorithm to solve the hybrid flow shop scheduling with multiprocessor tasks problems."

"Akkoyunlu et al. [54] proposed a harmony search algorithm for solving the hybrid flow shop scheduling with multiprocessor tasks problems."

Hybrid Flow Shop with Multiprocessor Task Scheduling

"The number of machines which the jobs require at each stage is indicated with $size_{ij}$."

"The number of jobs, stages, machines at each stage are known and fixed, All processors, machines and jobs are available through the planning horizon, The processing time of each job on each stage is given before, The processing times are included the setup times of the jobs, The objective is minimizing the makespan values."

"Let's denote the following variables [55]: Denote the following variables: n, number of jobs to be sequenced."

"m_i, number of machines at stage i. p_{ij}, processing time of job j on machine i. $size_{ij}$, number of processors required to process."

"Let's denote the following decision variables: SPT_{ij}, starting processing time of job j at stage i. C_{ij}, completion time of job j at stage i. C_{max}, makespan The mathematical formulation is given as follows [56]: s.t."

Memetic Global and Local Search Algorithm

"Mutation operator is essential to offer the opportunity to escape from trapping the local optima and thus to increase variability and diversity in the population to search different part of the solution area, by mutating genes in chromosomes with several different methods from the individuals randomly chosen."

"The steps of the proposed MGLS algorithm are given below: Adopt the parameter sizes of MGLS algorithm; such as population size, maximum iteration number, the mutation and crossover rate are determined by full factorial experimental design, Generate initial population consisting of random individual, including one individual generated by NEH algorithm; evaluate fitness of each individual in the population."

"Proceed to Step 4, Select parents by using tournament selection method, Generate new offspring with crossover operator and apply mutation procedure to the generated offspring, Perform local search using the best known solution, If the stopping criteria do not hold, go to Step 2; otherwise, stop the algorithm."

Computational Results

"These instances are divided into two types: type P and Q. The instances are combinations of n-job and k-stage (n = 10, 20, 50, 100; k = 2, 5, 8), and each combination contains 10 problems."

"The performance of the proposed MGLS algorithm is dependent on the selected parameter sets."

"A full factorial experimental design (FFED) is carried out to determine the best parameter set for each instance, solving a predetermined problem with each possible parameter combination."

"For each problem category, an instance is chosen out of 10 instances solved by using every possible parameter combination."

"The proposed MGLS algorithm found better solution than the others metaheuristic methods for HFSMT scheduling problems from the literature."

"The proposed MGLS algorithm is combined the global and local search methods."

"A FFED was carried out to determine the best parameter set for each instance, solving a predetermined problem with each possible parameter combination."

Conclusion and Future Research

"Hybrid flow shop with multiprocessor task problems, i.e., $Fk(Pm_1,\ldots,Pm_k)|size_{ij}|C_{max}$, is solved by proposed memetic global and local search algorithm."

"The performance of the proposed MGLS algorithm is dependent on the parameter sets such as crossover and mutation methods and rate, population and iteration size and parent selection rate."

"The Oğuz [57] benchmark HFSMT instances are solved by the proposed new MGLS algorithm with the best parameter sets."

"Besides, MGLS algorithm improved the best known solutions so far for 48 instances out of 240."

"The experiment results showed that the MGLS algorithm is a better approach to $Fk(Pm_1,\ldots,Pm_k)|size_{ij}|C_{max}$ than its predecessors in the literature."

Solving Technician Routing and Scheduling Problem Using Improved Particle Swarm Optimization [39]

This is a machine-generated summary of:

Pekel, Engin: Solving technician routing and scheduling problem using improved particle swarm optimization [39].

Published in: Soft Computing (2020).

Link to original: https://doi.org/10.1007/s00500-020-05333-5.

If you want to cite the papers, please refer to the original.

For technical reasons we could not place the page where the original quote is coming from.

Abstract-Summary "An improved particle swarm optimization (IPSO) algorithm is proposed to solve the technician routing and scheduling problem (TRSP)."

"The TRSP consists of the assignment of technicians into teams, the assignment of teams to tasks, the construction of routes, and the selection of the day on which a service is provided by considering the proficiency level of workers and the proficiency requirement of the task."

"The IPSO algorithm includes a particle swarm optimization (PSO) algorithm and one neighborhood operator."

"The computational results show that IPSO provides better solutions considering the branch-and-cut algorithm within reasonable computing time."

Introduction

"Several technicians must provide a specified proficiency to carry out various tasks at minimum cost while meeting resource constraints in the technician routing and scheduling problem (TRSP)."

"It also includes additional decisions that require to be considered, e.g., the assignment of technicians into teams, the assignment of teams to tasks, the construction of routes, and the selection of the day on which a service is provided [40]."

"The complexity of routing is an important part of many technician scheduling problems."

"Their paper aims to route technicians to execute tasks at various client locations to maximize the total profit associated with served customers, minus the operations costs."

"The authors apply the standard PSO and modified particle swarm optimization (MPSO), based on an intelligent optimization method, in their paper to solve the routing problem of a real case."

Mathematical Model of TRSP

"Each route starts and ends at the central depot, including a stream of tasks performed within a working day by a team."

"Team structures within the working day are not permitted."

"Various team configurations are enabled on different days."

"A technician works for a maximum of one team a day."

"Each day must have at least one scheduled team."

Methodology

"The neighborhood operator helps PSO to improve the quality of the solution by extending the search area."

"The PSO algorithm triggers the position of each particle to optimize the specified problem."

"One of them is the best local value that has been achieved in a given particle swarm so far, and the solution provides this value is stored as the best local solution."

"PSO algorithm meets the local and global best solutions from the initial solutions."

"The implementation of the PSO algorithm alone improves the initial solutions considerably but is insufficient to achieve better results."

"De and others [58] applied some neighborhood algorithms to provide local search before applying PSO algorithm."

"The IPSO algorithm proposed in this paper aimed to avoid the local optima trap by using neighborhood operator after the PSO algorithm."

Numerical Results

"The application of the IPSO algorithm to 29 test sets yields the best results in 24 test sets, while the branch-and-cut algorithm offers the best solution in 11 test sets."

"The IPSO algorithm provides a mean gap of 0.41 as a result of the application to 29 test sets, while the branch-and-cut algorithm offers an average gap of 2.73."

"It is advantageous to use the IPSO algorithm in TRSP problems considering both the average gap and the best number of solutions obtained although the IPSO algorithm takes a slightly higher time than the branch-and-cut algorithm."

"Both the IPSO algorithm and branch-and-cut algorithm are run 5 times."

"The IPSO algorithm gives the worst solution in the RC-105 sub-dataset with a 6.01% Gap."

"The branch-and-cut algorithm gives the worst solution in the RC-108 sub-data set with a 10.91% Gap."

Conclusion

"This paper proposed an improved PSO algorithm, called IPSO, to solve TRSP."

"Applying the IPSO algorithm to 29 test sets provides the best results in 24 test sets, while the branch-and-cut algorithm provides the best solution in 11 test sets."

"Of the application to 29 sample sets, the IPSO algorithm provides an average gap of 0.41, while the branch-and-cut algorithm provides an average gap of 2.73."

"The IPSO algorithm and the branch-and-cut algorithm solve 29 test sets in an average of 126 and 120 min, respectively."

"The computational results show that IPSO provides better solutions considering the branch-and-cut algorithm within reasonable computing time."

Multi-Objective Evolutionary Algorithm for Vehicle Routing Problem with Time Window Under Uncertainty [59]

This is a machine-generated summary of:

Tan, Fei; Chai, Zheng-yi; Li, Ya-lun: Multi-objective evolutionary algorithm for vehicle routing problem with time window under uncertainty [59].

Published in: Evolutionary Intelligence (2021).

Link to original: https://doi.org/10.1007/s12065-021-00672-0.

If you want to cite the papers, please refer to the original.

For technical reasons we could not place the page where the original quote is coming from.

Abstract-Summary "Vehicles route problems (VRP) are to arrange the optimal routes under the various requirements, and it is becoming significant in the logistics industry as electric commerce is rising."

"We formulate the robust multi-objective VRPTW (RMOVRPTW) model and propose a robust optimization algorithm based on MOEA/D (R-MOEAD-VRP) for simultaneously optimizing the total distance and the number of vehicles required for transport."

"For the new routes by reproduction, we use Monte-Carlo tests to check the feasibility of the routes after adding uncertainty."

"For the feasible routes, we calculate the solution robustness values based on the defined method."

"We consider both optimality and robustness to form a set of highly robust and relatively optimal solutions."

"Experimental results show that our proposed algorithm can bring more robust and non-dominated solutions under uncertainty and can achieve good performance."

Introduction

"The vehicle routing problem with time window constraints (VRPTW) [60] has become a research hotspot."

"Researchers began to focus on the multi-objective VRPTW (MOVRPTW) [61, 62] to minimize both the total travel distance and the number of vehicles simultaneously."

"For VRPTW under uncertainty, at first, many researches [63, 64] paid attention to single-objective VRPTW (SOVRPTW) under uncertainty to study the robust optimal solutions."

"There are very few robust optimization algorithms for multi-objective VRPTW under uncertainty."

"There is only one research work that Duan and others [65] proposed a PSO-based robust multi-objective optimization approach to solve the VRP with time windows under uncertainty."

"Current evolutionary algorithms for MOVRPTW rarely take into account the robustness in optimization processes, and therefore it often cannot handle the VRPTW of uncertainty."

"We propose a robust MOEA/D-based multi-objective evolutionary algorithm for VRPTW under uncertainty(R-MOEA/D-VRP)."

Background

"It is essential to exploit a robust multi-objective optimization algorithm for the MOVRPTW under uncertainty that optimizes both the total travel distance and the number of vehicles."

"In this paper, we study the VRPTW under uncertainty based on the robust multi-objective evolutionary algorithm."

"Discussion, either no studies on VRPTW with uncertainty, or robust evolutionary algorithms that can adapt to variable environment."

"The VRPTW under uncertainty was not investigated with a robust evolutionary algorithm."

"The motivation of this literature is to develop a robust MOEA/D algorithm to deal with the robustness and the optimal for the MOVRPTW under certainty."

"We propose a robust MOEA/D-based algorithm for VRPTW under uncertainty(R-MOEA/D-VRP)."

"Aimed at the RMOVRPTW model, we propose a robust MOEA/D-based algorithm for VRPTW under uncertainty (R-MOEAD-VRP) in order to simultaneously optimize the total travel distance and the number of vehicles required for transport."

The Model of RMOVRPTW Under Uncertainty

"For the model of RMOVRPTW under uncertainty, we adopt the general MOVRPTW stated by [60], adding the new form of uncertainty defined by Duan et al. [65]."

"For satisfying different requirements under uncertainty conditions over the total distance and the total number of vehicles, we present a multi-objective optimization algorithm to optimize the conflicting objective simultaneously."

The Proposed Algorithm

"In our proposed robust multi-objective optimization algorithm R-MOEA/D-VRP, at the beginning, the individual is encoded and the length of a individual is equal to the number of customers being served, with each location corresponding to one customer."

"Because we study the VRP with constraints, in this section we describe the method that transform the individual into a set of feasible routes according to the customer's requirement, vehicle capacity, etc. In previous studies [66, 67], when the constraints are exceeded, the chromosome is cut into several segments becoming multiple sub-paths, which will increase the number of dispatched vehicles and the cost of delivery companies."

"When constraints are violated, it doesn't generate new route but skip the current customer and continues to traverse through."

"Key techniques, the pseudo code of our proposed R-MOEAD-VRP algorithm is shown in algorithm 3."

Performance Evaluation

"We illustrate the superiority performance of our proposed algorithm R-MOEA/D-VRP on robust multi-objective VRPTW problems by comparing with several related algorithms in this section."

"The results show that R-MOEA/D-VRP yields more robust solutions at all conditions than the other algorithms."

"Speaking, our proposed algorithm R-MOEA/D-VRP can produce more robust solutions as the maximum disturbance increases, indicating that our proposed algorithm is more robust in response to uncertainty. (2) The non-dominated robust solutions We draw the non-dominated solutions produced by all algorithms under different disturbance degree for problem 1 and problem 3."

"The robust non-dominated solutions from R-MOEA/D-VRP can dominate other algorithms when disturbance is 2 and 3."

"The robust non-dominated solutions from R-MOEA/D-VRP can dominate the others when maximum disturance are 5 and 10."

"Our proposed algorithm R-MOEA/D-VRP can produce more robust non-dominated solutions."

[Section 6]

"We formulate the RMOVRPTW model and propose a robust optimization algorithm based on MOEA/D (R-MOEAD-VRP) for simultaneously optimizing the total transport distance and the number of vehicles under uncertainty."

"For the feasible routes, and then we calculate the solution robustness values based on the defined method."

"For the number of robust solutions, our algorithm R-MOEA/D-VRP generated more solutions than the other algorithms as the disturbance is increasing, indicating that our algorithm is more adapted to the uncertainty condition."

"The results show our algorithm can dominate the others and can get more solution at the same time under uncertainty."

"We will continue to improve the performance of R-MOEA/D-VRP by adding the additional measures."

"We also will consider the other disturbance resulting the R-MOEA/D-VRP more robust."

An Integrated Machine Scheduling and Personnel Allocation Problem for Large-Scale Industrial Facilities Using a Rolling Horizon Framework [68]

This is a machine-generated summary of:

Santos, Fernando; Fukasawa, Ricardo; Ricardez-Sandoval, Luis: An integrated machine scheduling and personnel allocation problem for large-scale industrial facilities using a rolling horizon framework [68].

Published in: Optimization and Engineering (2020).

Link to original: https://doi.org/10.1007/s11081-020-09542-7.

If you want to cite the papers, please refer to the original.

For technical reasons we could not place the page where the original quote is coming from.

Abstract-Summary "We present an optimization problem where machine scheduling and personnel allocation decisions are solved simultaneously."

"The machine scheduling consists of solving a variant of the job shop problem where jobs are allocated in batches and multitasking is allowed."

"The personnel allocation problem searches for the optimal allocation of human resources to support the facility operation."

"The results show that, overall improvements of up to 11.6% are achieved compared to the case where no personnel allocation decisions are considered within the scheduling problem."

Introduction

"We consider in this study a variant of the job shop problem where jobs comprise one or more units of material that can be allocated in batches to the machines that are capable of multitasking, i.e., machines are able to process multiple units of material from multiple jobs simultaneously."

"A stochastic version of the job shop problem was studied by Jamili [69], where the objective is to find the optimal scheduling using different robust approaches."

"In that study, each employee was modelled as a resource required by the machines and the classic flow shop problem was solved with this additional constraint."

"The approach proposed by Artigues et al. [70] combines methods based on integer linear programming and constraint programming for the integrated problem of job shop and personnel allocation that considered for the first time regulations, e.g. employees are not allowed to work for more than 8 h within a 24 h interval."

Mathematical Formulation

"Methods based on continuous time usually present higher complexity and are only able to evaluate optimal solutions for small-sized problems."

"In this representation, events are only allowed to happen at discrete time points, which leads to a trade-off between solution quality and computational cost: more time points lead to better quality solutions, at the expense of becoming more time consuming to solve the model."

"Lagzi et al. [71] compared formulations for the same variant of the job shop problem considered in this study using both continuous and discrete time representations."

"They concluded that the non-uniform discrete time representation provides the best model to solve medium/large size instances for the problem under consideration."

"We propose an extension to that formulation that integrates personnel allocation to the scheduling decisions using the non-uniform discrete time representation, which is described next."

Rolling Horizon Implementation

"If all information related to the scheduling problem over H is known in advance, the rolling horizon can then be applied as a decomposition technique [72–74]."

"Another common implementation of the rolling horizon arises in dynamic scheduling environments, where changes are allowed to happen over horizon H, e.g. arrival of new jobs or machine breakdowns [75, 76]."

"Our rolling horizon implementation considers a dynamic scheduling environment, where jobs arriving during horizon H may happen."

"We also use in our implementation the concept of idle horizons, which consists of horizons where no scheduling is performed in the facility but machines are still operating, e.g. machines running for rest of the day."

Computational Study

"The same Scheduling + Personnel model achieves only 12% throughput improvement for instances of group B, since these instances represent the facility operating under heavier loads and require all employees to be used, which averages down the gains obtained using employees with higher skill sets."

"The integrated approach achieves an improvement of 13.0%, 10.4% and 8.5% respectively for on the 10th, 20th and 30th day of the horizon compared to the Scheduling-Only model for instances of set C. The difference decreases to 12.2%, 6.6%, 4.8% and 7.8%, 4.0%, 3.1% for instances of sets B and A, respectively."

"The top three plots show the number of jobs processed using each algorithm after 30 days horizon for instances of sets A, B and C, respectively, while the bottom plots show the average computational time spent by each algorithm along the 30 days horizon for the same instances."

Concluding Remarks

"This study presents a solution approach for the integrated personnel allocation and machine scheduling problem in large-scale industrial facilities."

"The results show that, for a 30 days horizon, the integrated approach can improve the overall results up to 11.6% compared to the classic scheduling algorithm."

"We also implemented a sequential algorithm that first solves the personnel allocation and uses the solutions as input to the machine scheduling problem."

"As further research directions, we intend to conduct an economical analysis of the personnel allocation in the scheduling problem to understand how the resources such as employees and machines can be used to improve the facility overall profit."

Two-Stage Stochastic Days-Off Scheduling of Multi-Skilled Analysts with Training Options [77]

This is a machine-generated summary of:

Altner, Douglas S.; Mason, Erica K.; Servi, Les D.: Two-stage stochastic days-off scheduling of multi-skilled analysts with training options [77].

Published in: Journal of Combinatorial Optimization (2018).

Link to original: https://doi.org/10.1007/s10878-018-0368-5.

If you want to cite the papers, please refer to the original.

For technical reasons we could not place the page where the original quote is coming from.

Abstract-Summary "Motivated by a cybersecurity application, this paper studies a two-stage, stochastic days-off scheduling problem with (1) many types of jobs that require specialized training, (2) many multi-skilled analysts, (3) the ability to shape analyst skill sets through training decisions, and (4) a large number of possible future demand scenarios."

"We provide an integer linear program for this problem and show it can be solved with a direct feed into Gurobi with as many as 50 employees, 6 job types, and 50 demand scenarios per day without any decomposition techniques."

Introduction

"To help support these types of decisions, we study a stochastic days-off scheduling problem in an environment in which (1) there is a wide range of possible future demand scenarios, (2) there are several distinct job types, (3) employees are multi-skilled—meaning they are capable of performing more than one type of job depending on their training, and (4) planners can train individual employees to increase the set of jobs they can complete."

"We offer a two-stage stochastic integer programming model for a days-off scheduling problem with a unique combination of challenges: wide range of possible demand scenarios, many job types, many multi-skilled analysts, and the ability to increase individual employee skill sets with training; and we show this model can be solved without any decomposition techniques for reasonably sized instances."

Literature Review

"Several others have researched the challenges of deciding how to optimally multi-skill or cross-train a workforce—meaning, increasing the number of job types that specific employees can perform via training decisions or job routing policies—while also considering scheduling constraints and challenges (e.g., [78–81])."

"In the intersection of multi-skilled scheduling and optimization under uncertainty, Bodur and Luedtke study a two-stage multi-skilled employee scheduling problem with test cases with up to six job types and six employee types, each of which allow employees to complete two different types of jobs."

"Our work builds upon this great body of literature by tackling a unique combination of challenges: solving a two-stage stochastic days-off scheduling model with many job types, multi-skilled employees, and the ability to increase the job types that any individual employee can complete through first-stage training decisions."

Problem Statement

"We offer a model for determining scheduling and training decisions at a cybersecurity operations center (CSOC) with many distinct alert types (i.e., job types), many multi-skilled analysts, the ability to increase employee skills with training, and a large number of possible future demand scenarios."

"The purpose of the model is to help a planner with a given multi-skilled workforce—in this case, a set of individual employees each of whom is trained to complete one or more job types—decide (1) which days each employee should work during a future pay period, which we assume is two weeks or fourteen days; and (2) whether any employees should undergo training to increase the set of job types they can complete."

"For the employee capacity constraints, we assume each analyst can work up to 8 h during the days they are scheduled to work, and that they can only work on alert types they are trained for."

Solution Approach

"Gurobi 6.0 cannot find a solution within 40% of the best upper bound for problem instances with 50 employees, 6 alert types and 500 demand scenarios per day, even if we allow it to run overnight on an 8-processor virtual machine."

"For instances with 50 employees, 6 alert types, and 1500 demand scenarios per day, Gurobi struggles to produce an integer feasible solution even if it runs overnight on the same virtual machine."

"Our matheuristic iteratively solves smaller integer programs induced by freeing the decision variables for a small number of employees, and fixing all other decision variables to the values they assume in the current best known solution."

"The induced mixed integer program amounts to optimally revising the training and scheduling decisions for the subset of employees, while keeping those for all other employees fixed, and simultaneously determining the optimal assignment of jobs to all employees under each scenario."

Computational Experimentation

"The second subsection presents the results on smaller test cases, showing near optimal results were found in 1 min to 4 h. The third subsection describes how our local search method consistently produces solutions within 9% of the best bound on instances that are generally too large for a direct feed into a commercial IP solver."

"Each class of instances specifies different values to a few key dimensions to the test cases: the number of employees, the number of demand scenarios per day, and the number of alert types."

"Within a class, the instances differ by the composition of the employee productivity levels and initial skill sets as well as the volume and mix of alert types specified by each demand scenario for each day."

"We also test our local search heuristic on the large-B instances, which do not limit the number of skills that employees can be trained in."

Conclusions and Future Work

"We formulated this problem as a two-stage stochastic integer program and demonstrated this formulation can solve instances with as many as 50 employees, 6 alert types, and 50 demand scenarios per day by directly feeding the model into a commercial integer programming solver, without use of decomposition methods."

"Our model presently makes days-off scheduling decisions for a typical two-week pay period in the first stage, and considers how jobs might be assigned to employees for any particular day in the second stage."

"More work can be done to model employee productivity."

"Our model assumes there is no loss in productivity if employees are assigned three different types of alerts versus one type of alert."

"Perhaps that loss of efficiency from context-changing could be offset by some subtle productivity gains from allowing employees to switch between tasks when they prefer to work on a different type of job."

Multi-Objective Hybrid Flow Shop Scheduling with Variable Discrete Production Speed Levels and Time-of-Use Energy Prices [82]

This is a machine-generated summary of:

Schulz, Sven; Buscher, Udo; Shen, Liji: Multi-objective hybrid flow shop scheduling with variable discrete production speed levels and time-of-use energy prices [82].

Published in: Journal of Business Economics (2020).

Link to original: https://doi.org/10.1007/s11573-020-00971-5.

If you want to cite the papers, please refer to the original.

For technical reasons we could not place the page where the original quote is coming from.

Abstract-Summary "Energy costs play an important role in industrial production and are closely related to environmental concerns."

"A hybrid flow shop scheduling problem with variable discrete production speed levels is considered with the aim of minimizing both energy costs and total tardiness."

"Although lower speeds can reduce energy consumption, they also increase processing times, which counteract the objective of punctual delivery."

"The influence of variable discrete production speed levels on energy costs, energy consumption and punctual delivery as well as the interdependencies between these objectives are analysed in a numerical case study."

Introduction

"Energy costs are now being taken into account in many approaches of production planning and control and thus also in operative planning in the form of energy efficient scheduling (EES)."

"A general overview about different approaches to consider energy consumption in scheduling is given by Biel and Glock [83] or Gahm et al. [84]."

"Among the EES approaches, variable production speeds are probably one of the most promising methods in order to significantly reduce energy consumption [85]."

"High savings potential exists, for example, in pumps which have high energy consumption in injection moulding plants or for water supply in paper mills."

"Besides lower energy consumption a reduction in energy costs can also be achieved through, inter alia, the exploitation of time-dependent energy prices, which are also considered in this work."

"For that reason, this paper examines total tardiness and time depending energy costs as two separate objective functions using the ideas of multi-criteria optimization."

Related Literature

"Liu et al. [86] analyse a job shop problem with three different processing levels (idle, runtime, cutting), whereby the energy consumption is a deterministic value and can only be reduced by minimizing the non-processing time."

"In a later work Lei et al. [87] present a novel teaching-learning algorithm to minimize total energy consumption and total tardiness in a hybrid flow shop scheduling problem."

"Artigues et al. [88] minimize energy and power overrun costs in a parallel machine problem, and use maximum tardiness as decision criterion for the scheduling in the first step of a two-phase solution approach."

"The same objectives are considered by Wang et al. [89] in a batch scheduling problem under energy consumption uncertainties."

Problem Description

"Since we assume discrete integer production times, reducing the speed by one step leads to one additional processing time unit."

"If only energy costs were minimized, jobs would be processed at very low speeds and mainly times of low electricity prices would be used for production."

"In order to obtain a time-efficient schedule despite the energy cost optimization, we minimize the total tardiness of all jobs as a second objective."

"One alternative is to calculate all possible energy costs depending on job, stage, speed and starting time in advance."

"The eps-constraint method is a suitable approach for calculating all pareto optimal solutions (pareto front) or at least some NDSs by means of mathematical modelling."

"Since we limit the computing time of the models in the following, an optimal solution is not always found and the approaches can lead to different results."

Numerical Case Study

"For the considered problem minimum TT is 36 h with energy costs of 4360 €."

"The more TT increases the higher is the number of possible speed reductions and time shifts for TOU prices."

"Energy demand even increases and the total costs can be reduced by shifting the consumption to times of lower TOU prices."

"Since each problem instance has not only one optimal solution but different pareto optimal solutions, we first have to define a performance criterion to compare both approaches."

"For instances with 6 jobs, the average time of 0.39 h is reduced by 80% compared to the first model with 1.95 h. Interestingly the due date range R has no general influence on the used performance indicators and computation time."

Conclusion

"Two multi-objective MIP formulations are given for the HFS scheduling problem considering variable production speeds to reduce energy consumption at the expense of longer processing times."

"This is the first time that energy costs and tardiness are considered as objectives in a hybrid flow shop problem with the described properties."

"It can be stated, that energy costs can be reduced by shifting loads to times of lower energy prices without reducing consumption."

"Based on the second (sequence-dependent) model formulation, it can be shown that by calculating all energy cost scenarios in a pre-process, computation time can be enormously reduced."

"Regardless of objectives or which model is used, very good or even optimal solutions are found quickly by solver for the considered problems."

"It is appealing for future work to develop heuristic solutions in order to solve larger problems in reasonable computing time."

Multi-Phase Projects Selection and Scheduling Problem: A Multi-objective Optimization Approach [90]

This is a machine-generated summary of:

Alinezhad, Reza; Ansari, Ramin; Mahdikhani, Mahdi; Banihashemi, Sayyid Ali: Multi-phase Projects Selection and Scheduling Problem: A Multi-objective Optimization Approach [90].

Published in: Iranian Journal of Science and Technology, Transactions of Civil Engineering (2021).

Link to original: https://doi.org/10.1007/s40996-021-00721-9.

If you want to cite the papers, please refer to the original.

For technical reasons we could not place the page where the original quote is coming from.

Abstract-Summary "This research model the hybrid problem of selecting the optimal project portfolio and scheduling activities considering resource and budget constraints."

"A two-objective mixed-integer nonlinear programming model is presented that enables the selection and scheduling of phased projects."

"The proposed model has solved using two methods; augmented epsilon constraint and a meta-heuristic genetic algorithm based on non-dominated sorting."

"The results revealed that the augmented epsilon constraint method provided more favorable results and better efficiency in small-scale instances."

"By comparing the results of two methods of augmented epsilon-constraint and genetic algorithm in an example with five projects, six phases, and five time horizons (N5M6T5) showed that the NPF and NPS criteria in the augmented Epsilon constraint method are less than them in the genetic algorithm method and the DM and S criteria are more in the genetic algorithm method than them in augmented Epsilon constraint method."

"The most important practical application of this research is for project-based companies that seek to choose and schedule several projects."

Introduction

"Several investment projects are usually considered by an organization, so the management must decide which of the projects must be selected and for what durations each of them should be carried out."

"In the context of optimal management of the project portfolio, managers face with two main challenges: The first challenge is the decision on the selection of a set of projects among all available projects, and the second one is the scheduling according to which the selected projects need to be carried out [91]."

"The selection and scheduling aspects of the project are examined as two completely independent issues."

"This study developed a hybrid model for simultaneous modeling of two complex project selection and scheduling problems."

"After choosing the project and implementing the activities of the selected projects, this modeling can serve as a solution to existing problems."

Literature Review

"Gutjahr and others [92] developed a multi-objective optimization model for project portfolio selection that considered employee competence and their evaluation."

"The problem was stated as a bi-objective mixed-integer programming model containing two conflicting objectives to minimize the total cost of the multi-skilled workforce and obtain a sustainable schedule with minimum deviation from the projects' priorities."

"Chen and Askin [93] developed an implicit enumeration procedure to solve the two problems of project selection and task scheduling."

"Scheduling and staff assignment were optimized as sub-problems for a candidate set of selected projects."

"They used a mixed-integer non-linear programming (MINLP) formulation for the overall problem and then proposed heuristic solution techniques composed of a greedy heuristic for scheduling and staff assignment as well as alternative metaheuristics for project selection."

"Carazo et al. [94] solved a multi-objective binary programming model using a metaheuristic procedure based on a scatter search named SS-PPS (scatter search for project portfolio selection)."

Research Gap

"Against this background, this paper aims to address the project selection, scheduling, and resource allocation problem, for the first time, using an integrated process for projects composed of separate phases."

"Due to considering resource constraints in the problem, the proposed model exploits two objective functions of the maximum net present value (NPV) and minimum variance of resources."

"Based on the research questions and the identified research gap, this study strives to investigate and present the following objectives: Developing a mathematical model for the project selection, scheduling, and resource allocation problem using an integrated process Taking into account separate phases for available projects, with each one having separate costs and revenues Considering project resource leveling by maximizing the net present value Solving the model by mathematical and metaheuristic techniques for small-, medium-, or large-sized problems."

Mathematical Model and Methodology

"Constraint 9 indicates that the relationship among decision variables b, x, and y, which implies the value of variable x is equal to one from the execution time of phase j of project i to the completion time of that phase and will be equal to zero in other periods for that phase."

"Since there is an NP-hard problem and the number of decision variables increases exponentially when the problem becomes large-sized, a metaheuristic technique called the Non-Dominated Sorting Genetic Algorithm-II (NSGA-II) is utilized so that near-optimal solutions are provided within an acceptable time period."

"In this constraint, s is the number of equal parts divided by range (Z_2^{min}, Z_2^{max}), CSN is a parameter that is firstly equal to 1, and then one value is added to it during every problem-solving time period until it reaches value s. Now, upon determining the function of the mentioned method, the problem is solved in s iterations, and the outputs are stored separately in each solution."

Results and Outcomes

"To validate the model presented and its outputs, three different small-, medium-, and large-sized problems are solved, with the outputs of the two proposed solutions compared for each of them."

"In this medium-sized problem, there are 10 projects with a maximum of five phases for each project."

"Environmental parameters are the same as those in the previous problem, with the selection and scheduling of these projects supposed to be done with the initial budget of B0 = 2000 in time horizon = 10."

"For large-sized problems, in which more than 50 projects exist which are investigated on long time horizons, the only optimum solution method is the metaheuristic technique which delivers the Pareto front of the problem within an acceptable time."

"The results achieved from the two proposed methods for diverse small-, medium-, or large-sized problems are presented, which are evaluated by different metrics."

Results and Discussion

"Five evaluation metrics are presented in the following sections: The first metric is the amount of the solution time any method takes to solve the problem, which is one of the essential evaluation metrics for solution methods, especially in large-sized problems."

"The higher the number of Pareto solutions is, the more desirable the solution method will be."

"The lower the non-uniformity of the Pareto front is, the greater the uniformity will be, thereby leading to the more desirability of the solution method."

"The two metrics of time processing (TP) and spacing (S) have more desirable values in the non-dominated sorting genetic algorithm-II (NSGA-II) method than in the eps-constraint method."

"The value for the number of Pareto solutions (NPS) is roughly equal in the two methods, which indicates that the augmented epsilon constraint method has been more efficient for small- and medium-sized problems."

Conclusion

"A small example and an example with a larger dimension were solved using the two methods of augmented epsilon constraint and genetic algorithm based on non-dominant sorting to solve and validate the proposed model."

"There are some ideas suggested in the following part that could result in the development and improvement of the model by further investigations as follows: Adding new hypotheses and limitations to the model, such as the constraints of consumption resources by taking into consideration the interruptions present in carrying out projects' phases, etc.; developing and improving metaheuristic methods and comparing them with each other; considering and eliminating uncertainties in the model parameters; Employing other objective functions, such as quality or the satisfaction level, etc. This research examined the topic of selecting and scheduling projects in the optimal portfolio in consideration of budget and resource constraints."

Research Limitations/Lmplications

"This study used two objective functions of net present value and resource variance to select and schedule projects."

"Additional factors such as the level of risk associated with each project and the environmental effects resulting from its implementation affect project selection."

"The use of other problem-solving approaches and metaheuristic methods to increase the dimensions of the problem serves as a basis for further research in the future."

"The use of fuzzy and probable data can increase the accuracy of the results, whereas, in this study, definitive data were used."

Semi-Supervised Human Resource Scheduling Based on Deep Presentation in the Cloud [95]

This is a machine-generated summary of:

Lin, Yuanmo; Wang, Xinmei; Xu, Rongbin: Semi-supervised human resource scheduling based on deep presentation in the cloud [95].

Published in: EURASIP Journal on Wireless Communications and Networking (2020).

Link to original: https://doi.org/10.1186/s13638-020-01677-6.

If you want to cite the papers, please refer to the original.

For technical reasons we could not place the page where the original quote is coming from.

Abstract-Summary "To improve the effectiveness of human resource management and the degree of matching between jobs and staffs, we propose a novel auto-encoder neural network-based method in the cloud environment, which is a semi-automatic manner in the business process of human resource allocation."

"Our proposed method enables the modeling of semi-automatic human resource allocation process and can be used to facilitate optimized human resource allocation."

"Experimental results show that our method can bring significant improvements to personnel position matching and effectively enhance the efficiency of human resource allocation based on the cloud environment."

Introduction

"Accurate prediction of the required human resource for modern enterprises can help the enterprises find out whether the personnel positions are insufficient or too many, which will provide accurate guidance for rational allocation of human resources that has far-reaching significance for the sustainable development of enterprises [96]."

"By the convenience of using cloud computing resources, the proper allocation of enterprise human resources will contribute to the formation of meaningful economic activities and, thus, greatly improve the productivity of human resources [97]."

"The rational allocation of human resources is influenced by many factors such as scale, economic strength, culture, and the external environment of the enterprise."

"We adopt stack auto-encoder neural network to predict the required human resource and fully consider the characteristics of similarity between different staff, which combines the prior information of position fitness constraint to effectively realize the reasonable allocation of semi-supervised human resources."

Related Work

"This method collected the data from applicants through web interface and matched suitable jobs through a web-based human resource management system model by the network information server."

"Li et al. [98] analyzed the matching relationship between human resource management practice and organizational situation based on BP neural network model, which established a prediction model and received a good prediction effect."

"Most of them do not take into account the nonlinear influencing factors in the human resource allocation process as well as the specific relationship between personnel that results static distribution of cloud resources."

"We will propose a new deep nonlinear reconstruction model of human resources to solve the deficiencies in the traditional human resource allocation process based on cloud environment, which can construct the relationship between features and prediction results."

Human Resource Scheduling Based on Deep Representation in Cloud Environment

"In order to better predict unallocated employees for the most suitable job, we will use different attributes to construct the human resource network based on personnel relationship module and put forward a novel nonlinear reconstruction model about human resource scheduling in the cloud."

"Auto-encoder is mainly to study in the condition of a parameter $\theta = \{W_H,d_H,W_O,d_O\}$, which will reconstruct the original data to minimize the raw data M and the reconstruct the deviation between data O. So that a low dimensional nonlinear representation of H can be obtained."

"For a deep auto-encoder network, we train the first auto-encoder by reconstructing raw data in human relation matrix B and get a hidden layer representation $H^1 \in R(N \times t_1)$."

"That both person i and person j are suitable for the same post, there must be some similar attributes between persons i and j, which can be regarded as a priori information for human resource scheduling."

Experiment Results and Analysis

"The dataset of a given automobile manufacturing enterprise is used to test the performance of the human resource scheduling method proposed by this paper in the cloud environment this paper."

"Deep nonlinear reconstruction model for human resource is applied to train the first 9 subsets and the other subset is used as test data."

"As the deep nonlinear reconstruction model for human resource allocation, the stacked auto-encoder network consists of four auto-encoders."

"In order to evaluate the impact of post fitness constraints on our algorithm, 2000 experimental data are stacked and trained from four hidden layers in the auto-encoder."

"The deep nonlinear reconstruction model of human resource allocation based on the deep neural network proposed by this paper can achieve better degrees of job matching in cloud environment, which effectively and reasonably allocates human resources in automobile manufacturing enterprises and achieves the optimization of human resource allocation."

Conclusion and Future Work

"To effectively assist enterprises to achieve the optimized allocation of human resources, this paper proposes a deep nonlinear reconstruction model for human resource allocation in cloud environment."

"Our model can achieve semi-supervised post-personnel matching in human resource management."

"Experimental results show that the model proposed in this paper can achieve remarkable results in personnel position matching, which effectively improves the efficiency of human resource allocation."

"Our future work will try to apply the method proposed in this paper to the study of personnel arrangement process in projects for cloud resource allocation, considering that the working ability of personnel will directly affect the efficiency of project completion."

Multi-Phase Projects Selection and Scheduling Problem: A Multi-objective Optimization Approach [90]

This is a machine-generated summary of:

Alinezhad, Reza; Ansari, Ramin; Mahdikhani, Mahdi; Banihashemi, Sayyid Ali: Multi-phase Projects Selection and Scheduling Problem: A Multi-objective Optimization Approach [90].

Published in: Iranian Journal of Science and Technology, Transactions of Civil Engineering (2021).

Link to original: https://doi.org/10.1007/s40996-021-00721-9.

If you want to cite the papers, please refer to the original.

For technical reasons we could not place the page where the original quote is coming from.

Abstract-Summary "This research model the hybrid problem of selecting the optimal project portfolio and scheduling activities considering resource and budget constraints."

"A two-objective mixed-integer nonlinear programming model is presented that enables the selection and scheduling of phased projects."

"The proposed model has solved using two methods; augmented epsilon constraint and a meta-heuristic genetic algorithm based on non-dominated sorting."

"The results revealed that the augmented epsilon constraint method provided more favorable results and better efficiency in small-scale instances."

"By comparing the results of two methods of augmented epsilon-constraint and genetic algorithm in an example with five projects, six phases, and five time horizons (N5M6T5) showed that the NPF and NPS criteria in the augmented Epsilon constraint method are less than them in the genetic algorithm method and the DM and S criteria are more in the genetic algorithm method than them in augmented Epsilon constraint method."

"The most important practical application of this research is for project-based companies that seek to choose and schedule several projects."

Introduction

"Several investment projects are usually considered by an organization, so the management must decide which of the projects must be selected and for what durations each of them should be carried out."

"In the context of optimal management of the project portfolio, managers face with two main challenges: The first challenge is the decision on the selection of a set of projects among all available projects, and the second one is the scheduling according to which the selected projects need to be carried out [91]."

"The selection and scheduling aspects of the project are examined as two completely independent issues."

"This study developed a hybrid model for simultaneous modeling of two complex project selection and scheduling problems."

"After choosing the project and implementing the activities of the selected projects, this modeling can serve as a solution to existing problems."

Literature Review

"Gutjahr et al. [92] developed a multi-objective optimization model for project portfolio selection that considered employee competence and their evaluation."

"The problem was stated as a bi-objective mixed-integer programming model containing two conflicting objectives to minimize the total cost of the multi-skilled workforce and obtain a sustainable schedule with minimum deviation from the projects' priorities."

"Chen and Askin [93] developed an implicit enumeration procedure to solve the two problems of project selection and task scheduling."

"Scheduling and staff assignment were optimized as sub-problems for a candidate set of selected projects."

"They used a mixed-integer non-linear programming (MINLP) formulation for the overall problem and then proposed heuristic solution techniques composed of a greedy heuristic for scheduling and staff assignment as well as alternative metaheuristics for project selection."

"Carazo et al. [94] solved a multi-objective binary programming model using a metaheuristic procedure based on a scatter search named SS-PPS (scatter search for project portfolio selection)."

Research Gap

"Against this background, this paper aims to address the project selection, scheduling, and resource allocation problem, for the first time, using an integrated process for projects composed of separate phases."

"Due to considering resource constraints in the problem, the proposed model exploits two objective functions of the maximum net present value (NPV) and minimum variance of resources."

"Based on the research questions and the identified research gap, this study strives to investigate and present the following objectives: Developing a mathematical model for the project selection, scheduling, and resource allocation problem using an integrated process Taking into account separate phases for available projects, with each one having separate costs and revenues Considering project resource leveling by maximizing the net present value Solving the model by mathematical and metaheuristic techniques for small-, medium-, or large-sized problems."

Mathematical Model and Methodology

"Constraint 9 indicates that the relationship among decision variables b, x, and y, which implies the value of variable x is equal to one from the execution time of phase j of project i to the completion time of that phase and will be equal to zero in other periods for that phase."

"Since there is an NP-hard problem and the number of decision variables increases exponentially when the problem becomes large-sized, a metaheuristic technique called the Non-Dominated Sorting Genetic Algorithm-II (NSGA-II) is utilized so that near-optimal solutions are provided within an acceptable time period."

"In this constraint, s is the number of equal parts divided by range (Z_2^{min}, Z_2^{max}), CSN is a parameter that is firstly equal to 1, and then one value is added to it during every problem-solving time period until it reaches value s. Now, upon determining the function of the mentioned method, the problem is solved in s iterations, and the outputs are stored separately in each solution."

Results and Outcomes

"To validate the model presented and its outputs, three different small-, medium-, and large-sized problems are solved, with the outputs of the two proposed solutions compared for each of them."

"In this medium-sized problem, there are 10 projects with a maximum of five phases for each project."

"Environmental parameters are the same as those in the previous problem, with the selection and scheduling of these projects supposed to be done with the initial budget of B0 = 2000 in time horizon = 10."

"For large-sized problems, in which more than 50 projects exist which are investigated on long time horizons, the only optimum solution method is the metaheuristic technique which delivers the Pareto front of the problem within an acceptable time."

"The results achieved from the two proposed methods for diverse small-, medium-, or large-sized problems are presented, which are evaluated by different metrics."

Results and Discussion

"Five evaluation metrics are presented in the following sections: The first metric is the amount of the solution time any method takes to solve the problem, which is one of the essential evaluation metrics for solution methods, especially in large-sized problems."

"The higher the number of Pareto solutions is, the more desirable the solution method will be."

"The lower the non-uniformity of the Pareto front is, the greater the uniformity will be, thereby leading to the more desirability of the solution method."

"The two metrics of time processing (TP) and spacing (S) have more desirable values in the non-dominated sorting genetic algorithm-II (NSGA-II) method than in the eps-constraint method."

"The value for the number of Pareto solutions (NPS) is roughly equal in the two methods, which indicates that the augmented epsilon constraint method has been more efficient for small- and medium-sized problems."

Conclusion

"A small example and an example with a larger dimension were solved using the two methods of augmented epsilon constraint and genetic algorithm based on non-dominant sorting to solve and validate the proposed model."

"There are some ideas suggested in the following part that could result in the development and improvement of the model by further investigations as follows: Adding new hypotheses and limitations to the model, such as the constraints of consumption resources by taking into consideration the interruptions present in carrying out projects' phases, etc. Developing and improving metaheuristic methods and comparing them with each other. Considering and eliminating uncertainties in the model parameters. Employing other objective functions, such as quality or the satisfaction level, etc. This research examined the topic of selecting and scheduling projects in the optimal portfolio in consideration of budget and resource constraints."

Research Limitations/Lmplications

"This study used two objective functions of net present value and resource variance to select and schedule projects."

"Additional factors such as the level of risk associated with each project and the environmental effects resulting from its implementation affect project selection."

"The use of other problem-solving approaches and metaheuristic methods to increase the dimensions of the problem serves as a basis for further research in the future."

"The use of fuzzy and probable data can increase the accuracy of the results, whereas, in this study, definitive data were used."

References

1. Long Z, Wen X, Lan M, Yang Y (2021) Nursing rescheduling problem with multiple rescheduling methods under uncertainty. Complex Intell Syst. https://doi.org/10.1007/s40747-021-00554-z
2. Maenhout B, Vanhoucke M (2013) An integrated nurse staffing and scheduling analysis for longer-term nursing staff allocation problems. Omega 41(2):485–499
3. Bard JF, Purnomo HW (2005) Short-term nurse scheduling in response to daily fluctuations in supply and demand. Health Care Manag Sci 8(4):315–324
4. He F, Chaussalet T, Qu R (2019) Controlling understaffing with conditional value-at-risk constraint for an integrated nurse scheduling problem under patient demand uncertainty. Oper Res Perspect 6:100119
5. König S, Reihn M, Abujamra FG, Novy A, Vogel-Heuser B (2022) Flexible scheduling of diagnostic tests in automotive manufacturing. Flex Serv Manuf J. https://doi.org/10.1007/s10696-021-09438-3
6. Shi Y, Reich D, Epelman M, Klampfl E, Cohn A (2017) An analytical approach to prototype vehicle test scheduling. Omega 67:168–176. https://doi.org/10.1016/j.omega.2016.05.003
7. Spieckermann S, Gutenschwager K, Voß S (2004) A sequential ordering problem in automotive paint shops. Int J Prod Res 42(9):1865–1878. https://doi.org/10.1080/00207540310001646821
8. Alouneh S, Abed S, Al Shayeji MH, Mesleh R (2019) A comprehensive study and analysis on sat-solvers: advances, usages and achievements. Artif Intell Rev 52(4):2575–2601. https://doi.org/10.1007/s10462-018-9628-0
9. Weckenborg C, Kieckhäfer K, Spengler TS, Bernstein P (2020) The volkswagen pre-production center applies operations research to optimize capacity scheduling. INFORMS J Appl Anal 50(2):119–136. https://doi.org/10.1287/inte.2020.1029
10. Buergin J, Helming S, Andreas J, Blaettchen P, Schweizer Y, Bitte F, Haefner B, Lanza G (2018) Local order scheduling for mixed-model assembly lines in the aircraft manufacturing industry. Prod Eng 12(6):759–767. https://doi.org/10.1007/s11740-018-0852-x
11. Dörmer J, Günther HO, Gujjula R (2015) Master production scheduling and sequencing at mixed-model assembly lines in the automotive industry. Flex Serv Manuf J 27(1):1–29
12. Da Col G, Teppan E (2019) Google vs IBM: a constraint solving challenge on the job-shop scheduling problem. Electr Proc Theor Comp Sci 306:259–265. https://doi.org/10.4204/eptcs.306.30
13. Son TH, Van Lang T, Huynh-Tuong N, Soukhal A (2020) Resolution for bounded-splitting jobs scheduling problem on a single machine in available time-windows. J Ambient Intell Humaniz Comput. https://doi.org/10.1007/s12652-020-02162-0
14. Tamssaouet K, Dauzre-Prs S, Yugma C (2018) Metaheuristics for the job-shop scheduling problem with machine availability constraints. Comput Ind Eng 125:1–8. https://doi.org/10.1016/j.cie.2018.08.008
15. Wang S-J, Tsai C-W, Chiang M-C (2018) A high performance search algorithm for job-shop scheduling problem. Proc Comp Sci 141:119–126. https://doi.org/10.1016/j.procs.2018.10.157

16. Liu M, Yao X, Li Y (2020) Hybrid whale optimization algorithm enhanced with lvy flight and differential evolution for job shop scheduling problems. Appl Soft Comput 87:105954. https://doi.org/10.1016/j.asoc.2019.105954
17. Tahar DN, Yalaoui F, Chu C, Amodeo L (2006) A linear programming approach for identical parallel machine scheduling with job splitting and sequence-dependent setup times. Int J Prod Econ 99(12):63–73. https://doi.org/10.1016/j.ijpe.2004.12.007
18. Wang W-L, Wang H-Y, Zhao Y-W, Zhang L-P, Xu X-L (2013) Parallel machine scheduling with splitting jobs by a hybrid differential evolution algorithm. Comput Oper Res 40(5):1196–1206. https://doi.org/10.1016/j.cor.2012.12.007
19. Wang C, Liu C, Hai Zhang Z, Zheng L (2016) Minimizing the total completion time for parallel machine scheduling with job splitting and learning. Comput Ind Eng 97:170–182. https://doi.org/10.1016/j.cie.2016.05.001
20. Liu C, Wang C, Zhang Z-H, Zheng L (2018) Scheduling with job-splitting considering learning and the vital-few law. Comput Oper Res 90:264–274. https://doi.org/10.1016/j.cor.2017.02.011
21. Nguyen VH, Tuong NH, Nguyen HP, Nguyen TH (2013) Single-machine scheduling with splitable jobs and availability constraints. REV J Electron Commun 3(12):21–27. https://doi.org/10.21553/rev-jec.51
22. Nguyen VH, Tuong NH, Tran VH, Thoai N (2013) An milp-based makespan minimization model for single-machine scheduling problem with splitable jobs and availability constraints. In: 2013 international conference on computing, management and telecommunications (commantel). IEEE, Piscataway, pp 397–400. https://doi.org/10.1109/ComManTel.2013.6482427
23. Praveenchandar J, Tamilarasi A (2021) An enhanced load balancing approach for dynamic resource allocation in cloud environments. Wirel Pers Commun. https://doi.org/10.1007/s11277-021-09110-x
24. Li C, Li L (2017) Load-balancing based cross-layer elastic resource allocation in mobile cloud. Wirel Pers Commun 97:2399–2437. https://doi.org/10.1007/s11277-017-4615-3
25. Lin C, Chin H, Deng D (2014) Dynamic multiservice load balancing in cloud-based multimedia system. IEEE Syst J 8(1):225–234. https://doi.org/10.1109/JSYST.2013.2256320
26. Al-Rahayfeh A, Atiewi S, Abuhussein A, Almiani M (2019) Novel approach to task scheduling and load balancing using the dominant sequence clustering and mean shift clustering algorithms. Future Internet. https://doi.org/10.3390/fi11050109
27. Chitra Devi D, Rhymend Uthariaraj V (2016) Load balancing in cloud computing environment using improved weighted round robin algorithm for nonpreemptive dependent tasks. Sci World J. https://doi.org/10.1155/2016/3896065
28. Xiaolong X, Shucun F, Cai Q, Tian W, Liu W, Dou W, Sun X, Liu AX (2018) Dynamic resource allocation for load balancing in fog environment. Wirel Commun Mob Comput. https://doi.org/10.1155/2018/6421607
29. Rajagopal TKP, Venkatesan M, Rajivkannan A (2020) An improved efficient dynamic load balancing scheme under heterogeneous networks in hybrid cloud environment. Wirel Pers Commun 111:1837–1851. https://doi.org/10.1007/s11277-019-06960-4
30. Kim HY, Kim J (2017) An energy-efficient balancing scheme in wireless sensor networks. Wirel Pers Commun 94:17–29. https://doi.org/10.1007/s11277-015-3154-z
31. Krishana R, Laxmi V (2015) IEEE 802.11 WLAN load balancing for network performance enhancement. In 3rd international conference on recent trends in computing 2015 (ICRTC-2015), Ghaziabad, India
32. Zhou C-Q, Yang Y-W, Gong B, Li M-K, Li J, Wen R-Q (2018) The model design of mobile resource scheduling in large scale activities. Mobile Netw Appl. https://doi.org/10.1007/s11036-018-1023-1
33. Dai G, Qingli DA (2000) The study of combinatorial scheduling problem in emergency systems. Syst Eng Theory Pract 20(9):52–55
34. Gao SP, Liu SY (2003) Scheduling problem in multi-resource emergency systems based on the connection number. Syst Eng Theory Pract 6:17

35. Berrichi A et al (2010) Bi-objective ant colony optimization approach to optimize production and maintenance scheduling. Comput Oper Res 37(9):1584–1596
36. Xing L-N, Chen Y-W, Wang P, Zhao Q-S, Xiong J (2010) A knowledge-based ant colony optimization for flexible job shop scheduling problems. Appl Soft Comput 10:888–896. https://doi.org/10.1016/j.asoc.2009.10.006
37. Yang J, Zhuang Y (2010) An improved ant colony optimization algorithm for solving a complex combinatorial optimization problem. Appl Soft Comput 10(2):653–660
38. Pekel E (2022) A simple solution to technician routing and scheduling problem using improved genetic algorithm. Soft Comput. https://doi.org/10.1007/s00500-022-07072-1
39. Pekel E (2020) Solving technician routing and scheduling problem using improved particle swarm optimization. Soft Comput 24(24):19007–19015
40. Zamorano E, Stolletz R (2017) Branch-and-price approaches for the multiperiod technician routing and scheduling problem. Eur J Oper Res 257(1):55–68
41. Chen X, Thomas BW, Hewitt M (2016) The technician routing problem with experience-based service times. Omega 61:49–61
42. Damm RDB, Ronconi DP (2021) A multi-objective biased random-key genetic algorithm for service technician routing and scheduling problem. In: International conference on computational logistics. Springer, Cham, pp 471–486
43. Damm RB, Resende MG, Ronconi DP (2016) A biased random key genetic algorithm for the field technician scheduling problem. Comput Oper Res 75:49–63
44. Gao K, Huang Y, Sadollah A, Wang L (2019) A review of energy-efficient scheduling in intelligent production systems. Complex Intell Syst. https://doi.org/10.1007/s40747-019-00122-6
45. Lora AT, Riquelme JC, Ramos JLM, Santos JMR, Exposito AG (2003) Application of evolutionary computation techniques to the optimal short-term scheduling of the electrical energy production. Complex Intell Syst 3040:656–665
46. He Y, Liu F, Cao HJ, Li CB (2005) A bi-objective model for job-shop scheduling problem to minimize both energy consumption and makespan. J Cent S Univ Technol 12:167–171
47. Zhang R, Chiong R (2016) Solving the energy-efficient job shop scheduling problem: a multi-objective genetic algorithm with enhanced local search for minimizing the total weighted tardiness and total energy consumption. J Clean Prod 112:3361–3375
48. Engin BE, Engin O (2020) A new memetic global and local search algorithm for solving hybrid flow shop with multiprocessor task scheduling problem. SN Appl Sci. https://doi.org/10.1007/s42452-020-03895-5
49. Oğuz C, Ercan FM (1997) Scheduling multiprocessor tasks in a two stage flow shop environment. Comput Ind Eng 33:269–272
50. Oğuz C, Ercan MF, Cheng TCE, Fung YF (2003) Heuristic algorithms for multiprocessor task scheduling in a two stage hybrid flow shop. Eur J Oper Res 149:390–403
51. Şerifoğlu SFS, Ulusoy G (2004) Multiprocessor task scheduling in multistage hybrid flow shops: a genetic algorithm approach. J Oper Res Soc 55:504–512
52. Engin O, Yılmaz MK, Ceran G (2006) Using data mining to find patterns in genetic algorithm solutions to hybrid flow shop scheduling problems with multiprocessor task systems. In: Paper presented at international conference on modelling and simulation. AMSE, Konya, pp 1041–1045
53. Tseng CT, Liao CJ (2008) A particle swarm optimization algorithm for hybrid flowshop scheduling with multiprocessor tasks. Int J Prod Res 46(17):4655–4670
54. Akkoyunlu MC, Engin O, Büyüközkan K (2015) A harmony search algorithm for hybrid flow shop scheduling with multiprocessor task problems. In: Proceedings of the 6th international conference on modeling, simulation, and applied optimization. ICMSAO, Istanbul, pp 1–3
55. Engin B, Engin O (2018) Hybrid flow shop with multiprocessor task scheduling based on earliness and tardiness penalties. J Enterp Inf Manag 31(6):925–936
56. Ying KC, Lin SW (2018) Minimizing makespan for the distributed hybrid flowshop scheduling problem with multiprocessor tasks. Expert Syst Appl 92:132–141

57. Oğuz C (2006) Data for hybrid flow-shop scheduling with multiprocessor tasks. Koc University. http://home.ku.edu.tr/coguz/
58. De A, Wang J, Tiwari MK (2019) Hybridizing basic variable neighborhood search with particle swarm optimization for solving sustainable ship routing and bunker management problem. IEEE Trans Intell Transp Syst 21:986–997
59. Tan F, Chai Z-y, Li Y-L (2021) Multi-objective evolutionary algorithm for vehicle routing problem with time window under uncertainty. Evol Intel. https://doi.org/10.1007/s12065-021-00672-0
60. El-Sherbeny NA (2010) Vehicle routing with time windows: an overview of exact, heuristic and metaheuristic methods. J King Saud Univ Sci 22(3):123–131
61. Wang J, Zhou Y, Wang Y, Zhang J, Chen CLP, Zheng Z (2016) Multiobjective vehicle routing problems with simultaneous delivery and pickup and time windows: formulation, instances, and algorithms. IEEE Transact Cybernet 46(3):582–594
62. Wang J, Weng T, Zhang Q (2019) A two-stage multiobjective evolutionary algorithm for multiobjective multidepot vehicle routing problem with time windows. IEEE Transact Cybernet 49(7):2467–2478
63. Toklu NE, Gambardella LM, Montemanni R (2014) A multiple ant colony system for a vehicle routing problem with time windows and uncertain travel times. J Traffic Log Eng 2(1):52–58
64. Hu C, Lu J, Liu X, Zhang G (2018) Robust vehicle routing problem with hard time windows under demand and travel time uncertainty. Comput Oper Res 94:139–153
65. Duan J, He Z, Yen GG (2021) Robust multiobjective optimization for vehicle routing problem with time windows. IEEE Transact Cybernet 1:1. https://doi.org/10.1109/TCYB.2021.3049635
66. Zhang W, Yang D, Zhang G et al (2019) Hybrid multiobjective evolutionary algorithm with fast sampling strategy-based global search and route sequence difference-based local search for VRPTW. Expert Syst Appl 145:1151
67. Prins C (2004) A simple and effective evolutionary algorithm for the vehicle routing problem. Comput Oper Res 31(12):1985–2002
68. Santos F, Fukasawa R, Ricardez-Sandoval L (2020) An integrated machine scheduling and personnel allocation problem for large-scale industrial facilities using a rolling horizon framework. Optim Eng. https://doi.org/10.1007/s11081-020-09542-7
69. Jamili A (2016) Robust job shop scheduling problem: mathematical models, exact and heuristic algorithms. Expert Syst Appl 55:341–350
70. Artigues C, Gendreau M, Rousseau L, Vergnaud A (2009) Solving an integrated employee timetabling and job-shop scheduling problem via hybrid branch-and-bound. Comput Oper Res 36:2330–2340
71. Lagzi S, Fukasawa R, Ricardez-Sandoval L (2017a) A multitasking continuous time formulation for short-term scheduling of operations in multipurpose plants. Comput Chem Eng 97:135–146
72. Bassett MH, Pekny JF, Reklaitis GV (1996) Decomposition techniques for the solution of large-scale scheduling problems. AICHE J 42:3373–3384
73. Raman N, Talbot F (1993) The job shop tardiness problem: a decomposition approach. Eur J Oper Res 69:187–199
74. Singer M (2001) Decomposition methods for large job shops. Comput Oper Res 28:193–207
75. Chand S, Traub R, Uzsoy R (1997) Rolling horizon procedures for the single machine deterministic total completion time scheduling problem with release dates. Ann Oper Res 70:115–125
76. Fang J, Xi Y (1997) A rolling horizon job shop rescheduling strategy in the dynamic environment. Int J Adv Manuf Technol 13:227–232
77. Altner DS, Mason EK, Servi LD (2018) Two-stage stochastic days-off scheduling of multi-skilled analysts with training options. J Comb Optim. https://doi.org/10.1007/s10878-018-0368-5
78. Gnanlet A, Gilland WG (2014) Impact of productivity on cross-training configurations and optimal staffing decisions in hospitals. Eur J Oper Res 238:254–269

79. Henao C-A, Ferrer J-C, Muñoz J-C (2015) The impact of multi-skilling on personnel scheduling in the service sector: a retail industry case. J Oper Res Soc 66:1949–1959
80. Kilincli G, Zhang X (2017) Mathematical models and solution approach for cross-training staffscheduling at call centers. Comput Oper Res 87:258–269
81. Slomp J, Bokhorst JAC, Molleman E (2005) Cross-training in a cellular manufacturing environment. Comput Ind Eng 48:609–624
82. Schulz S, Buscher U, Shen L (2020) Multi-objective hybrid flow shop scheduling with variable discrete production speed levels and time-of-use energy prices. J Bus Econ. https://doi.org/10.1007/s11573-020-00971-5
83. Biel K, Glock CH (2016) Systematic literature review of decision support models for energy-efficient production planning. Comput Ind Eng 101:243–259
84. Gahm C, Denz F, Dirr M, Tuma A (2016) Energy-efficient scheduling in manufacturing companies: a review and research framework. Eur J Oper Res 248(3):744–757
85. Mecrow B, Jack A (2008) Efficiency trends in electric machines and drives. Foresight Sustainable Energy Management and the Built Environment Project. Energy Policy 36(12):4336–4341
86. Liu Y, Dong H, Lohse N, Petrovic S, Gindy N (2014) An investigation into minimising total energy consumption and total weighted tardiness in job shops. J Clean Prod 65:87–96
87. Lei D, Gao L, Zheng Y (2018) A novel teaching-learning-based optimization algorithm for energy-efficient scheduling in hybrid flow shop. IEEE Trans Eng Manag 65(2):330–340
88. Artigues C, Lopez P, Hait A (2013) The energy scheduling problem: industrial case-study and constraint propagation techniques. Int J Prod Econ 143(1):13–23
89. Wang J, Qiao F, Zhao F, Sutherland JW (2016) Batch scheduling for minimal energy consumption and tardiness under uncertainties: a heat treatment application. CIRP Ann 65(1):17–20
90. Alinezhad R, Ansari R, Mahdikhani M, Banihashemi SA (2021) Multi-phase projects selection and scheduling problem: a multi-objective optimization approach. Iran J Sci Technol Trans Civ Eng. https://doi.org/10.1007/s40996-021-00721-9
91. Liu SS, Wang CJ (2011) Optimizing project selection and scheduling problems with time-dependent resource constraints. Autom Constr 20(8):1110–1119
92. Gutjahr WJ, Katzensteiner S, Reiter P, Stummer C, Denk M (2010) Multi-objective decision analysis for competence-oriented project portfolio selection. Eur J Oper Res 205(3):670–679
93. Chen J, Askin RG (2009) Project selection, scheduling and resource allocation with time dependent returns. Eur J Oper Res 193(1):23–34
94. Carazo AF, Gómez T, Molina J, Hernández-Díaz AG, Guerrero FM, Caballero R (2010) Solving a comprehensive model for multiobjective project portfolio selection. Comput Oper Res 37(4):630–639
95. Lin Y, Wang X, Xu R (2020) Semi-supervised human resource scheduling based on deep presentation in the cloud. EURASIP J Wirel Commun Netw. https://doi.org/10.1186/s13638-020-01677-6
96. Dong-Jin Y, Jiao-Jiao W, Cheng-Fei L (2018) Approach to optimal staff assignment in workflows based on collaboration patterns. J Soft Comput 15:3340–3354
97. Jian S, Chang-Yong L, Jian X (2018) Trust-based multi-objectives task assignment model in cloud service system. J Comput Res Dev 55(6):1167
98. Ji CL, Zhao D-L, Ge B-F, Yang K-W, Chen Y-W (2017) A link prediction method for heterogeneous networks based on bp neural network. Physica A 495:17

Chapter 9
Optimization Methods in Machine Drawings

Introduction by the Editor

Topology optimization is very much necessary to generate feasible designs from basic design maintaining the functionalities. The efficient designs and 2D/3D drawings with several candidate solutions is becoming necessary in several viewpoints. A multi-component topology optimization method for structural assemblies that are made of components produced by die casting processes, where each part is guaranteed to be free from fully enclosed cavities and undercuts in the direction of die drawing are discussed. Moreover, 3D reconstruction of an indoor scene with large vertical span have been described. It represents a novel approach for 3D reconstruction of indoor scenes. It employed the 3D visual reconstruction model of indoor scene by the top-down image registration of point cloud data. The chapter also discussed a detailed optimization process of the combination of VR and VR-GIS. Moreover, two machine learning-based heuristic approaches to constructing sparse flexible designs that leverage a neural network to accurately and quickly predict the performance of large numbers of candidate designs.

The review highlights that the conventional visualization algorithms for small-work networks lead to a uniform clump of nodes or are restricted to a tree structure, making the network structure difficult to identify and analyze. The review underscores that if the original network is connected, the resulting drawing is required to be connected. Some smaller weight links are eliminated for the purpose of better visualization, the number of links that is to be deleted should be kept to a minimum. It concludes that the resulting weighted graph drawing should maintain the clustering structure of the original network. The study also suggests employing different force-directed methods, edge binding and more sophisticated visualization techniques. One of the other works detailed is associated with a multi-component topology optimization method for the structural assemblies that are made of components produced by die casting (MTO-D) processes. In this work, every part is guaranteed

A. J. Kulkarni (ed.), *Optimization Methods in Manufacturing Processes*,
https://doi.org/10.1007/978-981-96-5257-0_9

to be free from fully enclosed cavities and undercuts in the direction of mold drawings. This work assumes that there is no interior cavity in any component and the undercut features are all located on the parting line or the boundary of components. Moreover, the optimal model can be obtained by a slight moving of parting lines based on the optimized results for most situations.

In one of the reviews associated with micro-grid Planning in uncertainty domain, the observations and conclusions are derived that the selection of feasible renewable energy technology combination for optimal autonomous micro-grid planning with economic, technical, environmental and social objectives has been carried out by several researchers; however, uncertainty modelling and their respective dispatch strategies are required to be modelled. The work on the enhancement of light field disparity maps by reducing the silhouette effect and plane noise is critically reviewed and elaborated. The review concludes that the noise resilient seed growth algorithm managed to accurately estimate the position and orientation of planar surfaces in a light field, resulting in a significant reduction of low amplitude local noise in the resulting disparity maps, enabling the estimation of accurate normal maps. IN addition, the direct improvements to the method, replacing the matting-based optimisation method with an optimisation framework that takes into account the 4D structure of the light field may lead to more accurate results. Furthermore, an intelligent system that beautifies Chinese handwriting characters and physically writes them in a certain calligraphy style using a robotic arm is elaborated. It is claimed that the major contribution is the global optimization approach generating the optimal stroke for a robotic arm while retaining both the calligraphy and handwriting characteristics, in which a shape character vector is proposed to describe the shape of a stroke skeleton.

The work on complete 3D scene parsing from a image is discussed. The review elaborates that the representation encodes the layout of orthogonal walls and the extent of objects, modelled with CAD-like 3D shapes. The data-driven approach, generating sets of potential object regions, matching to regions in training images, as well as transferring and aligning associated 3D models have been utilized. It is important to mention that the representation encodes the layout of walls as 3D planes and all interpretable objects as 3D mesh models, parsing both the visible and occluded portion of the scene. The approach of levy flight and fitness distance balance-based coyote optimization algorithm is discussed in one of the works to improve the performance of automatic generation control of three different PV-based interconnected power systems. The algorithm is evaluated on the two-area PV-reheat thermal power system with PI controller, which is widely used in the literature, and the results are compared with the results obtained using genetic algorithm, firefly algorithm, and modified whale optimization algorithm.

Machine Generated Summaries

Disclaimer: The summaries in this chapter were generated from Springer Nature publications using extractive AI auto-summarization: An extraction-based summarizer aims to identify the most important sentences of a text using an algorithm and uses those original sentences to create the auto-summary (unlike generative AI). As the constituted sentences are machine selected, they may not fully reflect the body of the work, so we strongly advise that the original content is read and cited. The auto generated summaries were curated by the editor to meet Springer Nature publication standards. To cite this content, please refer to the original papers.

Machine generated keywords: object, image, node, network, power, depth, light, stateoftheart, user, region, structure, energy, computer, generation, design.

A Novel Centrality-Based Method for Visual Analytics of Small-World Networks [1]

This is a machine-generated summary of:

Lin, Chun-Cheng; Huang, Weidong; Liu, Wan-Yu; Wu, Sheng-Feng: A novel centrality-based method for visual analytics of small-world networks [1].

Published in: Journal of Visualization (2019).

Link to original: https://doi.org/10.1007/s12650-019-00582-5.

If you want to cite the papers, please refer to the original.

For technical reasons we could not place the page where the original quote is coming from.

Abstract-Summary “The network data that we need to deal with and make sense of are becoming increasingly large and complex.”

“Small-world networks are a type of complex networks whose underling graphs have small diameter, shorter average path length between nodes, and a high degree of clustering structures and can be found in a wide range of scientific fields, including social networks, sociology, computer science, business intelligence, and biology.”

“Conventional visualization algorithms for small-work networks lead to a uniform clump of nodes or are restricted to a tree structure, making the network structure difficult to identify and analyze.”

“Different from previous methods based on spanning trees, this method first generates a weighted planar sub-network based on the measurement of network centrality metrics.”

"A force-directed algorithm based on node-edge repulsion is then applied to visualize this sub-network into a proper layout for better understanding of the data."

"The experimental results show that compared to previous methods, the proposed method can be more effective in differentiating clusters and revealing relationship patterns among individual nodes and clusters in the network."

Introduction

"Because small-world networks possess a special structure, a lot of conventional visualization methods that aim to meet the above aesthetic criteria often cause nodes and links to mutually overlap, making the identification of the network structure difficult."

"To ensure that the backbone will lead to meaningful visualization results, whenever possible, setting weights should follow the following constraints [2]: If the original network is connected, the resulting drawing should be connected as well; while some smaller weight links are eliminated for the purpose of better visualization, the number of links that is to be deleted should be kept to a minimum; the resulting weighted graph drawing should maintain the clustering structure of the original network."

"Once the planar graph is expended with the maximum possible links, an improved force-directed method based on node-edge repulsion [3] is used to lay out the subgraph, in which the rest of links is then added back and drawn as curves to produce the final drawing."

Literature Review

"In graph theory and network analysis, centrality is an indicator for finding important nodes and links [4–7]."

"Closeness centrality Closeness centrality is a closeness that measures a node in relation to other nodes and is calculated based on the shortest path between two nodes."

"Betweenness centrality: In a network, if a node is located on the shortest path between many other nodes, this node can be considered as having an important position."

"The higher a node's control level is in a network means that it has a high betweenness centrality."

"The radiality centrality $C_r(s)$ for node s is calculated as follows:where D represents the network's longest path."

"Crnovrsanin et al. [8] used two analysis techniques (including centrality sensitivity analysis and modularity clustering analysis) and multiple visualization techniques (including edge bundling, centrality-based graph layout, radial layout, and parallel coordinates) to render social networks, and applied it in a real-life friendship network."

Methodology

"To maintain the linking structure within a cluster, this work uses low betweenness centrality links to draw the required planar graph, meaning that Step 2 of Algorithm 1 assigns higher weights to links with low link betweenness centrality scores."

"With the setting of weights, Step 2 of Algorithm 1 uses the PQ-tree algorithm of Junger et al. [9] to find a minimum-cost (maximum-weight) planar subgraph, which has the same number of nodes with the inputted graph, but has a smaller number of links."

"After the plane graph is obtained at Step 4, Step 5 adds back the links that were deleted in Step 2 to produce the final drawing of the inputted network."

"We observe that those links that had high centrality scores and were left out at Step 2 were mostly bridges between clusters."

Experiment Analysis

"In a non-directional graph, the cluster coefficient of node u is $2\ e_u/(k_u \cdot (k_u - 1))$, in which k_u is the number of neighbors that node u has, and e_u is the number of links that the neighbors have."

"The average cluster coefficient of a random graph with 515 nodes and 1070 links is approximately 0.0031."

"This means that the graph of Step 3 has the least number of bridge links, thus having the largest network diameter value."

"The peak for Step 5 curve appeared at 6 and was closer to the distribution of the original graph. (3) Cluster coefficient: This is the indicator used to measure the cluster relationship."

"Because the three graphs have different numbers of links, the stress centrality is normalized before making comparison."

"The artist cluster in the center is formed mostly by painters, in which the node with the most number of links is Picasso."

Application

"The semiconductor wafer production process often includes hundreds of processing stages."

"Because the semiconductor production process is very complex, it is difficult to find useful information from so much data."

"We hope to build a network that includes wafer information and use our visualization method to actually help analyze data of the semiconductor wafer manufacturing visually."

"For the links between the wafers, we used the aforementioned wafer processing characteristics to calculate the processing similarity between any two wafers."

"Because the processing similarity of any two wafers in the cluster is highly similar, the production time can be sped up and the production cost can be reduced."

"Because these types of wafers have lower similarity with other wafers or wafer processes, it may mean that these wafers have lower processing performance compared to other wafers and require improvement."

"We can use this graph to improve the wafer production procedure, find abnormal wafers and determine the cause."

Conclusion

"We first use betweenness centrality scores to choose links and produce the initial weighted planar subgraph."

"Very few of them have utilized planar graph drawing algorithms and force-directed methods based on node-link repulsion on small-world networks."

"We conducted a detailed basic parameter and centrality analysis on the structures of the original graph, the planar subgraph drawn with our method and the final graph."

"Based on the final visualization produced by our method, the analysis we conducted not only found the same information as that in van Ham and Wattenberg [2], but also successfully revealed more useful information."

"For future research directions, future studies may attempt to use different measurements or methods to produce the initial planar subgraphs."

"Further, different force-directed methods, edge binding, or other visualization techniques may also be used for visualization of the final graph to see if they could produce better results."

Multi-Component Topology Optimization for Die Casting (MTO-D) [10]

This is a machine-generated summary of:

Zhou, Hao; Zhang, Junyuan; Zhou, Yuqing; Saitou, Kazuhiro: Multi-component topology optimization for die casting (MTO-D) [10].

Published in: Structural and Multidisciplinary Optimization (2019).

Link to original: https://doi.org/10.1007/s00158-019-02317-4.

If you want to cite the papers, please refer to the original.

For technical reasons we could not place the page where the original quote is coming from.

Abstract-Summary "This paper presents a multi-component topology optimization method for the structural assemblies that are made of components produced by die casting (MTO-D) processes, where each part is guaranteed to be free from fully enclosed cavities and undercuts in the direction of mold drawing."

"The presence of undercut features and fully enclosed cavities are then evaluated for each mold half using the vector method with the numerically approximated density gradient."

"The optimized results show that each part is indeed moldable with no undercut in the drawing direction and has no fully enclosed cavity."

Introduction

"To realize manufacturability-driven synthesis of multi-component structures, the simultaneous optimization of topology and partitioning has been studied in both discrete, truss-like design domains [11] and continuum, pixel-like design domains [12, 13] by using non-gradient optimization algorithms."

"These works demonstrated the design solutions of multi-component assemblies with complex overall geometries but simple in each component, with the optimized trade-off between performance and manufacturing cost."

"This paper presents a multi-component topology optimization method for structural assemblies that are made of components produced by die casting processes (MTO-D), where each part is guaranteed to be free from fully enclosed cavities and undercuts in the direction of die drawing."

"Building upon our previous work on multi-component topology optimization for stamped sheet metal assemblies (MTO-S) [14], the proposed mathematical formulation incorporates new design variables specifying the location and orientation of the parting lines for each component."

Gradient-Based Multi-Component Topology Optimization

"This section reviews the continuously relaxed gradient-based multi-component topology optimization (MTO) framework, first proposed in Zhou and Saitou [14]."

"To enable the simultaneous exploration of topology and partitioning, two types of design variables are defined within a predefined design domain: the density variable representing the existence of structural material at each location in the design domain (similar to the SIMP formulation), and the component membership (vector) variables representing the membership of each location in the design domain to a component."

"Analogous to the relaxation of the binary concept of the existence of structural material (exist or not exist) at a design point to continuous density between 0 and 1, the binary concept of the belonging to a component (belong or not belong) at a design point is relaxed to take values between 0 and 1, allowing partial membership of a design point to a component."

"By adopting the dual-penalization power, the modified material interpolation scheme can help design variables converge to zero or one, leading to "black and white" topologies and partitioning with no fractional material and component membership."

Multi-Component Topology Optimization for Die Casting

"The die drawing directions and the parting lines play important roles in the moldability and the manufacturing costs of die cast components."

"In order to ensure that cast components can be removed from dies smoothly after solidification of metal (i.e., the part is moldable), die casting processes require the geometry of cast components to meet certain conditions: (1) interior cavities are not allowed in the structure and (2) the amount of undercut features in drawing direction should be minimized."

"In order to evaluate the moldability of the components during the optimization iterations, the finite elements that are members of a component (even fractionally) should be clustered into two, to the ones belonging to each mold half based on the parting line for the component, which is also a design variable for the proposed formulation."

Numerical Examples

"Considering the optimization accuracy and the complexity of the formulation model, the first derivatives of objectives and constraints are analytically derived with the assistance of the MATLAB symbolic math toolbox and their Hessians are numerically computed using a finite difference method."

"As there is no interior cavity in any component and the undercut features are all located on the parting line or the boundary of components, the optimal model can be obtained by a slight moving of parting lines based on the optimized results for most situations."

"The initial value for parting line direction may influence the optimization results."

"The optimization results show that the initial value of the parting line has no obvious influence on the overall topological structure and structure decomposition but has influence on parting line direction."

"The results indicate that with the increase of A^*, the volume fraction of the optimized structure increases while the structural compliance decreases, showing the trade-off between the two."

Conclusion

"This paper presented a multi-component topology optimization method for the assembly of components manufactured by die casting processes (MTO-D)."

"The method is based on our previous work on multi-component topology optimization for stamping (MTO-S) [14], with several new key developments."

"The vector-based, density-gradient method of moldability evaluation for monolithic topology optimization is extended to the multi-component framework."

"Based on the assumption of two-mold casting and straight parting lines, new design variables are introduced to represent the location and orientation of the parting lines for each component."

"Such extensions would include additional constraints on the maximum stress for joints, the influence of interfacial strength between different components, the cost model for welding cast components, relaxation to allow non-straight parting lines, relaxed constraint on undercut based on cost models for additional mold mechanisms such as side pins and core lifters, more general formulation for density gradient calculation, and finally a numerically efficient and robust 3D implementation."

Quad-Level MCDM Framework to Analyse Technology Combinations for Sustainable Micro-grid Planning in Uncertainty Domain [15]

This is a machine-generated summary of:

Thakkar, Nishant; Paliwal, Priyanka: Quad-Level MCDM Framework to Analyse Technology Combinations for Sustainable Micro-grid Planning in Uncertainty Domain [15].

Published in: Arabian Journal for Science and Engineering (2022).

Link to original: https://doi.org/10.1007/s13369-022-07019-6.

If you want to cite the papers, please refer to the original.

For technical reasons we could not place the page where the original quote is coming from.

Abstract-Summary "The selection of appropriate technology combination (TC) is one of the most important aspects of autonomous micro-grid (AMG) planning."

"This paper puts forward a Quad-level framework."

"The first level involves uncertainty modelling of generators and storage in probabilistic framework."

"Third level deals with the sustainability evaluation of each TC based on four main criteria and eleven sub-criteria."

"At the fourth level, the implementation of multi-criteria decision making approach is carried out using a combination of analytical hierarchy process and TOPSIS."

"The final ranking obtained through Borda count suggests that the combination of SPV, WTG and BST provides the best sustainable solution for AMG."

Introduction

"The four major pillars of sustainable energy production are described as follows [16]: Technical sustainability Economic sustainability Environmental sustainability Social sustainability Besides technical and economic considerations, environmental effects such as emission, and utilization of natural resources constitute important factors for sustainable AMG planning."

"Aberilla et al. [17] proposed a HOMER-based AMG framework and employed the MCDM approach for better resource optimization based on social, technological, and economic criteria."

"While different MCDM-based frameworks with different techniques have been developed for RET selection, among them several studies [18–21] utilized AHP and TOPSIS for weight assignment and ranking of RET based on different selected criteria and applications."

"On the basis of existing literature, following observations can be made: The selection of feasible RET combination for optimal AMG planning with economic, technical, environmental and social objectives has been carried out in various papers."

Level: 1—Uncertainty Modelling of TCs and their Respective Dispatch Strategies

"Uncertainty modelling of SPV involves two factors: The SIM comprises discretized states of solar irradiance and their respective probabilities for a specified time segment."

"Probabilistic battery storage model (PBSM) [22] has been used, which comprises of multiple states of battery state of charge (SOC) along with its respective probabilities."

"The charging and discharging of BST are dependent on four factors namely: present battery SOC, power availability from RET, hardware availability of RET and load."

"The DGT system is taken as a constant source of power supply, and the output power is considered to be equal to its rated value subjected to the hardware availability of generator."

"In those TCs where, both SPV and WTG type of RET is used, the power from SPV arrays is assigned first discharge priority (DP1) due to their low operation cost."

"The BST units are charged only when surplus power from RET is available after supplying the load."

[Section 3]

"To determine the optimal component size, a metaheuristic technique called as butterfly PSO is utilized [23] which has been briefly discussed in the later part of this section."

"The size of each component is determined so as to cater load demand of the system with best economy and defined reliability standard."

"The optimization tool (BFPSO) analyses different combinations of generators and storage units to come up with component sizes which provides minimum LCOE subjected to reliability constraints."

"Minimization of LCOE has been adopted as the objective function to obtain component size in AMG."

[Section 4]

"Four main sustainable criteria namely social, environmental, economic and technical along with 11 sub-criteria have been chosen to assess nine different TCs."

"The economic assessment lays the foundation of a sustainable project."

"Four different economic sub-criterion viz."

"The LCOE is an important economical parameter for cost comparison of different power generation technologies [22]."

"The ACC is the annual cost of a component that is distributed equally over the entire project lifespan."

"Water is required at different stages of power generation, starting from production of generators to the operational phase."

"The water footprint of a particular technology is defined as the amount of water consumed during the project lifespan [24]."

"a. With the installation of a new power generation project, additional jobs are created."

"The safety of workers is an important factor for successful implementation of any project [17]."

[Section 5]

"AHP assigns values based on both quantitative and qualitative criteria that making it a more striking and dependable method of decision making [25]."

"Determine the global weights between criteria and local weights between sub-criteria by using geometric average with relative acceptable valve (less than 0.1) of consistency ratio (CR) for each group."

"Calculate the final weights of each sub-criteria using integration of local weight of sub-criteria and global weight of criteria."

"local weight assessment of sub-criteria, global weight assessment of criteria and final weight assessment."

"The local weights of all sub-criteria are obtained by using AHP method."

"Final weights are determined by multiplying the local weights of sub-criteria and global weight of criteria."

"To determine sub-criteria weight in each PO, the global weight of criteria has been changed."

Results and Discussion

"The hardware availability model [22] is applicable to the generators such as SPV, WTG, BM, DGT in different TCs."

"In case of RET-based TC, such as solar and wind, the output power depends not only on hardware availability of generator but also on solar irradiance and wind velocity."

"The peak load of the AMG is considered as 100 kW. The BST is to be integrated with TCs based on solar and wind to provide firm support to supply load."

"In TC 3, the additional power generated from RET (SPV and WTG) is stored in BST."

"Based on load sharing proportions, it can be readily concluded that BST integration is essential for a reliable operation of AMG if dispatchable generators are not a component of TC."

"In TC 7 and TC 8, about 95% of load demand is supplied by the non-renewable technologies (from DGT in TC 7 and BM in TC 8) and the rest of 5% from SPV."

Conclusion

"The optimal planning of AMG is a complex task and involves evaluation of multiple criteria."

"The quad-level planning model presented in this paper adequately addresses each aspect of AMG planning."

"The major contributions of this paper can be enumerated as follows: The developed framework adequately addresses uncertainty parameter inherent to RET based DGs."

"The inclusion of system uncertainties improves the accuracy and reliability of AMG planning."

"Majority of existing papers initiate planning studies based on single/multi-criteria assessment."

"This paper initiates system planning based on optimal sizing and thereafter analyses each combination on different criteria."

"The important conclusions based on the study conducted in this paper are as follows: (1) The combination of renewable energy sources with battery storage technology provides the most reliable, economic and environmentally competent solution."

"AMG planning has to be site specific and governed by various socio-economic-environmental conditions."

Enhancement of Light Field Disparity Maps by Reducing the Silhouette Effect and Plane Noise [26]

This is a machine-generated summary of:

Lourenco, Rui M.; Tavora, Luis M. N.; Assuncao, Pedro A. A.; Thomaz, Lucas A.; Fonseca-Pinto, Rui; Faria, Sergio M. M.: Enhancement of light field disparity maps by reducing the silhouette effect and plane noise [26].

Published in: Multidimensional Systems and Signal Processing (2022).

Link to original: https://doi.org/10.1007/s11045-021-00807-7.

If you want to cite the papers, please refer to the original.

For technical reasons we could not place the page where the original quote is coming from.

Abstract-Summary "Among the existing ones, those based on the structure tensor have emerged as very promising to estimate disparity maps from Epipolar Plane Images."

"To address these problems, this work proposes a new method for improving disparity maps obtained from the structure-tensor approach by enhancing the silhouette and reducing the noise of planar surfaces in light fields."

"An edge-based approach is initially used for silhouette improvement through refinement of the estimated disparity values around object edges."

"A plane detection algorithm, based on a seed growth strategy, is used to estimate planar regions, which in turn are used to guide correction of erroneous disparity values detected in object boundaries."

"The proposed algorithm shows an average improvement of 98.3% in terms of median angle error for plane surfaces, when compared to regular structure-tensor-based methods, outperforming state-of-the-art methods."

Introduction

"This paper proposes a complete disparity estimation algorithmic pipeline, capable of achieving state-of-the-art results."

"Previous methods for LF disparity estimation based on the structure tensor [27, 28] suffer from known limitations, which consist in: (1) the enlargement of object boundaries where occlusions appear in transitions between distinct depths in the scene—this is known as the silhouette enlargement effect; (2) low amplitude local noise affecting planar regions of the scene—this can be characterised as an irregularity of surface normal maps."

"This work specifically addresses these two limitations of structure tensor-based methods by proposing methods to improve the accuracy of object silhouettes in disparity maps and to reduce the low amplitude local noise in planar regions of depth maps."

"This is a novel unified framework capable of minimising the inaccuracies of LF estimated disparity, thus improving the quality of the resulting depth maps."

Background on Disparity Estimation

"Tao et al. [29] achieved a good disparity estimation by combining defocus and correspondence cues and then propagating the results using Markov Random Fields."

"In Williem et al. [30], the problem of normal map regularisation is addressed directly by adding a novel optimisation step to the initial estimation introduced by Tao and others, achieving disparity maps that produce smooth surface normal maps."

"Schilling et al. [31] achieves excellent results by using a simple method based on Ransac line-fittings of the zero-crossings of the second order derivative of the EPIs to initialise a local optimisation method that directly enforces occlusion constraints."

"This method proves to be one of the best non-supervised methods at achieving good disparity in occlusion regions, however, it falls behind [30] when it comes to the accuracy of normal maps produced from the resulting disparity estimation."

Background on the Structure Tensor Method

"The structure tensor was proposed by Bigun [32] as a method for estimating the orientation of linear symmetry in images."

"Applying the tensor to overlapping windows of an EPI centered around each pixel is, thus, a good way of estimating the orientation of the linearly symmetric structures in the EPI and thus obtain an estimation of the disparity of the points represented in the EPI."

"Silhouette enlargement is one of the most noticeable artefacts created by disparity estimation methods using the structure tensor."

"Considering the disparity map for any view of the light field, this results in an enlargement of the occluding objects, as their disparity is spread into the surrounding occluded regions."

"The structure tensor does not enforce local smoothness in disparity maps."

"It is possible to observe that, in the regions where the ground truth normal map (c) has constant values, the normal vectors obtained from the structure tensor estimation appear as random (a)."

Proposed Method

"A plane fitting-based algorithm is applied in order to reduce the effects of local noise based on the orientation of planar surfaces in a light field image."

"A plane is estimated for each of these seeds and they grow by including neighboring points close to the estimated plane, in terms of perpendicular distance."

"An adapted RANSAC algorithm is used to estimate a plane based on three random candidate points."

"Every 10 iterations the same adapted RANSAC algorithm is run for all points currently considered as belonging to the plane, with the obtained plane estimate being considered as the new current estimate."

"Whenever two planes are considered similar, the plane area is merged by us of a logical 'or' operation on the binary map and RANSAC is run on all points to find a new estimate that better fits all points."

Results and Discussion

"The algorithm is compared with the state-of-the-art, where it proves slightly behind other methods in terms of the Mean Square Error metric, but mostly ahead in terms of median angle error in planar regions."

"The Plane Noise Reduction algorithm has an increased run-time for images with several non-contiguous planar regions."

"The proposed method is compared, in terms of mean square error (MSE) and median angle error in planar regions (MAE), to the Multi-Resolution (RM3DE) method by Neri et al. [33], the Spinning Parallelogram Operator (SPO) method [34], the OFSY method [35], a robust optimisation method (OBER), presented by Schilling et al. [31], and the attention-based neural network (AttNet) [36]."

"Those methods were selected for comparison, as the RM3DE, SPO and OBER methods are the non-neural network methods that provide the best results in terms of MSE in the literature, while the OFSY method is an optimization method that minimises the angle difference of a focus stack-based disparity estimation, which achieves the best results in terms of MAE."

Final Remarks

"A novel algorithm for disparity estimation was proposed to address known limitations of the traditional structure-tensor approach, namely silhouette enlargement and low amplitude local noise."

"A noise resilient seed growth algorithm managed to accurately estimate the position and orientation of planar surfaces in a light field, resulting in a significant reduction of low amplitude local noise in the resulting disparity maps, enabling the estimation of accurate normal maps."

"As for direct improvements to the method, replacing the matting-based optimisation method with an optimisation framework that takes into account the 4D structure of the light field may lead to more accurate results."

"Extending this method for multiple orientations, and not just the cross-hair views, would take advantage of the full light field volume and possibly achieve more accurate disparity results."

Intelligent Chinese Calligraphy Beautification from Handwritten Characters for Robotic Writing [37]

This is a machine-generated summary of:

Zhang, Xinyue; Li, Yuanhao; Zhang, Zhiyi; Konno, Kouichi; Hu, Shaojun: Intelligent Chinese calligraphy beautification from handwritten characters for robotic writing [37].

Published in: The Visual Computer (2019).

Link to original: https://doi.org/10.1007/s00371-019-01675-w.

If you want to cite the papers, please refer to the original.

For technical reasons we could not place the page where the original quote is coming from.

Abstract-Summary "It is a challenge for non-expert users to write visually pleasing calligraphy with his or her own unique style."

"We develop an intelligent system that beautifies Chinese handwriting characters and physically writes them in a certain calligraphy style using a robotic arm."

"To output aesthetically beautiful calligraphy with the user's style, we propose a global optimization approach to solve the minimization problem between the handwritten strokes and standard calligraphy strokes, in which a shape character vector is presented to describe the shape of standard strokes."

"Unlike existing systems that focus on the generation of digital calligraphy from handwritten characters, our system has the advantage of converting the user-input handwriting into physical calligraphy written by a robotic arm."

"The writing results show that our system can achieve visually pleasing calligraphy from various input handwriting while retaining the user's style."

Introduction

"Although Ma and Su [38] proposed a standard of aesthetics evaluation for robotic writing of Chinese calligraphy, it is challenging for the current writing results of robotic arms to meet the aesthetic requirements of calligraphy with a particular user's style because it needs to consider the complicated structure and transition of strokes when using ink and brush."

"We propose an intelligent Chinese calligraphy beautification method for robotic arms and build a system to generate calligraphy with a user's writing style from the input handwritten characters."

"This paper makes the following contributions regarding the writing of aesthetically pleasing Chinese calligraphy from handwriting: a global optimization approach to generate the optimal stroke for a robotic arm while retaining both the calligraphy and handwriting characteristics, in which a shape character vector (SCV) is proposed to describe the shape of a stroke skeleton; an intelligent system to physically write calligraphy using robotic arms."

Related Work

"The process of converting handwritten characters to calligraphy consists of four main parts: character writing and stroke extraction, stroke recognition, stroke beautification, and robot control."

"The second approach is more intuitive because the stroke information is determined while the user is writing a character."

"Although this approach could generate elegant calligraphy, the restrictions on input handwritten characters are too strict for non-expert users: It is difficult for them to write neat characters and adjust the corresponding parameters."

"Muller et al. [39] introduced a novel learning-based approach to write single strokes of calligraphy characters using a robotic arm, while the approach concentrated on writing single strokes rather than complete characters."

"Sun et al. [40] presented a local weighted linear regression approach to implement the mapping from stroke representation to brush trajectory, but it required human demonstrations and it is difficult to generate stylized calligraphy with the users' input."

Overview

"We propose an intelligent writing system to convert handwritten characters to aesthetically pleasing calligraphy for robotic writing."

"The target stroke should be close to the standard calligraphy in the database."

"It also should preserve the user's handwriting characteristics."

Handwriting Input and Recognition

"A CNN-based recognition approach was employed to identify the type of each stroke."

"We extract stroke coordinates directly from handwriting using an interface rather than from images, because it is easier to record the stroke coordinates and sequences."

"To generate a concise skeleton for robotic writing, every input stroke is resampled adaptively to form a polyline [41]."

"Black lines in the image refer to the handwritten strokes "Heng", "Shugou", and "Pie" in skeleton form."

"The obtained stroke points are used to generate the stroke image for subsequent CNN recognition."

"To avoid low recognition rate caused by image quality, the stroke images are preprocessed into a specified size."

"The previous studies showed that the rate of recognition is increased as the image size increases."

"Ten samples of each type of stroke were written by 20 different people, giving a total of 200 training images for each stroke."

Stroke Database Establishment

"To generate elegant calligraphy strokes for robotic writing, a unique parameter model was built for each type of stroke."

"As for parameter w, we approximate the relationship between an average stroke width and the descent of the robotic arm brush."

"To determine the relationship between w and h, we designed an experiment to collect the actual stroke width corresponding to every 1 mm descent of the brush tip."

"The average value of the three widths is assigned to w. The brush tip was lowered from 1 mm to 7 mm, and the corresponding stroke widths were 2, 3.2, 3.8, 4.5, 5.5, 6.2, and 6.8 mm."

"The parametric model for the 20 standard strokes we built is able to direct the motion of a robotic arm by controlling the three parameters to ensure that the brush follows the predetermined trajectory."

Calligraphy Beautification and Writing

"They have three limitations: (1) the input of the handwritten characters must be neat and beautiful, so it is not suitable for non-expert users. (2) It is difficult to preserve both the user's writing style and the standard calligraphy style. (3) The simulated calligraphy is inappropriate for robotic writing."

"For every stroke, an optimization function is employed to generate a target stroke for robot calligraphy."

"We control the skeleton points [42] rather than the contour points of each stroke."

"Because we have already obtained the skeleton points of the handwritten stroke and the standard calligraphy stroke, our goal is to find the optimal skeleton of the target strokes."

"The coordinates of the three-dimensional control points (x, y, z) of each stroke must be accurately transmitted to the robotic arm."

"We first expand the points of an optimized stroke and retrieve the corresponding points from the standard stroke database."

Results and Discussion

"With non-optimized writing results, the results of calligraphy writing after optimization showed some features of Kai-style calligraphy around the corner parts of the strokes."

"Robotic writing with optimized strokes and rigid segment replacement not only maintains the characteristics of handwritten characters better, but also preserves the Kai style of calligraphy."

"The results demonstrate that our calligraphy beautification approach for robotic writing preserves both the styles of user's handwriting and the style of Kai calligraphy."

"Our robotic calligraphy writing system is close to achieving the expected goal of generating visually pleasing calligraphy while preserving the user's writing style."

"Although we have achieved visually pleasing calligraphy writing using a robotic arm, our system still has three limitations: One limitation is the speed, which occurs because we have to manually tune the initial parameters for generating optimized strokes."

Conclusions and Future Work

"An intelligent system was built that consists of a simple interface for recording a user's handwriting and extracting individual strokes from a touch pad, and a robotic arm for physically writing the optimized calligraphy on paper."

"To preserve both the user's writing style and the style of the calligraphy, we segmented each stroke into rigid parts and flexible parts."

"We applied the style of standard Kai-style calligraphy to the rigid parts and applied the optimized strokes with the user's writing style to the flexible parts."

"The results of several experiments and a user study show that our robotic writing system not only maintains the user's unique writing style, but also preserves the Kai style of calligraphy."

Complete 3D Scene Parsing from an RGBD Image [43]

This is a machine-generated summary of:

Zou, Chuhang; Guo, Ruiqi; Li, Zhizhong; Hoiem, Derek: Complete 3D Scene Parsing from an RGBD Image [43].

Published in: International Journal of Computer Vision (2018).

Link to original: https://doi.org/10.1007/s11263-018-1133-z.

If you want to cite the papers, please refer to the original.

For technical reasons we could not place the page where the original quote is coming from.

Abstract-Summary "Our representation encodes the layout of orthogonal walls and the extent of objects, modeled with CAD-like 3D shapes."

"We parse both the visible and occluded portions of the scene and all observable objects, producing a complete 3D parse."

"Such a scene interpretation is useful for robotics and visual reasoning, but difficult to produce due to the well-known challenge of segmentation, the high degree of occlusion, and the diversity of objects in indoor scenes."

"We take a data-driven approach, generating sets of potential object regions, matching to regions in training images, and transferring and aligning associated 3D models while encouraging fit to observations and spatial consistency."

Introduction

"We propose an approach to recover a 3D model of room layout and objects from an RGBD image."

"We take a data-driven approach that proposes a set of potential object regions, matches each region to a similar region in training images, and transfers and aligns the associated labeled 3D models while encouraging their agreement with observations."

"The 3D labeling provides ground truth 3D scene representation with layout surfaces as 3D planar regions, furniture as CAD exemplars, and other objects as coarser polygonal shapes."

"We propose an approach to recover a 3D model of room layout and objects from an RGBD image, arguably producing the most detailed and complete 3D scene parses to date."

"To multiview 3D reconstruction methods, our approach recovers complete models from a limited viewpoint, attempting to use priors and recognition to infer occluded geometry, and parses the scene into individual objects and surfaces, instead of points, voxels, or contiguous meshes."

Related Work

"Room layout is often modeled as a 3D box (cuboid) [44–48]."

"Even when a depth image is available, a box layout is often used (e.g., [47]) due to the difficulty of parameterizing and fitting more complex models."

"3D objects are also often modeled as 3D boxes when estimating from RGB images [48–52] or RGB-D images [52]."

"Gupta et al. [53] fit shape models of 6 classes with poses to improve object detection."

"Xiang et al. [54] align 3D shapes from 100 categories to 2D images, contributing a large scale database for 3D object recognition."

"Our approach is efficient and light-weight compared to ScanNet by Dai et al. [55], which utilizes RGB-D video dataset to annotate instance-level semantic segmentations, since our model operates on single images instead of videos."

Detailed 3D Annotations for Indoor Scenes

"Each object and layout has a 2D segment labeling and a corresponding annotated 3D model, provided by Guo and Hoiem [56]."

"The 3D annotations use 30 models to represent 6 categories of furniture that are most common and use extruded polygons to label all other objects."

"To align name-space between datasets, we manually map all model class labels to the 633-class 3D object labels in NYUd v2 dataset."

"We also include the region's original coarse 3D annotation by Guo and Hoiem [56] in the model set M, so that we can preserve the original labeling if no provided CAD models are better fit in depth."

"We fit 3792 different CAD models to the object regions and through manual checking observe failures in only 10% of cases for automatic fitting."

Approach

"We also train and use a CNN with a Siamese network design to find the most similar 3D shape of a training object, based on region and depth features."

"We apply the above retrieval scheme to each of the 2000 region proposals in each image, obtaining shape similarity rank compared with all the training samples and 81-object class and non-object class probability for each region proposal."

"We tried using the estimated support height for each region proposal for aligning the related 3D shape models but observed a worse performance in the scene composition result."

"We train the support height prediction network with the ground truth support type for the region proposal and set the ground truth support height as the closest support height candidate that is within 0.15 m from the related 3D annotation's bottom."

Experiments

"We use both 2D and 3D quantitative measures to evaluate different aspects of our solutions: (1) instance segmentation and semantic segmentation performance induced by rendering; (2) label/depth accuracy of predicted layout surfaces; (3) voxel accuracy of occupied space and freespace."

"The depth accuracy of predicted layout surfaces and voxel accuracy of object occupancy are the most direct evaluations of our 3D model accuracy."

"We then evaluate on two metrics [57]: (1) scene completion with precision, recall and voxel-level IoU; (2) semantic scene completion with voxel-level IoU. Though our method is not optimized for voxel-based evaluation, we can still compare and evaluate to validate our approach on the scene-level 3D parsing performance."

"We can indirectly measure the effectiveness of our scene composition by comparing "Ours" to "GT-Region" (use the best fitting 3D object model and class for each ground truth 2D region)."

Discussions

"Our framework follows three steps: region proposal, classification and shape retrieval, and scene composition."

"For the region proposal step, we see improvements on applying the state-of-the-art RGBD region proposal method by Gupta et al. [58] instead of our previous Prim's based method."

"We observe less difference in overall performance if we use ground truth regions in our framework, which indicates the downstream step's robustness to errors from region proposals."

"The heavy occlusion characteristic of indoor scene makes the region classification harder, and accumulates errors to the retrieval step."

"To resolves the retrieval errors, we incorporate the scene composition step with combinatorial optimization to select a subset of shape candidates that fits the observation and scene regularity like no 3D overlap between shapes."

"Based on the previous discussions, we see the major sources of error of our framework are: (1) region classification error and (2) shape fitting error."

Conclusions

"We proposed an approach to predict a complete 3D scene model of individual objects and surfaces from a single RGBD image."

"We take a data-driven approach, generating sets of potential object regions, matching to regions in training images, and transferring and aligning associated 3D models while encouraging fit to observations and spatial consistency."

"We incorporate support inference to aid interpretation and propose a retrieval scheme that uses CNNs to classify regions and find objects with similar shapes."

"Compared to our earlier work [59], we demonstrate improvements due to better region proposals and use of CNNs to classify and retrieve similar shapes."

3D Reconstruction of Indoor Scenes via Image Registration [60]

This is a machine-generated summary of:

Li, Ce; Lu, Bing; Zhang, Yachao; Liu, Hao; Qu, Yanyun: 3D Reconstruction of Indoor Scenes via Image Registration [60].

Published in: Neural Processing Letters (2018).

Link to original: https://doi.org/10.1007/s11063-018-9781-0.

If you want to cite the papers, please refer to the original.

For technical reasons we could not place the page where the original quote is coming from.

Abstract-Summary "With the development of computer vision technologies, 3D reconstruction has become a hotspot."

"We aim at solving the problem of 3D reconstruction of an indoor scene with large vertical span."

"We propose a novel approach for 3D reconstruction of indoor scenes with only a Kinect."

"We achieve the 3D visual reconstruction model of indoor scene by the top-down image registration of point cloud data."

"This approach not only mitigates the sensor perspective restriction and achieves the indoor scene reconstruction of large vertical span, but also develops the fast algorithm of indoor scene reconstruction with large amount of cloud data."

"The experimental results show that the proposed algorithm has better accuracy, better reconstruction effect, and less running time for point cloud registration."

Introduction

"Until now, there is no method can easily get the color and depth information of 3D scene reconstruction at the same time."

"Newcombe [61] and Izadi [62] proposed a fusion method based on the Kinect data for 3D reconstruction."

"Choi et al. [63] proposed a global optimization method based on geometric registration which improved the accuracy of 3D reconstruction."

"Santos et al. [64] proposed a method that combined visual and depth information to compute more accurate convert matrices for 3D reconstruction of indoor scene."

"Considering that the traditional 3D reconstruction algorithm is time-consuming and not accurate, this paper proposed a 3D reconstruction approach for indoor scenes with only a Kinect."

"In order to solve the problem of 3D reconstruction of indoor scene that has large vertical span, we also proposed a method that based on the top and bottom registration."

The 3D Reconstruction of Indoor Scenes via ICP with Relative Coordinates

"The method includes the following four parts: (1) Data acquisition The color image and depth image of an indoor scene are acquired simultaneously by a Kinect sensor, and we fuse them to get the point cloud data. (2) Transformation matrix calculation We extract SIFT feature points in color images in adjacent frames, and then we calculate the transformation matrix between pair-wise adjacent frames based on feature matching which is treated as the initial matrix of ICP. (3) Point cloud registration We use the ICP algorithm to estimate the relative coordinate relationship of inter-frame point cloud data, and then we unite the center coordinates of multi-frame point cloud to obtain a multi-layer registration of 3D scene data."

"We make the inter-layer registration based on top and bottom layers. (4) Scene reconstruction We put the registration of 3D point cloud data and color texture information together to get the final visualization of 3D reconstruction model."

Experimental Results

"In order to verify the efficiency of the proposed algorithm, a comparison experiment is made between the frame registration before and after extraction points."

"We gain a better result of 3D reconstruction compared with Zhang [65] because we obtain a good initial transformation matrix based on the combination of SIFT and RANSAC and use relative coordinate ICP for data registration."

"In order to obtain the better outline of the 3D scene reconstruction, in this paper, the scene is got by annular shoots based on the top and bottom layers; each layer has twelve frames."

"Different from Zhang's method [65] which used each frame data for matching the scene, we implement image registration on the top and bottom layers and reconstruct the 3D scene hierarchically based on the interior of large buildings."

"Registration results of the five adjacent frames point cloud data use Zhang's method [65] and the proposed method, respectively."

Conclusions

"We proposed a new method of 3D reconstruction of indoor scene based on ICP."

"In order to quickly get the reconstruction result of the entire indoor scene, we propose a registration method based on the point cloud of the top and bottom layers."

"We match the point cloud data not only in the top layer but also in the bottom layer, which improves the visual effect of 3D reconstruction."

"The experimental results show that the proposed method can reduce the computational cost and can realize the 3D reconstruction of the indoor scene effectively."

Auxiliary Use and Detail Optimization of Computer VR Technology in Landscape Design [66]

This is a machine-generated summary of:

Shan, Pengyu; Sun, Wan: Auxiliary use and detail optimization of computer VR technology in landscape design [66].

Published in: Arabian Journal of Geosciences (2021).

Link to original: https://doi.org/10.1007/s12517-021-07131-1.

If you want to cite the papers, please refer to the original.

For technical reasons we could not place the page where the original quote is coming from.

Abstract-Summary "This paper analyzes the auxiliary use mode and detail optimization process of this technology in landscape design, and gives the comparative experiment of indoor and outdoor landscape design."

"Landscape design planning and design based on VR technology; secondly, the detailed optimization process of the combination of VR and VR-GIS is analyzed; finally, aiming at the problem of poor energy-saving effect of traditional methods, the energy-saving optimization design of data acquisition system is carried out."

"The auxiliary use and detail optimization in landscape design are given by using Lumion virtual reality (VR) platform."

"The experimental results show that the visual modeling error of this method is stable between 1 and 3%, the error is small and the stability is good; the average rendering time of batch objects is about 8.9 s, which is an efficient auxiliary use and detail optimization method for VR landscape design."

Landscape Planning and Design Based on Virtual Reality Assistant Technology

"In order to promote the organic combination of virtual reality (VR) technology and planning and design, so as to enhance the intuitiveness and authenticity of designers in space vision, so as to better carry out computer-aided scene design, this paper proposes to realize the digital design application of planning through VR technology."

"The advantages of VR technology aided planning and design are briefly analyzed; secondly, the VR modeling language tool (VRML) [67] used in VR technology aided design is introduced; finally, starting from the actual case, the overall structure of the project, the specific design process and the programming method are given by combining the drawing of CAD drawings and 3D character modeling."

"The VRML design process of virtual 3D scene is the process of placing 3D objects in a virtual space."

Detail Optimization of VR and VR-GIS

"After rendering by 3dstudio and other modeling tools, the unified data format is used to output the landscape reconstruction map [68]."

"In order to ensure the smooth implementation of the theory, firstly, the landscape image is obtained through overlay remote sensing, and then the VR development of the reconstructed image is completed by using the toolkit of Silicon Graphics GL graphics library."

"In the comprehensive consideration of the digital model designed in this paper, the cylindrical panorama is used to represent the landscape in the following process [69]."

"The construction of landscape 3D terrain model can be summarized as information digitization and DEM generation [70]."

"Through the previous section, we can get the three-dimensional terrain model and the reconstruction panorama of the landscape."

"F(t, s) is the established digital model of landscape reconstruction."

Energy Saving Design of Data Acquisition System and Virtual Image System

"The landscape image elevation and elevation diffusion weight coefficient of VR need to take absolute values, which are expressed as follows: Where, The central moment of the edge correlation of the imaging system is optimized as follows Where, Therefore, the final results of energy-saving calculation of 3D image system are as follows: Where, Through the aforementioned process, the energy saving process of imaging system is completed."

"Collection of landscape data to be modeled, import of 3D landscape model, detailed processing of VR scene simulation, and visualization rendering of VR landscape are the key steps of design."

"After the design of 3D visualization model of landscape is completed, the details of landscape design should be processed to show the rich, vivid and visual VR landscape simulation design effect The material processing of VR landscape design can be completed through the Edit button in Lumion software."

Simulation Experiment and Result Analysis

"This method is easy to use, rendering objects with high efficiency, good effect, has a strong plant library function; 3DSMAX VR simulation design method using process is more complex, difficult to understand, rendering effect is better, but the efficiency is low, and does not have the plant library function; SKETCHUP VR simulation design method is also easy to use, but the effect of rendering landscape is general and effective."

"The last experiment only consumes 0.2 s to render a single object; 3dsmax VR simulation design method is contrary to the method in this paper, with the expansion of the experiment, the rendering time increases continuously, and the longest time in five tests is as high as 8 s; SKETCHUP VR simulation design method in the whole test process shows the trend of first increase and then decrease, and the third experiment takes the most rendering time."

Conclusion

"This paper puts forward a kind of energy-saving design scheme of Forest Park Landscape Based on VR, and verifies the effectiveness of the design through the effective proof of the experiment."

"Lumion uses Irregular Triangulation model to create visualization effect for the imported landscape model; adds material and landscape effects to the visual landscape model to improve the details of landscape simulation design; uses get Value and set Value program to render landscape elements in batch, and this method is batch rendering The efficiency of the object is higher than that of a single object, which saves a lot of time for the designer, and outputs the visual image of VR Landscape after storage."

"Through the indoor and outdoor landscape design process, we verify the effectiveness of the VR Landscape energy-saving design, design the simulation contrast experiment, and fully prove the effectiveness of the forest park landscape energy-saving design based on VR through the experimental data analysis."

Levy Flight and Fitness Distance Balance-Based Coyote Optimization Algorithm for Effective Automatic Generation Control of PV-Based Multi-Area Power Systems [71]

This is a machine-generated summary of:

Tabak, Abdulsamed; Duman, Serhat: Levy Flight and Fitness Distance Balance-Based Coyote Optimization Algorithm for Effective Automatic Generation Control of PV-Based Multi-Area Power Systems [71].

Published in: Arabian Journal for Science and Engineering (2022).

Link to original: https://doi.org/10.1007/s13369-022-07004-z.

If you want to cite the papers, please refer to the original.

For technical reasons we could not place the page where the original quote is coming from.

Abstract-Summary "Levy flight and Fitness Distance Balance (FDB)-based coyote optimization algorithm (LRFDBCOA) is proposed to improve the performance of automatic generation control (AGC) of three different PV-based interconnected power systems."

"The achievement of the proposed algorithm is evaluated on the two-area PV-reheat thermal power system with PI controller, which is widely used in the literature, and the results are compared with the results obtained using genetic algorithm (GA), firefly algorithm (FA), and modified whale optimization algorithm (M-WOA) introduced in previous studies."

"The control performance of the proposed algorithm is demonstrated on the newly constructed two-area PV-non-reheat thermal power system with governor dead band (GDB) and two-area PV-multi-source power system with electrolyser and fuel cell."

"Both the performance of the proposed algorithm on AGC and the compatibility of the two newly tested systems with PID, TID and FOPID controllers are revealed."

Introduction

"Contributions of the study: (1) The compatibility of two new PV grid-based interconnected power systems with PID, TID and FOPID controllers is examined. (2) The LRFDBCOA algorithm was first applied to the optimal power flow (OPF) problem, which is one of the power systems planning problems."

"The performance of the proposed LRFDBCOA algorithm in finding the optimal parameters of the proposed controller structures has been tested in both linear and nonlinear interconnected power systems. (3) The hydrogen storage system, which has been used quite frequently as an energy store recently, has been included in both areas in test-system-3 (two-area PV-multi-source power system with electrolyser and fuel cell) and its effect on the interconnected system has been observed. (4) The control performance of the proposed algorithm against disturbances in the power system is tested. (5) Superior performance of proposed algorithm over compared algorithms is demonstrated."

Modelling of Power Systems

"First, two area PV-reheat thermal power plant is analysed."

"The last system analysed in the paper is the two-area PV-multi-source power system with electrolyser and fuel cell."

"An electrolyser and a fuel cell are added to both areas, which store energy in hydrogen during energy surplus and generate energy from hydrogen when needed."

"Hydrogen electrolyser (HE), fuel cell (FC) and hydrogen storage tank are the three main components that required to ensure continuity in energy supply and support energy generation resources."

"The third action is when the load needs electricity, the FC supports the grid by converting stored hydrogen into electricity."

"HE units generate H_2 as fuel for FC and are utilized for continuous power quality by absorption quickly changing load demands in the power system."

"FC stands out compared to other power generation systems with its advantages such as high efficiency, low pollution and no moving parts."

Utilized Control Structures

"The structure of PI, PID, TID and FOPID controllers, which are frequently used in the control of AGC, is briefly mentioned."

"The structure of the TID controller is similar to the PID controller."

"TID controller can be seen as a combination of fractional order and integer order controllers and parameter setting is as easy as PID controller."

"TID controller provides superiority to PID against disturbances in the output signal [72, 73]."

The Proposed LRFDBCOA Algorithm

"The fitness value of the newly created social status is calculated as follows."

"The jackals with the social conditions that could contribute the most to the pack's cultural trend were represented using the FDB selection method."

"These are expressed as fitness values and distance of the population from the best solution candidate (p_{best})."

"The steps of the FDB selection method [74–76] are identified as follows. (1) The proposed algorithm for solving an optimization problem consists of n number of solution candidates."

"W = 0.5 was set. (5) The n-dimensional S_p vector representing the FDB values of the solution candidates in the population is described as follows."

"After the S_p vector representing the FDB values of the solution candidates of the problem was created, individuals were selected from the population by using the roulette wheel method in the selection of the solution candidates to guide the search process."

Fitness Function

"Optimum tuning of controller parameters provides satisfactory control performance in AGC."

"A fitness function is needed to tune the controller parameters optimally."

"The coefficients must be chosen carefully as it affects the contribution of the components in the objective function."

Simulation Results

"The simulation results obtained by LRFDBCOA are compared with GA [77], FA [77] and modified whale optimization algorithm (M-WOA) [78] algorithms-based test system-1 with PI controller."

"Two-area PV-non reheat thermal power plant with governor dead band nonlinearities called test system-2 is evaluated to test the performance of proposed LRFDBCOA algorithm for commonly used controllers in power systems such as PID, TID and FOPID."

"The parameters of the TID controller in test system-2 are tuned with LRFDBCOA, GA and AOA, and the performance of the algorithms on the controller is compared."

"Two-area PV-multi-source with fuel cell energy storage structure called test system-3 is analysed to evaluate the achievement of proposed LRFDBCOA algorithm for commonly utilized controllers in the power systems such as PID, TID and FOPID."

"The parameters of the FOPID controller in the test system-3 are optimized with LRFDBCOA, AOA and GA and the results obtained by algorithms are compared."

Conclusion

"This paper proposes the LRFDBCOA algorithm to enhance the achievement of AGC of three different PV-based interconnected power systems."

"The two-area PV-reheat thermal power system with PI controller (test system-1), two-area PV-non-reheat thermal power system with governor dead band (GDB) nonlinearities (test system-2) and two-area PV-multi-source power system with electrolyser and fuel cell (test system-3) controlled by proportional–integral–derivative (PID), tilt-integral-derivative (TID) and fractional order PID (FOPID) controller are evaluated."

"The load frequency control of these two systems (test system-2 and test system-3) added to the literature is important."

"It has been shown that load frequency control can be performed successfully when the PV system is used together with a non-reheat thermal power plant with GDB or thermal/hydro/gas power plants."

Reverse Quantum Annealing Approach to Portfolio Optimization Problems [79]

This is a machine-generated summary of:

Venturelli, Davide; Kondratyev, Alexei: Reverse quantum annealing approach to portfolio optimization problems [79].

Published in: Quantum Machine Intelligence (2019).

Link to original: https://doi.org/10.1007/s42484-019-00001-w.

If you want to cite the papers, please refer to the original.

For technical reasons we could not place the page where the original quote is coming from.

Abstract-Summary "We investigate a hybrid quantum-classical solution method to the mean-variance portfolio optimization problems."

"We investigate several options to run the quantum computation optimally, ultimately discovering that the best results in terms of expected time-to-solution as a function of number of variables for the hardest instances set are obtained by seeding the quantum annealer with a solution candidate found by a greedy local search and then performing a reverse annealing protocol."

"The optimized reverse annealing protocol is found to be more than 100 times faster than the corresponding forward quantum annealing on average."

Introduction

"Several proposed applications of the D-Wave machine to portfolio optimization [80, 81], dealt with portfolios that were too small to evaluate the scaling of the chosen approach with the problem size."

"We extensively tune the D-Wave machine's runs using the hybrid method of reverse annealing which is available only on the most advanced models to date, and whose usage has never been published before on structured optimization problems."

"This observation spurred in the last few years a couple of proof-of-principle papers that were performing runs on the first-generation D-Wave machines solving discrete portfolio optimization problems [80, 81]."

"The discrete portfolio optimization problem is known to be NP-complete [82] and is much harder to solve than continuous mean-variance portfolio optimization problem, which can be solved efficiently by interior point algorithms for convex optimization."

Portfolio Optimization Beyond Markowitz

"The problem we are trying to solve here is construction of the optimal portfolio from the assets with known characteristics such as asset returns, volatilities, and pairwise correlations."

"From these returns, we calculated total realized return and realized volatility for each asset (which, obviously, differ from the their expected values μ and σ) and for the portfolio as a whole."

"With $\rho = 0.1$, $\mu = 0.075$, $\sigma = 0.15$, and the constant risk-free interest rate set at $r_0 = 0.015$, the expected Sharpe ratio for each asset in a portfolio is 0.4."

"The expected Sharpe ratio for the portfolio of N assets is significantly larger due to the diversification and low correlation between the assets, e.g., for a 48-asset portfolio, we would expect Sharpe ratio values from 0.5 (25th percentile) to 2.1 (75th percentile) with a mean around 1.4."

Quantum Annealing Hybrid Solver

"This forward quantum annealing procedure on DW2000Q can be ideally described as attempting to drive the evolution of the following time-dependent Hamiltonian: where the term multiplying A[t] is a Hamiltonian describing an independent collection of local transverse fields for each spin of the system [83]."

"At the end of the annealing run, A[t] = 0 and the system is projected on the computational basis by a measurement of each qubit magnetization."

"Recent research attempts to model accurately the forward annealing dynamics of DW2000Q and earlier models of D-Wave machines by means of ab initio reasoning have been moderately successful on carefully crafted problems [84]."

"The total duration of the selected reverse anneal protocol is $2\tau + \rho$ as opposed to τ for the forward anneal."

Experimental Results

"A common metric to benchmark performance of non-deterministic iterative heuristics against quantum annealing is Time-to-Solution (TTS)."

"In the case of portfolio optimization problem the solution (chromosome) is a vector $q = (q_1,\ldots,q_N)$ consisting of N elements (genes) that can take binary values $q_i \in\{0,1\}$."

"This simple procedure represents a statistical pre-characterization of the instance ensemble sensitivity to the parameters, and it is common for quantum annealing benchmarking initial studies of a problem [85–87]."

"With regard to reverse annealing, we should note that we don't pre-characterize all 30 instances, but only those that have not been solved by the greedy heuristics (Algorithm 1) that is used to initialize the reverse annealing procedure."

"The first observation considerably simplifies the parameter setting strategy when considering operational scenarios with a new unknown instance, while the second observation is consistent with recent results and interpretation of the annealing pause tested on native Chimera problems [88]."

Conclusion and Next Steps

"We did investigate a combinatorial optimization problem in Finance, identifying parameters and mappings that are related to real-world portfolio optimization approaches and that are compatible with the precision requirements of current D-Wave machines."

"Consistently with the rest of the literature, all reported times for runs on the DW2000Q have been calculated in terms of the annealing times τ (or ρ), ignoring all other necessary setup and iteration times, since they don't scale with the problem size and they could be reduced in the future as system integration of the chips improves."

"Research papers on reverse annealing and paused annealing are just beginning to appear [88, 89], and it seems that for hard problems it might be beneficial to implement a longer pause, larger than 100 ms, to gain a sizeable TTS advantage."

Sparse Flexible Design: A Machine Learning Approach [90]

This is a machine-generated summary of:

Chan, Timothy C. Y.; Letourneau, Daniel; Potter, Benjamin G.: Sparse flexible design: a machine learning approach [90].

Published in: Flexible Services and Manufacturing Journal (2022).

Link to original: https://doi.org/10.1007/s10696-021-09439-2.

If you want to cite the papers, please refer to the original.

For technical reasons we could not place the page where the original quote is coming from.

Abstract-Summary "For a general production network, state-of-the-art methods for constructing sparse flexible designs are heuristic in nature, typically computing a proxy for the quality of unseen networks and using that estimate in a greedy manner to modify a current design."

"This paper develops two machine learning-based approaches to constructing sparse flexible designs that leverage a neural network to accurately and quickly predict the performance of large numbers of candidate designs."

"We demonstrate how clinical constraints can be easily absorbed into the machine learning subroutine and how our sparse flexible treatment networks meet or beat the performance of those designed by state-of-the-art methods."

Introduction

"We provide a first look at how modern machine learning (ML) approaches can be used to solve a classical management science problem, namely the construction of sparse flexible designs."

"In the case of an asymmetric, unbalanced network, the focus tends to be on developing constructive approaches for designing sparse flexible networks [91–96]."

"This paper adds to the rich literature on process flexibility by developing a constructive, ML-based approach for sparse flexible design of general networks, which are not required to be balanced (same number of products and plants) or symmetric (identical plant capacities and i.i.d."

"We demonstrate how our ML-based approach can design sparse networks with comparable performance to Princes Margaret's existing, dense network."

"The ability to evaluate a large neighborhood of networks opens the door to new heuristic approaches for constructing sparse flexible designs."

Literature Review

"Feng et al. [94] solve a relaxation of a stochastic integer programming problem and use the optimal solution to determine the next arc to add."

"Chen et al. [96] develop an approach that adds multiples arcs at once to generate a final design that has good performance with high probability."

"Bello et al. [97] use a reinforcement learning approach to solve travelling salesman problem with up to 100 nodes."

"Khalil et al. [98] develop a framework to learn optimization problems over graphs by using a combination of reinforcement learning and graph embeddings."

"Larsen et al. [99] propose a framework for generating descriptions of optimal solutions (ranging from objective value to decision variable values) for a load planning problem in the railway industry."

"Chen et al. [100] propose a fixed-cost network design problem with unknown flow response by embedding a prediction model into the original network design problem and develop two solution approximation algorithms."

Model Development

"Sections, we develop a machine learning method for predicting the expected maximum flow of arbitrary network designs, and then develop a constructive heuristic for sparse network design that leverages this prediction model."

"To train a prediction model that can accurately approximate this function, we follow a supervised learning approach by using a set of randomly generated network designs (inputs) and their corresponding expected max flow values (outputs)."

"We consider two baseline predictors, a naive one that simply predicts the average value in the training set and a linear regression (LR) model that predicts the expected max flow based only on the number of arcs in the design."

"Suppose in a network with 10 plants and 10 products we greedily add arcs to reach an optimal design with 20 arcs, which we know to be the long chain."

Comparison Against Existing Methods

"In each test setting, the designs produced by the PS and PSRH heuristics are among the best performing ones."

"In test setting 1, the MDEP and PSRH heuristics significantly outperformed the others in expected max flow since they achieved the long chain."

"While no single heuristic dominated in all test settings and metrics, PSRH produced a network in each case that was highly competitive, if not the best."

"When examining the performance of the baselines, we notice the interesting result that for all three metrics and all three test settings, the PSRH heuristic completely dominates the average performance of the networks in the training set, and mostly dominates even the single best network in the training set."

"In test setting 1, this upfront cost allows the PSRH heuristic to evaluate over 150,000 network designs during a single run of the algorithm."

Radiation Therapy Case Study

"We then demonstrate the application of our ML-based heuristics in three computational case studies: (1) redesigning the patient group-LINAC assignment policies to form a sparse network with comparable performance as the existing network, (2) determining the value of operating a fleet of homogeneous LINACs, and (3) assessing the impact of an initial patient scheduling policy that spreads treatments more evenly across LINACs."

"With just 56 arcs—fewer than two arcs per patient group on average—the PSRH network achieves 99.94% of the expected max flow of the existing PM treatment network."

"Having additional ABC devices would make some LINACs more flexible with respect to addressing demand for ABC-based treatments, allowing other LINACs to specialize, and resulting in a net reduction in the total number of arcs compared to the current PM network design."

Discussion and Conclusion

"Focusing on the sparse flexible design problem, we develop ML-based algorithms for network design that offer comparable or better performance than state-of-the-art methods, which are based on deep mathematical insights."

"Although our ML-based heuristics show promise and possess notable computational efficiencies (e.g., the ability to make accurate, cost-free predictions for a huge number of candidate network designs) the approaches proposed in this paper are limited by the large number of arc combinations that are considered during construction of the network design."

"Methods that apply ML-based heuristics to solve combinatorial problems in a search like fashion need to strike the right balance between exploiting the use of a large number of predictions and not allowing the search space to explode in size."

"Future use cases of ML-based heuristics like the ones developed in this paper are likely best suited for problems that require repeated solution of optimization problems in an online fashion."

Design of Mutated Harmony Search Algorithm for Data Dissemination in Wireless Sensor Network [101]

This is a machine-generated summary of:

Poonguzhali, P. K.; Ananthamoorthy, N. P.: Design of Mutated Harmony Search Algorithm for Data Dissemination in Wireless Sensor Network [101].

Published in: Wireless Personal Communications (2019).

Link to original: https://doi.org/10.1007/s11277-019-06882-1.

If you want to cite the papers, please refer to the original.

For technical reasons we could not place the page where the original quote is coming from.

Abstract-Summary "Sensor based networks are a special category of a distributed network that is used to provide communication among the sensor nodes."

"The wireless sensor network consists of various classes of sensors like thermal, infrared, optical, and seismic for the measurement of temperature, heat, radiation, humidity which requires constant monitoring and detection in specific events."

"An optimal energy efficient cluster based routing algorithm is required for effective data diffusion."

"Conventional protocol like LEACH, HEED, PEGASIS protocol etc., fails to balance the network load and the coverage area when the sensor nodes are deployed in large scale."

"In harmony search algorithm (HSA) absence of gradient search leads the parameter search is in the local region where the required optimal solution remains outside the local region."

"HSA is heuristic algorithm uses random search with constant harmony memory consideration rate."

"This paper focuses on designing a Meta-heuristic optimized routing protocol for a distributed network using mutated harmony search algorithm (MHSA) to improve the energy efficiency by simultaneously analyzing the cluster patching."

"Cluster patching is examined for improving network coverage and connectivity thereby to optimize the energy distribution in WSN."

"MHSA is a refinement of heuristic algorithm in the global search by adjusting the harmony memory consideration rate HMCR."

"The computational time and the experimental results show the proposed MHSA gives 85% of connectivity improved that ensures the success of cluster formation for an increased number of nodes to increase increases the network lifetime when compared with existing algorithms."

Introduction

"Unequal Clustering Size model, Energy Efficient Clustering Scheme, Power-Efficient Gathering in Sensor Information Systems, Threshold sensitive Energy Efficient sensor Network protocol, Adaptive Threshold sensitive Energy Efficient sensor Network protocol, Two-Tier Data Dissemination, Concentric Clustering Scheme and Hierarchical Geographic Multicast Routing [102, 103] related with hierarchal routing tries to decrease the energy usage of sensor nodes during route or by optimizing some metrics."

"Suggested balance in modifying the harmony search parameters gives rise to energy efficiency optimization for selecting the cluster head."

"Harmony search (HS) is a music-based heuristic optimization method, experimented by Geem and others in 2001 as evolutionary optimization algorithm."

"Many researchers tried to contribute their work in Harmony search algorithm (HAS) for various applications to get optimized solution vector."

"Particle swarm optimization algorithm (PSOA) is a heuristic universal enhancement procedure and very efficient global search algorithm based on swarm intelligence."

Related Work

"An improvement in harmony search algorithm is focused to optimize the cluster head selection implemented simultaneously cluster patching is analyzed [104]."

"MHSA algorithm is computed at the Base station (BS).The parameters for the MHSA algorithm is derived from cluster and Node information table."

"The sensor nodes extract the information from ASG message to identify as CH or cluster member to switch over to data transmission module [105]."

"At the Sensor Node Level there are two modules Cluster set up module and Data transmission module."

"At the Base station level the Cluster Management Module is responsible for forming clusters by selecting the optimal cluster head by performing the Mutated HSA incorporated in Cluster Algorithm and the updated information is cached in the Node Information Table."

"Data Transmission Module consists of Data Aggregation cluster head aggregates the information send by member nodes and Data transmission modules."

Performance Evaluation of MHSA Protocol

"The performance evaluation of the Mutated Harmony Search Algorithm is studied with the five different quantitative metrics Fitness, Network Connectivity, Coverage rate, Execution time and Energy dissipation are analyzed and simulation results are compared and the performance proved the proposed algorithm is superior among existing HAS, PSO, GA."

"Concentrating on the coverage area and to maintain the network connectivity multi-hop communication between a data source CH and data sink BS is followed thereby reducing the energy consumption of the CH s. The coverage rate is a useful metric in terms of quality of service to monitor the sensing area covered by the sensate radius [106, 107]."

"Energy is dissipated in cluster head in various forms associated with energy consumption, aggregation while acquiring data from member nodes, and transmission process [108, 109]."

"Re-clustering begins in the next round, sensor nodes in the network will have sensed data all times the transmission process will come to an end only when the sensed data are aggregated by the cluster head and forwarded to BS."

Results and Discussion

"For improving the network coverage rate and network life time cluster patching is analyzed."

"Continuously increasing the sensor nodes 80–100 in sensing field the connectivity becomes 50% to that of the 85% for 10–20 nodes."

"Sensor nodes directly communicate with CH in Intra cluster communication."

"Percentage of active nodes is improved by increasing coverage area with node degree 'd' or cluster radius 'R'."

"If node degree d is increased it results in improving the transmission range and energy depletion of the nodes maximizes the percentage therefore for the desired coverage R value is changed as a result the output changed."

Conclusion

"The proposed Mutated Harmony Search Algorithm has several advantages when compared with other heuristic optimization techniques, setting the preliminary value for the solution vectors and it is free from complex calculus and divergence."

"The proposed optimized technique for choosing CH in a wide network with numerous sensor nodes can balance the energy consumption."

"Experimental results shows the execution time for the proposed MHSA is less that results MHSA to fast converge with best fitness value."

"In Mutated Harmony Search Algorithm PAR and BW range are imperative algorithm parameters in adjusting the solution vectors, and be likely essential in control the convergence rate in order to obtain best solution vectors."

"In step 3 the parameters PAR and HMCR are useful in locating the enhanced solutions in both global and local search area."

"In the heuristic algorithm pitch adjustment ratio and bandwidth values adjusted in initialization step and remain fixed for new solution vectors."

References

1. Lin C-C, Huang W, Liu W-Y, Wu S-F (2019) A novel centrality-based method for visual analytics of small-world networks. J Vis. https://doi.org/10.1007/s12650-019-00582-5
2. van Ham F, Wattenberg M (2008) Centrality based visualization of small world graphs. Comput Graph Forum 27:975–982
3. Bertault F (2000) A force-directed algorithm that preserves edge-crossing properties. Inf Process Lett 74:7–13

4. Bhandari A, Gupta A, Das D (2017) Betweenness centrality updation and community detection in streaming graphs using incremental algorithm. In: Proceedings of the 6th international conference on software and computer applications. ACM Press, New York, pp 159–164
5. Bonchi F, Morales GDF, Riondato M (2016) Centrality measures on big graphs: exact, approximated, and distributed algorithms. In: Proceedings of the 25th international conference companion on world wide web. ACM Press, New York, pp 1017–1020
6. Freeman LC (1979) Centrality in social networks conceptual clarification. Soc Networks 1:215–239
7. Opsahl T, Agneessens F, Skvoretz J (2010) Node centrality in weighted networks: generalizing degree and shortest paths. Soc Networks 32:245–251
8. Crnovrsanin T, Muelder CW, Faris R, Felmlee D, Ma K-L (2014) Visualization techniques for categorical analysis of social networks with multiple edge sets. Soc Networks 37:56–64
9. Junger M, Leipert S, Mutzel P (1998) A note on computing a maximal planar subgraph using PQ-trees. IEEE Trans Comput Aided Des Integr Circ Syst 17:609–612
10. Zhou H, Zhang J, Zhou Y, Saitou K (2019) Multi-component topology optimization for die casting (MTO-D). Struct Multidiscip Optim. https://doi.org/10.1007/s00158-019-02317-4
11. Lyu N, Saitou K (2005) Topology optimization of multicomponent beam structure via decomposition-based assembly synthesis. J Mech Des 127(2):170–183
12. Yildiz AR, Saitou K (2011) Topology synthesis of multicomponent structural assemblies in continuum domains. J Mech Des 133(1):011008
13. Guirguis D, Hamza K, Aly M, Hegazi H, Saitou K (2015) Multi-objective topology optimization of multi-component continuum structures via a Kriging-interpolated level set approach. Struct Multidiscip Optim 51(3):733–748
14. Zhou Y, Saitou K (2018) Gradient-based multi-component topology optimization for stamped sheet metal assemblies (MTO-S). Struct Multidiscip Optim 58(1):83–94
15. Thakkar N, Paliwal P (2022) Quad-level MCDM framework to analyse technology combinations for sustainable micro-grid planning in uncertainty domain. Arab J Sci Eng. https://doi.org/10.1007/s13369-022-07019-6
16. Danish MS, Senjyu T, Sabory NR, Danish SM, Ludin GA, Noorzad AS, Yona A (2017) Afghanistan's aspirations for energy independence: water resources and hydropower energy. Renew Energy 113:1276–1287. https://doi.org/10.1016/j.renene.2017.06.090
17. Aberilla JM, Schmid AG, Stamford L, Azapagic A (2020) An integrated sustainability assessment of synergistic supply of energy and water in remote communities. Sustain Prod Consump 22:1–21. https://doi.org/10.1016/j.spc.2020.01.003
18. Solangi Y, Tan Q, Mirjat N, Valasai G, Khan M, Ikram M (2019) An integrated Delphi-AHP and fuzzy TOPSIS approach toward ranking and selection of renewable energy resources in Pakistan. PRO. https://doi.org/10.3390/pr7020118
19. Rubio-Aliaga A, García-Cascales M, Sánchez-Lozano J, Molina-Garcia A (2021) MCDM-based multidimensional approach for selection of optimal groundwater pumping systems: design and case example. Renew Energy. https://doi.org/10.1016/j.renene.2020.08.079
20. Erdoğan M (2015) An integrated multi-criteria decision-making methodology based on type-2 fuzzy sets for selection among energy alternatives in Turkey. Iran J Fuzzy Syst 12:1–25
21. Balin A, Baraçli H (2017) A fuzzy multi-criteria decision making methodology based upon the interval type-2 fuzzy sets for evaluating renewable energy alternatives in Turkey. Technol Econ Dev Econ. https://doi.org/10.3846/20294913.2015.1056276
22. Paliwal P, Patidar N, Nema R (2014) A novel method for reliability assessment of autonomous PV-wind-storage system using probabilistic storage model. Int J Electr Power Energy Syst 55:692–703. https://doi.org/10.1016/j.ijepes.2013.10.010
23. Kamdar R, Paliwal P, Kumar Y (2020) A hybrid multi-agent-based BFPSO algorithm for optimization of benchmark functions. J Circ Syst Comput 29(7):2050112. https://doi.org/10.1142/S0218126620501121

24. Hoekstra AY, Chapagain AK (2007) Water footprints of nations: water use by people as a function of their consumption pattern. Water Resour Manag 21:35–48. https://doi.org/10.1007/s11269-006-9039-x
25. Kumar A, Singh A, Deng Y, He X, Kumar P, Bansal R (2019) Integrated assessment of a sustainable microgrid for a remote village in hilly region. Energy Convers Manag. https://doi.org/10.1016/j.enconman.2018.10.084
26. Lourenco RM, Tavora LMN, Assuncao PAA, Thomaz LA, Fonseca-Pinto R, Faria SMM (2022) Enhancement of light field disparity maps by reducing the silhouette effect and plane noise. Multidim Syst Sign Process. https://doi.org/10.1007/s11045-021-00807-7
27. Li J, Li ZN (2013) Continuous depth map reconstruction from light fields. In: IEEE international conference on multimedia and expo. IEEE, Piscataway, pp 1–6
28. Wanner S, Goldluecke B (2012) Globally consistent depth labeling of 4d light fields. In: IEEE conference on computer vision and pattern recognition. IEEE, Piscataway, pp 41–48
29. Tao MW, Hadap S, Malik J, Ramamoorthi R (2013) Depth from combining defocus and correspondence using light-field cameras. In: IEEE international conference on computer vision. IEEE, Piscataway, pp 673–680
30. Williem W, Park IK, Lee KM (2018) Robust light field depth estimation using occlusion-noise aware data costs. IEEE Trans Pattern Anal Mach Intell 40:2484–2497
31. Schilling H, Diebold M, Rother C, Jähne B (2018) Trust your model: light field depth estimation with inline occlusion handling. In: Conference on computer vision and pattern recognition. IEEE, Piscataway, pp 4530–4538
32. Bigun J (1987) Optimal orientation detection of linear symmetry. In: IEEE first international conference on computer vision. IEEE, Piscataway, pp 433–438
33. Neri A, Carli M, Battisti F (2015) A multi-resolution approach to depth field estimation in dense image arrays. In: IEEE international conference on image processing. IEEE, Piscataway, pp 3358–3362
34. Zhang S, Sheng H, Li C, Zhang J, Xiong Z (2016) Robust depth estimation for light field via spinning parallelogram operator. Comput Vis Image Underst 145:148–159
35. Strecke M, Alperovich A, Goldluecke B (2017) Accurate depth and normal maps from occlusion-aware focal stack symmetry. In: IEEE conference on computer vision and pattern recognition. IEEE, Piscataway, pp 2529–2537
36. Tsai YJ, Liu YL, Ouhyoung M, Chuang YY (2020) Attention-based view selection networks for light-field disparity estimation. AAAI Conf Artif Intell 34:12095–12103
37. Zhang X, Li Y, Zhang Z, Konno K, Hu S (2019) Intelligent Chinese calligraphy beautification from handwritten characters for robotic writing. Vis Comput. https://doi.org/10.1007/s00371-019-01675-w
38. Ma Z, Su J (2017) Aesthetics evaluation for robotic Chinese calligraphy. IEEE Trans Cogn Dev Syst 9(1):80–90
39. Mueller S, Huebel N, Waibel M, D'Andrea R (2013) Robotic calligraphy-learning how to write single strokes of Chinese and Japanese characters. In: 2013 IEEE/RSJ international conference on intelligent robots and systems. IEEE, Piscataway, pp 1734–1739
40. Sun Y, Qian H, Xu Y (2014) Robot learns Chinese calligraphy from demonstrations. In: 2014 IEEE/RSJ international conference on intelligent robots and systems. IEEE, New York, pp 4408–4413
41. Igarashi T, Matsuoka S, Tanaka TH (1999) A sketching interface for 3D freeform design. In: SIGGRAPH'99. ACM, New York
42. Yao F, Shao G, Yi J (2004) Trajectory generation of the writing-brush for a robot arm to inherit block-style Chinese character calligraphy techniques. Adv Robot 18(3):331–356
43. Zou C, Guo R, Li Z, Hoiem D (2018) Complete 3D scene parsing from an RGBD image. Int J Comput Vis. https://doi.org/10.1007/s11263-018-1133-z
44. Hedau V, Hoiem D, Forsyth D (2009) Recovering the spatial layout of cluttered rooms. IEEE, Piscataway

45. Flint A, Murray DW, Reid I (2011) Manhattan scene understanding using monocular, stereo, and 3D features. IEEE, Piscataway
46. Schwing AG, Urtasun R (2012) Efficient exact inference for 3D indoor scene understanding. IEEE, Piscataway
47. Zhang J, Kan C, Schwing AG, Urtasun R (2013) Estimating the 3D layout of indoor scenes and its clutter from depth sensors. IEEE, Piscataway
48. Zhang Y, Song S, Tan P, Xiao J (2014) Panocontext: a whole-room 3D context model for panoramic scene understanding. In: European conference on computer vision. Springer, Cham, pp 668–686
49. Hedau V, Hoiem D, Forsyth D (2010) Thinking inside the box: using appearance models and context based on room geometry. ACM, New York
50. Xiao J, Russell BC, Torralba A (2012) Localizing 3D cuboids in single-view images. NIPS, Islamabad
51. Zhao Y, Zhu SC (2013) Scene parsing by integrating function, geometry and appearance models. IEEE, Piscataway
52. Lin D, Fidler S, Urtasun R (2013) Holistic scene understanding for 3D object detection with RGBD cameras. IEEE, Piscataway
53. Gupta S, Arbeláez P, Girshick R, Malik J (2015) Aligning 3D models to RGB-D images of cluttered scenes. In: Proceedings of the IEEE conference on computer vision and pattern recognition. IEEE, Piscataway, pp 4731–4740
54. Xiang Y, Kim W, Chen W, Ji J, Choy C, Su H, Mottaghi R et al (2016) Objectnet 3D: a large scale database for 3D object recognition. European Conference Computer Vision (ECCV), Milano
55. Dai A, Chang AX, Savva M, Halber M, Funkhouser T, Nießner M (2017) Scannet: richly-annotated 3D reconstructions of indoor scenes. In: Proceedings computer vision and pattern recognition (CVPR). IEEE, Piscataway
56. Guo R, Hoiem D (2013) Support surface prediction in indoor scenes. IEEE, Piscataway
57. Song S, Yu F, Zeng A, Chang AX, Savva M, Funkhouser T (2017) Semantic scene completion from a single depth image. In: Proceedings of 30th IEEE conference on computer vision and pattern recognition. IEEE, Piscataway
58. Gupta S, Girshick R, Arbeláez P, Malik J (2014) Learning rich features from RGB-D images for object detection and segmentation. In: European conference on computer vision. Springer, Cham, pp 345–360
59. Guo, R., Zou, C., & Hoiem, D. (2015). *Predicting complete 3D models of indoor scenes.* arXiv preprint arXiv:1504.02437.
60. Li C, Lu B, Zhang Y, Liu H, Qu Y (2018) 3D Reconstruction of indoor scenes via image registration. Neural Process Lett. https://doi.org/10.1007/s11063-018-9781-0
61. Newcombe RA, Izadi S, Hilliges O, Molyneaux D, Kim D, Davison AJ, Kohi P, Shotton J, Hodges S, Fitzgibbon A (2011) KinectFusion: real-time dense surface mapping and tracking. In: IEEE international symposium on mixed and augmented reality. IEEE, Piscataway, pp 127–136
62. Izadi S, Kim D, Hilliges O, Molyneaux D, Newcombe R, Kohli P, Shotton J, Hodges S, Freeman D, Davison A (2011) KinectFusion: real-time 3D reconstruction and interaction using a moving depth camera. In: Proceedings of the 24th annual ACM symposium on user interface software and technology. ACM, New York, pp 559–568
63. Choi S, Zhou QY, Koltun V (2015) Robust reconstruction of indoor scenes. In: IEEE conference on computer vision and pattern recognition. IEEE, Piscataway, pp 5556–5565
64. Santos DRD, Basso MA, Khoshelham K, Oliveira ED, Pavan NL, Vosselman G (2016) Mapping indoor spaces by adaptive coarse-to-fine registration of RGB-D data. IEEE Geosci Remote Sens Lett 13(2):262–266
65. Zhang CH, Zhu RJ, Zhuang Y (2013) Indoor 3D scene reconstruction based on Kinect depth camera. Dalian Univ Technol 5:1–8

66. Shan P, Sun W (2021) Auxiliary use and detail optimization of computer VR technology in landscape design. Arab J Geosci. https://doi.org/10.1007/s12517-021-07131-1
67. Li Z, Cheng Y, Yuan Y (2018) Research on the application of virtual reality technology in landscape design teaching. Educ Sci Theory Pract 18(5):37. https://doi.org/10.12738/estp.2018.5.037
68. Li L, Chen F, Yu Y, Shen S (2020) Using 3D virtual VR technology to build rural landscape environment. In: International conference on applications and techniques in cyber security and intelligence. Springer, Cham, pp 313–319
69. Pan X, Hamilton AFDC (2018) Why and how to use virtual reality to study human social interaction: the challenges of exploring a new research landscape. Br J Psychol 109(3):395–417
70. Southall H, Marmion M, Davies A (2019) Adapting Jake Knapp's design sprint approach for AR/VR applications in digital heritage. In: Augmented reality and virtual reality. Springer, Cham, pp 59–70
71. Tabak A, Duman SL (2022) Flight and fitness distance balance-based coyote optimization algorithm for effective automatic generation control of PV-based multi-area power systems. Arab J Sci Eng. https://doi.org/10.1007/s13369-022-07004-z
72. Oshnoei S, Oshnoei A, Mosallanejad A, Haghjoo F (2021) Novel load frequency control scheme for an interconnected two-area power system including wind turbine generation and redox flow battery. Int J Electr Power Energy Syst 130:107033
73. Topno PN, Chanana S (2016) Automatic generation control using optimally tuned tilt integral derivative controller. In: 2016 IEEE 1st international conference on control, measurement and instrumentation, CMI 2016 206–210. Institute of Electrical and Electronics Engineers, Piscataway. https://doi.org/10.1109/CMI.2016.7413740
74. Kahraman HT, Aras S, Gedikli E (2020) Fitness-distance balance (FDB): a new selection method for meta-heuristic search algorithms. Knowl-Based Syst 190:105169
75. Aras S, Gedikli E, Kahraman HT (2021) A novel stochastic fractal search algorithm with fitness-distance balance for global numerical optimization. Swarm Evol Comput 61:100821
76. Guvenc U, Duman S, Kahraman HT, Aras S, Katı M (2021) Fitness-distance balance based adaptive guided differential evolution algorithm for security-constrained optimal power flow problem incorporating renewable energy sources. Appl Soft Comput 108:107421
77. Abd-Elazim SM, Ali ES (2018) Load frequency controller design of a two-area system composing of PV grid and thermal generator via firefly algorithm. Neural Comput & Applic 30(2):607–616
78. Khadanga RK, Kumar A, Panda S (2020) A novel modified whale optimization algorithm for load frequency controller design of a two-area power system composing of PV grid and thermal generator. Neural Comput & Applic 32(12):8205–8216
79. Venturelli D, Kondratyev A (2019) Reverse quantum annealing approach to portfolio optimization problems. Quantum Machine Intell. https://doi.org/10.1007/s42484-019-00001-w
80. Rosenberg G, Haghnegahdar P, Goddard P, Carr P, Wu K, López De Prado M (2016) Solving the optimal trading trajectory problem using a quantum annealer. IEEE J Select Top Signal Process 10(6):1053–1060
81. Marzec M (2016) Portfolio optimization: applications in quantum computing. In: Handbook of high-frequency trading and modeling in finance. Wiley, Hoboken, pp 73–106
82. Kellerer H, Mansini R, Speranza MG (2000) Selecting portfolios with fixed costs and minimum transaction lots. Ann Oper Res 99(1–4):287–304
83. Johnson MW, Amin MHS, Gildert S, Lanting T, Hamze F, Dickson N, Harris R, Berkley AJ, Johansson J, Bunyk P et al (2011) Quantum annealing with manufactured spins. Nature 473(7346):194
84. Boixo S, Smelyanskiy VN, Shabani A, Isakov S, Dykman M, Denchev VS, Amin MH, Smirnov AY, Mohseni M, Neven H (2016) Computational multiqubit tunnelling in programmable quantum annealers. Nat Commun 7:10327
85. Venturelli D, Mandra S, Knysh S, Ǵorman B, Biswas R, Smelyanskiy V (2015) Quantum optimization of fully connected spin glasses. Phys Rev X 5(3):031040

86. Venturelli D, Marchand DJJ, Rojo G (2015) Quantum annealing implementation of job-shop scheduling. arXiv: 1506.08479
87. Hamerly R, Inagaki T, McMahon PL, Venturelli D, Marandi A, Onodera T, Ng E, Langrock C, Inaba K, Honjo T (2018) Experimental investigation of performance differences between coherent ising machines and a quantum annealer. arXiv: 1805.05217
88. Marshall J, Venturelli D, Hen I, Rieffel EG (2018) The power of pausing: advancing understanding of thermalization in experimental quantum annealers. arXiv: 1810.05881
89. Ottaviani D, Amendola A (2018) Low rank non-negative matrix factorization with d-wave 2000q. arXiv: 1808.08721
90. Chan TCY, Letourneau D, Potter BG (2022) Sparse flexible design: a machine learning approach. Flex Serv Manuf J. https://doi.org/10.1007/s10696-021-09439-2
91. Mak HY, Shen ZJM (2009) Stochastic programming approach to process flexibility design. Flex Serv Manuf J 21(3–4):75–91
92. Chou MC, Chua GA, Teo CP, Zheng H (2011) Process flexibility revisited: the graph expander and its applications. Oper Res 59(5):1090–1105
93. Simchi-Levi D, Wei Y (2015) Worst-case analysis of process flexibility designs. Oper Res 63(1):166–185
94. Feng W, Wang C, Shen ZJM (2017) Process flexibility design in heterogeneous and unbalanced networks: a stochastic programming approach. IISE Trans 49(8):781–799
95. Yan Z, Gao SY, Teo CP (2018) On the design of sparse but efficient structures in operations. Manag Sci 64:3421–3445
96. Chen X, Ma T, Zhang J, Zhou Y (2019) Optimal design of process flexibility for general production systems. Oper Res 67(2):516–531
97. Bello I, Pham H, Le QV, Norouzi M, Bengio S (2016) Neural combinatorial optimization with reinforcement learning. arXiv preprint arXiv:161109940
98. Khalil E, Dai H, Zhang Y, Dilkina B, Song L (2017) Learning combinatorial optimization algorithms over graphs. In: Advances in neural information processing systems. MIT Press, Cambridge, pp 6348–6358
99. Larsen E, Lachapelle S, Bengio Y, Frejinger E, Lacoste-Julien S, Lodi A (2018) Predicting solution summaries to integer linear programs under imperfect information with machine learning. arXiv preprint arXiv:180711876
100. Chen S, Song JSJ, Wei Y (2020) Data-driven scalable e-commerce transportation network design with unknown flow response. Retrieved from SSRN 3590865
101. Poonguzhali PK, Ananthamoorthy NP (2019) Design of mutated harmony search algorithm for data dissemination in wireless sensor network. Wirel Pers Commun. https://doi.org/10.1007/s11277-019-06882-1
102. Abbasi AA, Younis M (2007) A survey on clustering algorithms for wireless sensor networks. Comput Commun 30:2826–2841
103. Jiang C, Yuan D, Zhao Y (2009) Towards clustering algorithms in wireless sensor networks: a survey. In: IEEE Wireless communications and networking conference. IEEE, Piscataway
104. Mahdavi M, Fesanghary M, Damangir E (2007) An improved harmony search algorithm for solving optimization problems. Appl Math Comput 188(2):1567–1579
105. Youssef M, Youssef A, Younis M (2009) Overlapping multi hop clustering for wireless sensor networks. IEEE Trans Parallel Distrib Syst 20:1844–1856
106. Sarvari H, Zamanifa K (2012) Improvement of harmony search algorithm by using statistical analysis. In: Artificial intelligence, vol 37. Springer, Cham, pp 181–215
107. Wang Y, Kelly BM, Li X (2013) On the network connectivity of wireless sensor networks following a random and non-uniform distribution. In: 2013 IEEE 9th international conference on wireless and mobile computing networking and communication (WiMob). IEEE, Piscataway

108. Zhang X (2007) Network lifetime optimization for heterogeneous sensor network with mixed communication. In: 2007 IEEE wireless sensor network communication and networking conference. IEEE, Piscataway
109. Rashid A, Khan F, Gul T, Khan S, Khalil FK (2018) Improving energy conservation in wireless sensor network using energy harvesting system. Int J Adv Comput Sci Apl 9:354–361

Chapter 10
Optimization Methods in Maintenance of Machines

Introduction by the Editor

The manufacturing systems must be maintained well to keep them working with full efficiency achieving maximum yield. In addition, the regular as well as preventive maintenance is necessary in order to minimize the downtime and eventually the cost of the finished product. The review of several contributions underscores that the optimal maintenance policy plays a vital role. There are several sectors serving the manufacturing industry such as energy sector, which also need to have optimal maintenance schedule. Efficient power generation of steam and electricity will be a key part of reliable systems usually used in industrial assets, including thermal power plants. The chapter reviews an optimal preventive maintenance strategy for a steam turbine in a thermal power plant with reliability cantered maintenance -oriented cost minimization. In addition, the work associated with ageing of the equipment, possible failures, and importantly the optimal schedule of equipment replacement is also discussed. The multi-objective optimization solution approach for the best sequence of jobs, optimal production rate and optimum preventive maintenance time is also discussed in very details. A Bayesian approach analysing data for probabilistic prediction of structure failures is also quite important application. Its applications are discussed with real world case studies.

The chapters mainly are around the maintenance policies as they play a significant role in helping to reduce costs by improving equipment reliability and decreasing the cost of replacement through preventing equipment from breakdowns and failures. The work reviewed carried out the reliability analysis and developed age replacement model. The case studies are associated with the main equipment of the thermal power plant for the analysis involving condenser steam turbine boiler condensate water pump. One of the works reviewed is the optimal preventive maintenance for systems having a continuous output and operating in a random environment. The work considers the systems that are operating in a random environment

A. J. Kulkarni (ed.), *Optimization Methods in Manufacturing Processes*,
https://doi.org/10.1007/978-981-96-5257-0_10

modelled by an external shock process. The model description and possible applications, expected quality of a system and optimal maintenance policy have been elaborated. A multi-objective optimization of energy-aware multi-product failure-prone manufacturing system is explored and elaborated. Two algorithms combining simulation and NSGA-II/PSO and simulation and NSGA-II/GA are employed in this work. Moreover, a single-machine and multi-product failure-prone manufacturing system was explored considering sequence of jobs where the processing times are stochastic and the purpose is the optimization of total tardiness, production cost and system reliability. Based on the appropriateness a suitable combination of the simulation approach and metaheuristic algorithms was employed optimizing the problems which have uncertainty and there is no analytical solution for them.

The diagnostics and prognostics of multi-mode failure scenarios in miter gates using multiple data sources and a dynamic Bayesian network has been critically reviewed and elaborated. The approach aims at an accurate simulation of the interaction between different failure features, and subsequently at the effective estimation and prediction of the damage states based on the generated structural physics. The work involves computer vision techniques to capture the cracks in the early stage while the strain sensor network detects the quoin block damage, resulting in load re-distribution within the whole structure. The contributions are implementation of a digital twin concept for a practical engineering problem with complicated degradation behaviors, which requires extensive model-based analysis to capture the interactions between multiple damages. In addition, the extension of the widely DBN framework to fuse information from strain gages and camera for damage diagnostics and failure prognostics of miter gates. The further objectives are to develop a joint production and maintenance control policy that minimizes the total cost composed of backlog, inventory holding, disposal, and maintenance costs. A critical review of the maintenance planning of electric distribution systems has been detailed. The review is with the view point of covering different criteria such as economic and reliability. It involves the maintenance planning of distribution systems and the reliability criterion in the maintenance planning of distribution systems.

Machine Generated Summaries

Disclaimer: The summaries in this chapter were generated from Springer Nature publications using extractive AI auto-summarization: An extraction-based summarizer aims to identify the most important sentences of a text using an algorithm and uses those original sentences to create the auto-summary (unlike generative AI). As the constituted sentences are machine selected, they may not fully reflect the body of the work, so we strongly advise that the original content is read and cited. The auto generated summaries were curated by the editor to meet Springer Nature publication standards. To cite this content, please refer to the original papers.

Machine generated keywords: maintenance, preventive, failure, preventive maintenance, repair, policy, interval, random, reliability, degradation, equipment, probability, breakdown, replacement, manufacturing system.

Reliability-Centered Maintenance of Rapier Dragline for Optimizing Replacement Interval of Dragline Components [1]

This is a machine-generated summary of:

Palei, Sanjay Kumar; Das, Smarak; Chatterjee, Snehamoy: Reliability-Centered Maintenance of Rapier Dragline for Optimizing Replacement Interval of Dragline Components [1].

Published in: Mining, Metallurgy & Exploration (2020).

Link to original: https://doi.org/10.1007/s42461-020-00226-5.

If you want to cite the papers, please refer to the original.

For technical reasons we could not place the page where the original quote is coming from.

Abstract-Summary "This paper proposes a preventive maintenance strategy for a dragline deployed in an opencast coal mine based on reliability-centered maintenance and failure-mode-effects analysis using real operational data."

"Reliability-centered maintenance replaced a cluster of critical failure components of the dragline while satisfying two specific conditions: (i) the components should be replaced at the earliest mean time to failure (MTTF) within the group and (ii) the time taken to replace all the parts should be equivalent to the maximum downtime within the group."

"Estimated threshold weightage factor identified twenty-six critical failure components for the preventive maintenance strategy, and clustering of components divided them into nine groups."

"In a group, the loss that is expected to occur by replacing some of the components before their failure, and the gain by reducing the overall maintenance downtime, has been explained through a cost-benefit analysis."

Introduction

"A suitable maintenance strategy plays a paramount role in the reliability and availability of draglines, which are mostly affected by downtimes because of unwanted failures [2], but it could be reduced by identifying and analyzing the faults [3]."

"A suitable maintenance strategy is needed immediately after the warranty period is over to maintain the reliability and availability of the equipment, and to decrease the portion of corrective maintenance that is significantly important to prevent the out-scheduled production due to the unexpected breakdowns."

"Preventive maintenance can be carried out at regular intervals for the equipment whose failure patterns are known or simply a reliability-centered maintenance (RCM)-based proactive maintenance action can be undertaken [4]."

"The RCM is a logical process for developing and/or optimizing the maintenance requirements of a capital equipment through an effective maintenance strategy by identifying the causes of failures and their effects [5]."

Methodology

"The MTTF for each component is calculated from the available failure data along with the cost and average downtime."

"The hourly revenue generated (R_g) can be expressed mathematically [6]: R_g = revenue generated by dragline per hour O = Overburden removed by dragline per year C.P. = coal price S.R. = stripping ratio D = number of working days per year H = average working hours per day L_i = loss of working hours of an ith part due to earlier replacement = $\sum(MTTF_i - MTTF_l)$ C_i = purchase cost of the ith component $MTTF_i$ = mean time to failure of the ith componentwhere $MTTF_i$ represents MTTF of the ith element of the cluster, and $MTTF_l$ represents the lowest value of MTTF in the cluster."

"The clustering method considers three parameters: MTTF, cost of the component, and mean downtime to cluster different components of the dragline."

Description of Case Study Mine

"The dragline considered for this study is a Rapier W2000 Walking Dragline, with a bucket capacity of 24 m^3, boom length 96 m, operating radius 81 m, dump height 48.1 m, and digging depth up to 67.5 m. Draglines are massive machines commonly used in opencast coal mining to strip overburden, revealing the targeted minerals for extraction [7]."

"A walking dragline can drag its tub along the ground using large eccentrically driven walking shoes at the side of the machine."

"The raw data obtained from the main workshop contains failure time data, maintenance data, and production data over 8 years (2008–2016), along with the cost of some of the components/spare-parts of the dragline."

"Based on the threshold weightage factor value, 21 components are not high risk and they can be repaired or replaced during scheduled inspection and these components were excluded from the preventive maintenance schedule."

Results and Discussion

"The acquisition cost is directly proportional to average downtime of dragline (there is an assumption that there are enough maintenance personnel to replace all the components in the cluster before D_{max} is reached, $\theta \propto$ Acquisition cost) due to component failure."

"D_{avg} = average downtime of dragline due to component failure F_f = frequency of failure of the component The data is normalized to bring the data sets on a comparable scale."

"The value of the mean downtime of dragline due to the failure of components lies within a specific range (i.e., within minimum and maximum mean downtime)."

"To show the effectiveness of the proposed clustering algorithm, for the maintenance schedule of the dragline components, the results from the proposed algorithm were compared with a traditional clustering algorithm."

"This result shows that the proposed clustering algorithm can effectively group the dragline components' failure data for preventive maintenance."

Summary

"Average downtime, the acquisition cost of components, MTTF, and failure types were the input parameters to the C++ program which delivered a preventive maintenance strategy, stating the time interval of replacement, duration of maintenance schedule and the associated profit."

"The RCM strategy divided the failure components into nine groups on the principle of clustering based on their MTTF."

"The replacement of the entire group of components was suggested at the lowest MTTF of the components within the group to reduce the overall downtime of the dragline system."

"Replacing all the components within the group at the given maximum downtime led to an overall improvement in dragline availability, thus fulfilling the overall goal of maintenance."

"In the first group, for example, the MTTF of the critical failure components such as the tooth adapter and tooth tip of dragline that is situated within close vicinity."

Optimum Preventive Maintenance Strategy for Turbine Based on Reliability Analysis and Mathematical Modeling [8]

This is a machine-generated summary of:

Eddouh, Yassine; Daya, Abdelmajid; Elotmani, Rabie: Optimum preventive maintenance strategy for turbine based on reliability analysis and mathematical modeling [8].

Published in: Life Cycle Reliability and Safety Engineering (2022).

Link to original: https://doi.org/10.1007/s41872-022-00206-2.

If you want to cite the papers, please refer to the original.

For technical reasons we could not place the page where the original quote is coming from.

Abstract-Summary "As energy consumption will gradually increase into the future, efficient power generation of steam and electricity will be a key part of reliable systems usually used in industrial assets, including thermal power plants."

"This need puts the thermal power plant equipment in a deteriorating condition which worsens the system's reliability."

"Maintenance policies play a significant role in helping to reduce costs by improving equipment reliability and decreasing the cost of replacement through preventing equipment from breakdowns and failures."

"This article presents the fundamental concepts of Reliability Centered Maintenance (RCM) aiming at its implementation in age-based preventive maintenance activities of critical equipment in thermal power plant."

"A Pareto analysis shows that the steam turbine is the most critical equipment for a thermal power plant."

"The age replacement model with the two-parameter Weibull failure distribution is used to model turbine performance."

"Based on the historical TBF database, results obtained under reliability analysis show that MTBF is 778 h, and the failure rate increases when the aging effect (time-related failures)."

Introduction

"Several published papers have investigated power plant equipment performance, reliability, and the implementation of preventive maintenance."

"Dhurvey and Soni [9] showed preventive maintenance of steam turbine used in thermal power plants by reliability investigation and failure modes and effects analysis."

"Preventive maintenance tasks are performed in a time-based schedule to avoid equipment failures."

"There are few studies that treat preventive maintenance optimization through age replacement models for power plant equipment. Verma et al. [10] presented an optimal time scheduling for carrying out multiple minor maintenance actions on a steam turbine, using Gamma distributions to model the wear/deterioration components. Sobamowo et al. [11] proposed an optimal preventive maintenance model using Weibull distribution to predict maintenance models for steam turbines."

"A case study based on a thermal power plant is shown in "Case study", where steam turbine performances are modeled using reliability analysis through Weibull distribution; hence, an optimal preventive replacement time and minimum cost of the intervention are figured out."

Reliability Analysis

"The reliability and availability concept when used to model equipment performance through Weibull distribution is analyzed."

"The Weibull distribution is one of the most widely used lifetime distributions in reliability engineering, as it offers considerable advantages due to its flexibility, which allows reliability data to be modeled, and it will be the distribution model used in this work."

"Reliability indicators of the steam turbine system have been found in a thermal power plant by using two-parameter Weibull distribution [12]."

"Kumar and Krishnan [13] presented the estimation of system reliability through the two-parameter Weibull distribution, they show that the Weibull distribution can be a decreasing, constant, or increasing failure rate. Sidibe et al. [14] performed a reliability and preventive maintenance analysis of Weibull's distributed lifetime, and they proved that there is an age interval where the preventive maintenance should be able to be delayed and that the shape parameter of the Weibull distribution controls the decision of the delay preventive maintenance."

Age Replacement Model

"The most common preventive maintenance policies are likely to be age replacement models."

"Under this model, replacement occurs at failure or at a time, T work, after its installation, whichever occurs first."

"The optimal strategy is defined by determining the optimal time of preventive replacement, which decreases the average total cost per time unit."

Case Study

"The main equipment of the thermal power plant selected for the analysis are: Condenser Steam turbine Boiler Condensate water pump Alternator The aim of this section is to find the critical equipment for the thermal power plant."

"Data have been collected from the maintenance section of the thermal power plant."

"The results help to provide that turbine is the first critical equipment in the thermal power plant, which needs to be monitored very closely."

"Data are analyzed and used to model the failure characteristics of the turbine."

"Estimation of Mean time between failures (MTBF): Reliability in MTBF time: The failure rate at MTBF time: Reliability analysis shows that the turbine reliability in normal conditions is 96.61%."

"The failure rate increased over time; the turbine is in the aging phase."

"The turbine has an increasing failure rate, so a minimized cost is achievable with a preventive maintenance strategy."

Conclusions

"Optimal preventive maintenance strategy for a steam turbine in a thermal power plant is presented, through the concept of RCM-oriented cost optimization."

"The analytical formulation of the expected cost rate for the proposed model has been obtained, considering the increasing failure rate of the steam turbine."

"Of this study, the following conclusions were drawn: Steam turbine is the critical equipment of thermal power plant Steam turbine has an increasing failure rate, and this is due to fatigue, wear, and corrosion."

"The optimum replacement schedule with an age replacement model is 1067 h. Moreover, the use of this maintenance policy has made a considerable contribution to decreasing the maintenance costs of the turbine when the costs of setting up maintenance actions are high."

"A replacement cost will be analyzed through Weibull and Gamma distribution with different shape parameter values."

Optimal Preventive Maintenance for Systems Having a Continuous Output and Operating in a Random Environment [15]

This is a machine-generated summary of:

Cha, Ji Hwan; Finkelstein, Maxim: Optimal preventive maintenance for systems having a continuous output and operating in a random environment [15].

Published in: TOP (2019).

Link to original: https://doi.org/10.1007/s11750-019-00508-2.

If you want to cite the papers, please refer to the original.
For technical reasons we could not place the page where the original quote is coming from.

Abstract-Summary "We consider systems that are operating in a random environment modeled by an external shock process."

"Performance of a system is characterized by a quality (output) function that is decreasing (due to degradation) in the absence of shocks."

"Shocks have a double impact, i.e., they affect the failure rate of a system directly and at the same time, and each shock contributes to the additional decrease in the quality function."

Introduction

"The performance of various engineering systems is often characterized not only by reliability characteristics such as availability and probability of failure-free operation in a given interval of time, but by characteristics of performance (output) as well."

"At many instances, it affects not only reliability characteristics that should be modeled by aging distributions of times to failure (e.g., Weibull), but the quality of performance as well."

"From the modeling point of view, they constitute a useful tool for describing an effect of environment on the reliability characteristics of engineering systems (see, e.g. [16–27] to name a few)."

"Shocks in our model have a double effect: they act directly on the failure rate that characterizes the time to failure of a system and at the same time on the quality function as well."

Model Description and Possible Applications

"To the described failure model that was studied in the literature, our system is characterized by a quality of performance function, which is also a stochastic process in the presence of shocks."

"An external shock along with the direct impact on the failure rate (hard failure) can act accordingly on these 'soft' processes, which is modeled in our paper in a simple, generalized way."

"A shock (e.g., electrical surge) acts on both parts increasing the failure rate of the first one and decreasing the accuracy of the gyro (and the system) performance."

"Other possible examples of relevant types of systems include, but are not limited to: In power-generating systems, shocks (e.g., lightning) can destroy some lines reducing overall transmission capacity and increasing the probability of an overall failure."

"Shocks (e.g., virus attacks and bugs initiation) can increase the failure rate of a software and at the same time decrease its computational capacity."

Expected Quality of a System

"The measure of our first interest iswhich is the expectation of a quality function of a system at time t (assuming that the quality is zero when a system is in the state of failure)."

"The following Proposition provides an explicit expression for this measure."

Optimal Maintenance Policy

"PM in general can be justified for aging systems to prevent the increasing risk of failure."

"The most common manifestation of stochastic aging is via the increasing failure rate."

"We will first prove that our failure model [10] results in the increasing failure rate of a system."

"This assumption describes the increasing susceptibility to shocks of a system with time, which is also a manifestation of the degradation processes."

Concluding Remarks

"Under the defined model, we obtain expressions for the expected quality function of a system and for the conditional expected quality function of a system, on condition that it is operable at a given time instant."

"It is shown that under certain assumption, the conditional quality function is initially decreasing (as expected), but then can increase for a certain period of time."

"The main focus of our paper is on the PM problem: to find an optimal replacement time that minimizes the long-run expected cost rate for the defined failure model and the quality function."

"It should be also noted that in our approach, we do not consider failures occurring when the quality function reaches some predetermined level/threshold as it is not set."

"A different mathematical model that takes into account accumulation of the shocks' impacts on the quality function should be developed."

Multi-objective Optimization of Stochastic Failure-Prone Manufacturing System with Consideration of Energy Consumption and Job Sequences [28]

This is a machine-generated summary of:

Amelian, S. S.; Sajadi, S. M.; Navabakhsh, M.; Esmaelian, M.: Multi-objective optimization of stochastic failure-prone manufacturing system with consideration of energy consumption and job sequences [28].

Published in: International Journal of Environmental Science and Technology (2018).

Link to original: https://doi.org/10.1007/s13762-018-1742-7.

If you want to cite the papers, please refer to the original.

For technical reasons we could not place the page where the original quote is coming from.

Abstract-Summary "Multi-objective optimization of energy-aware multi-product failure-prone manufacturing system is explored."

"The purpose is to determine the best sequence of jobs, optimal production rate and optimum preventive maintenance time for simultaneous optimization of three criterions of total weighted quadratic earliness and tardiness, system reliability and energy-consumption cost."

"Two novel algorithms by combining (1) simulation and NSGA-II/PSO and (2) simulation and NSGA-II/GA are proposed for solving this problem."

"Multi-product failure-prone manufacturing systems by considering sequence of jobs have not been explored in any paper and for the first time a new hedging point policy is presented for the mentioned problem."

Introduction

"Failure-prone manufacturing systems (FPMSs) are a subset of flexible systems in which stochastic breakdown and maintenance of machines is assumed."

"Since machines in flexible manufacturing systems can produce with different speeds, determining the optimal speed is highly important and has been recognized as the hedging point policy (HPP) in failure-prone manufacturing systems."

"Machine breakdown in manufacturing systems occurs randomly, determining the optimum preventive maintenance time is important."

"There are various approaches to determine the preventive maintenance time that are deterministic approach, stochastic approach and interval approach."

"A single-machine failure-prone stochastic scheduling problem is explored in order to determine the best sequence of jobs, the preventive maintenance time and the hedging point policy."

Literature Review

"Determining the optimal production rate in failure-prone manufacturing systems is very important and is known as the hedging point policy."

"Optimal production rate was determined in this survey by means of discrete event simulation and RSM methods to minimize holding and shortage costs."

"Khoury [29] extracted optimal production rate for a single-machine and single-product FPMS to minimize sum of holding and shortage costs when times between failures have exponential distribution and the repair time has general distribution via analytical method."

"Polotski et al. [30] determined optimal production rate for failure-prone manufacturing systems considering rework operations on the returned products from the quality control unit."

"Rivera-Gómez et al. [31] explored a single-machine and single-product failure-prone system through stochastic dynamic programming and combination of the simulation and RSM methods considering the rate of defective products which reenter the system."

Materials and Methods

"Chou et al. [32] investigated a deterministic single-machine problem considering sequence-dependent setup time and release time and the purpose was to minimize total weighted completion time."

"Kianfar and Moslehi [33] explored the deterministic single-machine problem in order to determine the best sequence of jobs to minimize total weighted quadratic earliness and tardiness."

"Ying et al. [34] proposed four MIP models to analyze the deterministic single-machine problem in order to minimize four purposes, i.e., mean tardiness, mean lateness, total flow time and maximum tardiness."

"Soroush [35] investigated the single-machine problem with stochastic processing times with the purpose of minimization of the number of tardy and early jobs."

"Cui and Lu [36] proposed an MIP model to explore joint production planning and preventive maintenance for a deterministic single-machine problem considering release time and nonresumable jobs."

Problem Description

"A single-machine and multi-product FPMS was explored considering sequence of jobs where the processing times are stochastic and the purpose is the optimization of total tardiness, production cost and system reliability."

"The below assumptions were considered in analyzing the problem: A machine can only process one job at a time."

"Increased number of preventive maintenance will increase the system reliability while lengthen the completion time."

"Production speed of jobs by the machine is variable, cost of production of each product is changed according to its processing time."

"The first function is total weighted quadratic earliness and tardiness penalties, the second function is system reliability, and the third function is total energy costs of production related to processing speed of each job."

Solution Methodology

"Combination of simulation and metaheuristic algorithms is an appropriate method to optimize problems which have uncertainty and there is no analytical solution for them."

"Given the stochastic processing times and machine breakdown as well as NP-hard nature of the problem, the combination of simulation and metaheuristic algorithms was used for solving the problem."

"There are two approaches in analysis of problems via combination of simulation and metaheuristic algorithms."

"The fitness function is extracted through simulation and methods such as RSM or neural network, and then the problem is optimized using metaheuristic algorithms."

"Simulation and metaheuristic algorithms are used jointly."

"To analyze the problem, combination of simulation and NSGA-II algorithm was used simultaneously."

Results and Discussion

"As quality of the obtained solutions from metaheuristic algorithms depends on the value of their parameters, tuning these parameters is important."

"There are different criterions like NPS, DM, MID and time to evaluate quality of solutions in multi-objective metaheuristic algorithms."

"As it is clear, a set of Pareto optimal solutions are obtained after the algorithm is executed."

"(4), one-time preventive maintenance has been used; therefore, reliability of the system is higher than solution (1)."

"In this solution, preventive maintenance has been used two times, and thus, reliability of the system is more than solutions (1) and (4)."

"In solutions (5) and (10), preventive maintenance has been used three times and the system in this state has the highest reliability."

"In solutions (5) and (6), preventive maintenance has been used three times but reliability of the system in solution (5) is more than solution (6)."

Conclusion

"A multi-product failure-prone system was explored in this study by considering the sequence of jobs and machine energy consumption."

"Decision variables which have been optimized via the proposed algorithm are: sequence of jobs, preventive maintenance time and production rate of each job."

"The concept of HPP was developed in this study for a multi-product failure-prone system given sequence of jobs."

Stochastic Modeling of Non-identical Redundant Systems with Priority, Preventive Maintenance, and Weibull Failure and Repair Distributions [37]

This is a machine-generated summary of:

Kumar, Ashish; Saini, Monika; Devi, Kuntal: Stochastic modeling of non-identical redundant systems with priority, preventive maintenance, and Weibull failure and repair distributions [37].

Published in: Life Cycle Reliability and Safety Engineering (2018).

Link to original: https://doi.org/10.1007/s41872-018-0040-1.

If you want to cite the papers, please refer to the original.

For technical reasons we could not place the page where the original quote is coming from.

Abstract-Summary "Analysis, a stochastic model has been designed for a non-identical redundant system having two units: one operative and other cold standby."

"Each unit has two possible states: either operative or failure."

"Original unit gets priority for repair over repair and preventive maintenance of duplicate unit."

"The repairs, preventive maintenance, and switches used to shift standby unit into operation are perfect."

Introduction

"Nakagawa and Osaki [38], Okumato and Osaki [39], Osaki and Okumato [40], Mokkadis et al. [41], Kishan and Jain [42], and Deswal and Malik [43] studied non-identical unit systems under a common assumption of constant failure and repair rates."

"Gupta et al. [44] carried out the cost–benefit analysis of a two dissimilar unit cold standby system with Weibull failure and repair laws."

"Priority to operation, over repair of other unit is given."

"Performance analysis of a non-identical unit system with priority, preventive maintenance, and arbitrary distribution in failure and repair disciplines has been done by Devi et al. [45]."

"By keeping above fact in mind, in the present paper, a stochastic model has been designed for a non-identical redundant system having two units: one operative and other cold standby."

"Original unit gets priority for repair over repair and preventive maintenance of duplicate unit."

"The repairs, preventive maintenance, and switches used to shift standby unit into operation are perfect."

Profit Analysis

"There must be an optimum poise between the reliability aspect of a product and its cost."

"The production cost also would increase with the requirement of greater reliability."

"The revenue and cost function leads to the profit function of a firm, as the profit is excess of revenue over the cost of production."

"The profit function in time t is given by the following: P(t) = Expected revenue in (0; t] – Expected total cost in (0; t]."

"By considering various cost and revenues, the profit incurred to the system model in steady state is given by the following relation: Weibull distribution plays the key role in reliability modeling due to its applicability to model both increasing and decreasing rates."

Numerical Results

"The system model is more profitable in which use of priority to repair of original unit over repair and preventive maintenance of duplicate unit jointly has been made over the system models in which either priority is not used or used in single repair discipline."

"Remark: Finally, we can conclude that a redundant system of non-identical units can be made more profitable and available for use by the following: Increasing its preventive maintenance rate and repair rate."

"Using the proper priority in repair disciplines."

Diagnostics and Prognostics of Multi-mode Failure Scenarios in Miter Gates Using Multiple Data Sources and a Dynamic Bayesian Network [46]

This is a machine-generated summary of:

Wu, Zihan; Fillmore, Travis B.; Vega, Manuel A.; Hu, Zhen; Todd, Michael D.: Diagnostics and prognostics of multi-mode failure scenarios in miter gates using multiple data sources and a dynamic Bayesian network [46].

Published in: Structural and Multidisciplinary Optimization (2022).

Link to original: https://doi.org/10.1007/s00158-022-03381-z.

If you want to cite the papers, please refer to the original.

For technical reasons we could not place the page where the original quote is coming from.

Abstract-Summary "This paper presents a novel online diagnostics and prognostics framework for structures subject to multiple failure modes and demonstrates the proposed method with a high-fidelity finite element model using multiple data sources (i.e., strain gages and images)."

"The approach aims at an accurate simulation of the interaction between different failure features, and subsequently at the effective estimation and prediction of the damage states based on the generated structural physics."

"A dynamic Bayesian network is used which incorporates different data sources to evaluate the structures under different kinds of deterioration mechanisms."

"In prognosis, the dynamic Bayesian network gives a probabilistic prediction of the remaining useful life of the structure based on the evolution of the failures."

Introduction

"Although FE analysis offers high interpretability [47–50], the separation in length scales between structural scale (e.g., at the scale of miter gates) and damage scale (e.g. at the scale of cracks) may increase numerical model discretization and add computational cost."

"In order to develop the framework for multiple failure modes and data sources, two main challenges need to be addressed, namely: (1) how to properly model different failure modes of miter gates; and (2) how to fuse multiple data sources for the model updating of the digital model."

"Within the context of large structures modeled as shells, such as a miter gate, the IGL method has been successfully used to connect shell global domains to solid local domains with welds [51]."

"The physics of two types of failure modes is represented in an FE model of the miter gate: the boundary condition loss represents the large-scale damage; the fatigue crack growth represents the small-scale local damage."

Modeling of Miter Gate Failures

"This compression causes the miter gate's quoin contact block to thrust into the lock wall contact block."

"The lengthening of this gap leads to a global re-distribution of the stress, which escalates crack evolution of the miter gates at different local regions."

"Besides the quoin block damage discussed above, fatigue cracks are a common form of miter gate damage due to the cyclic loads when the lock chambers are filled and emptied."

"An accurate crack analysis using the miter gate FE model is necessary to understand the behavior of such localized effect."

"Although the built-in Abaqus technology calculates SIF values through the contour integral method, crack analysis for the complicated and large-scale miter gate model is computationally expensive due to the fact that the crack can only be simulated with finely discretized solid elements in Abaqus."

Diagnosis and Prognosis of Miter Gates Using Multiple Data Sources and DBN

"Image-based observations enable computer vision techniques to capture the cracks in the early stage while the strain sensor network detects the quoin block damage, resulting in load re-distribution within the whole structure."

"We assume that there is uncertainty from noise in the two data sources, i.e., sensor noise and camera image quality; thus, a DBN is constructed which accommodates measurement uncertainty of observations along with probabilistic transitions of damage modes over time."

"Based on the surrogate modeling, functional representation, and probabilistic modeling of different nodes in the DBN, we now present the diagnostics and prognostics of miter gates with multiple failure modes using the DBN and multiple data sources (i.e. strain measurements and image monitoring data)."

"Failure prognostics is a process of predicting the remaining useful life (RUL) of structural assets based on all the information available at the current time step."

Case Study

"The following prognostic results are carried out based on considering crack and quoin block damage individually and considering two damage features jointly."

"The true RUL of the structure is 87.5 months based on quoin block damage."

"(before 35 months), the prediction slightly overestimates the RUL of the structure; in the second state (after 35 months), the model tends to be more conservative about the prediction as the predicted RULs are smaller than the true values."

"The prediction outperformed both results of using single failure threshold when the EOL is jointly determined from the two damage features."

"To understand the impact of different monitoring techniques (e.g., strain gages and camera images) on decisions related to maintenance planning, the optimal maintenance time and minimum CPUT are calculated based on the prognostics results using measurements from both monitoring techniques and from camera images only."

Conclusions

"An online diagnostic and prognostic framework that efficiently used multi-source data was proposed for structures with multiple failure modes."

"The contributions of this paper can be summarized as: (1) Implementation of a digital twin concept for a practical engineering problem with complicated degradation behaviors, which requires extensive model-based analysis to capture the interactions between multiple damages; (2) The extension of the widely DBN framework to fuse information from strain gages and camera for damage diagnostics and failure prognostics of miter gates; and (3) The investigation of the impact of using multiple structural health monitoring data sources (i.e. strain sensor and camera) on the final maintenance decision making process."

"The proposed framework provides a new approach of using DBN to incorporate multiple data sources for structures under different scales of failure modes."

Integrated Production and Maintenance Control Policies for Failure-Prone Manufacturing Systems Producing Perishable Products [52]

This is a machine-generated summary of:

Kaddachi, Rawia; Gharbi, Ali; Kenné, Jean-Pierre: Integrated production and maintenance control policies for failure-prone manufacturing systems producing perishable products [52].

Published in: The International Journal of Advanced Manufacturing Technology (2022).

Link to original: https://doi.org/10.1007/s00170-021-08273-y.

If you want to cite the papers, please refer to the original.

For technical reasons we could not place the page where the original quote is coming from.

Abstract-Summary "Ignoring this correlation effect may result in inaccurate values of the shelf-lives and inefficient control policies."

"The objective of this paper is to develop a joint production and maintenance control policy that minimizes the total cost composed of backlog, inventory holding, disposal, and maintenance costs."

"The proposed parametrized joint control policy combines a multi hedging point policy and an age-based preventive maintenance policy."

"The obtained results show that the correlation between machine degradation and shelf-life reduction has a major influence on the control parameters and that preventive maintenance interventions can lead to increasing the shelf-life of products and minimizing the total cost."

"The proposed joint control policy is then compared to three other polices for a wide range of system and cost data."

"The obtained results show that the proposed joint control policy outperforms the other polices in terms of total incurred costs."

Introduction

"In the case of failure-prone manufacturing systems producing perishable products, maintenance activities can play a crucial role into minimizing incurred costs as they can lead to increasing the shelf-life of products."

"It is important for manufacturing system managers to revise their strategies by adopting a joint maintenance and production control policy that take into consideration the characteristics of perishable products and the relationship between machine degradation and shelf-life reduction in order to minimize the total incurred cost."

"Research questions can be presented as follows: (1) How can we quantify the correlation between shelf-life reduction and machine degradation? (2) How does this correlation affect the maintenance and production control policies? (3) What is the best joint maintenance and production control policy to adopt to minimize the total incurred cost?"

"In this paper, we aim to develop a joint maintenance and production control policy for failure-prone manufacturing systems while taking into consideration the perishable nature of products and the related variation of the shelf-life."

Literature Review

"The work of Polotski et al. [53] deals with a production control problem of failure-prone manufacturing systems for perishable products having a deterministic shelf-life."

"There is no previous study that combines maintenance control policies with production control policies and take into consideration the perishable nature of products and failure prone manufacturing systems simultaneously in a stochastic and dynamic context."

"As can be noted, more studied are required to fully integrate maintenance and production activities in a joint control policy that incorporate a quantitative relationship between machine degradation and shelf-life reduction."

"Our objective is to fill the gap in the literature by proposing a joint maintenance and production control policy for a failure-prone manufacturing system producing perishable products having a random shelf-life that decreases as the machine degrades."

System Description and Problem Formulation

"The notations used in this study are defined as follows: The considered manufacturing system is composed of a single machine producing one type of finished products having a limited and random shelf-life."

"The shelf-life of the product is related to the degradation of the machine."

"When the machine degrades, the shelf-life decreases."

"This is expressed with the following equations:Here, we aim to establish a quantitative relationship between products' shelf-life and machine degradation represented by its age."

"It is also based on the work of [54] establishing the link between the quality of the products and machine degradation in function of the machine's age."

"The variability of the shelf-life increases with the increase of machine's age."

"In this study, we consider a dynamic and stochastic environment: random failures and repair times, random shelf-lives, evolutions of shelf-life variability, interactions between shelf-lives, and machine's age."

Proposed Joint Production and Maintenance Control Policy

"We propose a joint production and maintenance control policy for manufacturing systems subject to random failure and repair times producing perishable products where the shelf-life decreases with the machine's degradation."

"We propose to extend the classical (HPP) to a multi-hedging point policy (MHPP) for the case of perishable products."

"The objective of the proposed production control policy is to allow the system to satisfy demand during the non-operational times of the machine on one hand, and to minimize the number of perished products on the other hand."

"As for the maintenance policy, we propose an age-based PM policy."

Resolution Approach

"The structure of the proposed joint control policy is defined."

"The objective is to find the optimal control parameters that minimizes the total incurred cost."

"RSM is applied on the total cost obtained with the simulation to establish the effect of significant control parameters, their interactions, and their quadratic effects on the total incurred cost."

"This step aims to conduct a comparative study between the proposed joint control policy and other policies."

"The objective here is to provide the decision maker with insights to help him effectively implement the proposed joint control policy."

Simulation-Based Optimization Method

"A simulation-based optimization method is used in order to evaluate the economic performance of the proposed joint control policy."

"It combines simulation, DOE, and RSM in order to determine the optimal control parameters."

"The structure of the model consists of several events interacting with each other such as arrival of demand, the control policies, etc. The first block (block 1) is the initialization of input variables (demand rate, maximum production rate, shelf-life, control parameters, …)."

"The manufacturing system (block 2) prone to failure and repair times (block 5) is governed by the joint production and maintenance control policy."

"The preventive maintenance control policy (block 4) is responsible for providing changes in the state of the system based on the two control parameters: age of the machine and the level of inventory."

"To validate the simulation model, a numerical example was executed for a manufacturing system governed by the joint control policy PPMP."

Sensitivity Analyses

"By executing more PM actions, each time the mean value of shelf-life increases, and the variability decreases so less perished products, and the system risks less backlog."

"The control parameters are set in a way to avoid a more expensive cost of backlog."

"When the number of perished products increase, the less there is to satisfy demand which means more backlog."

"The system find itself increasing the value of the control parameters to protect itself from high costs especially those due to backlog."

"The system reacts in way to execute more PM actions in order to restore the machine to "As-good-as-New" condition thus increasing the shelf-life."

Comparative Study

"The second policy considered is from the literature and is a joint control policy that combine the classical HPP for production control and the age-based preventive maintenance [55] we note HPP_PM."

"The third joint policy used for comparison, it combines EPQ policy and the age-based preventive maintenance control policy we note EPQ_PM [56]."

"The first, is age replacement policy (ARP) and is based on the conducting a preventive maintenance at the time of failure or after a constant period of machine operation."

"These preventive maintenance strategies are combined with MHPP production policy and are noted respectively MHPP_ARP and MHPP_BRP."

"When applying PPMP, a better control is provided since it allows the system to track the quantity of each sub-stock at each age and adjust the production rate accordingly to minimize the risk of disposal and backlog."

Managerial Insights and Implementation

"Manufacturing systems dealing with products with a limited shelf-life face major challenges in terms of finding the best control policy that take into consideration the age of a product and the degradation of the machine to minimize the total cost."

"By implementing the proposed joint control policy, the manager is capable of deciding simultaneously on the production rate as a function of the quantity of inventory on the hand and on when to execute PM actions in order to minimize the total cost."

"To implement the proposed joint control policy, the manager is required to monitor the different inventory levels and the state of the machine (operational, under repair, or under PM interventions)."

"Our proposed joint control policy has a wide range of applications for different types of industries looking to establish an effective and dynamic plan for their maintenance and production activities under which the shelf-lives of the products is depend on the machine's age."

Conclusion

"The objective is to minimize a total cost composed of backlog, inventory holding, disposal, and maintenance, we propose a structure of a joint production and maintenance control policy."

"We propose an integrated production maintenance policy called perishable production maintenance policy (PPMP) that combines a production control policy of feedback nature with multiple hedging levels and, an age-based peplacement policy for maintenance control."

"The comparative study carried out demonstrates the advantage of the PPMP policy which allows executing PM actions to reduce the risk of backlog and disposal by increasing the shelf-life mean value and reducing shelf-life variability."

"A future study can incorporate random disposal costs depending on the shelf-life of the products."

"As we established in the literature review that there is a relationship between quality loss and products perishability, which means that this study can be utilized in the configurations of quality control policies."

Switching- and Hedging-Point Policy for Preventive Maintenance with Degrading Machines: Application to a Two-Machine Line [57]

This is a machine-generated summary of:

Magnanini, Maria Chiara; Tolio, Tullio: Switching- and hedging- point policy for preventive maintenance with degrading machines: application to a two-machine line [57].

Published in: Flexible Services and Manufacturing Journal (2019).

Link to original: https://doi.org/10.1007/s10696-019-09370-7.

If you want to cite the papers, please refer to the original.

For technical reasons we could not place the page where the original quote is coming from.

Abstract-Summary "Preventive Maintenance may be carried out and maintenance time reduced to the effective task duration, in order to prevent time waste."

"Companies and researchers have been focusing on policies able to mitigate the impact of Preventive Maintenance on system availability, by exploiting the knowledge about degradation profiles in machines and the joint information from the machine state and the buffer level."

"The mathematical proof of the optimal threshold-based control policy for Preventive Maintenance with inventory cost, maintenance cost, backlog cost is provided."

"The control policy is defined in terms of buffer thresholds and dependency of the thresholds on the degradation condition."

"The optimal control policy is proved to include a combination of switching points and hedging points, where the first ones activate the Preventive Maintenance for a given condition and the latter ones control the production rate in order to minimize the surplus."

"The results show that there exists cases in which the optimal policy is dominated by the effect of the hedging points or the switching points, alternatively."

Introduction

"Maintenance models, mostly, concern the analysis of machine degradation, in order to find the best policy to maximize the machine/system reliability."

"To control the degradation process the only solution is to force the critical item to stop in order to perform Preventive Maintenance."

"Maintenance function should take into account not only maintenance parameters, such as time to failure, time to repair, etc., but also parameters and variables from the other resources, such as machines and buffers, in order to reach a system optimum."

"The focus of this work is to prove the relation between a Preventive Maintenance policy based on the degradation condition of machines and the system dynamics, in order to minimize the overall costs."

Literature Review

"On, works by Kyriakidis and Dimitrakos [58] investigate the impact of Preventive Maintenance control policies based on buffer level and deterioration state for two-machine production systems."

"A joint Production and Preventive Maintenance strategy is investigated in the works by Aghezzaf et al. [59] which are based on non-linear problems."

"The iterative solution framework developed is very complex, it is then simplified in Aghezzaf and Najid [60] where the problem is approached through a non-linear mixed-integer program in the cases which each production line implements a cyclic preventive maintenance policy (time-based) and a linear mixed-integer program in the cases which noncyclic preventive maintenance policies are allowed (condition-based)."

"The problem of the optimal policy for Preventive Maintenance scheduling is analyzed in Ambani et al. [61], and more recently [62], investigates the impact of a Condition-based Maintenance policy on the performance evaluation of a two-machine line where machines are characterized by degrading behavior, by means of an analytical model."

Optimal Control Policy

"The resulting optimal control policy is characterized by switching points and hedging points: the first ones provide the thresholds and the degradation state for which the Preventive Maintenance needs to be activated immediately, the latter control the production rate of the upstream machine."

"The model should be reviewed and the control policy should be changed so that the production rate of the upstream machine could be slowed down, meaning that there is the need of an hedging point and not of a switching point."

"The ranking of hedging points and switching points provide an intuition about the criticalities of the system: if the switching point dominates the activation of the control policy, it is reasonable to assume that the system availability is critical, and therefore reducing the time spent on maintenance represents the key for the performance improvement."

Performance Evaluation Model

"The analytical performance evaluation model integrating the control policy is presented."

"The proposed model grounds on the work presented in Tolio and Ratti [63] for two-machine lines with generalized thresholds."

"In that paper, the authors present a stochastic flow model where machines may have different behavior for different levels of the buffer."

"It can be applied to model the proposed control policy."

"The model consists of a two-machine one-buffer system."

"The machines' states are discrete, while the parts are modeled by means of a continuous flow."

"The control policy is implemented on the upstream machine according to specified buffer levels."

Numerical Results

"The goal of the experimental campaign is to analyze the impact on the optimal control policy of the switching points and the hedging points, for different levels of degradation, duration of Preventive and Corrective Maintenance."

"The first group clusters the cases where both degrading states have high failure rates and the failure rates are increasing between the state 'as good as new' and the degraded one."

"The second group clusters the cases where the failure rates are increasing from a low failure rate in the state 'as good as new' to a high failure rate in the degraded state."

"If the degradation is fast, but the Corrective Maintenance is fast as well, the optimal policy is defined by a threshold-based switching point policy (Case 13)."

"When the Corrective Maintenance is fast, but the degradation is slow, there is not too much different for both activation points (Case 15)."

Conclusion

"A threshold-based control policy for Preventive Maintenance in a two-machine line is proposed."

"Switching points define the buffer level and the machine state for which the Preventive Maintenance should be activated."

"The switching points have a high impact on the policy optimization when the duration of the Preventive Maintenance is shorter enough than the duration of the Corrective Maintenance."

"The hedging points have a high impact on the policy optimization when the duration of the Preventive Maintenance is not enough convenient with respect to the duration of the Corrective Maintenance."

"The goal of the work is to provide the proof of the existence of both hedging points and switching points and their ranking in a threshold-based control policy."

"Threshold-based control policies for Preventive Maintenance based on the machine degradation state have been widely studied."

Preventive Maintenance Scheduling Optimization Based on Opportunistic Production-Maintenance Synchronization [64]

This is a machine-generated summary of:

Li, Li; Wang, Yong; Lin, Kuo-Yi: Preventive maintenance scheduling optimization based on opportunistic production-maintenance synchronization [64].

Published in: Journal of Intelligent Manufacturing (2020).

Link to original: https://doi.org/10.1007/s10845-020-01588-9.

If you want to cite the papers, please refer to the original.

For technical reasons we could not place the page where the original quote is coming from.

Abstract-Summary "This paper addresses a novel approach for integrating preventive maintenance (PM) into production planning of a complex manufacturing system based on availability and cost."

"The feasibility and benefits of this approach are investigated based on empirical study by using historical data from real manufacturing execution system and equipment maintenance system."

"Experimental results demonstrate the effectiveness of proposed approach, reduce personnel number while guarantee the maintenance tasks."

"The proposed approach is beneficial to improve the company's production efficiency."

Introduction

"The reason for using Non-periodic PM is that the equipment status changes due to real environmental factors such as order, maintenance, and repair."

"A good way to achieve these optimization objectives is to integrate the maintenance with the production for developing opportunistic maintenance preserving conjointly the product-production-equipment performances."

"Opportunistic maintenance methods focus on performing the several components maintenance at opportunities, such as equipment downtime due to failure and production plan."

"A novel model is proposed for optimizing the PM tasks of a complex re-entrant production system and preserving conjointly the product-production-equipment performances."

"The main contributions of this paper include: (1) uses historical data from actual MES and EMS to predict opportunities for maintenance tasks to have less impact on production based on availability, (2) applies improved ant colony algorithm to optimize staffing and obtains the least number of repairmen based on cost."

Problem Description

"When a machine becomes idle, it will decide which job in the buffer is to be processed or which PM task is to be implemented."

"This is possible to find some periods for each machine when the work-center have less workload for the opportunities to implement its maintenance tasks without or with less impact on the production."

"Information related to PM tasks include the PM type, PM cause, PM name, earliest start time, latest start time, best start time and duration, etc. That of CM tasks include the CM type, CM name, start time and duration, etc. Due to the processing flow of the jobs is re-entrant, it is difficult to calculate the required capacity of each machine with theoretical methods."

"Calculate the opportunities of each machine (shown in "Opportunities distribution and PM tasks acquisition" section)."

"The required capacity in this paper refers to the processing time period of the machine."

Proposed Approach Based on Availability and Cost

"Extract and analyze historical maintenance data, obtain cycle and duration of maintenance time, use improved ant colony algorithm to assign maintenance tasks to opportunities, and optimize maintenance personnel."

"Determine which machine take which period as their opportunities for maintenance according to optimization objectives."

"The PM task and PM duration that needs to be done on a certain day has been determined, and the opportunity time has also been obtained."

"Leave the largest percentage of opportunity time for the next PM scheduling, because the actual prediction is always error."

"The PM tasks and opportunity time of each machine are obtained, the unit is h (hour)."

"For PM task i, opportunity time i, j: Then, the start point is randomly distributed to each artificial ant."

"For different machines to be maintained at the same period, punish them to optimize the number of maintenance personnel."

Experiments and Results

"The steps of processing the data include data cleaning, sampling time selection, feature extraction (selection of feature extraction sources, dimension reduction), normalization and regression models."

"Sampling time refers to the sampling period, which determines the original input data of ELM algorithm."

"The final selected sampling time is 8:00 a.m. 2 days ago-8:00 a.m. Feature extraction source determines the original input features of the ELM algorithm."

"When it increases to a certain number, the effect of iterations doesn't change, then the certain number is the number of iterations. (1) Actual scheduling Actual scheduling is schedule that required capacity, opportunity and PM total from actual data."

"Compared with the previous routine scheduling, our algorithm optimizing the staffing and reducing the number of maintenance personnel. (2) Predictive scheduling Predictive scheduling is schedule that required capacity, opportunity and PM total obtained from historical data as mentioned earlier."

Discussion

"For most PM models, the data come from the simulation environment, and the actual production line data are rarely used into the algorithm development process, which causes gap between the model and the actual application."

"The reasons for this situation are: (1) the level of informatization of the factory is insufficient to obtain complete data, (2) the privacy of the factory, unable to provide data, (3) the actual data is not pure and missing, cannot be used, (4) the actual production line is too complicated to verify the model, etc. This paper ignores various complex assumptions and develops a predictive model based on actual production line data."

"This paper only uses a real data set to verify the model, and the model's prediction accuracy still cannot reach an error of less than 10% when all parameters are optimal at present, which can be improved in future research."

Conclusion

"A novel model is proposed to optimize the maintenance tasks and preserve conjointly the production-equipment performances for a re-entrant production system."

"Because the data comes from the real production line, which is greatly interfered by human beings and has great randomness, the prediction effect is not very ideal, this part can be improved in the future research."

"The second part to be improved is that although PM simulates the distribution of the real production line, it only simply selects the mean value for calculation, and later historical data can be selected for prediction."

Optimal Replacement Model for the Physical Component of Safety Critical Smart-World CPSs [65]

This is a machine-generated summary of:

Alemayehu, Temesgen Seyoum; Kim, Jai-Hoon; Cho, We-Duke: Optimal replacement model for the physical component of safety critical smart-world CPSs [65].

Published in: Journal of Ambient Intelligence and Humanized Computing (2021).
Link to original: https://doi.org/10.1007/s12652-021-03137-5.

If you want to cite the papers, please refer to the original.

For technical reasons we could not place the page where the original quote is coming from.

Abstract-Summary "The degradation of physical components of safety critical smart-world CPSs would deteriorate the performance of the smart system and lead to loss of human life with significant damage to properties."

"We focus on cost effective preventive replacement strategy that recovers failure of physical components of safety critical smart-world CPSs by mainly considering the deterioration state of the physical component on the availability of the system."

"In view of the degraded physical component of CPSs, we demonstrate the effects of involving maintenance actions during the deterioration state of physical components of CPSs on its availability."

"Since timely preventive replacement is crucial to support continuous and effective system operation, we compute the optimum time of the constant-interval of preventive replacement strategy for the physical components of CPSs."

"The effects of the shape and the scale parameters on the optimal preventive replacement interval of the physical component are also demonstrated."

Introduction

"This motivates us to study the effects of involving maintenance during the deterioration state of the physical component on the availability of safety critical smart-world CPSs, and to develop the optimum replacement interval for the physical component of safety critical smart-world CPSs to ensure its optimal operating condition, to avoid damage and catastrophic failures of the system, and also to prevent accidents to people in the system."

"Since timely preventive maintenance is necessary to support continuous and effective system operation, we compute the optimum time of the constant-interval of preventive replacement policy for the physical component of safety critical smart-world CPSs."

"The simplified optimum preventive replacement interval model proposed in this study can assist maintenance engineers to better make cost-effective maintenance decisions on the physical component of the safety critical smart-world CPSs."

Related Works

"Recovering failure of the physical components of the system can be performed by setting appropriate maintenance action or replacement techniques."

"Maintenance and replacement could be utilized as a recovery technique for the physical component of safety critical smart-world CPSs to avoid costly consequences of failures and guarantee the continuity of system operation."

"Preventive maintenance is designed to retain the healthy condition of an equipment by preventing failures due to wear-out using preventive actions, such as revisions, repairs and component replacements at predefined intervals."

"To that, it provides a more simplified and easier to use optimum preventive replacement interval model to better make cost-effective preventive replacement for the physical component of safety critical smart-world CPSs."

"It also demonstrates the effects of the shape and the scale parameters on the optimal preventive replacement interval of the physical component of the safety critical smart-world CPSs."

Effects of Involving Maintenance During the Deterioration State

"We demonstrated the effects of involving maintenance during the deterioration state of the physical component on the availability of the CPSs."

"Notice that no maintenance action is involved at the deterioration state D to return it to the original condition of the component (state N)."

"It can be noticed that performing maintenance action to revert the deterioration state to the normal state results a significant difference on availability."

"Availability of the system could further be incremented if we consider greater number of deterioration states and perform maintenance actions on each state."

"Performing maintenance action to revert the deterioration state to the normal state results a significant difference on the availability of the system."

Optimizing the Preventive Replacement Period

"As physical components of safety critical smart-world CPSs continue to wear-out, its failure rate increases with time."

"To restore the physical component to a state with a lower failure rate, preventive maintenance must be accomplished in a time-based manner."

"Preventive maintenance of a component is performed with the objective of minimizing the probability of failure that may occur during the operation of a system."

"An age-based preventive replacement policy is mainly used to provide an overall repairing services for simple equipment (single-units) at the time of failure (replacement), whereas the constant interval preventive replacement policy is mainly used to provide a preventive maintenance service for complex systems that are composed of a larger number of units within it."

"We mainly focus on computing the optimum time t_p of the constant-interval preventive replacement policy as it is suited for complex systems like safety critical smart-world CPS that comprises several units within it."

Numerical Computations

"In order to clearly illustrate the effects of failure cost on the optimum preventive replacement interval time, we consider different failure cost values for a failure: $C_f = \$50$, $C_f = \$100$, $C_f = \$500$ and $C_f = \$1000$."

"The optimal values: $t_{p\text{-opt-}50} = 45.59$, $t_{p\text{-opt-}100} = 38.37$, $t_{p\text{-opt-}500} = 25.64$ and $t_{p\text{-opt-}1000} = 21.56$ represent the optimal times that the preventive replacement should be carried out when the failure cost for a given failure is \$50, \$100, \$500 and \$1000 respectively."

"It is clear to observe that, the optimum preventive replacement interval time decreases as the cost of failure for a given failure increases."

"It can be noticed that as the value of the scale parameter η increases, the optimum preventive replacement interval also increases just as in the case of β."

"The right estimation of β and η is valuable to forecast the future behavior of the failure process and to decide the accurate optimum preventive replacement interval that maintain an equipment with minimal cost."

Conclusion

"As safety critical smart-world CPSs are composed of the cyber component integrated with the physical component, any failure of its components could result in a degradation of the physical state, which then cause major harm to human life with significant damage to properties."

"To avoid unexpected failures on safety critical smart-world CPSs due to degradation of its physical components, maintenance could be utilized as a recovery technique."

"Since an efficient and, at the same time, cost-effective maintenance is an important issue in maintaining the physical component of safety critical smart-world CPSs, we proposed and computed the optimum time of the constant-interval preventive replacement policy as it is suited for complex systems in a manner that minimizes cost."

"The simplified optimum preventive replacement interval model proposed in this paper enables maintenance practitioners to better make cost-effective maintenance decisions on the physical component of safety critical smart-world CPSs."

Optimization of Plant Location Problem in Interval Domain via Particle Swarm Optimization [66]

This is a machine-generated summary of:

Sahoo, Laxminarayan; Bhunia, Asoke Kumar: Optimization of plant location problem in interval domain via particle swarm optimization [66].

Published in: International Journal of System Assurance Engineering and Management (2021).

Link to original: https://doi.org/10.1007/s13198-021-01275-9.

If you want to cite the papers, please refer to the original.

For technical reasons we could not place the page where the original quote is coming from.

Abstract-Summary "This paper explores the concept of plant location problem in interval domain."

"In many formulations of plant location problem, it is assumed that parameters relating to the problems are precise valued, which is the traditional way to formulate the problem."

"We consider all the parameters of the problem are imprecise and impreciseness has been represented by interval numbers."

"Days, set up of different industries/plants as well as logistic centers plays an important role for developing the nation's economy and also significantly contribute to the economic and social development of specified regions in which they are located."

"Different types of optimization techniques have been advanced rapidly and applied widely to solve production planning and distribution problems."

"We have considered and solved plant location problem which is a most important optimization problem."

"The main goal of plant location problem is to setup a plant/industry in such a place that for which the total cost of transportations is minimum."

"In existing literature plant location problems have always been addressed in the deterministic domain i.e., all the parameters relating to the problems are fixed/ precise."

"Due to uncertainty/impreciseness, here we have presented the plant location problem in interval environment."

Introduction

"Plant location problem assumes careful thought of major business, including market demand, nature of the product, setup costs, transportation costs, production capacity, government policy etc. In particular, strategic decisions for plant location are the key for business successes in several Industries, viz."

"The plant location decision has often been formulated as a cost optimization problem."

"Given this assumption [67], formulated the problem of determination of plant location centre with least cost."

"Due to uncertainty of the social as well as political environments, parameters involving in plant location problems are not in deterministic."

"We have considered the least cost centre analysis approach for generalization purpose and expressed the uncertainty in terms of interval measure for cost and load parameters."

"We have formulated plant location problem minimizing total transportation cost considering interval valued transportation cost."

Formulation of the Problem

"Four cases may arise: Case 1: Cost and load parameters are intervals."

"Case 2: Cost parameters are intervals but load parameters are fixed/precise."

"Case 3: Cost parameters are fixed/ precise but load parameters are intervals."

"Case 4: All the parameters are fixed/precise."

"It is to be noted that positivity of the location parameters are the only constraints of the problem (3)."

Particle Swarm Optimization

"PSO is initialized with a population of random solutions and searches for optima by updating generations/iterations."

"In PSO, each single solution is a point in a search space, called a particle."

"In every generation/iteration, each particle is improved by two best positions (solutions)."

"To find out the efficiency and effectiveness of the proposed PSO, we have also solved the same problem by well-known heuristic method genetic algorithm (GA)."

"After initialization of GA parameters and bounds of the variables, chromosome representation is an important aspect of Genetic Algorithm."

"Whenever a chromosome is generated, the objective function is evaluated for the corresponding solution and the value is considered as fitness value of that chromosome."

"In a GA, selection process, crossover operation and mutation operation are performed in repeated way until a predefined termination condition is reached."

Numerical Example

"To illustrate how PSO and GA work for the plant location problem, we have solved a hypothetical example by Particle Swarm Optimization technique and Genetic Algorithm simultaneously."

"In this computation, the values of different parameters for PSO like, population size and maximum number of iteration/generations have been taken as 100 and 100 respectively."

"The values of different parameters for GA like, population size, maximum number of generation, probability of crossover and probability of mutation have been taken as 100, 100, 0.85 and 0.15."

"It can be claimed that the PSO algorithm is much better than that of GA for solving plant location problem."

Conclusions

"The study proposed a plant location problem by considering interval valued transportation cost and interval valued shifted load."

"These factors are (1) total transportation cost for transporting of raw materials from resource center and manufacturing goods to the markets (2) availability of raw materials with lower cost and (3) cost of land, building preparation and labour."

"Transport cost for transferring of produced/manufacturing items/products or raw materials may not be precise due to different situations of transportation."

"Considering the realistic assumptions, in this paper, we have formulated and solved plant location problem."

"With the help of interval arithmetic and interval order relations PSO and GA algorithms give best optimal solutions without involving any further assumptions for solving plant location problem in interval environment."

"For further work, one may solve the problem by minimizing the cost of land, labour and building preparation etc. with/without constraints."

Establishment of Maintenance Inspection Intervals: An Application of Process Mining Techniques in Manufacturing [68]

This is a machine-generated summary of:

Ruschel, Edson; Santos, Eduardo Alves Portela; Loures, Eduardo de Freitas Rocha: Establishment of maintenance inspection intervals: an application of process mining techniques in manufacturing [68].

Published in: Journal of Intelligent Manufacturing (2018).

Link to original: https://doi.org/10.1007/s10845-018-1434-7.

If you want to cite the papers, please refer to the original.

For technical reasons we could not place the page where the original quote is coming from.

Abstract-Summary "Well defined interval durations between maintenance inspections provide major support in achieving these targets."

"In order to establish the best interval length, process behavior, cycle times and related costs must be clearly known, and future estimates for these parameters must be established."

"The probability of a given activity occurring in the probabilistic model output establishes the forecast boundaries for predictive models, responsible for estimating process cycle times."

"Availability (uptime) and cost functions are mathematically defined and an iterative process is performed in the length of intervals between maintenance inspections until the time and costs wasted are minimized and the best interval duration is found."

Introduction

"Prior knowledge of cycle times and process behavior, in addition to future estimates of this data, are absolutely necessary in correctly determining the length of maintenance intervals."

"One of this paper's main contribution is to present a methodology capable of integrating probabilistic and predictive models fed with data from single databases, in addition to using this information to estimate possible future behaviors of the process."

"In order to solve this problem, the methodology proposed in this paper uses process mining techniques to handle shop-floor data and obtain information about activity durations, process deviations, cycle-time variations and the progress of the degradation rate."

"The possibility of continually feeding the models with new data allows for a much more dynamic visualization of the process and of its results and consequences for the different intervals defined between maintenance inspections."

Theoretical Background

"Mendes et al. [69] applied a periodic inspection model for redundant systems using Markov Chain's properties to minimize the total cost per cycle through optimizing the time between maintenance inspections."

"Jamshidi and Esfahani [70] present a mathematical model using genetic algorithms that integrate costs, operating time, work cycle duration and failure rates to reduce maintenance costs and increase component reliability."

"Changes in equipment behavior can be detected through the presented models, generating estimates of mean time to failures, cycle times, degradation rate, opportunistic maintenance inspections, among others."

"The present methodology integrates processes mining, probabilistic and predictive models that allow estimating future behaviors in two ways: (1) using real data of the process; (2) through simulations in the probabilistic model (Bayesian Network)."

"Further, the time series obtained in the present methodology can also be used to feed predictive models of other methods, providing greater support in determining maintenance intervals for a wide variety of equipment and types of processes."

Proposed Methodology

"Since several mean cycle times are calculated, one for each distinct group of units produced instead of one, any skewness or variance present in the process, although subtly attenuated, will be considered for calculation purposes in the predictive models."

"This approach is adapted from O'Connor [71] and De Almeida [72], considering that the occurrence of preventive maintenance inspection produces a similar effect to as-good-as-new, reducing both process cycle time and system degradation, returning the equipment to a better state."

"A probability distribution of the data input into the predictive models allows the DM to visualize process behavior trends and slopes, providing support for the decision to increase or reduce the length of the intervals between maintenance inspections with a view to adapting interval to address these trends and slopes."

"Any possible change in process behavior and cycle times will be perceived, allowing a new interval between maintenance inspections to be calculated more effectively."

Example of Application

"The event log showed a total of 3282 units produced and 59 maintenance inspections (activity Readjustment), which is equivalent to approximately 56 units produced between maintenance intervals."

"By minimizing the cost function, the interval between maintenance inspections corresponds to 56 units produced."

"The interval obtained from the original event log exactly corresponds to 56 units produced between two maintenance inspections."

"The optimal interval between maintenance inspections can be estimated by maximizing this function, weighted according to the DM's preferences."

"This evaluation allows the importance of the availability function related parameters presented in this paper to be pointed out, given their direct influence on the maintenance inspection interval balanced addressing both, equipment availability and process related costs."

"Although the total cost of units lost for the estimated interval is greater than the cost assumed as actual, the number of maintenance inspections is sufficiently low for the improvement to occur."

Conclusion

"Despite the use in different industry contexts of the vast amounts of data available on shop-floors contributing to the development of reliable prediction and diagnostic models, the lack of standardization for individual activity records requires that different sub-steps be added for each specific case, making it impossible to create a single guideline capable of addressing every situation."

"The available process mining techniques and their platforms are sufficiently effective to obtain the necessary information for the development of the present methodology, making possible its application due to the reduction of costs and resources."

"Future behavior of the process, cycle times, and probability of activities occurrence estimates, made possible by the methodology proposed in this paper, provide the necessary support so that the DM can define the best interval between maintenance inspections, since the process behavior and the activities occurrences are directly related to the equipment availability issues."

Maintenance Planning of Electric Distribution Systems: A Review [73]

This is a machine-generated summary of:

Trentini, Caroline; de Oliveira Guedes, Wanessa; de Oliveira, Leonardo Willer; Dias, Bruno Henriques; Ferreira, Vitor Hugo: Maintenance Planning of Electric Distribution Systems—A Review [73].

Published in: Journal of Control, Automation and Electrical Systems (2020).

Link to original: https://doi.org/10.1007/s40313-020-00663-z.

If you want to cite the papers, please refer to the original.

For technical reasons we could not place the page where the original quote is coming from.

Abstract-Summary "Electric distribution systems have the objective of supplying electricity with quality and reliability to the final consumers."

"In order to meet both criteria, efficient maintenance programs have a vital importance mainly due to the actual increase in the requirements for distribution service quality and in technologies related to electrical networks."

"This paper presents a comprehensive review on maintenance planning in electrical distribution systems covering different criteria such as economic and reliability."

Introduction

"These criteria are conflicting with each other requiring effective tools for the decision-making related to the maintenance actions [74] over a given planning horizon."

"These actions comprise asset management and maintenance scheduling planning."

"Maintenance consists of a set of actions that seek to maintain the system in a suitable functioning condition [75] and extends the lifespan of equipment or the average time to failure."

"In face of scheduled interruptions costs, maintenance actions should be optimized by utilities through a suitable cost–benefit analysis [76] and can be classified as predictive, preventive and corrective."

The Maintenance Planning of Distribution Systems

"The maintenance planning of EDS consists of a complex problem since it involves decision-making processes in environments of multiple options related to the definition of plans that include diverse actions over a planning horizon [77–80]."

"A trend for more planned maintenance actions is emerging for EDS due to contemporary management premises and operative requirements, as greater reliability of service."

"The RCM policy can be considered as an improvement over traditional preventive maintenance policies known as time-based maintenance (TBM) and condition-based maintenance (CBM) [81], since RCM considers both the probability of equipment failure and its consequences, which are not covered by CBM."

"The RCM policy is applied to evaluate maintenance plans for EDS in Afzali et al. [82], where an importance level is defined for main distribution feeders through a weighting index."

The Reliability Criterion in the Maintenance Planning of EDS

"This section presents works that addressed the reliability issue in the maintenance planning of EDS and comprises the main methods, reliability indexes and variables that have been used."

"Several failure modes can be considered to evaluate reliability of maintenance planning (MP), as done in Yin et al. [83]."

"Several indexes are distributed among the papers of literature, but it can be pointed out that few works consider diverse indexes in a comprehensive manner to plan maintenance actions in EDS [82, 84–88]."

"Works on maintenance planning of EDS that include the reliability criterion consider different equipment in the system, as transformers [89], breakers [90, 91], switches and distribution branches."

"It can be highlighted that transformers and breakers are the equipment most commonly considered in approaches to plan the EDS maintenance, due to their vital function in the distribution grid."

Main Contributions in the Subject of Maintenance Planning of EDS

"Research and development efforts to provide tools that can support the decision-making on the maintenance planning of EDS have been found in the literature because of the increasing relevance of this subject in the context of distribution networks [92–98]."

"In this sense, efforts to identify the system status as support to maintenance planning are presented in Endrenyi et al. [99] and Estava [100]."

"A weighting index is proposed by Afzali et al. [82] to obtain priority levels for main distribution feeders and equipment under the RCM policy."

"A decision-making process on maintenance of distribution branches to improve reliability is presented in Arya [95] by applying a weighted importance factor that classifies branches under their failure severity including load and distributed generation effects."

"Considering that the MP decision must be made within a suitable time, for instance on a monthly base, this decision must be supported by computational tools."

Objective Function

"Multi-objective problems, i.e., that have two or more indexes to be optimized, can be solved by mono-objective approaches when different indexes can be merged in a unique function."

"Given the conflicting nature of the core objectives in the EDS maintenance planning problem, i.e., the lowest cost with the highest reliability, multi-objective algorithms will be highly demanded in the future."

"Most mathematical programming and optimization approaches applied to the maintenance planning of EDS have constraints that can be equality or inequality, linear or nonlinear, or related to limits for decision variables."

"The modeling of a given criterion as objective function or constraint can be decided in function of the approach or software available to solve the MP problem."

"Linear programming algorithms like simplex and interior point require a full mathematical formulation of the optimization problem, which can be a very difficult task for some objectives (grid-related indexes) or constraints (reliability indexes)."

Conclusions and Remarks

"This review comprises works involving predictive, preventive and corrective maintenance, and it was organized by relevant aspects as indexes and probability functions, objectives, constraints and applied approaches within a comprehensive classification."

"Few papers have applied more proper approaches for multi-objective problems as, for instance, the Pareto method, which points out a lack of accurate methods in the literature and a promising research field."

"Some gaps were identified, and from them some future developments can be performed as: Development of comprehensive approaches that can consider several reliability indexes, random variables and equipment that are complementary for obtaining good trade-off EDS maintenance plans; Given the low number of references that consider a suitable trade-off between different criteria, mainly cost and reliability, by using a proper multiobjective approach, it can be pointed out as a lack and opens a range of possibilities for future investigations."

Analyzing the Productivity of Maintenance Systems Using System Dynamics Modeling Method [101]

This is a machine-generated summary of:

Esmaeili, Esmaeil; Karimian, Hasan; Najjartabar Bisheh, Mohammad: Analyzing the productivity of maintenance systems using system dynamics modeling method [101].

Published in: International Journal of System Assurance Engineering and Management (2018).

Link to original: https://doi.org/10.1007/s13198-018-0754-5.

If you want to cite the papers, please refer to the original.

For technical reasons we could not place the page where the original quote is coming from.

Abstract-Summary "Achievement of industries in production and offering products depends on different factors like having an appropriate maintenance system."

"Based on how the external parameters would influence the productivity of a maintenance system the researcher aims at designing a dynamic system model to analyze the productivity of a maintenance system using acquired information from a car device company by determination of the percentage of parameters and indexes and their influence on maintenance system productivity."

"The proposed model is simulated for a system with parameters of precise value according to fixed and variable maintenance policies which the results are represented."

"The results of different policies described by using computerized simulation of the prepared model on VENSIM software."

"In such environment a model of productivity evaluation and operation of maintenance seems required that in this research by analyzing influential factors on maintenance system productivity of a factory and by means of acquired samples, it was possible to measure the ratio of maintenance system operation condition according to the posed indexes and their influences on productivity operation."

"This situation shows that to the extent that positive indexes in the diagram are higher to the same extent maintenance system productivity of company increases."

Introduction

"Total proactive maintenance this method includes all features of the previous methods and focuses on the main agent of production or human resources who play key a role in maintenance, utilization and increasing efficiency."

"As maintenance activities have a lot of inputs and outputs, the maintenance productivity performance requires the measurement of the system using an integrated approach [102]."

"Measuring maintenance productivity as a strategic thinking element is necessary for services and building industry [103]."

"In their studies, Duffuaa and Al-Sultan found that optimization in using the resources of maintenance can decrease the price of operation and production or increase the productivity."

"Varelis et al. [104] developed and described a framework to study the dynamicity of maintenance system of a military airplane motors fleet using a comprehensive method."

Dynamic Assumptions

"Necessary resources for maintenance activities are limited and expensive [105]."

"Appropriate programming of preventive and corrective maintenance activities need to be consistent to boost the reliability and accessibility of the systems [106]."

"This limitation increases the maintenance expenses."

"It will focus on the factors that influence the effectiveness of preventive maintenance system [107]."

"The dynamic complex system is applied to undertake the maintenance activities."

"This paper used a simulated system to specify the rate of productivity of different maintenance policies in relation to influential coefficients on the maintenance system and the corrective and preventive activities and failures in systems."

"Preventive maintenance activities increase the effectiveness coefficient of labor that itself has a direct relationship with the quality of labour force and their performances and result in changes in maintenance system productivity."

Basic Framework and Principles Used in the Model

"Failures influence the level of productivity and system productivity affects the failures."

"When the system is engaged in preventive maintenance activities, there would not be any missing opportunity, because this system would never stop working and these types of activities (preventive maintenance) are considered at the programming phase and this factor is influenced by the failure and the average time required for each failure."

"In the creation of a defect which is a level variable, the equipment and their defect will directly affect the system productivity and is influenced by the factors such as the quality of parts, the quality of equipment, the quality and the productivity of corrective maintenance and the programmed quality and productivity."

"The last two factors will have a significant impact on the defects and the average time of each failure will affect them."

Parameters

"Personnel skill Maintenance agent the average of previous failures OEE level of access (equipment) level of performance (equipment) level of quality (equipment) OCE the quality level of labour exploitation of labour performance level of labour the opportunity coefficient capital cost (missed opportunity) sensitivity coefficient and its impact on preventive maintenance number of preventive maintenance in a year the average loss per failure preventive maintenance program the maximum allowable preventive maintenance the minimum allowable preventive maintenance the impact of improving the technology on failures the maximum allowable failure rate quality of parts quality of equipment quality and productivity of corrective maintenance programmed quality and productivity effort for corrective maintenance the proposed model is simulated for a system with parameters of precise value based on the fixed and changeable maintenance policies."

"The Maximum and minimum number of preventive maintenance in a year to visit the machinery and equipment is 6 and 2 times a year."

Circular Diagram of Cause and Effect

"Circular diagram of cause and effect is a suitable tool to depict the feedback loops and cause and effect relationships between variables."

"Relationships in this diagram are specified by pointers and the pointer starts from the variable of cause and moves towards the variable of the effect and describes the causal relationship."

Results of Model Simulation

"These are all external factors that create a stable and linear trend by influencing the flow rate."

"In the case of OCE, as this coefficient is related to external factors such as the rate of using the labor force, the access level and the quality rate, this flow rate affects the productivity of maintenance system through the three factors, and the OCE increases and causes some positive changes in the maintenance system productivity."

"The flow trend of the missed opportunity In this case, the average time of each failure and the average loss per failure related to the past activities were elicited and presented by statistics and external factors influencing the accumulated missed opportunities are presented."

"The rate of missed opportunity coefficient influenced by the external factor of capital cost enters into the diagram of accumulation of maintenance system productivity and changes it."

Conclusion

"The strength and weakness in either programming or machinery directly influences the productivity and profitability of productions."

"In such an environment, a model for evaluating productivity and maintenance operation is required."

"By analyzing influential factors contributing to maintenance system productivity of industry, it could be measure the ratio of maintenance system operation condition with regards to the indexes."

"It also be possible to measure their influences on the productivity operation."

Multiple Repair Scheduling of Markovian Deteriorated Concrete Structure [108]

This is a machine-generated summary of:

Shafiqul, Islam Md.: Multiple repair scheduling of Markovian deteriorated concrete structure [108].

Published in: Asian Journal of Civil Engineering (2021).
Link to original: https://doi.org/10.1007/s42107-021-00363-2.

If you want to cite the papers, please refer to the original.

For technical reasons we could not place the page where the original quote is coming from.

Abstract-Summary "Choice of suitable repair method at proper time can extend lifetime of infrastructure and can reduce overall cost, saves revenue of the government."

"Maintenance optimization is the most cost-effective solution for infrastructure management system, used by bridge management system (BMS) of several countries."

"This paper attempts to tabulate a schedule of multiple repair required by a Markov-deteriorated concrete structure."

"Challenges of incorporating different levels of cost and improvements to the facilities, imposed by the multiple repair at different age, were solved by the shortest path algorithm."

"Sensitivity on unequal capacity of failure reduction and non-identical cost incurred by multiple repair methods are presented to estimate maintenance schedule."

Introduction

"Selecting the optimum maintenance schedule that balances the improvement of the system and cost encountered by the maintenance plan at the same time is of significant importance."

"Costly and frequent maintenance may improve the system to a state of new at any time, and on the other hand, lower improvement to a system done by a maintenance plan may effectively reduce the operation cost."

"It used a life curve of 10 years which included different maintenance actions but applied in a single discrete time through its life of 10 years."

"Complicacy takes place where multiple maintenance actions are applied in multiple discrete times in structural lifetime what the author tried to include in this research."

"In this model, user can choose limiting value of failure probability to identify time to failure where maintenance action must be applied."

Methodology

"Right half of the flowchart explains the algorithm searching each of the possible paths to find optimal shortest route and continues until it reaches target life, restraining the failure below its limit, with the contribution of multiple repair methods."

"If the failure hits limit state, it uses repair option 1 and repair option 2 at the same discrete age to calculate required costs."

"Of computation, system will hit limiting failure again at two different discrete ages, and two different repair options will be applied at both the ages."

"If the failure reaches and exceeds the limit state, it requires having repair to come back to its operation state."

"It is found that minimizing the cost results the repair to take place at age before it reaches the limit state of failure."

Numerical Example 1

"The failure limit within which repair must be done is taken as 20."

"It means that the structure requires repair if the probability of failure crosses 20% and may be applied method 1, method 2, or method 3 (RM1, RM2, or RM3)."

"The reduction of failure done by repair method 1, 2, and 3 is Fr15 = 15%, FR10 = 10%, and Fr20 = 20%."

"It indicates that repair method 3 is able to take the structure back to its state as new."

Numerical Example 2 (Effect of Failure Reduction)

"Cost of failure is kept close to each other for both cases of repair methods to see the effect of improvements caused by repair."

"Case 1 consists of repair method 1, namely RM1, which has the power of failure reduction 6 (Fr6) costing CR1 = 45 and repair method 3, namely RM3, which has the power of failure reduction 18 (Fr18) costing CR3 = 50."

"Case 2 consists of repair method 2, namely RM2, which has the power of failure reduction 15 (Fr15) costing CR2 = 45 and repair method 3, namely RM3, which has the power of failure reduction 18 (Fr18) costing CR3 = 50."

Numerical Example 3 (Effect of Repair Cost)

"It indicates that the difference in improvement results from these two repairs is as high as 18–6 = 12%."

"In case of CR1 = 45 (difference in cost CR3-CR1 = 50–45 = 5 is close to each other), the scheduling is dominated by mixed type of repairs."

"If the failure reduction capacities' results from two different types of repair methods are close to each other and set as 18 and 15, the difference in cost does not influence the optimal solution."

"Increasing the performance improvement of the structure results from any repair methods can change the number of repair required and may bring change to the schedule."

"The cost of repair does not influence too much where the performance improvement ability results from the different repair methods are close to each other."

References

1. Palei SK, Das S, Snehamoy C (2020) Reliability-centered maintenance of Rapier dragline for optimizing replacement interval of dragline components. Mining Metall Explor. https://doi.org/10.1007/s42461-020-00226-5
2. Mohammadi M, Rai P, Gupta S (2016) Improving productivity of dragline through enhancement of reliability, inherent availability and maintainability. Acta Montan Slovaca 21:1–8. Retrieved April 3, 2020, from https://actamont.tuke.sk/pdf/2016/n1/1muhammadi.pdf
3. Demirel N, Golbasi O, Duzgun S, Kestel S (2013) System reliability investigation of draglines using fault tree analysis. In: Drebenstedt C, Singhal R (eds) Mine planning and equipment selection. Springer, Cham, pp 1151–1158. https://doi.org/10.1007/978-3-319-02678-7_112
4. Golbasi O (2015) Reliability-based maintenance optimization of walking draglines. The Graduate School of Natural and Applied Sciences, Middle East Technical University, Ankara. https://doi.org/10.1145/3132847.3132886
5. Ahuja IPS, Khamba JS (2008) Total productive maintenance: literature review and directions. Int J Qual Reliab Manag 25:709–756
6. Mishra A (2015) Reliability analysis of mining equipment using operational data. Indian Institute of Technology (Banaras Hindu University), Varanasi
7. Winstanley BYG, Usher K, Corke P, Dunbabin M, Roberts J (2007) Dragline automation—A decade of development. IEEE Robot Autom Mag 14:52–64. https://doi.org/10.1109/M-RA.2007.901315
8. Eddouh Y, Daya A, Elotmani R (2022) Optimum preventive maintenance strategy for turbine based on reliability analysis and mathematical modeling. Life Cycl Reliab Saf Eng. https://doi.org/10.1007/s41872-022-00206-2
9. Dhurvey S, Soni PK (2018) Preventive maintenance of steam turbine used in thermal power plant by reliability investigation and FMEA. Int J Sci Res 7(6):18–22
10. Verma AK, Srividya A, Rana A (2011) Optimal time scheduling for carrying out minor maintenance on a steam turbine. Int J Syst Assur Eng Manag 2(3):241–252
11. Sobamowo MG, Ismail SO, Ojolo SJ, Adebayo AA (2012) Optimal preventive maintenance models for steam turbines: a case study of egbin thermal station, Lagos, Nigeria. Int J Mech Comput Manuf Res 1(1):6–11
12. Kalaba DV, Đorđević ML, Kirin SD (2015) Determining the reliability function of the thermal power system in power plant "Nikola Tesla, Block B1". Therm Sci 19(3):793–800
13. Kumar AR, Krishnan V (2017) A study on system reliability in Weibull distribution. Int J Innov Res Electr Electron Instrum Control Eng 5:38–41
14. Sidibe I, Kathab A, Adjallah K (2013) Reliability and preventive maintenance analysis of Weibull distributed lifetime systems. IFAC Proc 46(7):306–311
15. Cha JH, Finkelstein M (2019) Optimal preventive maintenance for systems having a continuous output and operating in a random environment. TOP. https://doi.org/10.1007/s11750-019-00508-2

16. Castro IT (2013) An age-based maintenance strategy for a degradation-threshold shock-model for a system subjected to multiple defects. Asia-Pac J Oper Res 30:1350016–1350029
17. Castro IT, Caballe NC, Perez CJ (2015) A condition-based maintenance for a system subject to multiple degradation processes and external shocks. Int J Syst Sci 46:1692–1704
18. Montoro-Cazorla D, Pérez-Ocón R, Segovia MC (2009) Shock and wear models under policy n using phase-type distributions. Appl Math Model 33:543–554
19. Frostig E, Kenzin M (2009) Availability of inspected systems subject to shocks—a matrix algorithmic approach. Eur J Oper Res 193:168–183
20. Montoro-Cazorla D, Pérez-Ocón R (2011) Two shock and wear systems under repair standing a finite number of shocks. Eur J Oper Res 214:298–307
21. Montoro-Cazorla D, Pérez-Ocón R (2012) A shock and wear system under environmental conditions subject to internal failures, repair, and replacement. Reliab Eng Syst Saf 99:55–61
22. Montoro-Cazorla D, Pérez-Ocón R (2015) A reliability system under cumulative shocks governed by a BMAP. Appl Math Model 39:7620–7629
23. Huynh KT, Castro IT, Barros A, Berenguer C (2012) Modeling age-based maintenance strategies with minimal repairs for systems subject to competing failure modes due to degradation and shocks. Eur J Oper Res 218:140–151
24. van der Weide JAM, Pandey MD (2011) Stochastic analysis of shock process and modeling of condition-based maintenance. Reliab Eng Syst Saf 96:619–626
25. Nakagawa T (2007) Shocks and damage models in reliability theory. Springer, London
26. Finkelstein M, Cha JH (2013) Stochastic modelling for reliability (shocks, burn-in and heterogeneous populations). Springer, London
27. Mercier S, Pham HH (2016) A random shock model with mixed effect, including competing soft and sudden failures, and dependence. Methodol Comput Appl Probab 18:377–400
28. Amelian SS, Sajadi SM, Navabakhsh M, Esmaelian M (2018) Multi-objective optimization of stochastic failure-prone manufacturing system with consideration of energy consumption and job sequences. Int J Environ Sci Technol. https://doi.org/10.1007/s13762-018-1742-7
29. Khoury BN (2016) Optimal control of production rate in a manufacturing system prone to failure with general distribution for repair time. IEEE Trans Autom Control 61(12):4112–4117
30. Polotski V, Kenne JP, Gharbi A (2017) Production and setup policy optimization for hybrid manufacturing–remanufacturing systems. Int J Prod Econ 183:322–333
31. Rivera-Gómez H, Gharbi A, Kenné JP (2013) Joint production and major maintenance planning policy of a manufacturing system with deteriorating quality. Int J Prod Econ 146(2):575–587
32. Chou FD, Wang HM, Chang TY (2009) Algorithms for the single machine total weighted completion time scheduling problem with release times and sequence-dependent setups. Int J Adv Manuf Technol 43(7–8):810
33. Kianfar K, Moslehi G (2012) A branch-and-bound algorithm for single machine scheduling with quadratic earliness and tardiness penalties. Comput Oper Res 39(12):2978–2990
34. Ying KC, Lu CC, Chen JC (2016) Exact algorithms for single-machine scheduling problems with a variable maintenance. Comput Ind Eng 98:427–433
35. Soroush HM (2007) Minimizing the weighted number of early and tardy jobs in a stochastic single machine scheduling problem. Eur J Oper Res 181(1):266–287
36. Cui WW, Lu Z (2017) Minimizing the makespan on a single machine with flexible maintenances and jobs' release dates. Comput Oper Res 80:11–22
37. Kumar A, Saini M, Devi K (2018) Stochastic modeling of non-identical redundant systems with priority, preventive maintenance, and Weibull failure and repair distributions. Life Cycl Reliab Saf Eng. https://doi.org/10.1007/s41872-018-0040-1
38. Nakagawa T, Osaki S (1976) A summary of optimum preventive maintenance policies for two unit standby redundant system. Oper Res 20:171–187
39. Okumato K, Osaki S (1977) Optimum policies for a standby system with preventive maintenance. Oper Res Q 28:415–523

40. Osaki S, Okumato K (1977) Repair limit policies for a two unit standby redundant system with two phases repair. Microelectron Reliab 16:14–45
41. Mokkadis GS, Elias SS, Labib SW (1989) On a two dissimilar unit standby system with three modes and administrative delay in repair. Microelectron Reliab 29(4):511–515
42. Ram K, Divya J (2012) A two non-identical unit standby system model with repair, inspection and post-repair under classical and Bayesian view points. J Reliab Stat Stud 5(2):85–103
43. Deswal S, Malik SC (2015) Reliability measures of a system of two non-identical units with priority subject to weather conditions. J Reliab Stat Stud 8(1):181–190
44. Gupta R, Kumar P, Gupta A (2013) Cost–benefit analysis of a two dissimilar unit cold standby system with Weibull failure and repair laws. Int J Syst Assur Eng Manag 4(4):327–334
45. Devi K, Kumar A, Saini M (2017) Performance analysis of a non-identical unit system with priority and Weibull repair and failure laws. Int J Comput Appl 177(7):8–13
46. Wu Z, Fillmore TB, Vega MA, Hu Z, Todd MD (2022) Diagnostics and prognostics of multimode failure scenarios in miter gates using multiple data sources and a dynamic Bayesian network. Struct Multidiscip Optim. https://doi.org/10.1007/s00158-022-03381-z
47. Gravouil A, Moës N, Belytschko T (2002) Non-planar 3d crack growth by the extended finite element and level sets-part II: level set update. Int J Numer Methods Eng 53(11):2569–2586
48. Moës N, Gravouil A, Belytschko T (2002) Non-planar 3d crack growth by the extended finite element and level sets-part I: mechanical model. Int J Numer Methods Eng 53(11):2549–2568
49. Shi J, Chopp D, Lua J, Sukumar N, Belytschko T (2010) Abaqus implementation of extended finite element method using a level set representation for three-dimensional fatigue crack growth and life predictions. Eng Fract Mech 77(14):2840–2863
50. Xie M, Bott S, Sutton A, Nemeth A, Tian Z (2018) An integrated prognostics approach for pipeline fatigue crack growth prediction utilizing inline inspection data. J Press Vessel Technol 140(3):48
51. Li H, O'Hara P, Duarte C (2021) Non-intrusive coupling of a 3-d generalized finite element method and abaqus for the multiscale analysis of localized defects and structural features. Finite Elem Anal Des 193:103554
52. Kaddachi R, Gharbi A, Kenné J-P (2022) Integrated production and maintenance control policies for failure-prone manufacturing systems producing perishable products. Int J Adv Manuf Technol. https://doi.org/10.1007/s00170-021-08273-y
53. Polotski V, Kenne JP, Gharbi A (2021) Production control of unreliable manufacturing systems with perishable inventory. Int J Adv Manuf Technol 116:2473–2496
54. Ait-El-Cadi A, Gharbi A, Dhouib K, Artiba A (2021) Integrated production, maintenance and quality control policy for unreliable manufacturing systems under dynamic inspection. Int J Prod Econ 236:108140
55. Berthaut F, Gharbi A, Dhouib K (2011) Joint modified block replacement and production/inventory control policy for a failure-prone manufacturing cell. Omega 39(6):642–654
56. Widyadana GA, Wee HM (2012) An economic production quantity model for deteriorating items with preventive maintenance policy and random machine breakdown. Int J Syst Sci 43(10):1870–1882
57. Magnanini MC, Tolio T (2019) Switching- and hedging- point policy for preventive maintenance with degrading machines: application to a two-machine line. Flex Serv Manuf J. https://doi.org/10.1007/s10696-019-09370-7
58. Kyriakidis E, Dimitrakos TD (2006) Optimal preventive maintenance of a production system with an intermediate buffer. Eur J Oper Res 168(1):86–99
59. Aghezzaf EH, Jamali MA, Ait-Kadi D (2007) An integrated production and preventive maintenance planning model. Eur J Oper Res 181(2):679–685
60. Aghezzaf EH, Najid NM (2008) Integrated production planning and preventive maintenance in deteriorating production systems. Inf Sci 178(17):3382–3392
61. Ambani S, Meerkov SM, Zhang L (2010) Feasibility and optimization of preventive maintenance in exponential machines and serial lines. IIE Trans 42(10):766–777

62. Fitouhi MC, Nourelfath M, Gershwin SB (2017) Performance evaluation of a two-machine line with a finite buffer and condition-based maintenance. Reliab Eng Syst Saf 166:61–72
63. Tolio TA, Ratti A (2018) Performance evaluation of two-machine lines with generalized thresholds. Int J Prod Res 56(1–2):926–949
64. Li L, Wang Y, Lin K-Y (2020) Preventive maintenance scheduling optimization based on opportunistic production-maintenance synchronization. J Intell Manuf. https://doi.org/10.1007/s10845-020-01588-9
65. Alemayehu TS, Kim J-H, Cho W-D (2021) Optimal replacement model for the physical component of safety critical smart-world CPSs. J Ambient Intell Humaniz Comput. https://doi.org/10.1007/s12652-021-03137-5
66. Sahoo L, Bhunia AK (2021) Optimization of plant location problem in interval domain via particle swarm optimization. Int J Syst Assur Eng Manag. https://doi.org/10.1007/s13198-021-01275-9
67. Smykay EE, Bowersox DJ, Mossman FH (1961) Physical distribution management. Collier-Macmillan, London
68. Ruschel E, Santos EAP, Loures EFR (2018) Establishment of maintenance inspection intervals: an application of process mining techniques in manufacturing. J Intell Manuf. https://doi.org/10.1007/s10845-018-1434-7
69. Mendes AA, Coit DW, Ribeiro JLD (2014) Establishment of the optimal time interval between periodic inspections for redundant systems. Reliab Eng Syst Saf 131:148–165
70. Jamshidi R, Esfahani MMS (2015) Maintenance policy determination for a complex system consisting of series and cold standby system with multiple levels of maintenance action. Int J Adv Manuf Technol 78:1337–1346
71. O'Connor PDT (1985) Practical reliability engineering, 1st edn. Wiley, New York
72. De Almeida AT (2012) Multicriteria model for selection of preventive maintenance intervals. Qual Reliab Eng Int 28(6):585–593
73. Trentini C, Oliveira Guedes W, Oliveira LW, Dias BH, Ferreira VH (2020) Maintenance planning of electric distribution systems—A review. J Control Autom Electr Syst. https://doi.org/10.1007/s40313-020-00663-z
74. Abbasi E, Fotuhi-Firuzabad M, Abiri-Jahromi A (2009) Risk based maintenance optimization of overhead distribution networks utilizing priority based dynamic programming. IEEE Power Energy Soc General Meet 26:1–11
75. Caballé NC, Castro IT, Pérez CJ, Lanza-Gutiérrez JM (2015) A condition-based maintenance of a dependent degradation-threshold-shock model in a system with multiple degradation processes. Reliab Eng Syst Saf 134:98–109
76. Endrenyi J, Anders GJ, Da Silva AL (1998) Probabilistic evaluation of the effect of maintenance on reliability. An application to power systems. IEEE Trans Power Syst. https://doi.org/10.1109/59.667385
77. Dashti R, Yousefi S (2013) Reliability based asset assessment in electrical distribution systems. Reliab Eng Syst Saf 112:129–136
78. Lopez JC, Franco JF, Rider MJ, Romero R (2017) Optimal restoration/maintenance switching sequence of unbalanced three-phase distribution systems. IEEE Trans Smart Grid 9(6):6058–6068
79. Miloca SA, Volpi NMP, Yuan J, Pinto CLS (2015) Expansion planning problem in distribution systems with reliability evaluation: an application in real network using georeferenced database. Int J Electr Power Energy Syst 70:9–16
80. Pang N, Liu H, Huang S, Meng J (2016) Emergency rush repair task scheduling of distribution networks in large-scale blackouts. Int J Electr Power Energy Syst 82:373–381
81. Li F, Brown RE (2004) A cost-effective approach of prioritizing distribution maintenance based on system reliability. IEEE Trans Power Deliv 19(1):439–441
82. Afzali P, Keynia F, Rashidinejad M (2019) A new model for reliability-centered maintenance prioritisation of distribution feeders. Energy 171:701–709

83. Yin ML, Angus JE, Trivedi KS (2013) Optimal preventive maintenance rate for best availability with hypo-exponential failure distribution. IEEE Trans Reliab 62(2):351–361
84. Mirsaeedi H, Fereidunian A, Mohammadi-Hosseininejad SM, Lesani H (2017) Electricity distribution system maintenance budgeting: a reliability-centered approach. IEEE Trans Power Deliv 33(4):1599–1610
85. Mirsaeedi H, Fereidunian A, Mohammadi-Hosseininejad SM, Dehghanian P, Lesani H (2017) Long-term maintenance scheduling and budgeting in electricity distribution systems equipped with automatic switches. IEEE Trans Industr Inform 14(5):1909–1919
86. Mohammadnezhad-Shourkaei H, Abiri-Jahromi A, Fotuhi-Firuzabad M (2011) Incorporating service quality regulation in distribution system maintenance strategy. IEEE Trans Power Deliv 26(4):2495–2504
87. Yssaad B, Khiat M, Chaker A (2014) Reliability centered maintenance optimization for power distribution systems. Int J Electr Power Energy Syst 55:108–115
88. Wang J, Lu J, Bie Z, You S, Cao X (2014) Long-term maintenance scheduling of smart distribution system through a PSO-TS algorithm. J Appl Math. https://doi.org/10.1155/2014/694086
89. Koksal A, Ozdemir A (2016) Improved transformer maintenance plan for reliability centred asset management of power transmission system. IET Gener Transm Distrib 10(8):1976–1983
90. Abbasghorbani M, Mashhadi HR (2013) Circuit breakers maintenance planning for composite power systems. IET Gener Transm Distrib 7(10):1135–1143
91. Abbasghorbani M, Mashhadi HR, Damchi Y (2014) Reliability-centred maintenance for circuit breakers in transmission networks. IET Gener Transm Distrib 8(9):1583–1590
92. Briš R, Byczanski P, Goňo R, Rusek S (2017) Discrete maintenance optimization of complex multi-component systems. Reliab Eng Syst Saf 168:80–89
93. Carnero MC, Gómez A (2017) Maintenance strategy selection in electric power distribution systems. Energy 129:255–272
94. Mazidi P, Bobi MAS (2017) Strategic maintenance scheduling in an islanded microgrid with distributed energy resources. Electr Power Syst Res 148:171–182
95. Arya R (2016) Ranking of feeder sections of distribution systems for maintenance prioritization accounting distributed generations and loads using diagnostic importance factor (DIF). Int J Electr Power Energy Syst 74:70–77
96. Piasson D, Bíscaro AA, Leão FB, Mantovani JRS (2016) A new approach for reliability-centered maintenance programs in electric power distribution systems based on a multiobjective genetic algorithm. Electr Power Syst Res 137:41–50
97. Wang Y, Liu C, Shahidehpour M, Guo C (2016) Critical components for maintenance outage scheduling considering weather conditions and common mode outages in reconfigurable distribution systems. IEEE Trans Smart Grid 7(6):2807–2816
98. Yumbe Y, Miyakoshi M, Kondo M, Arao T, Furukawa N (2016) Evaluation of optimization method for inspection scheduling of power distribution facilities using maintenance data accumulated by power utility. IEEE Trans Power Deliv 32(2):696–702
99. Endrenyi J, Aboresheid S, Allan RN, Anders GJ, Asgarpoor S, Billinton R et al (2001) The present status of maintenance strategies and the impact of maintenance on reliability. IEEE Trans Power Syst 16(4):638–646
100. Estava NM (1987) Use of extreme value theory to evaluate effectiveness maintenance programs in electrical distribution systems. IEEE Trans Power Deliv 2(4):1271–1275
101. Esmaeili E, Karimian H, Najjartabar Bisheh M (2018) Analyzing the productivity of maintenance systems using system dynamics modeling method. Int J Syst Assur Eng Manag. https://doi.org/10.1007/s13198-018-0754-5
102. Liyanage JP, Kumar U (2003) Towards a value-based view on operations and maintenance performance management. J Qual Maint Eng 9(4):333–350
103. Parida A, Kumar U (2009) Maintenance productivity and performance measurement. In: Handbook of maintenance management and engineering. Springer, Cham

104. Varelis AG, Stamboulis YA, Adamides ED (2002) A life-cycle system dynamics model of aircraft-engine maintenance. In: 20th system dynamics conference, vol 25. System Dynamics Society, Littleton, pp 412–428
105. Bengtsson M (2004) Condition based maintenance in technical systems. Ann Oper Res 25(7):45–53
106. Honkaen T (2004) Modeling industrial maintenance system and the effects of automatic condition monitoring. Helsinki University of Technology, Espoo
107. Reilly M (2001) Design of computerized maintenance management system for radionuclide monitoring. University of Virginia, Virginia
108. Shafiqul I (2021) Multiple repair scheduling of Markovian deteriorated concrete structure. Asian J Civ Eng. https://doi.org/10.1007/s42107-021-00363-2

The manufacturer's authorised representative in the EU is Springer Nature Customer Service Centre GmbH, Europaplatz 3, 69115 Heidelberg, Germany. If you have any concerns regarding our products, please contact ProductSafety@springernature.com

Printed and bound by CPI Group (UK) Ltd, Croydon, CR0 4YY
06/10/2026
02228459-0007